Oxford Graduate Texts in Mathematics

Series Editors

R. Cohen   S. K. Donaldson
S. Hildebrandt   T. J. Lyons
M. J. Taylor

OXFORD GRADUATE TEXTS IN MATHEMATICS

Topology: A Geometric Approach

Terry Lawson

Mathematics Department, Tulane University,
New Orleans, LA 70118

OXFORD

UNIVERSITY PRESS

*This book has been printed digitally and produced in a standard specification
in order to ensure its continuing availability*

OXFORD
UNIVERSITY PRESS

Great Clarendon Street, Oxford OX2 6DP
United Kingdom

Oxford University Press is a department of the University of Oxford.
It furthers the University's objective of excellence in research, scholarship,
and education by publishing worldwide. Oxford is a registered trade mark of
Oxford University Press in the UK and in certain other countries

© Oxford University Press, 2003

First published 2003
First published in paperback 2006
Reprinted 2012

British Library Cataloguing in Publication Data
Data available

Library of Congress Cataloging in Publication Data
Data available

ISBN 978-0-19-920248-5

Preface

This book is intended to introduce advanced undergraduates and beginnning graduate students to topology, with an emphasis on its geometric aspects. There are a variety of influences on its content and structure. The book consists of two parts. Part I, which consists of the first three chapters, attempts to provide a balanced view of topology with a geometric emphasis to the student who will study topology for only one semester. In particular, this material can provide undergraduates who are not continuing with graduate work a capstone experience for their mathematics major. Included in this experience is a research experience through projects and exercise sets motivated by the prominence of the Research Experience for Undergraduate (REU) programs that have become important parts of the undergraduate experience for the best students in the US as well as VIGRE programs. The book builds upon previous work in real analysis where a rigorous treatment of calculus has been given as well as ideas in geometry and algebra. Prior exposure to linear algebra is used as a motivation for affine linear maps and related geometric constructions in introducing homeomorphisms. In Chapter 3, which introduces the fundamental group, some group theory is developed as needed. This is intended to be sufficient for students without a prior group theory course for most of Chapter 3. A prior advanced undergraduate level exposure to group theory is useful for the discussion of the Seifert–van Kampen theorem at the end of Chapter 3 and for Part II.

Part I provides enough material for a one-semester or two-quarter course. In these chapters, material is presented in three related ways. The core of these chapters presents basic material from point set topology, the classification of surfaces, and the fundamental group and its applications with many details left as exercises for the student to verify. These exercises include steps in proofs as well as application of the theory to related examples. This style fosters the highly involved approach to learning through discussion and student presentation which the author favors, but also allows instructors who prefer a lecture approach to include some of these details in their presentation and to assign others. The second method of presentation comes from chapter-end exercise sets. Here the core material of the chapter is extended significantly. These exercise sets include material an instructor may choose to integrate as additional topics for the whole class, or they may be used selectively for different types of students to individualize the course. The author has used them to give graduate students and undergraduates in the same course different types of assignments to assure that undergraduates get a well-balanced exposure to topology within a semester while graduate students get exposure to the required material for their PhD written examinations. Finally, these chapters end with a project, which provides a research experience that draws upon the ideas presented in the chapter. The author has used these projects as group projects which lead to the students involved writing a paper and giving a class presentation on their project.

Part II, which consists of Chapters 4–6, extends the material in a way to make the book useful as well for a full-year course for first-year graduate students with no prior exposure to topology. These chapters are written in a very different style, which is motivated in part by the ideal of the Moore method of teaching topology combined with ideas of VIGRE programs in the US which advocate earlier introduction of seminar and research activities in the advanced undergraduate and graduate curricula. In some sense, they are a cross between the chapter-end exercises and the projects that occur in Part I. These last chapters cover material from covering spaces, CW complexes, and algebraic topology through carefully selected exercise sets. What is very different from a pure Moore method approach is that these exercises come with copious hints and suggested approaches which are designed to help students master this material while at the same time improving their abilities in understanding the structure of the subject as well as in constructing proofs. Instructors may use them with a teaching style which ranges from a pure lecture–problem format, where they supply key proofs, to a seminar–discussion format, where the students do most of the work in groups or individually. Class presentations and expository papers by students, in groups or individually, can also be a component here. The goal is to lay out the basic structure of the material in carefully developed problem sets in a way that maximizes the flexiblity of the instructor in utilizing this material and encourages strong involvement of students in learning it.

We briefly outline what is covered in the text. Chapter 1 gives a basic introduction to the point set topology used in the rest of the book, with emphasis on developing a geometric feel for the concepts. Quotient space constructions of spaces built from simpler pieces such as disks and rectangles is stressed as it is applied frequently in studying surfaces. Chapter 2 gives the classification of surfaces using the viewpoint of handle decompositions. It provides an application of the ideas in the first chapter to surface classification, which is an important example for the whole field of manifold theory and geometric topology. Chapter 3 introduces the fundamental group and applies it to many geometric problems, including the final step in the classification of surfaces of using it to distinguish nonhomeomorphic surfaces. In Chapter 3, certain basic ideas of covering spaces (particularly that of exponential covering of the reals over the circle) are used, and Chapter 4 is concerned with developing these further into the beautiful relationship between covering spaces and the fundamental group. Chapter 5 discusses CW complexes, including simplicial complexes and Δ-complexes. CW complexes are motivated by earlier work from handle decompositions and used later in studying homology. Chapter 6 gives a selective approach to homology theory with emphasis on its application to low-dimensional examples. In particular, it gives the proof (through exercise sets) of key results such as invariance of domain and the Jordan curve theorem which were used earlier. It also gives a more advanced approach to the concept of orientation, which plays a key role in Chapter 2.

The coverage in the text differs substantially in content, order, and type from texts at a similar level. The emphasis on geometry and the desire to have a balanced one-semester introduction leads to less point set topology but a more thorough application of it through the handlebody approach to surface classification. We also move quickly enough to allow a significant exposure to algebraic topology through the fundamental group within the first semester. The extensive exercise sets, which form the core of developing the more advanced material in the text, also foster more flexibility in how the text can be used. When individual parts are counted, there are more than a thousand exercises in the text. In particular, it should serve well as a resource for independent study and projects outside of the standard course structure as well as allow many different types of courses.

There is an emphasis on understanding the topology of low-dimensional spaces which exist in three-space, as well as more complicated spaces formed from planar pieces. This particularly occurs in understanding basic homotopy theory and the fundamental group. Because of this emphasis, illustrations play a key role in the text. These have been prepared with LaTeX tools pstricks and xypic as well as using figures constructed using Mathematica, Matlab, and Adobe Illustrator.

The material here is intentionally selective, with the dual goals of first giving a good one-semester introduction within the first three chapters and then extending this to provide a problem-oriented approach to the remainder of a year course. We wish to comment on additional sources which we recommend for material not covered here, or different approaches to our material where there is overlap. For a more thorough treatment of point set topology, we recommend Munkres [24]. For algebraic topology, we recommend Hatcher [13] and Bredon [5]. All of these books are written at a more advanced level than this one. We have used these books in teaching topology at the first- and second-year graduate levels and they influenced our approach to many topics. For some schools with strong graduate students, it may be most appropriate to use just the first three chapters of our text for undergraduates and to prepare less prepared graduate students for the graduate course on the level of one of the three books mentioned. In that situation, some of the projects or selected exercises from Chapters 4–6 could be used as enhancements for the graduate course.

The book contains as an appendix some selected solutions to exercises to assist students in learning the material. These solutions are limited in number in order to maximize the flexibility of instructors in using the exercise sets. Instructors who are adopting this book for use in a course can obtain an Instructor Solutions Manual with solutions to the exercises in the book in terms of a PDF file through Oxford University Press (OUP). The LaTeX files for these solutions are also available through OUP for those instructors who wish to use them in preparation of materials for their class. Please write to the following address, and include your postal and e-mail addresses and full course details including student numbers:

Marketing Manager
Mathematics and Statistics
Academic and Professional Publishing
Oxford University Press
Great Clarendon Street
Oxford OX2 6DP, UK

Contents

II Covering Spaces, CW Complexes and Homology

List of Figures

Part I

A Geometric Introduction to Topology

1

Basic point set topology

1.1 Topology in $\mathbb{R}^n$

Topology is the branch of geometry that studies "geometrical objects" under the equivalence relation of homeomorphism. A *homeomorphism* is a function $f : X \to Y$ which is a bijection (so it has an inverse $f^{-1} : Y \to X$) with both f and f^{-1} being continuous. One of the prime aims of this chapter will be to enhance our understanding of the concept of continuity and the equivalence relation of homeomorphism. We will also discuss more precisely the "geometrical objects" in which we are interested (called *topological spaces*), but our viewpoint will primarily be to understand more familiar spaces better (such as surfaces) rather than to explore the full generalities of topological spaces. In fact, all of the spaces we will be interested in exist as subspaces of some Euclidean space $\mathbb{R}^n$. Thus our first priority will be to understand continuity and homeomorphism for maps $f : X \to Y$, where $X \subset \mathbb{R}^n$ and $Y \subset \mathbb{R}^m$. We will use bold face x to denote points in $\mathbb{R}^k$.

One of the methods of mathematics is to abstract central ideas from many examples and then study the abstract concept by itself. Although it often seems to the student that such an abstraction is hard to relate to in that we are frequently disregarding important information of the particular examples we have in mind, the technique has been very successful in mathematics. Frequently, the success is rooted in the following idea: knowing less about something limits the avenues of approach available in studying it and this makes it easier to prove theorems (if they are true). Of course, the measure of the success of the abstracted idea and the definitions it suggests is frequently whether the facts we can prove are useful back in the specific situations which led us to abstract the idea in the first place. Some of the most important contributions to mathematics have been made by those who have figured out good definitions. This is difficult for the student to appreciate since definitions are usually presented as if they came from some supreme being. It is more likely that they have evolved through many wrong guesses and that what is presented is what has survived the test of time.

It is also quite possible that definitions and concepts which seem so right now (or at least after a lot of study) will end up being modified at a later stage.

We now recall from calculus the definition of continuity for a function $f : X \to Y$, where X and Y are subsets of Euclidean spaces.

Definition 1.1.1. f is *continuous at* $x \in X$ if, given $\epsilon > 0$, there is a $\delta > 0$ so that $d(x, y) < \delta$ implies that $d(f(x), f(y)) < \epsilon$. Here d indicates the Euclidean *distance function* $d((x_1, \ldots, x_k), (y_1, \ldots, y_k)) = ((x_1 - y_1)^2 + \cdots + (x_k - y_k)^2))^{1/2}$. We say that f is *continuous* if it is continuous at x for all $x \in X$.

It will be convenient to have a slight reformulation of this definition. For $z \in \mathbb{R}^k$, we define the *ball* of radius r about z to be the set $B(z, r) = \{y \in \mathbb{R}^k : d(z, y) < r\}$ If C is a subset of $\mathbb{R}^k$ and $z \in C$, then we will frequently be interested in the intersection $C \cap B(z, r)$, which just consists of those points of C which are within distance r of z. We denote by $B_C(x, r) = C \cap B(x, r) = \{y \in C : d(y, x) < r\}$. Our reformulation is given in the following definition.

Definition 1.1.2. $f : X \to Y$ is *continuous at* $x \in X$ if given $\epsilon > 0$, there is a $\delta > 0$ so that $B_X(x, \delta) \subset f^{-1}(B_Y(f(x), \epsilon))$. f is *continuous* if it is continuous at x for all $x \in X$.

Exercise 1.1.1. Show that the reformulation Definition 1.1.2 is equivalent to the original Definition 1.1.1. This requires showing that, if f is continuous in Definition 1.1.1, then it is also continuous in Definition 1.1.2, and vice versa.

We reformulate in words what Definition 1.1.2 requires. It says that a function is continuous at x if, when we look at the set of points in X that are sent to a ball of radius ϵ about $f(x)$, no matter what $\epsilon > 0$ is given to us, then this set always contains the intersection of a ball of some radius $\delta > 0$ about x with X. This definition leads naturally to the concept of an open set.

Definition 1.1.3. A set $U \subset \mathbb{R}^k$ is *open* if given any $y \in U$, then there is a number $r > 0$ so that $B(y, r) \subset U$. If X is a subset of $\mathbb{R}^k$ and $U \subset X$, then we say that U is *open in* X if given $y \in U$, then there is a number $r > 0$ so that $B_X(y, r) \subset U$.

In other words, U is an open set in X if it contains all of the points in X that are close enough to any one of its points. What our second definition is saying in terms of open sets is that $f^{-1}(B_Y(y, \epsilon))$ satisfies the definition of an open set in X containing x; that is, all of the points in X close enough to x are in it. Before we reformulate the definition of continuity entirely in terms of open sets, we look at a few examples of open sets.

Example 1.1.1. $\mathbb{R}^n$ is an open set in $\mathbb{R}^n$. Here there is little to check, for given $x \in \mathbb{R}^n$, we just note that $B(x, r) \subset \mathbb{R}^n$, no matter what $r > 0$ is.

Example 1.1.2. Note that a ball $B(x, r) \subset \mathbb{R}^n$ is open in $\mathbb{R}^n$. If $y \in B(x, r)$, then if $r' = r - d(y, x)$, then $B(y, r') \subset B(x, r)$. To see this, we use the triangle inequality for the distance function: $d(z, y) < r'$ implies that

$$d(z, x) \leq d(z, y) + d(y, x) < r' + d(y, x) = r.$$

Figure 1.1 illustrates this for the plane.

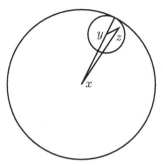

Figure 1.1. Balls are open.

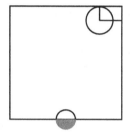

Figure 1.2. Open and closed rectangles.

Example 1.1.3. The inside of a rectangle $R \subset \mathbb{R}^2$, given by $a < x < b, c < y < d$, is open. Suppose (x, y) is a point inside of R. Then let $r = \min(b - x, x - a, d - y, y - c)$. Then if $(u, v) \in B((x, y), r)$, we have $|u - x| < r, |v - y| < r$, which implies that $a < u < b, c < v < d$, so $(u, v) \in R$. However, if the perimeter is included, the rectangle with perimeter is no longer open. For if we take any point on the perimeter, then any ball about the point will contain some point outside the rectangle. We illustrate this in Figure 1.2.

Example 1.1.4. The right half plane, consisting of those points in the plane with first coordinate positive, is open. For given such a point (x, y) with $x > 0$, then if $r = x$, the ball of radius r about (x, y) is still contained in the right half plane. For any $(u, v) \in B((x, y), r)$ satisfies $|u - x| < r$ and so $x - u < x$, which implies $u > 0$.

Example 1.1.5. An interval (a, b) in the line, considered as a subset of the plane (lying on the x-axis), is not open. Any ball about a point in it would have to contain some point with positive y-coordinate, so it would not be contained in (a, b). Note, however, that it is open in the line, because, if $x \in (a, b)$ and $r = \min(b - x, x - a)$, then the intersection of the ball of radius r about x with the line is contained in (a, b). Of course, the line itself is not open in the plane. Thus we have to be careful in dealing with the concept of being open in X, where

X is some subset of a Euclidean space, since a set which is open in X need not be open in the whole space.

Exercise 1.1.2. Determine whether the following subsets of the plane are open. Justify your answers.

(a) $A = \{(x, y) : x \geq 0\}$,

(b) $B = \{(x, y) : x = 0\}$,

(c) $C = \{(x, y) : x > 0 \text{ and } y < 5\}$,

(d) $D = \{(x, y) : xy < 1 \text{ and } x \geq 0\}$,

(e) $E = \{(x, y) : 0 \leq x < 5\}$.

Note that all of these sets are contained in A. Which ones are open in A?

We now give another reformulation for what it means for a function to be continuous in terms of the concept of an open set. This is the definition that has proved to be most useful to topology.

Definition 1.1.4. $f : X \to Y$ is *continuous* if the inverse image of an open set in Y is an open set in X. Symbolically, if U is an open set in Y, then $f^{-1}(U)$ is an open set in X.

Note that this definition is not local (i.e. it is not defining continuity at one point) but is global (defining continuity of the whole function). We verify that this definition is equivalent to Definition 1.1.2. Suppose f is continuous under Definition 1.1.2 and U is an open set in Y. We have to show that $f^{-1}(U)$ is open in X. Let $\boldsymbol{x}$ be a point in $f^{-1}(U)$. We need to find a ball about $\boldsymbol{x}$ so that the intersection of this ball with X is contained in $f^{-1}(U)$. Now $\boldsymbol{x} \in f^{-1}(U)$ implies that $f(\boldsymbol{x}) \in U$, and U open in Y means that there is a number $\epsilon > 0$ so that $B_Y(f(\boldsymbol{x}), \epsilon) \subset U$. But Definition 1.1.2 implies that there is a number $\delta > 0$ so that $B_X(\boldsymbol{x}, \delta) \subset f^{-1}(B_Y(f(\boldsymbol{x}), \epsilon)) \subset f^{-1}(U)$, which means that $f^{-1}(U)$ is open in X; hence f is continuous using Definition 1.1.4.

Suppose that f is continuous under Definition 1.1.4 and $\boldsymbol{x} \in X$. Let $\epsilon > 0$ be given. We noted above that a ball is open in $\mathbb{R}^k$ and the same proof shows that the intersection of a ball with Y is open in Y. Since $B_Y(f(\boldsymbol{x}), \epsilon)$ is open in Y, Definition 1.1.4 implies that $f^{-1}(B_Y(f(\boldsymbol{x}), \epsilon))$ is open in X. But the definition of an open set then implies that there is $\delta > 0$ so that $B_X(\boldsymbol{x}, \delta) \subset f^{-1}(B_Y(f(\boldsymbol{x}), \epsilon))$; hence f is continuous by Definition 1.1.2.

Before continuing with our development of continuity, we recall from calculus some functions which were proved to be continuous there. It is shown in calculus that any differentiable function is continuous. This includes polynomials, various trigonometric and exponential functions, and rational functions. Certain constructions with continuous functions, such as taking sums, products, and quotients (where defined), are shown to give back continuous functions. Other important examples are inclusions of one Euclidean space in another and projections onto Euclidean spaces (e.g. $P(x, y, z) = (x, z)$). Also, compositions of continuous functions are shown to be continuous. We re-prove this latter fact with the open-set definition.

Suppose $f : X \to Y$ and $g : Y \to Z$ are continuous. We want to show that the composition $gf : X \to Z$ is continuous. Let U be an open set in Z. Since g is continuous, $g^{-1}(U)$ is open in Y; since f is continuous, $f^{-1}(g^{-1}(U))$ is open in X. But $f^{-1}(g^{-1}(U)) = (gf)^{-1}(U)$, so we have shown that gf is continuous. Note that in this proof we have not really used that X, Y, Z are contained in some Euclidean spaces and that we have our particular definition of what it means for a subset of Euclidean space to be open. All we really are using in the proof is that in each of X, Y, Z, there is some notion of an open set and the continuous functions are those that have inverse images of open sets being open. Thus the proof would show that even in much more general circumstances, compositions of continuous functions are continuous. We pursue this in the next section.

1.2   Open sets and topological spaces

The notion of an open set plays a basic role in topology. We investigate the properties of open sets in X, where X is a subset of some $\mathbb{R}^n$. First note that the empty set is open since there is nothing to prove, there being no points in it around which we have to have balls. Also, note that X itself is open in X since given any point in X and any ball about it, then the intersection of the ball with X is contained in X. This says nothing about whether X is open in $\mathbb{R}^n$.

Next suppose that $\{U_i\}$ is a collection of open sets in X, where i belongs to some indexing set I. Then we claim that the union of all of the U_i is open in X. For suppose $\boldsymbol{x}$ is a point in the union, then there must be some i with $\boldsymbol{x} \in U_i$. Since U_i is open in X, there is a ball about $\boldsymbol{x}$ with the intersection of this ball with X contained in U_i, hence contained in the union of all of the U_i.

We now consider intersections of open sets. It is not the case that arbitrary intersections of open sets have to be open. For example, if we take our sets to be balls of decreasing radii about a point $\boldsymbol{x}$, where the radii approach 0, then the intersection would just be $\{\boldsymbol{x}\}$ and this point is not an open set in X. However, if we only take the intersection of a finite number of open sets in X, then we claim that this finite intersection is open in X. Let $U_1, \ldots, U_p$ be open sets in X, and suppose $\boldsymbol{x}$ is in their intersection. Then for each i, $i = 1, \ldots, p$, there is a radius $r_i > 0$ so that the intersection of X with the ball of radius r_i about $\boldsymbol{x}$ is contained in U_i. Let r be the minimum of the r_i (we are using the finiteness of the indexing set to know that there is a minimum). Then the ball of radius r is contained in each of the balls of radius r_i, and so its intersection with X is contained in the intersection of the U_i. Hence the intersection is open.

The properties that we just verified about the open sets in X turn out to be the crucial ones when studying the concept of continuity in Euclidean space, and so the natural thing mathematicians do in such a situation is to abstract these important properties and then study them alone. This leads to the definition of a topological space.

Definition 1.2.1. Let X be a set, and let $\mathcal{T} = \{U_i \colon i \in I\}$ be a collection of subsets of X. Then $\mathcal{T}$ is called a *topology* on X, and the sets U_i are called the *open sets* in the topology, if they satisfy the following three properties:

(1) the empty set and X are open sets;

(2) the union of any collection of open sets is open;

(3) the intersection of any finite number of open sets is open.

If $\mathcal{T}$ is a topology on X, then $(X, \mathcal{T})$, or just X itself if $\mathcal{T}$ is made clear by the context, is called a *topological space*.

Our discussion above shows that if X is contained in $\mathbb{R}^n$ and we define the open sets as we have, then X with this collection of open sets is a topological space. This will be referred to as the "standard" or "usual" topology on subsets of $\mathbb{R}^n$ and is the one intended if no topology is explicitly mentioned. Note that Definition 1.1.4 makes sense in any topological space. We use it to define the notion of continuity in a general topological space. Our proof above that the composition of continuous functions is continuous goes through in this more general framework. As we said before, the spaces that we are primarily interested in are those that get their topology from being subsets of some Euclidean space. Nevertheless, it is frequently useful to use the notation of a general topological space and to give more general proofs even though we are dealing with a very special case. We will also use quotient space descriptions of subsets of $\mathbb{R}^n$, which will require us to use topologies more generally defined than those of $\mathbb{R}^n$ and its subsets.

One of the important properties of $\mathbb{R}^n$ and its subsets as topological spaces is that the topology is defined in terms of the Euclidean distance function. A special class of topological spaces are the metric spaces, where the open sets are defined in terms of a distance function.

Definition 1.2.2. Let X be a set and $d : X \to \mathbb{R}$ a function. d is called a *metric* on X if it satisfies the following properties:

(1) $d(x, y) \geq 0$ and $= 0$ iff $x = y$;

(2) $d(x, y) = d(y, x)$;

(3) $d(x, z) \leq d(x, y) + d(y, z)$ (triangle inequality).

The metric d then determines a topology on X, which we denote by $\mathcal{T}_d$, by saying a set U is open if given $x \in U$, there is a ball $B_d(x, r) = \{y \in X : d(x, y) < r\}$ contained in U. $(X, \mathcal{T}_d)$ (or more simply denoted (X, d)) is then called a *metric space*.

To verify that the definition of a topology on a metric space does indeed satisfy the three requirements for a topology is left as an exercise. The proof is essentially our proof that Euclidean space satisfied those conditions. Also, it is easy to verify that the usual distance function in $\mathbb{R}^n$ satisfies the conditions of a metric.

From the point of view of some forms of geometry, the particular distance function used is very important. From the point of view of topology, the important idea is not the distance function itself, but rather the open sets that it determines. Different metrics on a set can determine the same open sets. For

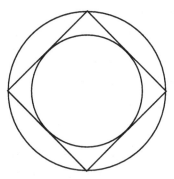

Figure 1.3. Comparing balls.

an example of this, let us consider the plane. Let d denote the usual Euclidean metric in the plane and let $d'((x,y),(u,v)) = |x - u| + |y - v|$. We will leave it as an exercise to verify that d' is a metric. We will use a subscript to indicate the metric being used when determining balls and open sets. As illustrated in Figure 1.3, balls in the metric d' look like diamonds. We show that these two metrics determine the same open sets. Since the open sets are determined by the balls and each type of ball is open, it is enough to show that if $B_d(z,r)$ is a ball about z, then there is a number r' so that $B_{d'}(z,r') \subset B_d(z,r)$, and conversely, that each ball $B'_d(z,r')$ contains a ball $B_d(z,r)$. First suppose that we are given a radius r for a ball $B_d(z,r)$. We need to find a radius r' so that $B_{d'}(z,r') \subset B_d(z,r)$. Note that we want $|x_1 - u_1| + |x_2 - u_2| < r'$ to imply that $(x_1 - u_1)^2 + (x_2 - u_2)^2 < r^2$. But if $r' = r$, then this will be true as can be seen by squaring the first inequality. For the other way, given a ball $B_{d'}(z,r')$, we need to find a ball $B_d(z,r)$ within it. Here $r = r'/2$ will work: $(z_1 - u_1)^2 + (z_2 - u_2)^2 < (r')^2/4$ implies that $|z_1 - u_1| < r'/2, |z_2 - u_2| < r'/2$, and so $d'(z,u) < r'$. As Figure 1.3 suggests, we could actually take $r = r'/\sqrt{2}$. This figure shows the inclusions $B_d(z, r/\sqrt{2}) \subset B_{d'}(z,r) \subset B_d(z,r)$.

From the topological point of view, the best value of r given r' is not really of much importance; it is just the existence of an appropriate r. The existence can be seen geometrically.

Exercise 1.2.1. Verify that the definition of an open set for a metric space satisfies the requirements for a topology.

Exercise 1.2.2. Verify that d' is a metric.

We give two examples of a metric space besides the usual topology on a subset of $\mathbb{R}^n$. For the first example, we take as a set $X = \mathbb{R}^n$, but define a metric d by $d(x,y) = 1$ if $x \neq y$, and $d(x,x) = 0$. It is straightforward to check that this satisfies the conditions for a metric. Then a ball $B\left(x, \frac{1}{2}\right) = \{x\}$, so one point sets are open. Hence every set, being a union of one-point sets, will be open. The topology on a set X where all sets are open is called the *discrete topology*.

The next example is of no special importance to us here, but similar constructions are very important in analysis. The points in our space will be continuous functions defined on the interval $[0, 1]$. We can then define the distance between two such functions to be $d(f, g) = \int_0^1 |f(x) - g(x)| \, dx$. We leave it as an exercise to check that this satisfies the definition of a metric.

Exercise 1.2.3. Show that the above definition of the distance between two functions does satisfy the three properties required of a metric. This depends on the fact, which you may assume in your argument, that the integral of a nonnegative continuous function is positive unless the function is identically 0.

We give an example of a topological space which is not a metric space. To define a topology on a set, we have to give a collection of subsets of the set (which we will call open sets) and then verify that they satisfy the three properties required of open sets in a topology. The simplest example of a nonmetric space is to take any set X with more than one point and define the open sets by saying that the only open sets are ϕ and X. This topology is called the *indiscrete topology* on X. For a slightly more complicated example, we will take our set to be the set with three points $\{a, b, c\}$ and then define the following sets to be open: $\phi, \{a, b\}, \{a, b, c\}$. We may verify that this collection of open sets does satisfy the three required properties: the empty set and the whole space are open, unions of open sets are open, and finite intersections of open sets are open. Of course, this is just one of many possible topologies on the three-point set. In order to get a better feeling for the requirements of a topology, we will leave it as an exercise to find some more topologies on this set.

Exercise 1.2.4.

   (a) Find five different topologies for the set $\{a, b, c\}$.

   (b) Find all the possible topologies on the set $\{a, b, c\}$.

How do we know that the topology that we put on $\{a, b, c\}$ does not arise from some metric? The answer lies in a separation property that any metric space possesses and our topology does not. Given any two distinct points x, y in a metric space, there is some distance $r = d(x, y)$ between them. Then the ball of radius $r/2$ about x does not intersect the ball of radius $r/2$ about y and vice versa. Hence there are two disjoint open sets, one of which contains x and the other y. But this is not true for the points a and b in the topology given above, since every open set which contains b also contains a. The same argument shows that the indiscrete topology on any set X with at least two points does not come from a metric. A topological space X is called *Hausdorff* if given $x, y \in X$ there are disjoint open sets U_x, U_y with $x \in U_x$, $y \in U_y$. The argument above says a metric space is Hausdorff, and our examples are shown not to arise from a metric since they are not Hausdorff.

We look at some specific examples of continuous functions. The inclusion of a subset B of A into A will always be continuous, where $A \subset \mathbb{R}^n$. For if $B \subset A, i : B \to A$ is the inclusion, and if U is an open set in A, then $i^{-1}(U) = B \cap U$. We need to see why $B \cap U$ is open in B if U is open in A. Let $\boldsymbol{x} \in B \cap U$. Then U open in A means that there is a ball $B(\boldsymbol{x}, r)$ with $B(\boldsymbol{x}, r) \cap A \subset U$.

Since $B(x, r) \cap B \subset B \cap U$, $B \cap U$ is open in B. Note that this proof would work equally as well in any metric space as long as we use the same metric for the subset. In a general topological space, we have to specify how we get the topology on the subset from the topology on the original set.

Definition 1.2.3. Suppose A is a topological space and $B \subset A$. A set $V \subset B$ is open in the *subspace topology* on B iff V is the intersection of B with an open set in the whole space A; that is, V is open in B iff $V = U \cap B$, where U is open in A.

It is straightforward to show that an inclusion map is continuous when the subset has the subspace topology. From now on, we will assume that a subset is given the subspace topology unless otherwise stated. The topology on a subset of $\mathbb{R}^n$ coming from using the usual metric is a special case of the subspace topology.

Exercise 1.2.5. For $X \subset \mathbb{R}^n$, show that the usual topology on X is the same as the subspace topology.

Here is another useful construction for continuous functions. Suppose that $f : A \to B$ is continuous and C is a subset of B which contains the image of f. Then we may regard f as a function from A to C. This function, which we denote by f_C, is still continuous when C is given the subspace topology. For if we take an open set V of C, it will have the form $V = U \cap C$, where U is open in B. Then $f_C^{-1}(V) = f^{-1}(U)$ is open since U is open and f is continuous.

Putting these last two constructions together and using the fact that compositions of continuous functions are continuous shows that if we start with a function f from $\mathbb{R}^n$ to $\mathbb{R}^m$ which we already know is continuous, such as a polynomial, and then restrict the function to a subset A and restrict the range to a subset B which contains $f(A)$, then this new function with restricted domain and range will be continuous.

For many constructions involving continuous functions, it is more convenient to work with the concept of closed sets rather than open sets.

Definition 1.2.4. A set $C \subset X$ is said to be *closed* if its complement $X \backslash C$ is open.

From their definition, the closed sets are completely determined by the open sets and vice versa. From the three properties that the open sets satisfy, we can deduce three properties that the closed sets must satisfy:

(1) the empty set and X are closed sets;

(2) the intersection of any collection of closed sets is closed;

(3) the union of any finite number of closed sets is closed.

Critical for verifying these properties from the properties of open sets are DeMorgan's laws regarding complements:

(1) $X \backslash \cup_i A_i = \cap_i (X \backslash A_i)$;

(2) $X \backslash \cap_i A_i = \cup_i (X \backslash A_i)$.

First, the empty set and the whole space X will be closed since their complements (X and the empty set) are open. Second, any intersection of closed sets will be closed since the complement of the intersection will be the union of the individual complements, and thus will be open since the union of open sets is open. Finally, any finite union of closed sets will be closed since the complement of the finite union will be the intersection of the individual complements and so will be open since the finite intersection of open sets is open. It is possible to define a topology in terms of the concept of closed sets and work with closed sets instead of open sets. The most familiar example of a closed set is the closed interval $[a, b]$. We leave it as an exercise to show that it is closed.

Exercise 1.2.6. Show that $[a, b]$ is a closed set in $\mathbb{R}$. Show that a rectangle (including the perimeter) is a closed set in $\mathbb{R}^2$.

Exercise 1.2.7. Show that $[a, b)$ is neither open nor closed in $\mathbb{R}$.

We now prove a couple of useful propositions involving the concept of closed sets. Each proposition follows from corresponding statements involving open sets by taking complements.

Proposition 1.2.1. $f : A \to B$ *is continuous iff the inverse images of closed sets are closed.*

Proof. Suppose f is continuous and C is a closed subset in B. Then $B \backslash C$ is open and $f^{-1}(C) = A \backslash f^{-1}(B \backslash C)$ is closed since it is the complement of an open set in A. The converse follows similarly and is left as an exercise. $\square$

Exercise 1.2.8. Complete the proof above by proving the converse.

Proposition 1.2.2. *If $A \subset X$ has the subspace topology, then $D \subset A$ is closed in A iff $D = A \cap E$, where E is closed in X.*

Proof. By the definition of the subspace topology, the open sets in A are the intersections of A with the open sets in X. What we are trying to prove here is a similar statement for closed sets. Suppose D is closed in A. Then $D = A \backslash F$, where F is open in A. Then $F = A \cap G$, where G is open in X. Hence, if $E = X \backslash G$, then E is closed in X and

$$D = A \backslash F = A \backslash (A \cap G) = A \cap (X \backslash G) = A \cap E.$$

The converse is left as an exercise. $\square$

Exercise 1.2.9. Complete the proof above by proving the converse.

Exercise 1.2.10. Suppose A is a closed subset of X. Then $D \subset A$ is closed in A (with the subspace topology) iff D is closed in X.

Definition 1.2.5. The *closure* of a set $A \subset X$, denoted $\bar{A}$, is the intersection of all closed sets containing A. The *interior* of A, denoted int A, is the union of all open sets contained in A. A point in int A is called an *interior point* of A. The *boundary* of A, denoted Bd A, is $\bar{A} \cap \overline{X \backslash A}$. A point in Bd A is called a *boundary point* of A.

Exercise 1.2.11. Show that $\bar{A}$ is closed and int A is open.

To find $\bar{A}$ in examples, it is useful to have another characterization. Note that a point x is *not* in $\bar{A}$ exactly when there is a closed set C containing A which does not contain x. But this means that $X \backslash C$ is an open set containing x which is disjoint from A, or, equivalently, is contained in $\text{int}(X \backslash A)$. Thus $\bar{A}$ consists of points of A and points not in A that have the property that every open set about them intersects A. Since points of A also have that property, points of $\bar{A}$ can be characterized in that every open set about them intersects A nontrivially. The description of $X \backslash \bar{A}$ above can also be rephrased as saying $X \backslash \bar{A} = \text{int}(X \backslash A)$. Using the definition of Bd A and the reformulation of $\bar{A}$, we can characterize points of Bd A as those points where every open set intersects both A and $X \backslash A$.

As an example, we determine $\bar{A}$, int A, and Bd A for $A = \{(x,y) \colon x > y > 0\}$. First note that this set is open since it is the intersection of the two open sets, $A_1 = \{(x,y) \colon x - y > 0\}$ and $A_2 = \{(x,y) \colon y > 0\}$. The sets A_1 and A_2 are open since they are the inverse images of $(0, \infty)$ under the continuous functions $x - y$ and y, respectively. Thus int $A = A$. The closure is found from A by adding the rays $x = y$ and $y = 0$ within the first quadrant. These points are in the closure since every open ball about a point in them will intersect A. The set $B = \{(x,y) \colon x \geq y \geq 0\}$ is closed since it is the intersection of two closed sets, $B_1 = \{x,y) \colon x - y \geq 0\}$ and $B_2 = \{(x,y) \colon y \geq 0\}$. These sets are closed since they are the inverse images of $[0, \infty)$ under the continuous functions $x - y$ and y, respectively. Thus $\bar{A} = B$. The set $X \backslash A$ is closed since its complement is open. Thus $\overline{X \backslash A} = X \backslash A$. Hence Bd $A = \bar{A} \cap \overline{X \backslash A} = \{(x,y) \colon x \geq 0, x = y\} \cup \{(x,y) \colon x \geq 0, y = 0\}$.

Exercise 1.2.12. Find $\bar{A}$, int A, and Bd A for the following sets A in $\mathbb{R}^2$:

(a) $\{(x,y) \colon x \geq 0,\ y \neq 0\}$;

(b) $\{(x,y) \colon x \in \mathbb{Q},\ y > 0\}$;

(c) $\{(x,y) \colon x^2 + y^2 < 1\}$.

Exercise 1.2.13. Show that $\bar{A} = \text{Int } A \cup \text{Bd } A$ and $\text{Int } A \cap \text{Bd } A = \emptyset$.

We will now prove a piecing lemma, which is very useful in verifying that certain functions which are constructed by piecing together continuous functions are themselves continuous.

Lemma 1.2.3 (Piecing lemma). *Suppose $X = A \cup B$, where A and B are closed subsets of X. Let $f \colon X \to Y$ be a function so that the restrictions of f to A and B (given the subspace topology) are each continuous (another way of saying this is that the compositions of f with the inclusions of A and B into X give continuous functions). Then f is continuous.*

Proof. Let $C \subset Y$ be closed. Our hypothesis then says that $A \cap f^{-1}(C)$ is closed in A and $B \cap f^{-1}(C)$ is closed in B. But Exercise 1.2.10 then says that these two sets are in fact closed in X since A and B are assumed to be closed subsets of X. Then $f^{-1}(C) = (A \cap f^{-1}(C)) \cup (B \cap f^{-1}(C))$ is closed since it is the union of two closed sets. $\qquad \square$

Exercise 1.2.14. Prove the analog of Lemma 1.2.3 where the word closed is replaced by the word open. Give an example to show that the conclusion that f is continuous is not true without some hypothesis on the sets A, B.

We will give many examples of continuous functions in the next section constructed by piecing together continuous functions defined on closed subsets. We state the definition of a homeomorphism and give the relevant version of the piecing lemma for homeomorphisms.

Definition 1.2.6. A *homeomorphism* is a bijection (1–1 and onto) between topological spaces so that the map and its inverse are both continuous. If $f : X \to Y$ is a homeomorphism, then we will say X is *homeomorphic* to Y, denoted $X \cong Y$.

Homeomorphism gives an equivalence relation on topological spaces, as it satisfies the three conditions of an equivalence relation: (1) reflexivity—the identity $1_X : X \to X$ has continuous inverse 1_X; (2) symmetry—if $f : X \to Y$ has continuous inverse $g : Y \to X$, then g has f as its continuous inverse; (3) transitivity—if $f : X \to Y$ has continuous inverse f^{-1}, and $g : Y \to Z$ has continuous inverse g^{-1}, then $gf : X \to Z$ has continuous inverse $f^{-1}g^{-1}$. A topologist looks at homeomorphic spaces as being essentially the same. One of the fundamental problems of topology is to decide when two topological spaces are homeomorphic. One technique for solving this problem (more successful in showing that spaces are not homeomorphic than in showing that they are homeomorphic) is to find properties of spaces which are preserved by homeomorphisms. We will study two such properties in this chapter, compactness and connectedness. Later we will study an invariant that is associated to any topological space called the fundamental group of the space. It has the property that homeomorphic topological spaces have isomorphic fundamental groups, and thus it may be used to distinguish between topological spaces up to homeomorphism.

We state our lemma for piecing together homeomorphisms. It follows from the piecing lemma in a straightforward manner, and we leave the proof as an exercise.

Lemma 1.2.4 (Piecing lemma for homeomorphisms). *Suppose that* $X = A \cup B, Y = C \cup D$, *where* A, B *are closed in* X, *and* C, D *are closed in* Y. *Let* $f : A \to C$ *and* $g : B \to D$ *be homeomorphisms, and suppose that the restrictions of* f *and* g *to the intersection* $A \cap B$ *agree as maps into* Y. *Define* $h : X \to Y$ *by* $h|A = f$ *and* $h|B = g$ *(or we could start with* h *and define* f *and* g *just by restricting them to* A *and* B*). If* h *is a bijection (this just requires that the only points that are in the image of both* f *and* g *are the points in the image of* $A \cap B$*), then* h *is a homeomorphism.*

Exercise 1.2.15. Prove the piecing lemma for homeomorphisms.

1.3 Geometric constructions of planar homeomorphisms

We now look at some geometric constructions which give continuous functions and homeomorphisms. For simplicity, we will restrict our domain space to the plane, although these constructions have analogues for other $\mathbb{R}^n$.

Our first example is a rotation. If a point in the plane is given by $r(\cos\theta, \sin\theta)$, then a rotation by an angle ϕ sends this to $r(\cos(\theta + \phi), \sin(\theta + \phi))$. One way of seeing that this is continuous is to note that distances between points are unchanged by this map. A map between metric spaces which leaves the distance between any two points unchanged is continuous; we leave this as an exercise.

Exercise 1.3.1.

(a) Show that any map from $\mathbb{R}^2$ to $\mathbb{R}^2$ which leaves distances between points unchanged (i.e. $d(f(\boldsymbol{x}), f(\boldsymbol{y})) = d(\boldsymbol{x}, \boldsymbol{y})$) is continuous.

(b) Generalize this to show that $f : (X, d) \rightarrow (Y, d')$ with $d'(f(x), f(y)) \leq Kd(x, y), K > 0$, is continuous.

That a rotation does in fact preserve distances can be checked using trigonometric formulas and the distance formula in the plane. Another way of seeing that a rotation by ϕ is continuous is to note that it is given by a linear map, $\boldsymbol{x} \rightarrow A\boldsymbol{x}$, where $\boldsymbol{x}$ represents a point in the plane as a column vector and A is the 2×2 matrix

$$\begin{pmatrix} \cos\phi & -\sin\phi \\ \sin\phi & \cos\phi \end{pmatrix}.$$

For a rotation, A is an orthogonal matrix, which means that it preserves the Euclidean inner product between vectors, and hence preserves distances between points. Multiplication by any matrix can be shown to give a continuous map. This is usually shown indirectly in advanced calculus courses by noting that a linear map is differentiable (it gives its own derivative) and that differentiable maps are continuous. It could also be shown directly using part (b) of Exercise 1.3.1 and the inequality $|A\boldsymbol{x} - A\boldsymbol{y}| \leq \|A\| \, |\boldsymbol{x} - \boldsymbol{y}|$ shown in linear algebra. Note that a rotation is reversible; after rotating a point by an angle θ, we can get back to our original point by rotating by an angle $-\theta$. From the matrix point of view, the matrix A is invertible. Either way may be used to show that rotation represents a homeomorphism from the plane to itself.

Another familiar geometric operation which gives a continuous map (and a homeomorphism) is a translation, $T_v(\boldsymbol{x}) = \boldsymbol{x} + \boldsymbol{v}$. This is seen to be continuous either directly from the definition or by the fact that it preserves distances between points. Its inverse is translation by $-\boldsymbol{v}$, and so it gives a homeomorphism.

Of course, we could rotate about some other point besides the origin. This also preserves distances and so can be shown to give a homeomorphism. Note that a rotation by angle ϕ about the point $\boldsymbol{x}$ is the composition of a translation by $-\boldsymbol{x}$ to send $\boldsymbol{x}$ to the origin, then a rotation of angle ϕ about the origin, and finally a translation by $\boldsymbol{x}$ to send the origin back to $\boldsymbol{x}$. A composition of homeomorphisms will give a homeomorphism, since a composition of continuous maps is continuous and the inverse of gf, given that g and f have inverses, is $f^{-1}g^{-1}$.

Another geometric construction which gives a homeomorphism is a reflection through a line. That this gives a homeomorphism follows from the fact that it is its own inverse and that it preserves distances between points. Alternatively, reflections through lines passing through the origin are given by multiplication by orthogonal matrices, and other reflections are conjugate to these using translations which move the line to one passing through the origin.

We may reinterpret the equivalence relation of congruence of triangles frequently studied in high school in terms of these three types of homeomorphisms: translations, rotations, and reflections. Suppose two triangles T_1, T_2 are congruent. Then they have corresponding sides A_1, B_1, C_1 and A_2, B_2, C_2, which are of the same length, and the angles between corresponding sides are the same. Let $\boldsymbol{v}_1$ be the vertex between A_1 and B_1 and $\boldsymbol{v}_2$ the vertex between A_2 and B_2. First translate the plane so that the vertex $\boldsymbol{v}_2$ is sent to $\boldsymbol{v}_1$. Now rotate about $\boldsymbol{v}_1$ so that the side A_2 lies along the side A_1. Now either the two triangles will agree or we can get from shifted triangle T_2 to T_1 by reflecting through the line going through side A_1. Thus two triangles are congruent if we can get from one to the other by a composition of translations, rotations, and reflections. Note that each type of map used above preserves distances between points. A map from the plane to itself which preserves distances between points is called a *rigid motion* or an *isometry*. In general, the term *isometry* is used for a map between metric spaces which preserves distance between points and their images.

It can be shown that any rigid motion of the plane is just a composition of translations, rotations, and reflections. We outline this argument. Starting with a rigid motion f, we get a new rigid motion g from f by translating by $-f(\boldsymbol{0})$: $g = T_{-f(\boldsymbol{0})}f$. Then $g(\boldsymbol{0}) = \boldsymbol{0}$. Now we use the relation of the dot product with the distance function $\langle \boldsymbol{x} - \boldsymbol{y}, \boldsymbol{x} - \boldsymbol{y} \rangle = d(\boldsymbol{x}, \boldsymbol{y})^2$ to show that $g(\boldsymbol{0}) = \boldsymbol{0}$ and $d(g(\boldsymbol{x}), g(\boldsymbol{y})) = d(\boldsymbol{x}, \boldsymbol{y})$ implies that $\langle g(\boldsymbol{x}), g(\boldsymbol{y}) \rangle = \langle \boldsymbol{x}, \boldsymbol{y} \rangle$. Thus g will send unit vectors to unit vectors and orthogonal vectors to orthogonal vectors. In particular, $\boldsymbol{q}_1 = g(\boldsymbol{e}_1), \boldsymbol{q}_2 = g(\boldsymbol{e}_2)$ are orthogonal unit vectors. If Q denotes multiplication by the orthogonal matrix $(\boldsymbol{q}_1 \ \boldsymbol{q}_2)$ with column vectors $\boldsymbol{q}_1, \boldsymbol{q}_2$, then Q is a rotation or reflection, and $h = Q^{-1}g$ is a rigid motion which preserves $\boldsymbol{0}, \boldsymbol{e}_1, \boldsymbol{e}_2$. Then h can be shown to be the identity by using the relation $\boldsymbol{v} = \langle \boldsymbol{v}, \boldsymbol{e}_1 \rangle \boldsymbol{e}_1 + \langle \boldsymbol{v}_2, \boldsymbol{e}_2 \rangle \boldsymbol{e}_2$.

Exercise 1.3.2. Fill in the details of the argument sketched above to show that a rigid motion in the plane is the composition of rotations, reflections, and translations.

Another familiar geometric relation is the similarity of triangles. If two triangles are similar, their angles will correspond exactly, but corresponding side lengths need not be equal but only have to have some common ratio k. If T_1 and T_2 are similar, we may use a rigid motion to align them so that sides A_1 and A_2 lie on the same line, as do the sides B_1 and B_2. Then the shifted T_2 will be sent to T_1 by a map that takes a line through $\boldsymbol{v}_1$ and sends the line to itself by

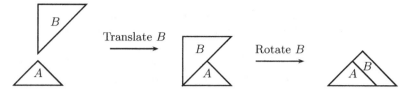

Figure 1.4. Similarity transformation.

shrinking or expanding along the line by a factor of k (in terms of the distance to v_1). This last map may be described as a composition of a translation of v_1 to the origin, multiplication of a vector by k, and then a translation of the origin back to v_1. The multiplication by k gives a continuous map, and its inverse is given by multiplication by $1/k$, so it gives a homeomorphism. We illustrate the first three steps in a similarity in the Figure 1.4. In this figure, no reflection was necessary as part of the rigid motion.

We have seen that congruences and similarities are both examples of homeomorphisms. In geometry, a triangle and a rectangle are distinguished from one another by the number of sides, and two triangles, although possibly not similar, still are seen to have the same "shape". We will see below that the inside of a triangle and the inside of a rectangle are in fact homeomorphic. Thus what is meant when one says that two triangles have the same shape and a triangle and a rectangle do not? It means that we are looking at the triangle and rectangle through "affine linear eyes".

There is a standard triangle $\Delta(e_0, e_1, e_2)$ with vertices $e_0 = 0$, e_1, e_2. Each point in it can be expressed as $(\lambda_1, \lambda_2) = \lambda_1 e_1 + \lambda_2 e_2$, with $\lambda_1, \lambda_2 \geq 0$ and $0 \leq \lambda_1 + \lambda_2 \leq 1$. We define $\lambda_0 = 1 - \lambda_1 - \lambda_2$, and then we can write $(\lambda_1, \lambda_2) = \lambda_0 e_0 + \lambda_1 e_1 + \lambda_2 e_2$, where $\lambda_0, \lambda_1, \lambda_2 \geq 0$ and $\lambda_0 + \lambda_1 + \lambda_2 = 1$. Now suppose we have another triangle with vertices e_0, v_1, v_2, where v_1, v_2 are linearly independent. If $V = (v_1 \ v_2)$, then multiplication by V is a linear transformation which gives a homeomorphism between $\Delta(e_0, e_1, e_2)$ and $\Delta(e_0, v_1, v_2)$. If three points a_0, a_1, a_2 satisfy the property that $v_1 = a_1 - a_0$, $v_2 = a_2 - a_0$ are linearly independent, then we say that a_0, a_1, a_2 are *affinely independent*. This is equivalent to $\lambda_1 a_0 + \lambda_1 a_1 + \lambda_2 a_2 = 0$, $\lambda_0 + \lambda_1 + \lambda_2 = 0$ implying $\lambda_0 = \lambda_1 = \lambda_2 = 0$. If a_0, a_1, a_2 are affinely independent, then there is a triangle $\Delta(a_0, a_1, a_2)$ with vertices a_0, a_1, a_2. Translation by a_0 gives a homeomorphism between $\Delta(e_0, v_1, v_2)$ and $\Delta(a_0, a_1, a_2)$, where $v_1 = a_1 - a_0$, $v_2 = a_2 - a_0$. The composition of multiplication by V and the translation then gives a map, called an *affine linear* map, which is a homeomorphism between the standard triangle $\Delta(e_0, e_1, e_2)$ and $\Delta(a_0, a_1, a_2)$. This affine linear map A has the property that $\lambda_0 e_0 + \lambda_1 e_1 + \lambda_2 e_2$ is sent to $\lambda_0 a_0 + \lambda_1 a_1 + \lambda_2 a_2$. In particular, this means that the triangle $\Delta(a_0, a_1, a_2)$ is characterized as the points $\lambda_0 a_0 + \lambda_1 a_1 + \lambda_2 a_2$ where $\lambda_i \geq 0$, $\lambda_0 + \lambda_1 + \lambda_2 = 1$. If $\Delta(b_0, b_1, b_2)$ is another triangle with affinely independent vertices b_0, b_1, b_2, then there is an affine linear map B sending the standard triangle to it. Then $C = BA^{-1}$ gives an affine linear map sending $\Delta(a_0, a_1, a_2)$ to $\Delta(b_0, b_1, b_2)$. Thus any two triangles in the plane are homeomorphic via a canonical affine linear map, and the image of a triangle under an affine linear map will be another triangle. In particular, there is no affine linear map sending a triangle to a rectangle. Affine linear maps from one triangle $\Delta(a_0, a_1, a_2)$ to another triangle $\Delta(b_0, b_1, b_2)$ are determined completely by the map on the vertices $a_i \to b_i$ and the affine linearity condition $\sum \lambda_i a_i \to \sum \lambda_i a_i$.

Exercise 1.3.3.

(a) Show that $a_1 - a_0$, $a_2 - a_0$ are linearly independent iff $\lambda_0 a_0 + \lambda_1 a_1 + \lambda_2 a_2 = 0$, $\lambda_0 + \lambda_1 + \lambda_2 = 0$ implies $\lambda_0 = \lambda_1 = \lambda_2 = 0$.

(b) Show that if a_0, a_1, a_2 are affinely independent, then $\lambda_1 a_0 + \lambda_1 a_1 + \lambda_2 a_2 = \mu_1 a_0 + \mu_1 a_1 + \mu_2 a_2$ with $\sum \lambda_i = \sum \mu_i = 1$ implies $\mu_i = \lambda_i, i = 0, 1, 2$.

(c) Show that any finite composition of translations and linear maps in the plane can be written as a single composition TL, where T is a translation and L is a linear map.

(d) Show that any composition M of translations and linear maps satisfies $M(\sum_{i=1}^{k} \lambda_i a_i) = \sum_{i=1}^{k} \lambda_i M(a_i)$ when $\sum_{i=1}^{k} \lambda_i = 1$. Conversely, show that if M satisfies this condition for any three affinely independent points, then M is a composition of a translation and a linear map, so is an affine linear map.

(e) Show that an affine linear map sending a_i to b_i will always send a line segment $\overline{a_0 a_1}$ to the line segment $\overline{b_0 b_1}$ via $(1-t)a_0 + ta_1 \to (1-t)b_0 + tb_1$, $0 \le t \le 1$.

Triangles and rectangles are not equivalent under invertible affine linear maps. A triangle and a rectangle are homeomorphic, however. Moreover, the homeomorphism may be taken to be "piecewise linear". If a_0, a_1, a_3 are the vertices of the triangle and b_0, b_1, b_2, b_3 are the vertices of the rectangle, then we can divide the rectangle into two triangles $B_1 = \Delta(b_0, b_1, b_2)$, $B_2 = \Delta(b_0, b_2, b_3)$ by introducing the edge $\overline{b_0 b_2}$ (see Figure 1.5). We can also introduce a vertex a_2 in the triangle at the midpoint of $\overline{a_1 a_3}$ and then an edge $\overline{a_0 a_2}$. Now the triangle is divided into two triangles, $A_1 = \Delta(a_0, a_1, a_2)$ and $A_2 = \Delta(a_0, a_2, a_3)$. The

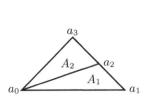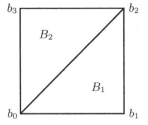

Figure 1.5. PL homeomorphism between a triangle and a rectangle.

map sending a_i to b_i can be extended affine linearly on triangles to give maps sending A_i to B_i. Figure 1.5 shows how the triangle and square are subdivided. This defines a homeomorphism between the triangle and the rectangle. That it is a homeomorphism follows from the piecing lemma for homeomorphisms. Note that on the triangles A_1, A_2, the map is affine linear $(\sum_i \lambda_i a_i \to \sum_i \lambda_i b_i)$. Our homeomorphism is an example of a *piecewise linear (PL) homeomorphism* of planar regions—the domain and range are divided into triangles, and the homeomorphism is an affine linear homeomorphism on each triangle.

We can generalize the argument above to show that any two convex polygonal regions in the plane are homeomorphic. By a polygonal region R, we mean a region that is bounded by a *closed polygonal path*; that is, $f([0, n])$, where $f : [0, n] \to \mathbb{R}^2$ with f affine linear on $[i, i + 1]$, $f(i) = x_i$, $x_0 = x_n$ and $f(a) = f(b)$ implies $a = b$ or $\{a, b\} = \{0, n\}$. The region R is called *convex* if R lies on one side of each line $\overline{x_i x_{i+1}}$ or, equivalently, line segments joining two points of R are in R. The region R bounded by P is then given by the union of line segments joining points in P. The idea of the proof that two convex polygonal regions are homeomorphic is to divide each region into the same number of triangles and then send the triangles to each other consistently. Our argument above with a triangle and a rectangle is the simplest case of this procedure.

Exercise 1.3.4.

   (a) Construct a PL homeomorphism between a square and a hexagon.

   (b) Show that any two convex polygonal regions are homeomorphic via a PL homeomorphism.

So far all of our examples of homeomorphisms have been piecewise linear. Here is an example of one that is not. The unit disk $D^2 = \{(x, y) : x^2 + y^2 \le 1\}$ is homeomorphic to the square $S = \{(x, y) : |x| \le 1, |y| \le 1\}$ (hence to any convex polygonal region). The homeomorphism may be described geometrically as follows. Each ray from the origin intersects D^2 and S in a line segment. The intersection with D^2 is sent linearly to the intersection with S.

We can verify that this is a homeomorphism by deriving a formula for it. This is somewhat tedious, however, so we will give a geometrical explanation, leaving the verification based on this as an exercise. We describe some corresponding open sets from our construction. Given a point x inside the disk which is not the center, we get $f(x)$ by first forming the circle about the center on which x lies, then forming the square which circumscribes this circle, and then sending x to the the point $f(x)$ on the intersection of the perimeter of this square and the ray through x. The region between two circles is then sent to the region between the corresponding squares. The basic open sets inside the circle are given by the region between two circles, which lie between two lines of angles $\theta = \theta_1$, $\theta = \theta_2$, as well as disks about the center. For the inside of a square, the basic open sets are given by regions between two smaller squares, again limited by the same two radial lines, as well as small squares about the center. Our map gives a correspondence between these basic open sets about x and $f(x)$ as pictured in Figure 1.6. At the center, a small disk about the center corresponds to a small

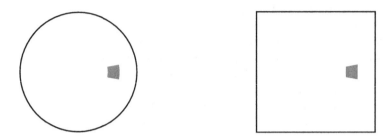

Figure 1.6. Basic open sets for disk and square.

rectangle about the center. From these facts, we can verify that the map is a homeomorphism.

Exercise 1.3.5. Use the geometrical facts cited above to verify that our construction gives a homeomorphism. You will need to use the fact that any open set about a point contains one of the basic open sets as described above.

Note that this homeomorphism sends the boundary circle to the perimeter of the square. In fact, if the homeomorphism of the circle to the perimeter of the square is given by $x \to f(x)$, then our homeomorphism is just $tx \to tf(x)$, $0 \leq t \leq 1$. We are using the convexity of each region to realize the region as the "cone" on its boundary and extending the homeomorphisms of boundaries by "coning".

This same idea could be used to give a homeomorphism between the unit disk and the inside of an ellipse, for example.

Exercise 1.3.6. Write down a formula for a homeomorphism between the unit disk D^2 and the ellipse $E = \{(x,y): x^2 + y^2/4 \leq 1\}$, and check whether it satisfies $f(tx, ty) = tf(x, y)$, $0 \leq t \leq 1$.

In the exercise above and the preceding example, there is a common idea. We take two subspaces in the plane A, B and points $p \notin A$, $q \notin B$. Then we form spaces pA, qB from taking the line segments joining p to points of A and line segments joining q to points of B. The set A is chosen so that each point in pA lies on a unique line segment from p to a unique point of A (and similarly for qB). In the case of the inside of the circle and the inside of the square, A is the circle and B is the square. For the disk and the inside of the ellipse, A is the circle and B is the ellipse. In both cases, $p = q = 0$. Then we take a homeomorphism $f : A \to B$, and then get a homeomorphism $F : pA \to qB$ by sending $(1 - t)p + ta$ to $(1 - t)q + tf(a)$. That F turns out to be a homeomorphism depends on pA, qB having the appropriate types of corresponding basic open sets. This can be rephrased in terms of the notion of a quotient topology, which we will study in Section 1.7. The construction of F from f is called *coning*.

We have seen many examples of different regions in the plane that turn out to be homeomorphic. Each of the regions so far has been homeomorphic to a disk. An important problem of topology is to characterize all regions in the plane that are homeomorphic to the disk. The homeomorphism would send the circle to a homeomorphic image—this is called a *simple closed curve*. Thus, a region R homeomorphic to a disk would have to be "bounded" by a simple closed curve.

The Jordan curve theorem and the Schönflies theorem combine to say that, if C is a simple closed curve in the plane, then it "bounds" a region R, and the homeomorphism $f : S^1 \to C$ extends to a homeomorphism between the unit disk D and R. The Jordan curve theorem says that the complement of the curve separates into two open connected pieces, one of which is bounded and the other of which is unbounded. It says the curve is the boundary of each piece. The Schönflies theorem then says that the bounded piece is homeomorphic to a disk and the unbounded piece is homeomorphic to the complement of a closed disk. We discuss connectedness in Section 1.6 and have a project to prove both theorems in the polygonal case in Section 1.8. A full proof of the Jordan curve theorem and it's generalization, the Jordan separation theorem, is given in terms of homology in Section 6.14 (see Theorems 6.14.2 and 6.14.6). The full proof of the Schönflies theorem can be found in [22]. A proof of the generalization of the Schönflies theorem to higher dimensions for locally flat embeddings is given in [5] based on the proof by Morton Brown [6].

A natural question would be to ask for examples of regions in the plane that are not homeomorphic to a disk. A simple example would be an annulus (see Figure 1.7), which is the region enclosed between two circles. There are two ways of seeing that this is not homeomorphic to a disk. One way is to compare their boundaries. The annulus has two circles as boundary and the disk has one. Of course, we have to understand why one circle is not homeomorphic to two circles (this can be based on the concept of connectedness, which we will study later) and why a homeomorphism between the annulus and the disk must restrict to a homeomorphism between their boundaries. A justification of the last fact actually leads us to the other reason that they are not homeomorphic. This involves the ideas surrounding the fundamental group of a space. Intuitively speaking, there is a circle (the middle circle) in the annulus which cannot be deformed continuously to a point, but every circle in the disk may be deformed

Figure 1.7. Annulus.

to a point (just contract the whole disk to its center and see what happens to the circle). This idea is responsible for a large number of applications and is pursued in Chapter 3. The classification of regions in the plane up to homeomorphism is a special case of the classification of surfaces with boundary. This latter topic is pursued in Chapter 2.

1.4   Compactness

We now discuss the property of compactness. We will discuss this in the context of a general topological space, but will specialize to metric spaces or subspaces of $\mathbb{R}^n$ on occasion.

Definition 1.4.1. Let X be a topological space. A subset $A \subset X$ is said to be *compact* if whenever A is contained in a union of open sets U_i (called an *open cover* of A), then A is contained in the union of a finite subcollection of these open sets (called a *finite subcover*).

This can be rephrased in terms of the open sets of A in the subspace topology by saying that whenever A is written as the union of a collection of open sets in A, then it may be written as the union of a finite number of these open sets.

One of the prime reasons that compactness is important as a topological concept is that it is preserved by continuous maps.

Proposition 1.4.1. *Let $f : X \to Y$ be continuous and X compact. Then the image set $f(X)$ is compact.*

Proof. Let $\mathcal{V} = \{V_i\}$ be an open cover of $f(X)$. Then $\mathcal{U} = \{U_i\} = \{f^{-1}(V_i)\}$ is an open cover of X. Since X is compact, there is a finite subcover $U_{i(1)}, \ldots, U_{i(k)}$ of X. Then the corresponding open sets $V_{i(1)}, \ldots, V_{i(k)}$ give a finite subcover of $f(X)$.   □

In particular, this implies that if two sets are homeomorphic, then either both are compact or both are not compact. A property that is invariant under homeomorphisms is called a *topological invariant*. Thus compactness is a topological invariant.

Let us look at some examples.

Example 1.4.1. The real line $\mathbb{R}$ is not compact since it can be written as the union of intervals $U_k = (-k, k)$ where k ranges over the integers, and it cannot be written as a union of a finite subcollection of these open sets. The same idea will show that, for a subset of $\mathbb{R}$ to be compact, it must be bounded (i.e. contained in a large interval). For if it is not, then we can use the collection $\{(-k, k)\}$ to cover the subset, and it cannot be contained in any finite subcollection of these. We leave it as an exercise to generalize this to subsets of metric spaces.

Exercise 1.4.1. A set A of a metric space is said to be *bounded* if it is contained in some ball $B(x, r)$. Show that a subset of a metric space which is compact must be bounded.

Example 1.4.2. A finite set $A = \{a_1, \ldots, a_k\} \subset X$ is compact. For if it is contained in a union of open sets U_i, then there must be some set $U_{i(j)}$ in the collection which contains a_j. Thus $U_{i(1)}, \ldots, U_{i(k)}$ gives a finite subcover of A.

Exercise 1.4.2. Show that a finite union of compact sets is compact.

Exercise 1.4.3. Decide whether or not the following subsets of $\mathbb{R}$ are compact:

  (a) $A = \{1/n : n \in \mathbb{N}\}$;
  (b) $B = \{0\} \cup A$;
  (c) $(0, 1]$.

We have seen that $\mathbb{R}$ is not compact, but $\mathbb{R}$ is closed as a subset of itself. Thus a closed set does not have to be compact. A compact set does not have to be closed in a general topological space, either. For example, the two-point space, where the only open sets are the empty set and the space itself, has either of its points as a compact subset, but that point is not a closed set with this topology. However, if we are dealing with subsets of Euclidean space and the standard topology, then compact sets are closed. We will give a proof in the more general situation of a metric space.

Proposition 1.4.2. *In a metric space, compact sets are closed.*

Proof. Let X be a metric space and A a compact subset of X. To show that A is closed, we have to show that its complement is open. Let $x \in X \backslash A$; we need to find a ball about x that does not intersect A. Let y be a point of A. Then we can find disjoint balls $B(y, r(y))$ and $B(x, r(x))$. The union of the $B(y, r(y))$ over all y in A will contain A; since A is compact, there is a finite subcollection of these balls which covers A. Then the intersection of the corresponding balls about x will be an open set about x which does not intersect the union of the subcollection, and hence does not intersect A. $\qquad\square$

The crucial property of a metric space X which we used here was that a metric space is Hausdorff.

Definition 1.4.2. X is called *Hausdorff* if given $x, y \in X$ then there are disjoint open sets U_x, U_y with $x \in U_x, y \in U_y$.

Exercise 1.4.4. Show that in a Hausdorff space, compact sets are closed. (Hint: In a general topological space, a set U will be open if given $x \in U$, then there is an open set $U(x)$ with $x \in U(x) \subset U$; for then we can write U as the union of the sets $U(x)$ as x ranges over the points of U, and the union of open sets is open. With this criterion for a set to be open, the proof in the metric case can be modified to prove the result.)

The next proposition allows us to deduce that certain sets are compact by knowing that they are closed subsets of a compact set.

Proposition 1.4.3. *Let X be compact and let A be closed in X. Then A is compact.*

Proof. Suppose that $\mathcal{U} = \{U_i\}$ is a collection of open sets of X whose union contains A. Then the U_i together with $X \backslash A$ is a collection of open sets whose union is X, and so some finite subcollection will contain X. Since no points of A are contained in $X \backslash A$, then the U_i in this subcollection will contain A.   □

We combine the propositions connecting compact and closed sets to prove the following very useful proposition that certain bijections between sets are homeomorphisms.

Proposition 1.4.4. *Let $f : X \to Y$ be a bijection (i.e. 1–1 and onto). Suppose that f is continuous, X is compact, and Y is Hausdorff. Then f is a homeomorphism.*

Proof. Since f is a bijection, it has an inverse $f^{-1} : Y \to X$. To see that f is a homeomorphism, we need to see that f^{-1} is continuous. We use here the characterization of a continuous function as one which has the inverse image of a closed set being closed. Let C be a closed set in X. Then X compact implies that C is compact. But $(f^{-1})^{-1}(C) = f(C)$ is the image of a compact set, and so is compact. In a Hausdorff space, a compact set is closed, so $f(C)$ is closed as required.   □

In the proof above, we showed that if $f : X \to Y$ is continuous, X is compact, and Y is Hausdorff, then f sends closed sets to closed sets. A map which sends closed sets to closed sets is called a *closed map*. When f is invertible, then f^{-1} being continuous is the same thing as f being a closed map.

This proposition would no longer be true if we removed the hypothesis that X is compact. For example, consider the function $f : [0, 1) \to S^1$ given by $f(x) = (\cos 2\pi x, \ \sin 2\pi x)$.

Exercise 1.4.5. Show that the function f defined above is a bijection that is continuous but is not a homeomorphism. (Hint: Consider the open set $\left[0, \frac{1}{2}\right) \subset [0, 1)$ and its image.)

We begin studying some basic compact sets in the reals. We first show that a closed interval $[a, b]$ is compact in the usual topology of the line. This proof is based on the least upper bound property of the real numbers, which we now review. A subset $A \subset \mathbb{R}$ is said to have an *upper bound* u if $a \leq u$ for all $a \in A$. u is called the *least upper bound* of A if it is a upper bound and it is less than or equal to any other upper bound. The *least upper bound property* of the real numbers asserts that any nonempty subset of the reals with an upper bound has a least upper bound. This property does not hold for the rationals; for example, the set of rational numbers with square less than 2 has an upper bound, but does not have a least upper bound. As a subset of the reals, the least upper bound would be $\sqrt{2}$. We can think of the reals as being formed from the rationals by adding to the rationals all the least upper bounds of subsets of the rationals that are not already in the rationals.

Theorem 1.4.5. *The closed interval $[a, b]$ is compact.*

Proof. Suppose that we have an open cover $\mathcal{U} = \{U_i\}$ of $[a, b]$. Consider the set $A = \{x \in [a, b]: [a, x]$ has a finite subcover$\}$. We intend to show that $A = [a, b]$. First note that A is not empty since some U_i contains a, and thus must contain some interval $[a, b_1]$, for $b_1 > a$. Since b is an upper bound for A, the set A must have a least upper bound, which we will call u. We want to show that $u = b$ and that $b \in A$. Suppose first that $u < b$. Since $u \in [a, b]$, there must be some element of the cover, which we will call $U_{i(u)}$, which contains u. Now $U_{i(u)}$ contains some interval $[u_1, u_2]$, where $a < u_1 < u < u_2 < b$. Since u is the least upper bound for A, there must be an element $a_1 \in A$ with $u_1 < a_1 \leq u$ (if not, then u_1 would be a smaller upper bound, contradicting the choice of u as the least upper bound). But then $[a, a_1]$ is covered by a finite number of the U_i and thus so is $[a, u_2]$ (just use those that cover $[a, a_1]$ together with $U_{i(u)}$). But this contradicts u being an upper bound for A, since now $u_2 \in A$. Thus the least upper bound must be b. Now choose an element $U_{i(b)}$ of the cover which contains b, and choose u_1 with $[u_1, b] \subset U_{i(b)}$. Then b being the least upper bound for A implies that there is an element $a_1 \in A$ with $u_1 < a_1 \leq b$. But $[a, a_1]$ is covered by a finite number of the U_i and $[a_1, b]$ is contained in $U_{i(b)}$, so $[a, b]$ is contained in a finite number of the elements of the cover, showing that it is compact. $\square$

As a corollary, we can now characterize the compact sets in the line.

Corollary 1.4.6. *$A \subset \mathbb{R}$ is compact iff it is closed and bounded.*

Proof. If it is compact, then it must be bounded by Exercise 1.4.1 and closed by Proposition 1.4.2. Conversely, suppose that it is closed and bounded. Since it is bounded, it is contained in some closed interval $[a, b]$. Since it is closed as a subset of the line, it will also be closed in $[a, b]$. But this makes it a closed subset of a compact space, and so it is compact. $\square$

Exercise 1.4.6. Analogous to the definition of least upper bound is that of greatest lower bound. Give a definition of greatest lower bound for a set $A \subset \mathbb{R}$ and use the least upper bound property to show that a set with a lower bound must have a greatest lower bound.

Exercise 1.4.7. Give an example of a closed, bounded subset A of a metric space X that is not compact. (Hint: Consider the metric space X itself to be a bounded noncompact subset of $\mathbb{R}$ and $A = X$.)

For $\mathbb{R}$ we extract an important property of a closed bounded set.

Proposition 1.4.7. *A compact subset A of $\mathbb{R}$ has a largest element M and a smallest element m; that is, $m \leq a \leq M$ for all $a \in A$.*

Proof. We show that it has a largest element; the proof for a smallest element is analogous. Since A is compact, it is bounded, and so has a least upper bound u. We claim that $u \in A$, and hence u will be the largest element of A. Suppose that u is not in A; then we claim that A could not be closed. For every interval about u has to contain an element of A in order for u to be the least upper bound of A. But this means that the complement of A is not open; hence A is not closed. $\square$

Now we give an application of this to analysis.

Proposition 1.4.8. *Let $f \colon X \to \mathbb{R}$ be continuous and X compact. Then f assumes a maximum (and minimum) on X; that is, there are $x, y \in X$ with $f(x) \leq f(z) \leq f(y)$ for all $z \in X$.*

Proof. To say that f assumes a maximum just means that $f(X)$ has a largest element. But X compact and f continuous means that $f(X)$ is compact and so has a largest element. □

When X is a closed interval, this is the familiar theorem from calculus that a continuous function assumes a maximum and a minimum on a closed interval.

1.5   The product topology and compactness in $\mathbb{R}^n$

We wish to generalize our characterization of compact sets in $\mathbb{R}$ to show that a subset of $\mathbb{R}^n$ is compact iff it is closed and bounded. The only missing ingredient from our proof above is knowing that a cube $[a_1, b_1] \times [a_2, b_2] \times \cdots \times [a_n, b_n]$ is compact. This can be proved inductively if we can show that the product of compact sets in a product of Euclidean spaces is compact. To do this most efficiently, we need to discuss the notion of a product topology on the product $X \times Y$ of two topological spaces.

Suppose that X and Y are topological spaces and consider their product $X \times Y = \{(x, y) \colon x \in X, \ y \in Y\}$. We will define a topology on $X \times Y$ by saying that a set $W \subset X \times Y$ is open if given any $(x, y) \in W$, then there are open sets U in X and V in Y so that $(x, y) \in U \times V \subset W$. In particular, products of open sets will be open, and the general open set will be a union of products of open sets. It is not difficult to verify that this definition of open sets does satisfy the three requirements for a topology, which is called the *product topology*.

Exercise 1.5.1. Verify that open sets as defined above satisfy the three properties required of a topology.

Now the product topology in the plane is not defined in exactly the same way as the usual metric topology, but it does give the same topology; that is, it gives the same collection of open sets. To see this, first note that if a set W is open in the plane in the usual metric topology, and (x, y) is a point of A, then there is a small ball about (x, y) that is contained in W. But inside this ball we can find a rectangle that is a product of intervals which contains (x, y). Hence W is open in the product topology. Conversely, suppose W is open in the product topology, and $(x, y) \in W$. Then there is a product $U \times V$ of open sets (which we may choose to be intervals) with $(x, y) \in U \times V \subset W$. Then the rectangle $U \times V$ is contained in W. But then we can find a ball contained in the rectangle and containing (x, y), so W is open in the metric topology (see Figure 1.3). Inductively, a similar argument shows that the metric topology on $\mathbb{R}^n$ arises as the inductive product of n copies of $\mathbb{R}$ using the product topology.

Thus to show that a product of closed intervals is compact in $\mathbb{R}^n$, it suffices to show that the product of compact sets is compact in the product topology. We first need a preliminary lemma on product spaces.

Proposition 1.5.1. *Suppose X and Y are topological spaces and let $X \times Y$ have the product topology. Then the inclusions $i_x : Y \to X \times Y, i_x(y) = (x, y)$, $i_y : X \to X \times Y$, $i_y(x) = (x, y)$, are continuous. Moreover, each projection $p_X : X \times Y \to X$, $p_X(x, y) = x$, $p_Y : X \times Y \to Y$, $p_Y(x, y) = y$, is continuous. In particular, the map $X \to X \times \{y\}$ given from i_y by restricting the range, and $Y \to \{x\} \times Y$ given from i_x similarly, are homeomorphisms, where $X \times \{y\}$ and $\{x\} \times Y$ are given the subspace topology.*

Proof. We first show that i_x is continuous; the proof is analogous for i_y. Let W be an open set in the product topology on $X \times Y$, and let $y \in i_x^{-1}(W)$. Then $(x, y) \in W$, so there are open sets U, V with $(x, y) \in U \times V \subset W$. Then $y \in V \subset i_x^{-1}(W)$, so $i_x^{-1}(W)$ is open (using the hint in Exercise 1.4.4). We now show that p_X is continuous; the proof for p_Y is analogous. Let U be an open set in X. Then $p_X^{-1}(U) = U \times Y$, which is an open set in the product topology. Finally, note that i_x and p_Y are inverses to one another (when properly restricted) and so give homeomorphisms between Y and $\{x\} \times Y$; similarly, i_y and p_X give homeomorphisms between X and $X \times \{y\}$. $\square$

We now show that the product of compact spaces is compact.

Theorem 1.5.2 (Tychanoff). *Suppose X and Y are compact. Then the product $X \times Y$ is compact.*

Proof. Let $\mathcal{W} = \{W_i\}$ be an open cover of the product. Fix $x \in X$ and consider the set $\{x\} \times Y$. It is homeomorphic to Y, so it is compact. Thus there are a finite number of the W_i, which we will denote by $W_{i_{x,1}}, \ldots, W_{i_{x,k}}$, which cover $\{x\} \times Y$. Let $W_x = W_{i_{x,1}} \cup \cdots \cup W_{i_{x,k}}$. Then for each $y \in Y$, select an open set

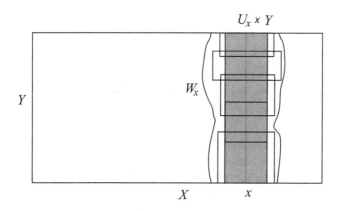

Figure 1.8. A tube $U_x \times Y \subset W_x$.

$U_y \times V_y$ with $(x, y) \in U_y \times V_y \subset W_x$. Then this gives an open cover of $\{x\} \times Y$, and so there is a finite subcover $U_{y_1} \times V_{y_1}, \ldots, U_{y_p} \times V_{y_p}$. Let $U_x = \bigcap_{j=1}^{p} U_{y_j}$. Note that $\{x\} \times Y \subset U_x \times Y \subset W_x$. The set $U_x \times Y$ is sometimes called a *tube* about $\{x\} \times Y$ inside W_x. This is illustrated in Figure 1.8. As x varies over X, the sets U_x give an open cover of X and so there is a finite subcover $U_{x(1)}, \ldots, U_{x(r)}$. Then $U_{x(1)} \times Y, \ldots, U_{x(r)} \times Y$ will cover $X \times Y$, and so will the corresponding $W_{x(i)}$. But each $W_{x(i)}$ is the union of a finite number of sets in our original cover, and so we will get a covering by a finite number of sets in our original cover. $\square$

The Tychanoff theorem holds for infinite products as well, and it is closely related to the axiom of choice. See [24] for a discussion and proof in this context.

Now we are ready to characterize the compact sets in $\mathbb{R}^n$.

Theorem 1.5.3 (Heine–Borel). *A subset of $\mathbb{R}^n$ is compact iff it is closed and bounded.*

Proof. We showed that compact implies closed and bounded in a metric space. Suppose A is closed and bounded. Then A will be a closed subset of some large cube (which is compact) and hence will be compact. $\square$

We now wish to introduce another form of compactness, sequential compactness, and show that it is equivalent to compactness in a metric space. In general, these concepts are not equivalent but counterexamples are rather sophisticated. In the course of doing so, we will also introduce the concept of the Lebesgue number of a cover, and show that compact metric spaces have Lebesgue numbers, a fact which will be very useful to us in Chapter 3.

Definition 1.5.1. A sequence in X is a function $s : \mathbb{N} \to X$, where $\mathbb{N}$ denotes the natural numbers. We usually denote $s(n)$ by s_n and the sequence by $\{s_n\}$. A subsequence s' of a sequence s is a sequence formed by taking the composition $s' = sj$, where $j : \mathbb{N} \to \mathbb{N}$ is order preserving ($a < b$ implies $j(a) < j(b)$). It is usually denoted by s_{n_i} where $n_i = j(i)$. A sequence is said to converge to x if given an open set U about x, there is a natural number N so that $n > N$ implies $s_n \in U$.

Definition 1.5.2. X is called *sequentially compact* if every sequence in X has a convergent subsequence.

We wish to give a criterion for a sequence to have a subsequence which converges to x. If a subsequence converges to x, then the definition of convergence implies that for any open set U containing x, there are an infinite number of values of n so that $s_n \in U$. We show the converse is true in a metric space.

Proposition 1.5.4. *Suppose X is a metric space and $x \in X$. If $\{s_n\}$ is a sequence so that for any ball about x, the ball contains an infinite number of the s_n (this means that there are an infinite number of n so that s_n is in the ball), then there is a subsequence which converges to x.*

Proof. Choose n_1 so that s_{n_1} is contained in the ball of radius 1 about x. Since there are an infinite number of the s_n in the ball of radius $\frac{1}{2}$ about x, we can find n_2 so that $n_2 > n_1$ and $s_{n_2} \in B\left(x, \frac{1}{2}\right)$. Inductively, we then use the same idea to choose $n_1 < n_2 < n_3 < \cdots$ so that $s_{n_j} \in B(x, 1/j)$. This will give us our convergent subsequence. We leave the details as an exercise. $\square$

Exercise 1.5.2. Fill in the details in the proof above.

Proposition 1.5.5. *In a metric space, compactness implies sequential compactness.*

Proof. We prove the contrapositive. Suppose X is not sequentially compact and s_n is a sequence with no convergent subsequence. If there are only a finite number of distinct values s_n, then some value must be repeated infinitely often and we can use this to get a constant, hence convergent, subsequence. Thus we may assume that there are an infinite number of distinct values s_n. For each $x \in X$, there is no subsequence which converges to x. By the criterion above, there is an open set U_x about x which contains only a finite number of the s_n. But a covering of X by these balls, one for each x, can have no finite subcover, since a finite subcover would have to contain only a finite number of the values of the sequence (which are infinite in number), and hence could not contain all of X. $\square$

The proof above does not need the full strength of the metric space hypothesis, just the existence for each x of a sequence of open sets U_n with $U_{i+1} \subset U_i$ about x so that any open set about x contains some U_i. This property is called first countability and is pursued in Exercises 1.9.39–1.9.41 at the end of the chapter.

Exercise 1.5.3. Show that if $\{s_n : n \in \mathbb{N}\}$ is finite, then the sequence has a convergent subsequence.

We now show that in a metric space, sequential compactness implies compactness. To prove this, we introduce the concept of the Lebesgue number of a cover. Let A be a subset of the metric space (X, d). Consider $D_A = \{d(a_1, a_2) : a_1, a_2 \in A\}$. If D_A is bounded from above, define $d_A = \sup D_A$. We will call d_A the *diameter* of the set A.

Definition 1.5.3. A covering $\mathcal{U} = \{U_i\}$ of a metric space is said to have *Lebesgue number* $\delta > 0$ if every set $A \subset X$ of diameter less than δ is contained in some element of the covering.

Proposition 1.5.6. *Let X be a metric space which is sequentially compact. Then every open covering of X has a Lebesgue number.*

Proof. We prove the contrapositive: if there is an open cover with no Lebesgue number, then there is a sequence with no convergent subsequence. Let $\mathcal{U} = \{U_i\}$ be an open cover with no Lebesgue number. Then there is a sequence of sets $\{A_n\}$ with the diameter of A_n less than $1/n$ which are not contained in any element of the cover. Choose $a_n \in A_n$. Then we claim that $\{a_n\}$ is a sequence with no convergent subsequence. Suppose there were a subsequence $\{a_{n_k}\}$ which converges to a point x, and choose an element U_p of the cover containing x.

Choose m large enough so that $B(x, 1/m) \subset U_p$, and choose $k_1 \geq 2m$ so that if $k \geq k_1$, $a_{n_k} \in B(x, 1/2m)$. Then if $a \in A_{n_k}$, $d(a, x) \leq d(a, a_{n_k}) + d(a_{n_k}, x) < 1/2m + 1/2m = 1/m$. But this means $A_{n_k} \subset U_p$, which is a contradiction.   □

Proposition 1.5.7. *In a metric space, sequential compactness implies compactness.*

Proof. A metric space is *totally bounded* if given $\epsilon > 0$, we can cover X by a finite number of balls of radius ϵ. We first show that X sequentially compact implies that it is totally bounded. We show this by proving the contrapositive. Suppose X cannot be covered by a finite number of balls of radius ϵ. Let $x_1 \in X$. Since $B(x_1, \epsilon)$ does not cover X, choose $x_2 \notin B(x_1, \epsilon)$. Since $B(x_1, \epsilon) \cup B(x_2, \epsilon)$ does not cover X, we may choose $x_2 \notin B(x_1, \epsilon) \cup B(x_2, \epsilon)$. Inductively, we can choose a sequence $\{x_n\}$ in this manner with $x_{n+1} \notin \bigcup_{k=1}^{n} B(x_k, \epsilon)$. Since $d(x_n, x_k) \geq \epsilon$ for $k < n$, any ball of diameter ϵ can contain at most one x_n, so the sequence can have no convergent subsequence.

Now suppose X is sequentially compact and $\mathcal{U} = \{U_i\}$ is an open cover. By Proposition 1.5.6 we can find a Lebesgue number δ for this cover. By the above argument, there is a cover of X by a finite number of balls of radius $\delta/3$. But each such ball will be of diameter less than δ, so it will lie in an element of our original cover, $B(x_k, \delta/3) \subset U_{i(k)}, k = 1, \ldots, n$. Then $U_{i(1)}, \ldots, U_{i(n)}$ give a finite subcover of our original cover.   □

Definition 1.5.4. Let $(X, d_X), (Y, d_Y)$ be metric spaces. Then $f : X \to Y$ is said to be *uniformly continuous* if given $\epsilon > 0$ there exists a $\delta > 0$ such that for $x_1, x_2 \in X, d_X(x_1, x_2) < \delta$ implies $d_Y(f(x_1), f(x_2)) < \epsilon$.

Exercise 1.5.4. Show that f uniformly continuous implies f is continuous, but construct an example to show that the converse does not hold.

Exercise 1.5.5. Let $f : X \to Y$ be a continuous map of the compact metric space (X, d_X) to the metric space (Y, d_Y). Show that f is uniformly continuous. (Hint: Use the Lebesgue number of the covering $\{f^{-1}(B(y, \epsilon/2))\}_{y \in Y}$ of X.)

1.6   Connectedness

We next want to discuss the concept of connectedness. The definition is given in terms of its negation, as it is easier to say what we mean by a space not being connected.

Definition 1.6.1. A topological space X is called *separated* if it is the union of two disjoint, nonempty open sets. A subset $A \subset X$ is *separated* if A is separated as a topological space, using the subspace topology. A set is called *connected* if it is not separated.

Exercise 1.6.1. Show that a space X is connected iff the only subsets of X which are both open and closed are $\emptyset$ and X.

We rephrase the conditions for a subset $A \subset X$ to be separated or connected in terms of open sets in X.

Proposition 1.6.1.

(a) $A \subset X$ *is separated iff there are two open sets* $U, V \subset X$ *so that* $A \subset U \cup V$, $U \cap V \cap A = \emptyset$, $U \cap A \neq \emptyset, V \cap A \neq \emptyset$.

(b) *A set* $A \subset X$ *is connected iff whenever* U, V *are open sets in* X *so that* $U \cap V \cap A = \phi, A \subset U \cup V$, *then* $A \subset U$ *or* $A \subset V$.

Proof. We only prove (a), leaving (b) as an exercise. Suppose A is separated. Then there are two disjoint nonempty sets U', V' which are open in A so that $A = U' \cup V'$. Since U', V' are open in A, there are open sets U, V in X with $U' = U \cap A$, $V' = V \cap A$. Since U' and V' are disjoint, we have $U \cap V \cap A = \emptyset$. Since $A = U' \cup V'$, we have $A \subset U \cup V$. This proves one direction. For the other direction, given U, V with $A \subset U \cup V$, $U \cap V \cap A = \emptyset, U \cap A \neq \emptyset$, $V \cap A \neq \emptyset$, then defining $U' = U \cap A$, $V' = V \cap A$ gives two nonempty sets U', V' which are open in A and show that A is separated. □

Exercise 1.6.2. Deduce (b) from (a).

Example 1.6.1. We use Proposition 1.6.1 to describe some examples of separated sets. The first example we give is the union of two points in the line $X = \{0, 1\}$. To see that this is separated, we choose $U = (-0.1, 0.1), V = (0.9, 1.1)$. A similar example would be to let $Y = [0, 1] \cup [2, 3]$. Then we could choose $U = (-0.1, 1.1), V = (1.9, 3.1)$. Our final example may be somewhat less intuitive. The rationals $\mathbb{Q}$ in the line are separated. Here we can choose $U = (-\infty, \sqrt{2}), V = (\sqrt{2}, \infty)$. We will show that the $\mathbb{R}$ itself is connected, so the missing irrational numbers were crucial in separating the rational ones. Note that the openness condition in the definition is crucial. For example, you cannot get an interval being separated by dividing it into two pieces, say $[0, 2] = [0, 1] \cup (1, 2]$. The problem is that to get an open set U about $[0, 1]$ you have to include points greater than 1 and so it will not be disjoint with an open set about $(1, 2]$.

We first investigate connectedness for subsets of the line. Consider the following property:

(*) If $x, y \in A \subset \mathbb{R}$, then the interval $[x, y] \subset A$.

Proposition 1.6.2. *Any connected set in the line satisfies* (*) *or, equivalently, any set that does not satisfy* (*) *is separated.*

Proof. If A does not satisfy (*), then there are points x, y, z with $x < y < z$ and $x, z \in A$ and $y \notin A$. But then A is separated by the two open sets $(-\infty, y)$ and (y, ∞). □

What are the sets that satisfy (*)? The next proposition says that they are just the intervals, rays, and $\mathbb{R}$.

Proposition 1.6.3. *A set* $A \subset \mathbb{R}$ *satisfies* (*) *iff it is an interval, a ray, or* $\mathbb{R}$.

Proof. It is straightforward to see that an interval, ray, or $\mathbb{R}$ satisfies (*). Suppose A satisfies (*). There are a number of cases to consider; we will only consider one of the cases and leave the completion of the proof as an exercise. We consider the case where A is bounded both from above and below. Let a be the greatest lower bound and b the least upper bound of A. This implies $A \subset [a,b]$. We will show that $(a,b) \subset A$. Let c be a point in (a,b). Since a is the greatest lower bound, there is an element $e \in A$ with $a \le e < c$. Similarly, b being the least upper bound implies that there is an element $f \in A$ with $c < f \le b$. But (*) implies that $[e,f] \subset A$ and so $c \in A$. Hence $(a,b) \subset A$. But $A \subset [a,b]$, so there are four possibilities for $A: (a,b), [a,b), (a,b], [a,b]$, each of which is an interval. The other cases one has to consider are when A is not bounded on one side or the other or both.   □

Exercise 1.6.3. Complete the proof of the proposition by considering the other cases.

Our previous two propositions say that the only possibilities for connected sets in $\mathbb{R}$ are intervals, rays, and $\mathbb{R}$. We now show that they are connected.

Proposition 1.6.4. *Any interval, ray, or $\mathbb{R}$ is connected.*

Proof. We will just give the proof for a closed interval $[a,b]$, and leave the other cases for the reader. In Proposition 1.6.1 we re-expressed the condition of connectivity by saying that a set is connected if, whenever it is contained in the union of two open sets U, V with $U \cap V \cap A = \emptyset$, then it is entirely contained in one of the two sets. Suppose that $[a,b]$ is contained in the union of two open sets U, V with $U \cap V \cap [a,b] = \emptyset$. Assume that $a \in U$. To show that $[a,b]$ is connected, we must show that $[a,b] \subset U$. Analogously to the proof that $[a,b]$ is compact, we form the set $A = \{x \in [a,b]: [a,x] \subset U\}$. Since U is open, we see that A contains some $x > a$. Since A is bounded, it must have a least upper bound u. We first claim that $u \in U$. If not, then $u \in V$ and so there is $u_1 < u$ with $[u_1, u] \subset V$ since V is open. But u being the least upper bound of A means that there is $c \in A$ with $u_1 < c \le u$. But then $c \in U \cap V \cap A$, which is a contradiction. If $u \ne b$, we can find an interval $[u_1, u_2] \subset U$, with $u_1 < u < u_2$, and so $[a, u_2] \subset U$, contradicting the choice of u as an upper bound. Thus we must have $[a,b] \subset U$, and so $[a,b]$ is connected.   □

Exercise 1.6.4. Show that $\mathbb{R}$ is connected.

The three preceding propositions together yield the following theorem.

Theorem 1.6.5. *The connected sets in $\mathbb{R}$ are intervals, rays, and $\mathbb{R}$.*

Here is a useful proposition about connectedness, which could be used to show that $\mathbb{R}$ is connected, knowing that $[a,b]$ is connected.

Proposition 1.6.6. *Suppose that A_i is a collection of connected subsets of a topological space X so that they all have at least one point a in common. Then the union $A = \cup_i A_i$ is connected.*

Proof. Suppose $A \subset U \cup V$, where $U \cap V \cap A = \emptyset$, and suppose further that $a \in U$. Then we have to show that $A \subset U$. But each A_i being connected will imply that $A_i \subset U$, so $A \subset U$. □

Exercise 1.6.5. Use the proposition above to deduce that $\mathbb{R}$ is connected from the fact that a closed interval is connected.

Unfortunately, there is no nice characterization of connected subsets of other Euclidean spaces as there is for compact subsets, although the above proposition is very useful in recognizing connected sets.

We prove that connectedness is preserved under continuous maps, and hence gives another topological invariant for a space up to homeomorphism.

Proposition 1.6.7. *The continuous image of a connected space is connected.*

Proof. Suppose $f : X \to Y$ is continuous and X is connected. Suppose $f(X) \subset U \cup V$, where $U \cap V \cap f(X) = \emptyset$ and U, V are open. Then $X \subset f^{-1}(U) \cup f^{-1}(V)$, and $X \cap f^{-1}(U) \cap f^{-1}(V) = \emptyset$. Now f continuous implies that $f^{-1}(U)$ and $f^{-1}(V)$ are open, and so X connected means that X is contained in one of these, say $f^{-1}(U)$. Hence $f(X) \subset U$, and so $f(X)$ is connected. □

Since the continuous image of a connected set is connected, so is a homeomorphic image. Hence connectedness is also a topological invariant. This fact could be used to show, for example, that two disjoint intervals could not be homeomorphic to one interval.

A somewhat more intuitive property than connectedness is path connectedness.

Definition 1.6.2. A space X is called *path connected* if, given $x, y \in X$, there is a continuous map $f : [0, 1] \to X$ (called a *path* in X) with $f(0) = x$ and $f(1) = y$. We say that the path connects x to y.

There is an equivalence relation generated by this definition as follows: we say $x \sim y$ if there is a path connecting x to y. The constant path at x shows $x \sim x$. That $x \sim y$ implies $y \sim x$ can be seen by composing a path connecting x to y with a self homeomorphism of $[0, 1]$ which is order reversing; usually one uses the linear map $\alpha(t) = 1 - t$, but any order reversing homeomorphism will work. That $x \sim y, y \sim z$ implies $x \sim z$ involves reparametrizing the paths and lying them end on end. Geometrically, we just traverse the path connecting x to y and then traverse the path from y to z. However, to get a parametrized path from the two paths involves reparametrizing them so that their domains fit together nicely. For example, we can compose f with $\alpha(t) = 2t$, so $f\alpha(0) = f(0) = x, f\alpha(\frac{1}{2}) = f(1) = y$. Then we could similarly reparametrize g with an affine linear map $\beta : [\frac{1}{2}, 1] \to [0, 1]$ and define the path connecting x to z by making it $f\alpha$ on $[0, \frac{1}{2}]$ and $g\beta$ on $[\frac{1}{2}, 1]$. We leave the details as an exercise.

Exercise 1.6.6. Show that the relation $x \sim y$ as defined above is an equivalence relation.

The equivalence classes under this equivalence relation are called the *path components* in X. For example, if $X = [0,1] \cup [2,3]$, then the intervals $[0,1]$ and $[2,3]$ would be the path components. A set is path connected iff it has only one path component.

We show that path connectedness is preserved by continuous maps, hence, by homeomorphisms, so is a topological invariant.

Proposition 1.6.8. *Suppose X is path connected and $f : X \to Y$ is a continuous map. Then $f(X)$ is path connected.*

Proof. Let $u = f(x), v = f(y)$ be points of $f(X)$. Since X is path connected, there is a path α connecting x and y. Then $f\alpha$ is a path connecting u and v. □

The basic relationship between the two forms of connectivity is given by the following proposition.

Proposition 1.6.9. *If X is path connected, then X is connected.*

Proof. Pick a point $x \in X$, and for each point $y \in X$, choose a path connecting x to y. The images of these paths are all connected since they are images of connected sets under continuous maps, and each of them contains x. Their union (as we let y range over all of the points of X) is all of X, and so we get that X is connected by applying Proposition 1.6.6. □

It is not the case that a connected set has to be path connected. Here is an example of a set in the plane, called the *topologist's sine curve*, which is connected but is not path connected. Our set is based on the $\sin 1/x$ curve. Figure 1.9 shows a global and a local view (near a point on the y-axis) of its graph. It is the union of two sets, A and B. Here A is just the graph of $\sin 1/x$, where $0 < x \leq 1$, and B is the segment along the y-axis where the y-coordinate ranges from -1 to 1. To see that $A \cup B$ is connected, the idea is that if it were contained in a union $U \cup V$ of open sets with no points in both U and V, then since A and B are each connected (being the images of connected sets under continuous maps), each would have to lie entirely in one of the sets. Suppose that $B \subset U$. Then since U is open, we can show that at least one point of A must also lie in U. Since A is connected, then all of A must also lie in U and so $A \cup B$ lies in U. That $A \cup B$ is not path connected is based on the idea that there can be no path connecting a point of A to a point of B. The basic idea is to use the fact that a small ball about a point in B will intersect A in an infinite number of disjoint arcs, and to show that for $A \cup B$ to be path connected, we would have to be able to connect points in different arcs while staying in such a ball, which is impossible. Verification of the details are left as Exercises 1.9.44–1.9.46 at the end of the chapter.

We now consider some examples of path connected, hence connected, sets in Euclidean spaces.

Example 1.6.2. As our first example, note that $\mathbb{R}^n$ is path connected. We can take a straight line path connecting any two points $\boldsymbol{x}, \boldsymbol{y}, f(t) = (1-t)\boldsymbol{x} + t\boldsymbol{y}$. By analogous reasoning, any convex set (a set where straight lines joining any two

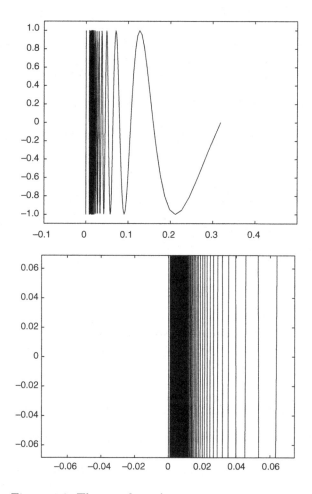

Figure 1.9. The topologist's sine curve—two views.

points in the set lie in the set) is path connected. This contains balls and cubes, for example.

Example 1.6.3. The unit sphere $S^n \subset \mathbb{R}^{n+1}$ is path connected, $n \geq 1$. The best way to see this is to show that $\mathbb{R}^{n+1}\backslash\{0\}$ is path connected, and then show that there is a continuous map from $\mathbb{R}^{n+1}\backslash\{0\}$ onto S^n. To see that $\mathbb{R}^{n+1}\backslash\{0\}$ is path connected, note that if $x, y \in \mathbb{R}^{n+1}$ and the straight line joining them does not pass through 0, then it may be used to give a path connecting them as before. If it does pass through 0, then choose a point z that is not on this line (here we use $n \geq 1$). Then the straight line from x to z together with the straight line from z to y may be used to give a path from x to y. We can get a continuous map from $\mathbb{R}^{n+1}\backslash\{0\}$ onto S^n by projecting along lines through the origin. Precisely, this map is given by the formula, $P(x) = x/|x|$, where $|x|$ denotes the length of x.

Exercise 1.6.7. Show that a union of path connected sets with a point in common is path connected. (Hint: Let z be the common point. Then show that given x, y in the union, we can find a path that joins them by using paths in individual path connected spaces that join x to z and join z to y.)

Although a connected set need not be path connected, here is a situation where that is true.

Proposition 1.6.10. *Let A be an open subset of $\mathbb{R}^n$. If A is connected, then A is path connected.*

Proof. We show that A has only one path component, hence is path connected. Note that each path component P is open in A, since each point has a ball about it contained in A and each point of the ball can be connected to the center by a straight line path. If A had more than one path component, let P_1 be a path component and P_2 be the union of the other path components. Then P_1, P_2 give a separation of A into two disjoint, nonempty open sets, a contradiction.   □

We conclude this section on connectedness by proving a version of the intermediate value theorem.

Proposition 1.6.11 (Intermediate value theorem). *Suppose that $f : X \to \mathbb{R}$ is a continuous function and X is connected. Let $c = f(x_1)$ and $d = f(x_2)$ and suppose that $c < e < d$. Then there is $x_3 \in X$ with $f(x_3) = e$.*

Exercise 1.6.8. Prove the intermediate value theorem using the fact that $f(X)$ is connected and our characterization of connected sets in the line.

This theorem is encountered in calculus when X is a closed interval $[a, b]$. In this context, it says that a continuous function must assume every value between $f(a)$ and $f(b)$. Another way of stating this is to say that the closed interval with end points $f(a), f(b)$ is a subset of $f([a, b])$. By combining compactness and connectedness, we can describe completely what the image of a closed interval under a continuous map to the reals must be. Since it must be connected, it has to be an interval, a ray, or all of the reals. Since it must be compact, the only possible choice is a closed interval. The end points of this interval will be the minimal value and the maximal value of the function. We state this as a proposition.

Proposition 1.6.12. *If $f : [a, b] \to \mathbb{R}$ is continuous, then $f([a, b]) = [m, M]$ where m, M are the minimal and maximal values of the function.*

Exercise 1.6.9. Show that the letter T is not homeomorphic to the letter O. (Hint: Consider what happens when a point is removed from each letter and the corresponding connectivity properties.)

Exercise 1.6.10. Show that S^1 is not homeomorphic to $\mathbb{R}$ by showing $S^1 \backslash \{x\}$ is not homeomorphic to $\mathbb{R} \backslash \{y\}$.

Exercise 1.6.11. Show that two disjoint concentric circles in the plane are not homeomorphic to one circle.

1.7 Quotient spaces

We discuss the notion of a quotient space, which is also called an identification space. We will be using quotient spaces extensively in Chapter 2 when we study surfaces.

Definition 1.7.1. Suppose X, Y are topological spaces, and we have a surjective map $q : X \to Y$. Then we say Y has the *quotient topology with respect to* (X, q) if $U \subset Y$ is open iff $q^{-1}(U) \subset X$ is open. Y is then called a *quotient space* of X and q is called a *quotient map*.

A simple example of a quotient map is the map $q : \mathbb{R} \to S^1$ where $q(t) = e^{2\pi i t}$. The arcs in the circle which provide a basis for its topology have as their inverse images the unions of disjoint intervals in the reals.

Note that the map q is continuous when Y has the quotient topology. For whenever $U \subset Y$ is open, the definition of the quotient topology requires that $q^{-1}(U)$ has to be open.

Exercise 1.7.1. Suppose $q : X \to Y$ and Y has the quotient topology with respect to (X, q). Show $C \subset Y$ is closed iff $q^{-1}(C) \subset X$ is closed.

Quotient spaces often arise by starting with some known space X and then forming Y from X by identifying certain points of X, usually by means of an equivalence relation we put on points of X. The map q then sends a point $x \in X$ to the equivalence class of all points that are identified with x. In this context, Y is sometimes called an *identification space* and the quotient map q is called an *identification map*. The equivalence class containing x is denoted by $[x]$ and the map sending a point to its equivalence class is denoted by $q(x) = [x]$. A simple, but quite important, example comes from starting with $X = [0, 1]$, and then making 0 equivalent to 1 the only nontrivial equivalence relation. The quotient space then can be imagined geometrically by taking a piece of string and then joining the end points to get a circle up to homeomorphism for the quotient space Y.

Suppose $f : X \to Z$ is a continuous function and $Y = X/\!\sim$ is formed from X by identifying points in X within the same equivalence class, $q : X \to Y, q(x) = [x]$. Then f induces a map $\bar{f} : Y \to Z$ if whenever x_1 is equivalent to x_2 then $f(x_1) = f(x_2)$; that is, identified points are sent to the same point by f. We define $\bar{f}$ by $\bar{f}([x]) = f(x)$. This is well defined because, if we choose $[x_1] = [x_2]$, then $x_1 \sim x_2$ and $f(x_1) = f(x_2)$. We are defining $\bar{f}$ by $\bar{f}q(x) = \bar{f}([x]) = f(x)$. We call $\bar{f}$ *the map induced by* f. The quotient topology is set up so that f continuous implies $\bar{f}$ is continuous. For if U is an open set in Z, then to check that $\bar{f}$ is continuous, we verify that $\bar{f}^{-1}(U)$ is open in Y. To check this, we use the quotient map $q : X \to Y, q(x) = [x]$. Then $\bar{f}^{-1}(U)$ is open in Y iff $q^{-1}(\bar{f}^{-1}(U))$ is open in X. Since $\bar{f}q = f$, the condition is that $f^{-1}(U)$ is open, which is true since f is continuous.

Figure 1.10. Saturated open sets $q^{-1}(U)$ about $[0]$ for $[0,1]$ and $\mathbb{R}$.

When a quotient space is formed by identifying points, it is difficult to picture the equivalence classes directly and the open sets in the quotient space. What we can do, however, is picture their inverse images within the space X. The sets $q^{-1}(U)$ are open sets that are *saturated* with respect to the equivalence relation. This means that if $x \in q^{-1}(U)$ and $x \sim y$, then $y \in q^{-1}(U)$. For a simple example, consider $X = [0,1]$ with the only nontrivial equivalence being $0 \sim 1$. Then the basis for the topology of $X/\sim$ will have inverse images being open intervals in $(0,1)$ and also sets of the form $[0,a) \cup (b,1]$ for $0 < a < b < 1$. The last set is a saturated open set that contains the equivalence class $\{0,1\}$. A related example uses $X' = \mathbb{R}$ and forms the quotient space using the equivalence relation $x \sim x+n, n \in \mathbb{Z}$. A typical equivalence class is $\{\ldots, x-2, x-1, x, x+1, x+2, \ldots\}$. A basic open set U about this point will have inverse image $q^{-1}(U) = \cup_{n \in \mathbb{Z}}(x + n - \epsilon, x + n + \epsilon)$, where $\epsilon < \frac{1}{2}$. This is just an interval about x together with all of its translates by integers. See Figure 1.10.

We prove some elementary propositions about quotient spaces. The first proposition formalizes our last observation about induced maps.

Proposition 1.7.1. *Let Y be a quotient space of X with quotient map $q \colon X \to Y$. Let $g \colon Y \to Z$ be a map. Then g is continuous iff the composition gq is continuous.*

Proof. If g is continuous, then the composition is continuous since the composition of continuous functions is continuous. Conversely, suppose the composition is continuous and $U \subset Z$ is an open set. Look at $g^{-1}(U)$. To see that it is open, we have to show that $q^{-1}(g^{-1}(U))$ is open. But $q^{-1}(g^{-1}(U)) = (gq)^{-1}(U)$, so it is open since gq is continuous. □

Proposition 1.7.2. *Suppose Y is a quotient space with respect to (X,q) and Y' is a quotient space with respect to (X',q'). Let $f \colon X \to X', \bar{f} \colon Y \to Y'$ be maps with $q'f = \bar{f}q$. We also could express this by saying that the following diagram is commutative.*

$$
\begin{array}{ccc}
X & \xrightarrow{\ f\ } & X' \\
{\scriptstyle q}\big\downarrow & & \big\downarrow{\scriptstyle q'} \\
Y & \xrightarrow{\ \bar{f}\ } & Y'
\end{array}
$$

Then $\bar{f}$ is continuous if f is continuous.

Proof. To show $\bar{f}$ is continuous, we need to show that $\bar{f}q$ is continuous, by Proposition 1.7.1. But $q'f = \bar{f}q$ and f,q' continuous imply $q'f$ is continuous. □

Proposition 1.4.4 has a nice application for quotient spaces.

Proposition 1.7.3. *Suppose $f: X \to Y$ is a surjective continuous map, X is compact and Y is Hausdorff. Define an equivalence relation on X by saying $u \sim v$ iff $f(u) = f(v)$; the equivalence classes are the inverse images $f^{-1}(y)$. Then the induced map $\bar{f}: X/\sim \to Y$ is a homeomorphism.*

Proof. Proposition 1.7.1 implies that $\bar{f}$ is continuous. It is a bijection since we are identifying points in X which map to the same point. Since $X/\sim$ is the continuous image of the compact space X by the quotient map $q: X \to X/\sim$, we have that $X/\sim$ is compact. Then Proposition 1.4.4 implies that $\bar{f}$ is a homeomorphism. $\qquad\square$

We now apply these propositions to the quotient spaces $Y = [0,1]/\sim$ and $Y' = \mathbb{R}/\sim$. Consider the map $f: [0,1] \to S^1$ given by $f(t) = (\cos 2\pi t, \sin 2\pi t)$. This is a continuous surjection and the only nontrivial inverse image is $f^{-1}\{(1,0)\} = \{0,1\}$. Thus if we form the quotient space Y from the interval $X = [0,1]$ by identifying 0 with 1, then Proposition 1.7.3 implies that the induced map $\bar{f}$ is a homeomorphism.

We could instead start with $X' = \mathbb{R}$ and identify x with $x + n, n \in \mathbb{Z}$ to form the quotient space Y'. We claim that Y' is also homeomorphic to the circle. We start with the same map p, now considered as a map from the reals. It determines a map $\bar{p}: Y' \to S^1$ by $\bar{p}[t] = (\cos 2\pi t, \sin 2\pi t)$. This is well defined since $(\cos 2\pi(t + n), \sin 2\pi(t + n)) = (\cos 2\pi t, \sin 2\pi t)$ and is continuous, by Proposition 1.7.1. Note that it is onto since both q and p are. It is also 1–1, since $\bar{p}[t] = \bar{p}[t']$ implies $(\cos 2\pi t, \sin 2\pi t) = (\cos 2\pi t', \sin 2\pi t')$. But this only happens if $t = t' + n$ for some integer n; hence $[t] = [t']$. To see that $\bar{p}$ is in fact a homeomorphism, we can no longer use Proposition 1.7.3 since $\mathbb{R}$ is not compact. We need to see that its inverse $\bar{p}^{-1}$ is continuous. But this is equivalent to $(\bar{p}^{-1})^{-1}(U) = \bar{p}(U)$ being open when U is open; that is, $\bar{p}$ sends open sets to open sets. Since $\bar{p}(U) = pq^{-1}(U)$, this condition is equivalent to p sending saturated open sets to open sets. But p is an *open map*; that is, it sends open sets to open sets. Hence $\bar{p}$ is a homeomorphism from $\mathbb{R}/\sim$ to S^1.

We state, for future use, the principle used in the last example.

Proposition 1.7.4. *Suppose $f: X \to Y$ is a surjective continuous map. Define an equivalence relation on X by saying $u \sim v$ iff $f(u) = f(v)$; the equivalence classes are the inverse images $f^{-1}(y)$. Then the induced map $\bar{f}: X/\sim \to Y$ is a homeomorphism exactly when f sends saturated open sets $q^{-1}(U)$ to open sets. In particular, it is a homeomorphism if f is an open map.*

Since each of Y, Y' is homeomorphic to S^1, they are homeomorphic to each other. We now show this more directly. Let $q: X \to Y$, $q': X' \to Y'$ be the identification maps. Define $i: X \to X'$ by inclusion. Since $[i(0)] = [0] = [1] = [i(1)]$, i induces a map $\bar{i}: Y \to Y'$ defined by $\bar{i}([x]) = [i(x)]$. Thus we have a commutative diagram (i.e. $\bar{i}q = q'i$) by definition. Thus $\bar{i}$ is continuous since i is. Next note that $\bar{i}$ is 1–1 since $\bar{i}q$ is except for $0,1$, and $[0] = [1]$ in Y. $\bar{i}$ maps onto Y' since any $[y] \in Y'$ is represented by a y between 0 and 1. We leave it as

Figure 1.11. Cylinder and torus as quotient spaces of the square.

an exercise to construct an inverse for $\bar{i}$ and to prove it is continuous.

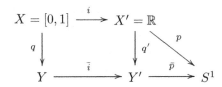

Exercise 1.7.2. Construct an inverse for $\bar{i}$ and show that it is continuous. (Hint: Consider the discontinuous function from X' to X defined by sending x to $x - [x]$, where $[x]$ denotes the greatest integer in x; i.e. the unique integer satisfying $[x] \leq x < [x] + 1$.)

Consider the product of the circle with itself. This space is called a *torus* and will be studied in more depth in Chapter 2. From our description of the circle as a quotient space, we may give a description of $S^1 \times S^1$ as a quotient space. We take the product $\mathbb{R} \times \mathbb{R}$ and make the following identifications: $(s, t) \sim (s + m, t + n)$, where $m, n \in \mathbb{Z}$. An alternate description would be to take $[0, 1] \times [0, 1]$ and identify $(0, t)$ with $(1, t)$ and $(s, 0)$ with $(s, 1)$. A pictorial description is given in Figure 1.11. It is supposed to indicate that we identify the edges labeled a and the edges labeled b. Geometrically, we can think of gluing the edges labeled a together to form a cylinder (the b edges becoming circles) and then gluing the two circles together to get a torus.

Exercise 1.7.3. Describe basic open sets in the quotient space $[0, 1] \times [0, 1]/(0, t) \sim (1, t), (s, 0) \sim (s, 1)$ about each of the points $[(0, 0)], [(\frac{1}{2}, 0)], [(0, \frac{1}{2})]$, and $[(\frac{1}{2}, \frac{1}{2})]$. Describe the inverse image $q^{-1}(U)$ of each of these basic open sets.

Exercise 1.7.4. Show that the quotient space formed from a square by identifying all of the points in the bottom edge of the square to each other is homeomorphic to a triangle. (Hint: Start with the map from the rectangle to the triangle preserving y-levels and sending the bottom edge of the rectangle to the bottom vertex of the triangle. See Figure 1.12, where the bottom line that is to be collapsed to a point is thickened, as is the image point.)

Figure 1.12. Triangle as a quotient space of the square.

We now discuss quotient spaces that are formed from two disjoint sets by identifying certain points in one of the sets with points in the other by means of a function. Suppose A and B are disjoint topological spaces. Then the union of A and B can be regarded as a topological space by saying a set is open iff it is the union of an open set in A with an open set in B. We will denote the union with this topology as $A \bigsqcup B$, and call it the disjoint union. Frequently, we will perform this construction when A and B are not disjoint. In this case we will regard them as disjoint by distinguishing points by saying the point comes from A or it comes from B. This is the reason for our terminology "disjoint union"—we want to emphasize that we are regarding the two sets as disjoint. Now suppose K is a closed subset of B and f is a homeomorphism from K onto a closed subset $f(K)$ of A. Then we may form the quotient space $A \cup_f B = (A \bigsqcup B)/x \sim f(x), x \in K \subset B$, formed from the disjoint union by identifying $x \in K$ with $f(x) \in f(K)$.

Proposition 1.7.5. *Let $g : A \cup_f B \to C$ be a map induced from continuous functions $g_A : A \to C$, $g_B : B \to C$ with $g_A f = g_B | K$. That is, if $x \in A \subset A \cup_f B$, then $g(x) = g_A(x)$, and if $x \in B, g(x) = g_B(x)$. Then g is continuous.*

Proof. To show g is continuous, we have to show that the composition $gq : A \bigsqcup B \to A \cup_f B \to C$ is continuous. But the topology on the disjoint union is just the union of the topologies on A and B. Since the restriction of this composition to A, B is just g_A, g_B, respectively, it is continuous. $\qquad \square$

Proposition 1.7.6. *Let $A_i \cup_{f_i} B_i = A_i \bigsqcup B_i / x \sim f_i(x)$, $x \in K_i \subset B_i$ be the quotient space of $A_i \cup B_i$ coming from identifying $x \in K_i \subset B_i$ with $f(x) \in A_i$ via a homeomorphism $f_i : K_i \to f(K_i)$, $i = 1, 2$. Suppose there are homeomorphisms $F_A : A_1 \to A_2, F_B : B_1 \to B_2$ with $F_B(K_1) = K_2$ and $F_A f_1 = f_2 F_B$. Then the map $F : A_1 \cup_{f_1} B_1 \to A_2 \cup_{f_2} B_2$ given by $F(x) = F_A(x)$ if $x \in A_1$ and $F(x) = F_B(x)$ if $x \in B_1$ is a homeomorphism.*

Exercise 1.7.5. Prove Proposition 1.7.6.

Given a topological space X and closed subsets A, B with $A \cup B = X$, we can regard X as a quotient space of $A \bigsqcup B$ using $\mathrm{id} : A \cap B \subset B \to A \cap B \subset A, \mathrm{id}(x) = x$. For the inclusion maps give a map $q : A \bigsqcup B \to X$; this induces $A \cup_{\mathrm{id}} B \to X$, which is a bijection. To see that it is a homeomorphism just requires showing X has the quotient topology. A set C in X is closed iff $C \cap A$ and $C \cap B$ are closed since $C = (C \cap A) \cup (C \cap B)$, and A and B are assumed closed. The quotient topology on X from $(A \bigsqcup B, q)$ comes from requiring C to be closed iff $q^{-1}(C)$ is closed in $A \cup B$; that is, $C \cap A$ and $C \cap B$ are closed. Thus X does have the quotient topology and so $\bar{q}$ is a homeomorphism.

Suppose now we have homeomorphisms $h_A : A \to A', h_B : B \to B'$. Then Proposition 1.7.6 implies $X = A \cup_{id} B \simeq A' \cup_f B'$, where $f : h_B(A \cap B) \to h_A(A \cap B)$ is $f(x) = h_A h_B^{-1}(x)$. We will use this in situations where we can choose A', B' to be particularly nice spaces such as disks or rectangles.

As an example, consider the annulus $X = \{(x_1, x_2) : 1 \le x_1^2 + x_2^2 \le 2\}$. We can first break X up into $A = X \cap \{(x_1, x_2) : x_2 \le 0\}$ and $B = X \cap \{(x_1, x_2) : x_2 \ge 0\}$. We will give a number of different descriptions of the annulus as a quotient space (see Figure 1.13). The variety of descriptions given below illustrate that a space may arise as a quotient space in many different ways. The simplest description comes from using $f : [-1, 1] \times [1, 2] \to X, f(s, t) = (t \cos \pi s, t \sin \pi s)$. The first coordinate s is used to wrap the interval around the circle (giving the angle up to a factor of π), and the second coordinate t measures the distance from the origin. This map sends $(-1, t)$ and $(1, t)$ to the same point $(-t, 0)$ and is otherwise 1–1. Thus f induces a homeomorphism between the quotient space $Q_1 = [-1, 1] \times [1, 2]/(-1, t) \sim (1, t)$ and the annulus X. We could replace the interval $[1, 2]$ by the homeomorphic interval $[-1, 1]$ and thus identify Q_1, and hence X, to the quotient space $Q_2 = [-1, 1] \times [-1, 1]/(-1, t) \sim (1, t)$. We will think of this as the standard description of the annulus as a quotient of the square $D^1 \times D^1$, where we are identifying the left-hand boundary interval to the right hand boundary interval. We depict this identification and the corresponding image on the annulus by labeling the identified edges with the letter a.

We now split the interval $[-1, 1]$ into two intervals $[-1, 0]$ and $[0, 1]$ and think of it as a quotient of the disjoint union by identifying the two copies of 0. Using

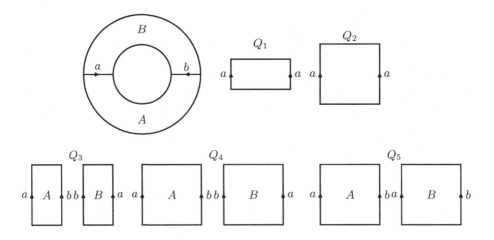

Figure 1.13. Expressing the annulus as a quotient space.

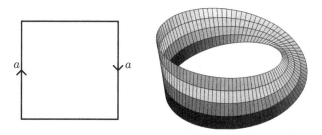

Figure 1.14. Möbius band.

this, the inclusion gives maps of the disjoint union $[-1,0] \times D^1 \bigsqcup [0,1] \times D^1$ to $D^1 \times D^1$. This map then induces a homeomorphism between the quotient space $Q_3 = [-1,0] \times D^1 \cup_f [0,1] \times D^1$ and Q_2, where $f(-1,t) = (1,t), f(0,t) = (0,t)$. This identification of the two copies of $0 \times D^1$ is labeled with b, as is its image in the annulus. Now by identifying $[-1,0]$ and $[0,1]$ with D^1 using the order preserving affine linear maps, we can re-express Q_3 as the quotient $Q_4 = D^1 \times D^1 \cup_g D^1 \times D^1$, where $g(-1,t) = g(1,t), g(1,t) = (-1,t)$. As another description, form a quotient $Q_5 = D^1 \times D^1 \cup_h D^1 \times D^1$, where $h(-1,t) = (-1,t), h(1,t) = (1,t)$; that is, h is the identity on the identified edges. The homeomorphism $F: Q_5 \to Q_4$ is induced from the map that sends the left-hand copy of $D^1 \times D^1$ to itself via the identity, and sends the right-hand copy of $D^1 \times D^1$ to itself via $(s,t) \to (-s,t)$. We use Proposition 1.7.6 to see that this induces a homeomorphism from Q_5 to Q_4.

Descriptions such as the last ones will be very useful to us in Chapter 2, where we study surfaces. We will decompose a surface into a number of pieces, each of which is homeomorphic to D^2 or $D^1 \times D^1$ and then think of the surface as a quotient space of the disjoint union of these nice pieces. The structure of the surface will be contained in the pieces involved and how they are glued together.

Here is another example. The Möbius band B is formed from a rectangular strip by identifying the ends after making a half twist as in Figure 1.14. More formally, $B = D^1 \times D^1/(-1,t) \sim (1,-t)$. We might also write this as a quotient space formed from two rectangles by splitting $D^1 = [-1,0] \cup [0,1]$ to form $Q'_1 = [-1,0] \times D^1 \cup_k [0,1] \times D^1$, with $k(-1,t) = (1,-t), k(0,t) = (0,t)$. By identifying $[-1,0]$ and $[0,1]$ with D^1, we can re-express this as a quotient space $D^1 \times D^1 \cup_p D^1 \times D^1$, with $p(-1,t) = (1,t), p(1,t) = (-1,-t)$.

Exercise 1.7.6. Consider the space X formed from two copies of $R = D^1 \times D^1$ by identifying $\{-1,1\} \times D^1$ to itself via d with $d(1,y) = (1,-y)$ and $d(-1,y) = (-1,-y)$; that is, $X = R \cup_d R$. Construct a homeomorphism between X and the annulus.

Exercise 1.7.7. Suppose $X = A \cup_g B, Y = A \cup_{g'} B$, where $g, g': K \subset B \to g(K), g'(K) \subset A$ are homeomorphisms. Suppose $(g')^{-1}g: K \to K = h|K$, where

$h: B \to B$ is a homeomorphism. Show that the identity on A and h on B piece together to give a homeomorphism from X to Y.

Exercise 1.7.8. Show that the Möbius band can also be described as a quotient space $D^1 \times D^1 \cup_f D^1 \times D^1$, with $f(-1,t) = (-1,t)$, $f(1,t) = (1,-t)$.

Exercise 1.7.9. Identify all points in the lower half of the circle to each other. Show that the resulting quotient space $S^1/\sim$ is homeomorphic to S^1. (Hint: Find a continuous map from the circle to the circle which sends the lower half of the circle to a point and is 1–1 elsewhere.)

Exercise 1.7.10. Put an equivalence relation on the unit disk by making all points on the boundary circle equivalent to each other. Show that the resulting quotient space $D^2/\sim$ is homeomorphic to S^2. (Hint: Send diameters to great circles with the origin going to the south pole and the boundary circle going to the north pole.)

1.8 The Jordan curve theorem and the Schönflies theorem

In this section we outline proofs of the Jordan curve theorem and the Schönflies theorem for polygonal curves. The section is essentially in the form of a project to fill in the details of the outline to prove these results. The proofs of these theorems in the polygonal case will provide us with many opportunities to apply the concepts from the chapter in justifying geometric steps in the argument.

We start by carefully stating these theorems in their general versions.

Definition 1.8.1. A *simple closed curve* in the plane is a function $f: S^1 \to \mathbb{R}^2$ which is a homeomorphism onto its image. The image $C = f(S^1) \subset \mathbb{R}^2$ is sometimes also called a simple closed curve when the parametrization is not important. Alternatively, a simple closed curve in the plane can be given as a map $f: [a,b] \to \mathbb{R}^2$, with $f(x) = f(y)$ for $x \neq y$ iff $\{x,y\} = \{a,b\}$ so that when the quotient space $[a,b]/a \sim b$ is identified with S^1, the induced map $\bar{f}$ is a homeomorphism onto its image.

Theorem 1.8.1 (Jordan curve theorem). *Let $C = f(S^1)$ be a simple closed curve in the plane. Then $\mathbb{R}^2 \backslash C$ is the disjoint union of two open sets A, B so that each is path connected. Moreover, one of these sets A is bounded and the other B is unbounded. Also, C is the boundary of each of these sets.*

Theorem 1.8.2 (Schönflies theorem). *Let $C = f(S^1)$ be a simple closed curve in the plane and $\mathbb{R}^2 \backslash C = A \cup B$ as given by the Jordan curve theorem, with A bounded. Then there is a homeomorphism of the plane to itself which sends the open unit disk to A and the closed unit disk to $A \cup C$.*

The Jordan curve theorem was first stated as a theorem by Camille Jordan (1838–1932) in his Cours d'Analyse in the late nineteenth century. His original proof was very complicated and was found to have gaps, which required considerable effort to fill in. Modern proofs use homology theory, where the separation

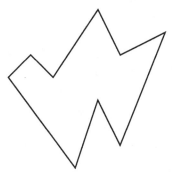

Figure 1.15. A polygonal simple closed curve.

part of the theorem is expressed by saying that $H_0(\mathbb{R}^2\backslash C)$ is the free abelian group on two generators. H_0 measures the path components of a space; the two generators correspond to A and B. The difficulty, in general, has to do with the very wild nature a simple closed curve may have. If the curve is restricted somewhat, then the theorem becomes much easier. The Schönflies theorem was proved in 1908.

In this section we will only look at the case of a polygonal simple closed curve, which is the image of a map $p : [0, n] \to \mathbb{R}^2$ where, on each subinterval $[k, k+1]$, the map is an affine linear map onto a line segment L_k determined by the points (called *vertices*) $p(k) = v_k$ and $p(k + 1) = v_{k+1}$. We assume that $p(0) = p(n)$ but $p(a) \neq p(b)$ if $a \neq b$ unless $\{a, b\} = \{0, n\}$. Note that the quotient space $[0, n]/0 \sim n$ is homeomorphic to S^1 and p determines a map $\bar{p} \colon S^1 \to \mathbb{R}^2$ as in the original definition of a simple closed curve. Figure 1.15 shows an example of a polygonal simple closed curve and the bounded region which it bounds. We will assume that adjacent segments in C do not lie on the same line.

We give an outline of the proofs of these theorems, giving the major steps with illustrations when appropriate.

Step 1. Show that both theorems are unaffected by composing $f \colon S^1 \to \mathbb{R}^2$ with an affine linear homeomorphism (sends lines to lines) $H \colon R^2 \to \mathbb{R}^2$. Use this to show that we can reduce the theorems to the case that no segment in the polygonal curve is horizontal, which we will assume from now on.

Step 2. There are two types of points in C, *edge points* in $p(k, k + 1)$, and vertices, which are the points $p(k)$. The vertices can be divided into two types, *regular vertices* and *special vertices*. The special vertices are those which are a local maxima or local minima for the y-coordinate on C, and the regular vertices are the others. Figure 1.16 shows neighborhoods of each type of point, and smaller *regular neighborhoods* within these which consist of nearby parallel line segments. Show that such neighborhoods exist for each type of point in C.

Step 3. Consider a horizontal line at height y_0. Suppose it intersects C in k points (not counting any special vertices). Show that there is a number ϵ so that horizontal curves at height between $y_0 - \epsilon$ and $y_0 + \epsilon$ intersect C in l points besides

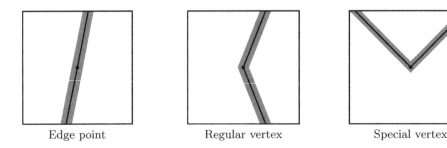

| Edge point | Regular vertex | Special vertex |

Figure 1.16. Nice neighborhoods.

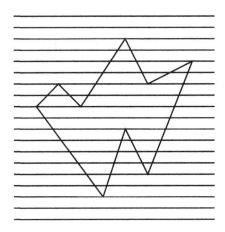

Figure 1.17. How lines intersect C.

special vertices, where $k \equiv l \mod 2$. It is necessary to consider line segments in C which are missed at height y_0 as well as special vertices at height y_0. Use this fact to show that the function that sends y to the number of points of C mod 2 that are not special vertices at height y is a continuous function from $\mathbb{R}$ to $\{0, 1\}$. Show that the horizontal line at height y intersects C in an even number of points which are not special vertices. See Figure 1.17 for an illustration.

Step 4. For each $(x, y) \notin C$, define $I(x, y)$ to be 0 if there are an even number of points of C (not counting special vertices) at height y to the left of (x, y), and equal to 1 when there are an odd number of such points. Show that I is continuous, and that the sets $A = I^{-1}(\{1\})$ and $B = I^{-1}(\{0\})$ are disjoint open sets with $\mathbb{R}^2 \backslash C = A \cup B$.

Step 5. Figure 1.18 shows a regular neighborhood of the curve C consisting of parallel polygonal curves near C. Show that C has such a regular neighborhood $N(C)$. Show that that $N(C)$ is homeomorphic to the annulus $S^1 \times [\frac{1}{2}, \frac{3}{2}] \subset \mathbb{R}^2$ enclosed between the circles of radii $\frac{1}{2}$ and $\frac{3}{2}$ with C corresponding to $S^1 \times \{1\}$.

Figure 1.18. A regular neighborhood.

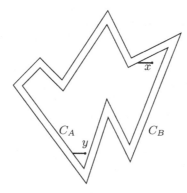

Figure 1.19. Using C_A to connect $x, y \in A$.

In particular, show that $N(C) \backslash C$ consists of two sets which are path connected but that $N(C) \backslash C$ is not path connected.

Step 6. Show that $\operatorname{Bd} N(C) = C_A \cup C_B$, where $C_A \subset A$ and $C_B \subset B$ are parallel polygonal curves to C. Use the curves C_A and C_B to show that each of A and B are path connected. See Figure 1.19 for a motivating example of such a path connecting two points $x, y \in A$ that uses C_A. Use $N(C)$ to show that $\bar{A} = A \cup C$, $\bar{B} = B \cup C$.

Step 7. Use the fact that C is compact to show that $\bar{A}$ is compact and $\bar{B}$ is not compact.

These steps then complete the proof of the polygonal version of the Jordan curve theorem. We now outline an approach to proving the polygonal Schönflies theorem. Our starting point is the setup from the polygonal Jordan curve theorem above.

For our proof of the Schönflies theorem, we first need to modify C slightly before we give our argument, but in a way that does not change the validity of the Schönflies theorem.

Step 1. Show that there is a homeomorphism $h : \mathbb{R}^2 \to \mathbb{R}^2$ which is the identity outside a regular neighborhood of C so that the special vertices all occur at

Figure 1.20. Moving a vertex.

different y-values. The idea is depicted in Figure 1.20, where we push the vertex vertically, keep the boundary of the part of the regular neighborhood of two adjacent edges that come together at the vertex fixed, and extend this to a PL map of the regular neighborhood. This allows an extension via the identity outside of the regular neighborhood to get a homeomorphism of $\mathbb{R}^2$ which displaces the vertex slightly. Show that this modification does not affect the validity of the Schönflies theorem. Figure 1.20 shows the original piece of the curve and the displaced piece in a regular neighborhood.

Step 2. Use compactness to show that there is the minimal value m assumed by C and the maximal value M assumed by C and the A lies between the $y = m$ and $y = M$. Moreover, show that there is a minimal special vertex at height m and a maximal special vertex at height M. Show that there are an even number of special vertices, half of which are local minima and half local maxima.

Step 3. Show that if there are just two special vertices, then A is homeomorphic to a triangle, and that the homeomorphism can be chosen to fix pointwise a small subtriangle at the bottom of A and is the identity outside a large rectangle containing A. Use induction on the number V of vertices, with starting point $V = 3$. Your homeomorphism should be expressible as a composition of homeomorphisms which are the identity outside a small neighborhood of a triangle which is being worked on. At each step a triangle is added or removed from A where two of its sides are on C and the third side is not. The interior of the triangle will lie entirely in A or entirely in B. The argument should show the existence of such triangles. The requirement that there are no horizontal lines occurring in any intermediate steps may require working on two adjacent triangles in a single reduction step. As a hint, we illustrate an example of a complete reduction of such a region to a triangle in Figure 1.21. The dotted lines show intermediate triangles being used and the numbering shows new edges in C as it is homeomorphed to the bottom triangle.

Step 4. The general argument is by induction on the ordered pairs (V, S), where V denotes the total number of vertices and S is the number of special vertices. The ordering is lexicographic ordering: $(V_1, S_1) < (V_2, S_2)$ iff (1) $V_1 < V_2$, or (2) $V_1 = V_2$ and $S_1 < S_2$. The starting point for the induction is $(3, 2)$

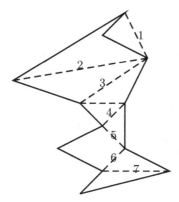

Figure 1.21. Homeomorphing A to a triangle.

and the way it works is to either keep the total number of vertices the same and reduce the number of special vertices by 2, or reduce the number of total vertices. Consider the lowest special vertex (minimal y-value) v_m which is a local maximum. There are a number of cases to consider. A useful concept to look at is the position of the first two vertices on the segments moving downward from v_m and look for ways to homeomorph $\mathbb{R}^2$ to simplify the image of C so that it has one fewer regular vertex. All of your homeomorphisms should fix a small triangle at the bottom of A and the region outside a large rectangle containing A. In fact, at each step the homeomorphism should just fix everything outside a region near a triangle on which you are working. Some steps may require composing a couple of these as well as introducing new vertices to avoid horizontal edges. Figure 1.22 illustrates how C is deformed to a C' with a single local maximum and local minimum. Intermediate triangles being used are indicated with dotted lines.

1.9 Supplementary exercises

Definition 1.9.1. A collection $\mathcal{B} = \{B_i : i \in I\}$ of subsets of X is called a *basis* and its elements are called *basis elements* if the following properties are satisfied:

(a) every $x \in X$ is contained in some B_i;

(b) if $x \in B_i \cap B_j, B_i, B_j \in \mathcal{B}$, then there is a basis element B_k with $x \in B_k \subset B_i \cap B_j$.

The topology $\mathcal{T}_\mathcal{B}$ determined by the basis $\mathcal{B}$ is defined as follows: a set $U \subset X$ is open if, for every $x \in U$, there is a basis element B_i with $x \in B_i \subset U$.

The first six problems concern the concept of a basis and the topology which it determines.

Exercise 1.9.1. Verify that $\mathcal{T}_\mathcal{B}$ satisfies the three properties required of a topology.

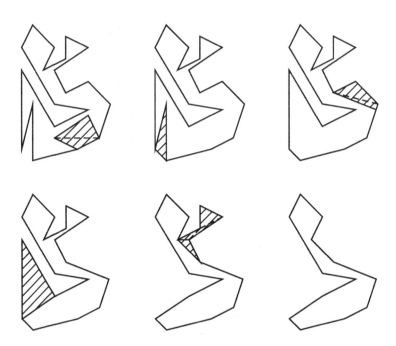

Figure 1.22. Removing excess special vertices.

Exercise 1.9.2. Show that the set of balls $\{B(x,r), \ x \in X, \ r > 0\}$ is a basis for the topology of a metric space X; that is, show that it is a basis and the topology it determines is the metric topology.

Exercise 1.9.3. Show that any open set in $\mathcal{T}_{\mathcal{B}}$ is a union of basis elements. (Hint: For $x \in U$, choose a basis element $B_{i(x)}$ with $x \in B_{i(x)} \subset U$.)

Exercise 1.9.4. Suppose the topology for Y is determined by a basis. Show that $f : X \to Y$ is continuous iff, for each basis element B of $Y, f^{-1}(B)$ is open.

Exercise 1.9.5. Show that the open intervals give a basis for the topology of $\mathbb{R}$.

Exercise 1.9.6. Combine the last two exercises to show that a map to $\mathbb{R}$ is continuous iff, for each open interval $I \subset \mathbb{R}$, we have $f^{-1}(I)$ is open. Formulate and prove an analogous statement for maps to a metric space in terms of balls.

Exercise 1.9.7. Let X be a metric space with metric d and let A be a subset of X. Show that the metric topology on A given by d is the same as the subspace topology.

Exercise 1.9.8. Suppose $A \subset B \subset X$, and B has the subspace topology. Show that, if A is open in B and B is open in X, then A is open in X.

Exercise 1.9.9. If $x, y \in \mathbb{R}^2$, let $d(x,y) = |x_1 - y_1|$. Which of the three properties of a metric does d satisfy?

Exercise 1.9.10. On $\mathbb{R}^2\backslash\{0\}$, write each point in polar coordinates as (r, θ), where $0 \leq \theta < 2\pi$. Define $d((r_1, \theta_1), (r_2, \theta_2)) = |r_1 - r_2| + |\theta_1 - \theta_2|$. Show that this gives a metric on $\mathbb{R}^2\backslash\{0\}$ but the topology formed is not the usual topology.

Definition 1.9.2. A point x is called a *limit point* of a set A if every open set U containing x intersects $A\backslash\{x\}$ in a nonempty set; that is, $U \cap (A - \{x\}) \neq \emptyset$ for U open, $x \in U$. Denote the limit points of A by A'.

Exercise 1.9.11. Show that a set A is closed iff it contains all of its limit points.

Exercise 1.9.12. Find the limit points of the following subsets of $\mathbb{R}$: $(a)(0, 1)$; $(b)\mathbb{Q}$, the rationals; (c) $\{1/n\colon n \in \mathbb{N}\}$.

Exercise 1.9.13. Let X be a metric space and $A \subset X$. Show that x is a limit point of A iff every ball $B(x, r)$ contains infinitely many points of A.

Exercise 1.9.14. Show that, in Hausdorff space X with subset A, x is a limit point of A iff every open set containing x contains infinitely many points of A.

Exercise 1.9.15.

  (a) Show that if C is a closed set containing A, then $\bar{A} \subset C$.

  (b) Show that if V is an open set contained in A, then $V \subset \text{int } A$.

Exercise 1.9.16. Show that A is closed iff $A = \bar{A}$, and A is open iff $A = \text{int } A$.

Exercise 1.9.17. Show that $\bar{A} = A \cup A'$.

Exercise 1.9.18.

  (a) Show that in $\mathbb{R}^n$ we have $\overline{B(z, r)} = \{x : d(z, x) \leq r\}$ and Bd $B(z, r) = \{x : d(z, x) = r\}$.

  (b) By using the discrete topology, show that this does not hold generally in a metric space.

Definition 1.9.3. A topological space X is called *limit point compact* if every infinite set has a limit point.

Exercise 1.9.19. Show that a compact space is limit point compact.

Exercise 1.9.20. Show that if X is a metric space, then X is limit point compact iff it is sequentially compact iff it is compact.

Exercise 1.9.21. Show that if $f : X \rightarrow Y$ is continuous at x and x_n is a sequence converging to x, then $f(x_n)$ converges to $f(x)$.

Exercise 1.9.22. Show that a map $f : (X, d) \rightarrow (Y, d')$ between metric spaces is continuous at x iff for every sequence x_n which converges to x, the sequence $f(x_n)$ converges to $f(x)$.

Exercise 1.9.23. Show that, in a Hausdorff space, the limit of a sequence is well defined; that is, if x_n converges to x and to y, then $x = y$.

Exercise 1.9.24. Show that, in a Hausdorff space, if x_n converges to x, then x is the only limit point of the set of all of values $\{x_n\colon n \in N\}$. Is the converse true? Give a proof or counterexample.

Exercise 1.9.25. Show that a finite set in a Hausdorff space is closed.

Definition 1.9.4. A Hausdorff space is called *regular* if, given $x \in X$ and a closed set C with $x \notin C$, then there are disjoint open sets U and V with $x \in U$ and $C \subset V$. A Hausdorff space is called *normal* if, whenever C, D are disjoint closed subsets, then there are disjoint open sets U, V with $C \subset U, D \subset V$.

Exercise 1.9.26. Show that a compact Hausdorff space is regular. (Hint: Use the fact that a closed subset of a compact space is compact. Get an open cover of the closed, hence compact, set C where there is an open set V_y for each point $y \in C$ with $y \in V_y$ and an open set U_y so that $x \in U_y$ with $U_y \cap V_y = \emptyset$.)

Exercise 1.9.27. Suppose $x \in U \subset X$, where X is regular and U is open. Show that there is an open set V with $x \in V \subset \bar{V} \subset U$. (Hint: Consider the point x and the disjoint closed set $X \backslash U$.)

Exercise 1.9.28. Show that a compact Hausdorff space is normal. (Hint: Apply the conclusion of Exercise 1.9.26 to pairs x, D, where $x \in C$ and use the compactness of C.)

Exercise 1.9.29. Suppose $C \subset U \subset X$, where X is normal, C is closed and U is open. Then show that there is an open set V with $C \subset V \subset \bar{V} \subset U$. (Hint: C and $X \backslash U = D$ are disjoint closed sets.)

Exercise 1.9.30. Show that a metric space is normal. (Hint: If C, D are disjoint closed subsets, then cover C by balls $B(c, r(c))$ disjoint from D and cover D by balls $B(d, r(d))$ disjoint from C. Then show that $U = \cup_{c \in C} B(c, r(c)/2)$ and $V = \cup_{d \in D} B(d, r(d)/2)$ are disjoint open sets containing C and D.)

Definition 1.9.5. Let C be a subset of a metric space X. For each $x \in X$, define the *distance from x to C* by $d(x, C) = \inf\{d(x, y) : y \in C\}$.

Exercise 1.9.31. Show that $\{x : d(x, C) = 0\} = \bar{C}$. Show that if C is closed and $x \notin C$, then $d(x, C) > 0$.

Exercise 1.9.32. Show that $C_\epsilon = \{y : d(y, C) < \epsilon\}$, where $\epsilon > 0$, is an open set containing C.

Exercise 1.9.33. Show that the function $f : X \to \mathbb{R}$ given by $f(x) = d(x, C)$ is continuous. (Hint: Look at the inverse image of an interval.)

Exercise 1.9.34. Suppose C, D are disjoint closed sets in a metric space X. Use the notion of the distance from a point to a set to define $d(C, D)$ in terms of $d(x, D)$ for $x \in C$. Give an example to show that this distance could be 0. Show that if X is compact, then the distance must be positive.

The next two exercises give versions of Urysohn's lemma and the Tietze extension theorem for metric spaces.

Exercise 1.9.35. Suppose the (X, d) is a metric space and A, B are disjoint closed sets. Show that the function

$$f(x) = \frac{d(x, A) - d(x, B)}{d(x, A) + d(x, B)}$$

is a continuous real-valued function $f : X \to [-1, 1]$ with $f^{-1}\{-1\} = A$, $f^{-1}\{1\} = B$. The existence of a function from X to $[-1, 1]$ which is -1 on A and 1 on B is called Urysohn's Lemma and holds in the more general situation of a normal space.

Exercise 1.9.36. Suppose X is a metric space and $C \subset X$ is a closed subset, with a continuous function $f : C \to \mathbb{R}$. This exercise leads you through a proof that there is a continuous extension $F : X \to \mathbb{R}$. This result is called the Tietze extension theorem and holds in the more general situation of a normal space.

(a) Reduce to the case where f is bounded by considering the composition of f with a homeomorphism from $\mathbb{R}$ to $(-1, 1)$.

(b) Because of (a), we assume from now on that $f : X \to [-M, M]$. We inductively define continuous maps from X to $[-M, M]$ which give better and better approximations to f on C. Let $A_1 = f^{-1}([-M, -M/3])$, and $B_1 = f^{-1}([M/3, M])$. Show that A_1, B_1 are disjoint closed subsets of X. Apply Exercise 1.9.35 to show that there is a continuous map $g_1 : X \to [-M/3, M/3]$ with $A_1 \subset g_1^{-1}\{-M/3\}$, and $B_1 \subset g_1^{-1}\{M/3\}$. Show that $|f(x) - g_1(x)| \le 2M/3$ on C.

(c) Repeat the construction in (b) applied to $h_1 = f - g_1$ defined on C to construct $g_2 : X \to [-2M/9, 2M/9]$ so that

$$h_1^{-1}\left(\left[\frac{-2M}{3}, \frac{-2M}{9}\right]\right) \subset g_2^{-1}\left(\left\{\frac{-2M}{9}\right\}\right),$$
$$h_1^{-1}\left(\left[\frac{2M}{9}, \frac{2M}{3}\right]\right) \subset g_2^{-1}\left(\left\{\frac{2M}{9}\right\}\right)$$

and $|f(x) - g_1(x) - g_2(x)| \le 4M/9$.

(d) Use induction to construct a sequence of maps $g_n : X \to [-2^{n-1}M/3^n, 2^{n-1}M/3^n]$ so that if $h_n(x) = f(x) - g_1(x) - \cdots - g_n(x)$, then $|h_n(x)| \le 2^n M/3^n$ on C and $|g_n(x)| \le 2^{n-1}M/3^n$.

(e) Define $g(x) = \sum_{n=1}^{\infty} g_i(x)$. Show that $g(x)$ converges uniformly to a continuous function $g : X \to [-M, M]$ which is an extension of f; that is, $g(x) = f(x)$ for $x \in C$.

Exercise 1.9.37. Urysohn's lemma states that for a normal space X with disjoint closed sets A, B, there is a continuous function $f : X \to [-1, 1]$ with $A \subset f^{-1}\{-1\}$ and $B \subset f^{-1}\{1\}$. The Tietze extension theorem states that for a normal space X with a closed subset C and a continuous function $f : C \to \mathbb{R}$, there is a continuous function $g : X \to \mathbb{R}$ with $f(x) = g(x)$ for $x \in C$. Show that Urysohn's lemma is equivalent to the Tietze extension theorem. (Hint: Use the argument in Exercise 1.9.36 to show that Urysohn's lemma implies the Tietze extension theorem.)

Definition 1.9.6. A set is *countable* if it can be put in 1–1 correspondence with the natural numbers $\mathbb{N}$ or is finite. For example, the rationals $\mathbb{Q}$ and n-tuples of rationals $\mathbb{Q}^n = \{(r_1, r_2, \ldots, r_n) : r_i \in \mathbb{Q}\}$ are countable. A space X is called *first* *countable* if, for each $x \in X$, there is a countable basis of open sets containing x; that is, there is a collection $\{B_n : n \in \mathbb{N}\}$ of open sets containing x so that, if U is an open set containing x, then there is a set $B_k \subset U$. This is called a *neighborhood basis*. A space X is called *second countable* if there is a countable basis for the topology of X. A space X is called *countably compact* if every countable open cover has a finite subcover. A metric space is called *separable* if there is a countable set $\{x_n : n \in \mathbb{N}\}$ so that every open set contains at least one x_n (we say $\{x_n\}$ is *dense* in X and $\{x_n\}$ is a *countable dense subset*).

Exercise 1.9.38. Show that $\mathbb{Q}$ is dense in $\mathbb{R}$, and hence $\mathbb{R}$ is separable.

Exercise 1.9.39. Show that a metric space is first countable.

Exercise 1.9.40. Show that, if X is first countable and x is a limit point of C, then there is a sequence $x_i \in C$ which converges to x.

Exercise 1.9.41. Show that if X is first countable, Hausdorff, and compact, then X is sequentially compact. (Hint: Adapt the proof given in Section 1.5 for metric spaces.)

Exercise 1.9.42. Show that a metric space is separable iff it is second countable. (Hint: If it is separable, use balls about the countable dense subset to get a countable basis. A countable number of countable sets is still countable. If it is second countable, select a countable set by choosing one point from each basis element.)

Exercise 1.9.43. Show that compactness implies countable compactness, and that the converse holds in a second countable space. (Hint: In a second countable space show that for each open covering $\{U_i\}_{i \in I}$ there is a covering by basis elements so that each basis element is contained in an element of the given covering $\{U_i\}$.)

Consider the space $X = A \cup B$, where $A = \{(x, \sin 1/x) : 0 < x \leq 1\}$ and $B = \{(0, y) : -1 \leq y \leq 1\}$. The space X is called the *topologist's sine curve*. The next three problems show that X is connected but not path connected. See Figure 1.9.

Exercise 1.9.44. Show that $A \cup B = \bar{A}$.

Exercise 1.9.45. Show that the closure of a connected set is connected, thus implying that $A \cup B$ is connected.

Exercise 1.9.46. We show here that $A \cup B$ is not path connected. Suppose $A \cup B$ were path connected. Let $f : [0,1] \to A \cup B$ be a path with $f(0) = (0,0)$ and $f(1) = (1, \sin 1)$. Consider the set $S = \{t : f([0,t]) \in B\}$. Let $p : \mathbb{R}^2 \to \mathbb{R}$ be $p(x,y) = x$.

    (a) Show that S is nonempty, is bounded from above, and has a least upper bound $u < 1$.

    (b) Show that $u \in S$.

(c) Show that there is a neighborhood $N_{f(u)}$ of $f(u)$ consisting of an infinite number of separated arcs, and a neighborhood N_u of u with $f(N_u) \subset N_{f(u)}$.

(d) Show that there is $u_1 > u$, $u_1 \in N_u$ with $pf(u_1) > 0$.

(e) Show that there are disjoint open sets $U, V \subset N_{f(u)}$ with $f(u) \in U$, $f(u_1) \in V$, and $U \cup V = N_{f(u)}$; that is, $N_{f(u)}$ is separated by U, V.

(f) By looking at $f|[u, u_1]$, arrive at a contradiction.

Definition 1.9.7. A topological space X is called *locally path connected at x* if for each open set V containing x, there is a path connected open set U with $x \in U \subset V$. It is called *locally path connected* if it is locally path connected at each $x \in X$. It is called *locally connected at x* if, for each open set V containing x, there is a connected open set U with $x \in U \subset V$. It is called *locally connected* if it is locally connected for each $x \in X$.

Exercise 1.9.47. Show that a locally path connected space is locally connected.

Exercise 1.9.48. Show that the topologist's sine curve is not locally connected.

Exercise 1.9.49. Show that an open set in $\mathbb{R}^n$ is locally path connected.

Exercise 1.9.50. Show that the path components of a locally path connected space are open sets.

Exercise 1.9.51. Show that if X is locally path connected and connected, then X is path connected. (Hint: Modify the proof that a connected open set in $\mathbb{R}^n$ is path connected.)

Exercise 1.9.52. Give an example of a path connected space which is not locally path connected.

Exercise 1.9.53. Define an equivalence relation on X by $x \sim y$ if there is a connected set containing both x and y. The equivalence classes are called the *components* of X.

(a) Verify that this is an equivalence relation.

(b) Show that each component is connected, that any two components are equal or disjoint, and that the union of the components is X.

(c) Show that any connected subset of X intersects at most one component and is a subset of that component.

(d) Show that each path component is contained in a component, and that a component is a disjoint union of path components.

(e) Show that a component is a closed set.

Exercise 1.9.54.

(a) Show that a space is locally connected iff, for each open set U, each component of U is open in X.

(b) Show that a space is locally path connected iff, for each open set U, each path component of U is open in X.

Exercise 1.9.55. A collection $\mathcal{D}$ of subsets of X is said to satisfy the *finite intersection property* (F.I.P.) if for every finite subcollection $\{D_1, \ldots, D_k\}$ of $\mathcal{D}$, the intersection $D_1 \cap \cdots \cap D_k \neq \emptyset$. Show that X is compact iff for every collection $\mathcal{D}$ of closed sets satisfying the F.I.P., the intersection of all of the elements of $\mathcal{D}$ is nonempty. (Hint: If not, consider the covering of X by the complements $\{X \backslash D_i\}$.)

Exercise 1.9.56. Let (X, d) be a compact metric space, and $f: X \to X$ continuous. $x \in X$ is called a *fixed point* of f if $f(x) = x$. f is called a *contraction* if there is a number $a < 1$ such that $d(f(x), f(y)) \leq ad(x, y)$ for all $x, y \in X$. Show that a contraction has a unique fixed point. This result is known as the *contraction mapping principle* and plays a key role in analysis. (Hint: Consider $\cap f^n(X)$. where f^n denotes the n-fold composition of f with itself and use the finite intersection property from the previous exercise.)

Definition 1.9.8. X is called *locally compact at x* if there is an open set U and a compact set C with $x \in U \subset C$. It is called *locally compact* if it is locally compact at each $x \in X$.

Exercise 1.9.57. Show that a compact space is locally compact.

Exercise 1.9.58. Show that $\mathbb{R}^n$ is locally compact.

Exercise 1.9.59. Show that if X is Hausdorff and locally compact at x, then there is an open set U containing x so that $\bar{U}$ is compact.

Exercise 1.9.60. Suppose X is a locally compact Hausdorff space. Show that if $x \in U \subset X$, U open, then there exists an open set V containing x such that $\bar{V}$ is compact and $\bar{V} \subset U$. (Hint: Use the preceding exercise and the argument of Exercise 1.9.27.)

The following exercise leads through the construction of the one-point compactification of a locally compact Hausdorff space.

Exercise 1.9.61. Suppose that X is a locally compact Hausdoff space. Form a new space X^+, called the *one-point compactification* of X as $X^+ = X \cup \{p\}$, the disjoint union of X and an added point p. A set $U \subset X^+$ is called open if (1) U is an open set in X, or (2) $p \in U$ and $X^+ \backslash U$ is a compact set in X.

(a) Show that this definition of open set satisfies the three properties of a topology.

(b) Show that the subspace topology on $X \subset X^+$ is the same as its usual topology.

(c) Show that X^+ is a compact Hausdorff space.

Exercise 1.9.62. Show that if X is a locally compact Hausdorff space and Y is a compact Hausdorff space with $Y \backslash \{y_0\} \simeq X$, then $Y \simeq X^+$.

Exercise 1.9.63.

   (a) Show that the one-point compactification of $\mathbb{R}$ is homeomorphic to S^1.

   (b) Show that the one-point compactification of $\mathbb{R}^2$ is homeomorphic to S^2. (Hint: Use projection from the point $p = (0,0,1)$ to get a homeomorphism from $S^2 \backslash \{p\}$ to $\mathbb{R}^2$.)

   (c) Show that the one-point compactification of $\mathbb{R}^n$ is homeomorphic to S^n.

Exercise 1.9.64. Show that, if X is compact, the one-point compactification of X is $X^+ = X \bigsqcup \{p\}$, where the set $\{p\}$ is an open set in the disjoint union.

Exercise 1.9.65. Consider the space $\mathbb{R}^\infty$, which is the product of a countably infinite number of copies of $\mathbb{R}$. This is given the product topology with basis the sets which are product of a finite number of intervals with a product of copies of the reals:
$$(a_1, b_1) \times \cdots \times (a_n, b_n) \times \mathbb{R} \times \cdots \times \mathbb{R} \times \cdots .$$
Show that this space is not locally compact. (Hint: Look at basic open sets and show that they do not have compact closure.)

Exercise 1.9.66.

   (a) Show that the following two subsets of $\mathbb{R}$ are not homeomorphic: $A = \{1/n : n \in \mathbb{N}\} \cup \{0\}$, $B = \mathbb{N}$.

   (b) Show that B^+ is homeomorphic to A.

Exercise 1.9.67. Show that ∞ is not homeomorphic to O (Hint: Consider where the crossing point of ∞ could go under a homeomorphism.)

The next four exercises concern the homeomorphism type of the letters of the alphabet. In each case, assume the letter is written as given below in the sans serif style, with no adornments:

ABCDEFGHIJKLMNOPQRSTUVWXYZ

Exercise 1.9.68. Show that the letter X is not homeomorphic to the letter Y, but that the letter Y is homeomorphic to the letter T.

Exercise 1.9.69. Construct a homeomorphism between the letter D and the letter O.

Exercise 1.9.70. Prove that the letter A is not homeomorphic to the letter B.

Exercise 1.9.71. Group the letters of the alphabet into equivalence classes so that equivalent letters are homeomorphic and nonequivalent letters are not homeomorphic.

Exercise 1.9.72. Prove that if an open set $U \subset \mathbb{R}^2$ is path connected, then any two points in U can actually be connected by a polygonal path in U.

Exercise 1.9.73. For each of the following subsets of $\mathbb{R}^2$ indicate which of the following properties it possesses, namely, (i) compact; (ii) connected; (iii) path connected; (iv) open; (v) closed:

(a) $A = \{x_1, x_2\} : x_1 \geq 0, 4 < x_2 \leq 8\}$;
(b) $B = \{(x_1, x_2) : x_1^2 + x_2^2 = 25\}$;
(c) $C = A \cap B$;
(d) $D = \{(x_1, x_2) : x_1^2 + x_2^2 < 1\}$;
(e) $E = \bar{D}$.

Exercise 1.9.74. Show that the torus is not homeomorphic to an open set in $\mathbb{R}^2$. (Hint: Use the properties of compactness and connectedness.)

Exercise 1.9.75. Show that if $A \subset S^1$ with $A \neq S^1$, then A is not homeomorphic to S^1.

Exercise 1.9.76. Put an equivalence relation $\sim$ on $\mathbb{R}^2$ by saying that two points are equivalent if they both lie on the circle of radius r about the origin. Show that $\mathbb{R}^2 / \sim$, with the quotient topology, is homeomorphic to $[0, \infty)$.

Exercise 1.9.77. Identify all points in the lower hemisphere of the sphere S^2. Show that the resulting quotient space $S^2 / \sim$ is homeomorphic to S^2.

Exercise 1.9.78. Identify points on the boundary circles of an annulus A between the circles of radius 1 and radius 2 that lie on the same ray from the origin. Show that the resulting quotient space $A / \sim$ is homeomorphic to the torus T^2.

Exercise 1.9.79. Identify the points on the outer circle of an annulus A to one-point and the points on the inner circle to a (different) point. Show that $A / \sim$ is homeomorphic to S^2. Describe what space you would get if you identified the points on both circles to a single point.

Exercise 1.9.80. Consider the quotient space X formed from two copies of $D^1 \times D^1$ using $f : \{-1, 1\} \times D^1 \to \{-1, 1\} \times D^1$ by $f(-1, x) = (1, -x)$, $f(1, y) = (-1, y)$, $X = (D^1 \times D^1) \cup_f (D^1 \times D^1)$. Decide whether X is homeomorphic to the annulus or the Möbius band, and prove your assertion.

Exercise 1.9.81. Show that the upper hemisphere of S^2 is homeomorphic to D^2 and similarly for the lower hemisphere. Use this to show S^2 is homeomorphic to $D^2 \cup_g D^2$ for $g : S^1 \to S^1$ and determine g.

Exercise 1.9.82. Show that $D^2 \cup_f D^2$, where $f : K \to K$ is $f(x) = x$, and $K = \{(x_1, x_2) \in S^1 : x_1 \geq 0\}$ is homeomorphic to D^2. (Hint: First choose a homeomorphism h from D^2 to $D^1 \times D^1$ where $h(K) = \{(x_1, x_2) \in D^1 \times D^1 : x_1 = 1\}$. Use this to get a homeomorphism $D^2 \cup_f D^2 \simeq (D^1 \times D^1) \cup_g (D^1 \times D^1)$, where $g : h(K) \to h(K)$ is $g(x) = x$.)

Exercise 1.9.83. Construct a homeomorphism between a square and a diamond.

Exercise 1.9.84. Construct a homeomorphism between the two regions in Figure 1.23.

Figure 1.23. Annular regions.

Figure 1.24. Star.

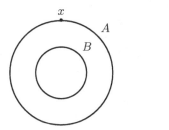
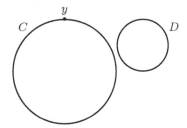

Figure 1.25. Two pairs of circles.

Exercise 1.9.85. Show that any two rectangles in the plane are homeomorphic.

Exercise 1.9.86. Construct a homeomorphism between the inside of a square and the star in Figure 1.24.

Exercise 1.9.87. Show that there is no homeomorphism of the plane to itself which sends the unit circle to itself and sends $(0,0)$ to $(2,0)$.

Exercise 1.9.88. Construct an example of a simple closed curve in the plane where a horizontal line intersects the curve in an infinite number of points but the curve contains no horizontal line segments.

Exercise 1.9.89. Show that the complement of two disjoint polygonal simple closed curves in the plane consists of three disjoint open, path connected sets.

The next three problems concern Figure 1.25.

Exercise 1.9.90. Show that there is a homeomorphism sending $A \cup B$ to $C \cup D$.

Exercise 1.9.91. Show that any homeomorphism sending $A \cup B$ to $C \cup D$ which sends $x \in A$ to $y \in C$ must send A homeomorphically to C and B homeomorphically to D.

Exercise 1.9.92. Show that there does not exist a homeomorphism of the plane sending $A \cup B$ to $C \cup D$. (Hint: Consider the regions bounded by A and C.)

Exercise 1.9.93. Consider a polygonal path P that is not closed and does not intersect itself. Show that $\mathbb{R}^2 \backslash P$ is path connected using a polygonal path. (Hint: Use induction on the number of segments in the path.)

Exercise 1.9.94. Show directly that the triangle with vertices $(0,0), (1,0), (0,1)$ separates the plane into two nonempty disjoint open path connected sets, one of which is bounded and the other not.

Exercise 1.9.95. Consider the shaded region in Figure 1.26 (which is not homeomorphic to a disk). Analyze how the region changes as we move upward past the special vertices.

Figure 1.26. A polygonal annular region.

Figure 1.27. A curvy disk.

Exercise 1.9.96. Show that the region in Figure 1.26 is homeomorphic to the region R enclosed between the squares $[-1, 1] \times [-1, 1]$ and $[-2, 2] \times [-2, 2]$. Do this both by a direct argument and by breaking the each region into two regions to which we can apply the polygonal Schönflies theorem.

Exercise 1.9.97. Describe a homeomorphism between the region in Figure 1.27 and a disk.

2

The classification of surfaces

2.1 Definitions and construction of the models

In this chapter we discuss surfaces and classify them up to homeomorphism. A surface is a topological space which locally looks like a piece of the plane, such as a sphere or the exterior of a donut (a torus). Before specializing to surfaces, we begin by introducing the concept of an *n-manifold*, where a surface is a 2-manifold. Manifolds constitute one of the primary areas of study in topology. Our study of surfaces will introduce us to many of the key ideas in manifold theory in a fairly concrete geometric setting.

Definition 2.1.1. A topological space M is an n-manifold if

(1) there is an *embedding* (a homeomorphism onto its image with the subspace topology) of M into $\mathbb{R}^N$ for some N;

(2) given $x \in M$, there is a neighborhood U of x and a homeomorphism h from U onto an open set in $\mathbb{R}^n$.

Condition (1) turns out to be equivalent (in the presence of (2)) to either requiring M to be a separable metric space or requiring M to be a second countable, Hausdorff space (see the supplementary exercises of Chapter 1 and [24, 5]). We will mainly concern ourselves with condition (2), regarding (1) as a technicality to rule out certain pathological examples. Condition (2) is sometimes phrased as requiring M to be *locally homeomorphic to* $\mathbb{R}^n$ or, if n is clear from the context, requiring M to be *locally Euclidean*. In condition (2), we may require $h(U) = \mathbb{R}^n$, and not just an open set in $\mathbb{R}^n$. For if $h(U)$ is open in $\mathbb{R}^n$, there is a smaller neighborhood V of x with $h(V) = B(\boldsymbol{y}, r), \boldsymbol{y} = h(x)$. But there is a homeomorphism $g : B(\boldsymbol{y}, r) \to \mathbb{R}^n$ and so $gh : V \to \mathbb{R}^n$ is a homeomorphism. An n-manifold is said to be of *dimension n*. By a *surface* we mean a 2-manifold.

Exercise 2.1.1.

(a) Verify that $g : (-1, 1) \to \mathbb{R}$, $g(x) = x/(1 - x^2)$, is a homeomorphism and that it restricts to a homeomorphism of $[0, 1)$ onto $[0, \infty)$.

(b) Construct a homeomorphism of $B(\boldsymbol{y}, r)$ onto $B(\boldsymbol{0}, 1)$, where $\boldsymbol{y}, \boldsymbol{0} \in \mathbb{R}^n$, $r > 0$.

(c) Construct a homeomorphism from $k : B(\boldsymbol{0}, 1) \to \mathbb{R}^n$. (Hint: Use the analog of (a).)

The classification problem for n-manifolds seeks a collection $M_i, i \in I$, of n-manifolds so that each n-manifold is homeomorphic to one of the M_i, and M_i is not homeomorphic to M_j for $i \neq j$. Also, a procedure should be given for deciding which M_i a given n-manifold is homeomorphic to. Frequently, the class of n-manifolds under consideration is restricted in some way. A common restriction is to compact, connected n-manifolds, and we will only give the classification of compact, connected surfaces in this chapter. The problem of classifying compact, connected 1-manifolds turns out to be relatively simple. A proof will be outlined in Exercises 2.9.1–2.9.8 that any compact, connected 1-manifold is homeomorphic to the circle. For compact, connected 3-manifolds, the problem is still unsolved. For n-manifolds, $n \geq 4$, the problem has been shown to be undecidable in a precise logical sense. Nevertheless, some more restricted classification problems have been solved and have constituted some of the most fruitful areas of research in topology in the last 50 years.

Surfaces constitute a familiar example of manifolds from advanced calculus. Many surfaces arise there as solution sets of some equation. For example, the 2-sphere S^2 is the solution to $F(x, y, z) = x^2 + y^2 + z^2 - 1 = 0$. Higher-dimensional manifolds also arise as solution sets to equations. Besides an open set U in $\mathbb{R}^n$, the simplest n-manifold is the graph of a function $f : U \subset \mathbb{R}^n \to \mathbb{R}^p$, $\Gamma(f) = \{(\boldsymbol{x}, f(\boldsymbol{x})): \boldsymbol{x} \in U\}$. It is homeomorphic to U by projection to its first n coordinates. If $F : \mathbb{R}^{n+k} \to \mathbb{R}^k$ is a differentiable function so that on $F^{-1}(\boldsymbol{0})$ the matrix of partial derivatives has rank k at each point, then $F^{-1}(\boldsymbol{0})$ is an n-manifold. This is shown by using the implicit function theorem to show that a neighborhood of $\boldsymbol{x} \in F^{-1}(\boldsymbol{0})$ is the graph of a function defined on an open set in a hyperplane determined by n of its coordinates.

We will also need the more refined notion of an *n-manifold with boundary*. First let $\mathbb{H}^n = \{(x_1, \ldots, x_n) \in \mathbb{R}^n: x_n \geq 0\}$ and $\partial \mathbb{H}^n = \{(x_1, \ldots, x_n) \in \mathbb{R}^n: x_n = 0\}$.

Definition 2.1.2. An *n-manifold with boundary* is a topological space M so that

(1) there is an embedding of M into $\mathbb{R}^N$ for some N;

(2) given $x \in M$, there is a neighborhood U of x and a homeomorphism h of U onto an open set in $\mathbb{H}^n$.

Again, (1) may be replaced by requiring M to be either separable metric or second countable, Hausdorff, and (2) may be refined by requiring that either $h(U) = \mathbb{R}^n$ or $h(U) = \mathbb{H}^n$. For if $h(\boldsymbol{x}) = \boldsymbol{y} \notin \partial \mathbb{H}^n$, then there will be a smaller neighborhood V with $h(V) = B(\boldsymbol{y}, r)$, and then $kgh : V \to \mathbb{R}^n$ will be a homeomorphism as before. We leave it as an exercise to modify this argument in the case where $h(x) \in \partial \mathbb{H}^n$.

Exercise 2.1.2. Show that if $h : U \to \mathbb{H}^n$ is a homeomorphism onto an open set in $\mathbb{H}^n$ and $h(x) \in \partial\mathbb{H}^n$, then there is a smaller open set V about x and a homeomorphism $h' : V \to \mathbb{H}^n$.

Definition 2.1.3. In a manifold with boundary, those points x with $h(x) \notin \partial\mathbb{H}^n$ are called *interior points* and those points with $h(x) \in \mathbb{H}^n$ are called *boundary points*. The collection of all interior points is called the *interior* of M, and is denoted by int M. The collection of all the boundary points is called the *boundary* of M, and is denoted by ∂M. int M is an n-manifold (without boundary) and ∂M is an $(n - 1)$-manifold (without boundary).

We now quote some basic results related to these definitions. These will be proved in Chapter 6 using homology theory. First, an open set in $\mathbb{R}^n$ is not homeomorphic to an open set in $\mathbb{R}^m$ for $m \neq n$. Second, if $x \in \partial\mathbb{H}^n$ and U is an open set in $\mathbb{H}^n$ about x, then U is not homeomorphic to an open set in $\mathbb{R}^n$ (or even any $\mathbb{R}^m$).

Exercise 2.1.3. Assuming the two results quoted above, show that the dimension of a manifold is well defined and that an interior point cannot be a boundary point as well, so the concepts of interior point and boundary point are well defined.

Exercise 2.1.4. Suppose U is an open set about $0 \in [0, \infty) = \mathbb{H}^1$ and V is an open set in $\mathbb{R}$. Show that U is not homeomorphic to V. (Hint: Note that there is an interval $[0, c)$ contained in U which is path connected after 0 is removed.)

Exercise 2.1.5. Show that the boundary of an n-manifold is either empty or an $(n - 1)$-manifold. (Hint: Use the fact that a point is either an interior point or a boundary point but not both.)

Exercise 2.1.6. Suppose M, N are n-manifolds with boundary. Show that if $h : M \to N$ is a homeomorphism, then $h|\partial M$ is a homeomorphism from ∂M to ∂N.

There is a stronger statement about open sets in $\mathbb{R}^n$ from which some of the above statements can be deduced. This is the invariance of domain property, which we will also prove in Chapter 6.

Theorem 2.1.1 (Invariance of domain). *Suppose U is an open subset of $\mathbb{R}^n$ and $f : U \to \mathbb{R}^n$ is 1-1 and continuous. Then f is an open map; that is, it maps open sets to open sets.*

Exercise 2.1.7. Apply Theorem 2.1.1 to prove the following version for n-manifolds: If M^n, N^n are n-manifolds and $f : M^n \to N^n$ is 1-1 and continuous, then f is an open map. Show that if we also assume that M^n is compact, connected and N^n is connected, then f must be a homeomorphism.

Exercise 2.1.8. Deduce the fact that the dimension of an n-manifold is well defined from the invariance of domain.

Exercise 2.1.9. Prove that the invariance of domain holds when $n = 1$.

In order to classify compact, connected surfaces with boundary, we need to construct some examples which will be our basic building blocks.

The simplest example of a surface with boundary is the closed disk D^2. It is compact since it is closed and bounded, and path connectivity follows by using straight line paths. The boundary will be the circle S^1. To see that points on the circle are boundary points, note first that if $x, y \in S^1$, there is a rotation (hence homeomorphism) of D^2 sending x to y (hence a neighborhood of x to a neighborhood of y). Thus it is sufficient to exhibit one point x on the circle with a neighborhood homeomorphic to an open set in $\mathbb{H}^2$. But there is a homeomorphism of D^2 onto a rectangle R which sends the circle to the perimeter of R. For a point on the interior of the bottom edge of the rectangle, there is a neighborhood homeomorphic to an open set in $\mathbb{H}^n$, and so all points of the boundary circle (and thus also on the perimeter of the rectangle) have the required neighborhood. Note that this implies that the corner points on the perimeter have appropriate neighborhoods. The following exercise asks the reader to show this directly.

Exercise 2.1.10. Show directly that $\mathbb{H}^2_+ = \{(x_1, x_2): x_1 \geq 0, x_2 \geq 0\}$ is homeomorphic to $\mathbb{H}^2$. (Hint: Use polar coordinates to define the map.)

In the following examples, we will frequently refer to the boundary of certain surfaces without verifying precisely that the points involved are boundary points. Frequently, a direct verification is possible using the ideas for the disk above together with Exercise 2.1.10, and you are encouraged to convince yourself that the boundary is as indicated.

The 2-sphere S^2 is defined as $\{(x_1, x_2, x_3) \in \mathbb{R}^3: x_1^2 + x_2^2 + x_3^2 = 1\}$. It is compact since it is closed and bounded in $\mathbb{R}^3$. Path connectivity was shown in Chapter 1. To see that it is locally homeomorphic to $\mathbb{R}^2$, we show the stronger fact that if $p \in S^2$, then $S^2 \backslash \{p\}$ is homeomorphic to $\mathbb{R}^2$. We first do this for the special points $p = (0, 0, 1)$, the north pole N, and $p = (0, 0, -1)$, the south pole S. Our technique is standard and is called *stereographic projection* (see Figure 2.1). For each $x \in S^2 \backslash \{N\}$, consider the line through N and x. It intersects the plane $\mathbb{R}^2 \subset \mathbb{R}^3$ in some point, which we call $h_N(x)$. This gives us a map $h_N: S^2 \backslash \{N\} \to \mathbb{R}^2$. We will show that this is a homeomorphism by exhibiting formulas for h_N and h_N^{-1}. First note, however, that h_N sends the upper hemisphere $S^2_+ = \{(x_1, x_2, x_3) \in S^2: x_3 \geq 0\}$ to the exterior of the unit disk in $\mathbb{R}^2$ and the lower hemisphere $S^2_- = \{(x_1, x_2, x_3) \in S^2: x_3 \leq 0\}$, to the unit disk.

To derive a formula for h_N, it is useful to restrict h_N to the plane determined by N, x, and $\mathbf{0}$. Let $r^2 = x_1^2 + x_2^2$, $x = (x_1, x_2, x_3)$, and $h_N(x) = (a, b)$, $s^2 = a^2 + b^2$. Then by similar triangles we get $r/(1 - x_3) = s$ or $s/r = 1/(1 - x_3)$. By projection of these two triangles onto the planes $x_2 = 0$ and $x_1 = 0$, we also get the equalities $x_1/(1 - x_3) = a, x_2/(1 - x_3) = b$. Thus $h_N(x) = (x_1/(1 - x_3), x_2/(1 - x_3))$. To get a formula for h_N^{-1} we solve for x_1, x_2, x_3 in terms of a, b. Using our three equations together with $x_3^2 = 1 - r^2$, algebraic manipulation yields

$$x_1 = \frac{2a}{1 + s^2}, \qquad x_2 = \frac{2b}{1 + s^2}, \qquad x_3 = \frac{s^2 - 1}{1 + s^2}.$$

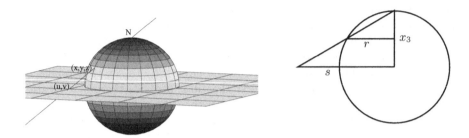

Figure 2.1. Stereographic projection.

Thus $h_{\boldsymbol{N}}^{-1}(a,b) = (2a/(1+s^2), 2b/(1+s^2), s^2 - 1/(1+s^2))$. The continuity of $h_{\boldsymbol{N}}$ and $h_{\boldsymbol{N}}^{-1}$ follows from the algebraic nature of their formulas in terms of rational functions.

The homeomorphisms from $S^2 \backslash \{\boldsymbol{p}\}$ to $\mathbb{R}^2$ for other $\boldsymbol{p}$ follow similarly by projecting onto the plane perpendicular to the line through $\boldsymbol{0}$ and $\boldsymbol{p}$ and then identifying that plane with $\mathbb{R}^2$. A derivation similar to the one above shows that if we project from the south pole, we get

$$h_{\boldsymbol{S}}(\boldsymbol{x}) = \left(\frac{x_1}{1+x_3}, \frac{x_2}{1+x_3} \right), \qquad h_{\boldsymbol{S}}^{-1}(a,b) = \left(\frac{2a}{1+s^2}, \frac{2b}{1+s^2}, \frac{1-s^2}{1+s^2} \right).$$

Note that $h_{\boldsymbol{S}}^{-1}$ sends the unit disk to the upper hemisphere and $h_{\boldsymbol{N}}^{-1}$ sends the unit disk to the lower hemisphere and each maps the circle to itself via the identity. Let D_i^2 denote a copy of the unit disk, $i = 1, 2$. Form the quotient space $D_1^2 \cup_{\mathrm{id}} D_2^2$ by identifying corresponding points of the circle in the two copies of the disk via the identity. Then $h_{\boldsymbol{N}}^{-1}, h_{\boldsymbol{S}}^{-1}$ fit together to give a homeomorphism from $D_1^2 \cup_{\mathrm{id}} D_2^2$ to S^2.

The representation of the sphere as two copies of the unit disk glued together along their boundary is a very useful one in topology and can be given for higher-dimensional spheres by an analogous construction. The identification of the two copies of S^1 by the identity is unnecessary in the following sense. If $f : S^1 \to S^1$ is any homeomorphism, we may form $D_1^2 \cup_f D_2^2$, where $\boldsymbol{x} \in S_1^1$ is identified to $f(\boldsymbol{x}) \in S_2^1$ with the quotient topology. Now f extends to a homeomorphism of D^2 via $F(t\boldsymbol{x}) = tf(\boldsymbol{x})$, $0 \le t \le 1$. We can construct a homeomorphism $h : D_1^2 \cup_f D_2^2 \to D_1^2 \cup_{\mathrm{id}} D_2^2$ by sending D_1^2 to D_1^2 via the identity and D_2^2 to D_2^2 via F. The piecing lemma for homeomorphisms shows that this is a homeomorphism.

Our second model surface will be the torus, which we denote by T. Its most convenient description is as a product space, $T = S^1 \times S^1$. A geometric realization occurs in $\mathbb{R}^3$ by taking a circle in the right $x_2 x_3$-half-plane and revolving it about the x_3-axis. That it is a surface is left as an exercise.

Exercise 2.1.11. Show that the product of an m-manifold M and n-manifold N is an $(m+n)$-manifold $M \times N$. Conclude that T is a compact, connected surface.

We give a description of T analogous to our description of S^2 as the union of two disks glued along their boundaries. First note that S^1 is homeomorphic to $D_1^1 \cup_{\mathrm{id}} D_2^1$ where D_i^1 denotes a copy of the interval $[-1, 1]$. The proof is left as an exercise.

Exercise 2.1.12. Construct a homeomorphism between the quotient space $D_1^1 \cup_{\mathrm{id}} D_2^1$ and the circle S^1.

Thus $T = S^1 \times S^1$ is homeomorphic to

$$(((D_1^1 \times D_1^1) \cup_f (D_1^1 \times D_2^1)) \cup_g (D_2^1 \times D_1^1)) \cup_h (D_2^1 \times D_2^1),$$

where f, g, h indicate that certain points in the boundary of each product are identified via homeomorphisms to points in the boundary of the space preceding it. We now make these identifications more explicit. The boundary of $D_1^1 \times D_1^1$ is $\{-1\} \times D_1^1 \cup \{1\} \times D_1^1 \cup D_1^1 \times \{-1\} \cup D_1^1 \times \{1\}$. The map f identifies copies of $D_1^1 \times \{-1\}$ and $D_1^1 \times \{1\}$. See Figure 2.2 to see how the first two pieces form an annulus within the front half of the torus.

Note that the boundary of $D_1^1 \times D_1^1 \cup_f D_1^1 \times D_2^1$ is $\{\pm 1\} \times (D_1^1 \cup D_2^1) = \{\pm 1\} \times S^1$, which is the union of two circles. Now $D_2^1 \times D_1^1$ has boundary $\{-1\} \times D_1^1 \cup \{1\} \times D_1^1 \cup D_2^1 \times \{-1\} \cup D_2^1 \times \{1\}$. The map g identifies $\{-1\} \times D_1^1 \cup \{1\} \times D_1^1$ in this boundary with $\{-1\} \times D_1^1 \cup \{1\} \times D_1^1$ in the boundary of $(D_1^1 \times D_1^1) \cup_f (D_1^1 \times D_2^1)$. The boundary of $((D_1^1 \times D_1^1) \cup_f (D_1^1 \times D_2^1)) \cup_g (D_2^1 \times D_1^1)$ is $D_2^1 \times \{-1\} \cup D_2^1 \times \{1\} \cup \{-1\} \times D_2^1 \cup \{1\} \times D_2^1$, which is also the boundary of $D_2^1 \times D_2^1$. The map h now identifies these two boundaries. See Figure 2.3 for views of the first two pieces and then the first three pieces within the torus.

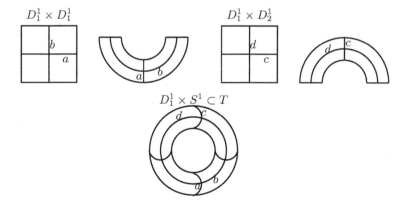

Figure 2.2. Decomposition of front half of the torus.

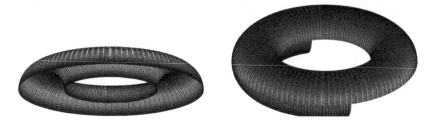

Figure 2.3. Views of one-half and three-fourths of the torus.

2.2   Handle decompositions and more basic surfaces

Recall that the last piece $D_2^1 \times D_2^1 = [-1,1] \times [-1,1]$ of the torus is homeomorphic to the unit disk D^2. Our identifications are occurring on the boundary of $D_2^1 \times D_2^1$, which is homeomorphic to the circle. We could also think of $D_1^1 \times D_1^1$ as a disk D^2. Our decomposition of $S^1 \times S^1$ is then better expressed as

$$(((D^0 \times D^2) \cup_f (D^1 \times D^1)) \cup_g (D^1 \times D^1)) \cup_h (D^2 \times D^0),$$

where D^0 is a point and $D^i \times D^j$ is attached to the boundary of the space preceding it by an embedding of $\partial D^i \times D^j$.

Definition 2.2.1. We will call $D^i \times D^j$ an *i-handle*. We say that it has *index i* and denote it by h^i, with subscripts used to distinguish different handles of the same index. A *handle decomposition* of a surface (with boundary) is a decomposition of the form

$$h_1^0 \cup \cdots \cup h_{k_0}^0 \cup h_1^1 \cup \cdots \cup h_{k_1}^1 \cup h_1^2 \cup \cdots \cup h_{k_2}^2,$$

where $k_0 \geq 1$ and $\partial D^i \times D^j$ is identified with a homeomorphic image in the boundary of the space (which is a surface with boundary) preceding it in the decomposition. A surface (possibly with boundary) with a handle decomposition is called a *handlebody*.

As an alternate description, a handlebody H is built up inductively from a disk by attaching handles of nondecreasing index $X_0 \subset X_1 \subset \cdots X_k \subset X_{k+1} \subset X_p = H$, where $X_{k+1} = X_k \cup h^i$, with $\partial D^i \times D^j \subset h^i$ identified to a homeomorphic image in ∂X_k. In the case of the 2-sphere, we have exhibited a handle decomposition with one 0-handle and one 2-handle. For the torus, we have given a handle decomposition with one 0-handle, two 1-handles, and one 2-handle.

Note that the above definition assumes implicitly that each X_k is a surface with boundary. This will be proved later in the chapter as a step in proving the classification theorem. Basically, the idea is that attaching a 1-handle to a surface

Figure 2.4. Attaching a 1-handle.

with boundary gives a surface with boundary. This is clear geometrically if we know that there is a "nice neighborhood" of the part of the boundary where the handle is attached and the handle is attached in a "nice" manner. See Figure 2.4 for a picture of how the boundary changes when a 1-handle is attached. Two arcs that were part of the boundary where the 1-handle is attached are no longer in the boundary, but the two complementary arcs of the boundary of the 1-handle become part of the boundary.

The index of a handle indicates the way that the handles are attached in forming the surface inductively. The 0-handles just consist of disjoint disks. The 1-handles are also homeomorphic to disks, but in terms of the handle decomposition they are best thought of as rectangles $D^1 \times D^1$. They are attached to the surface already formed by identifying $\{-1\} \times D^1$ and $\{1\} \times D^1$ with arcs in the boundary circles. The 2-handles are best thought of as disks D^2 and they are attached to the preceding surface by identifying their boundary circle with a circle in the boundary of the surface. Attaching a 2-handle can be regarded as filling in a hole in the surface. There is no choice in how a 0-handle is added to a surface (just as a disjoint disk) and little choice for the 2-handles (essentially only which boundary circle one fills in). The interesting operation in forming the surface is attaching the 1-handles, and the bulk of our work in classifying surfaces is to understand this.

The existence of a handle decomposition on a compact surface follows from the existence of a triangulation or a differentiable structure on the surface. Examples are given in Section 2.8 to show how these structures give rise to handle decompositions. We will see through examples how handle decompositions arise naturally for surfaces with boundary in 3-space that come from embeddings of the circle in 3-space (knots). In 1925, Radó [29] first proved that surfaces could be triangulated as PL manifolds, and it was later shown that two PL surfaces were homeomorphic iff they were PL homeomorphic [27]. Moreover, all surfaces arise as differentiable surfaces, and they are homeomorphic iff they are diffeomorphic. These results also hold for 3-manifolds, but are not true in higher dimensions—understanding the distinctions between topological manifolds, PL manifolds, and differentiable manifolds have provided some of the most fascinating research problems in topology. Some of the early

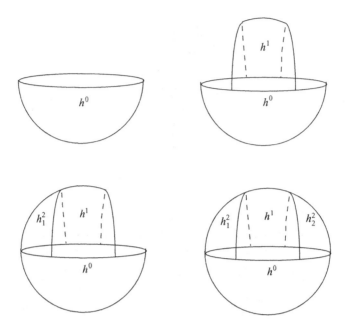

Figure 2.5. Another handle decomposition of the sphere.

foundational papers involving both surfaces and higher-dimensional manifolds include [23, 4, 8, 26, 33]. See [3, 22, 10, 25] for more expository presentations.

Note that a surface may have many different handle decompositions. The ones we have exhibited so far are minimal in that they have the minimal number of handles of each index necessary for a handle decomposition of each surface. We illustrate by a picture another handle decomposition of the sphere in Figure 2.5. This is formed with one 0-handle, one 1-handle (starting off like a torus with an annulus), and then two 2-handles. The role of the first 2-handle is to close a hole in the annulus to get a disk. Thus the effect of the first three handles is to give a complicated handle decomposition of a disk. The remaining 2-handle then completes the disk to a sphere, as in the standard handle decomposition of the sphere.

Our next two model surfaces are best described in terms of handle decompositions. Neither of them can be embedded in $\mathbb{R}^3$, although one of their fundamental building blocks—the Möbius band—can be. The first is the projective plane, which we denote by P. The projective plane has a handle decomposition with one 0-handle, one 1-handle, and one 2-handle. The 0-handle may be regarded as $[-1, 0] \times D^1$ and the 1-handle as $[0, 1] \times D^1$. The 1-handle is attached to the 0-handle by identifying the two copies of $0 \times D^1$ via the identity and identifying $(1, y)$ with $(-1, -y)$. The space obtained so far is the Möbius band, which we had looked at previously in Section 1.7 as an example of a quotient space. It can be formed by taking a rectangular strip of paper and joining the ends after

$[-1,0] \times D^1$ $[0,1] \times D^1$

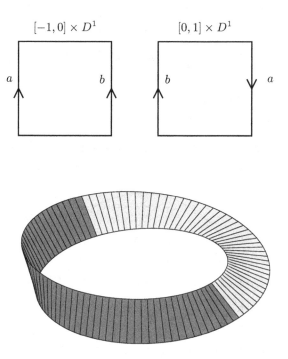

Figure 2.6. Handle decomposition of Möbius band.

making a half twist (see Figure 2.6). Its boundary is one circle, which comes from $[-1,1] \times \{\pm 1\}$ by identifying $(1,-1)$ with $(-1,1)$ and identifying $(1,1)$ with $(-1,-1)$. This is homeomorphic to a circle since it is the union of two intervals with their respective end points identified.

The Möbius band has the interesting property that it is "one-sided". You can run your finger around a path on the Möbius band to get from one "side" to the other without crossing the edge. Practical application of this property is made in conveyor belts so that they wear out evenly. Another description of this property, which is called *nonorientability*, is given by establishing near a point a direction of counterclockwise rotation as being a *positive orientation* at that point. Then there is a path in the Möbius band so that if we carry this orientation consistently along the path, we will return to the initial point with the opposite orientation (clockwise) from which we started. See Figure 2.7 for an illustration of this in the quotient space model. It is also useful to make a paper model of the Möbius band and confirm this property in the paper model.

It is worthwhile to experiment with a model of the Möbius band to gain some intuitive feeling for the property of nonorientability. You should also construct a model of the cylinder by taking a rectangular strip and joining the ends without making a twist. For each model, try cutting it lengthwise down the middle and also lengthwise one-third of the way across. Try to predict what will happen

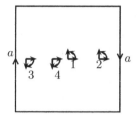

Figure 2.7. Orientation-reversing path.

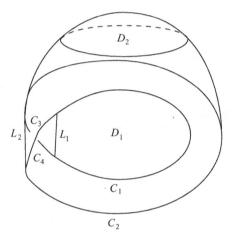

Figure 2.8. Decomposition of P.

in each case before you cut it. You may find more experiments and further manifestations of nonorientability discussed in [2]. We will study orientability more thoroughly in Section 2.4. It plays a key role in the classification of surfaces. We will give a more advanced treatment of orientability from the viewpoint of homology in Chapter 6.

Returning to the projective plane, note that the boundary of the Möbius band is a circle. The projective plane is formed from the Möbius band by attaching a 2-handle $D^2 \times D^0 = D^2$ to the Möbius band by identifying the boundary circle S^1 with the boundary circle of the Möbius band. The projective plane may not be embedded as a subspace of $\mathbb{R}^3$ so that it is difficult to draw a picture of it as in the case of the 2-sphere and the torus.

Here is a way to see how to embed P in $\mathbb{R}^4$. Consider the boundary circle C of the Möbius band. We can find an embedded disk in $\mathbb{R}^3$ with boundary C as follows—first draw lines L_1, L_2 connecting points of C, breaking C into C_1, C_2, C_3, C_4 as indicated in Figure 2.8.

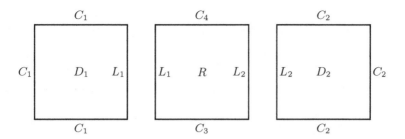

Figure 2.9. Forming a disk from three disks.

Now $C_1 \cup L_1$ bounds a disk D_1 and $C_2 \cup L_2$ bounds a disk D_2 that lies above the paper (like the upper hemisphere of the 2-sphere bounding the equator). Then $L_1 \cup C_3 \cup L_2 \cup C_4$ bounds a twisted rectangular strip R. To see that D_1, D_2, and L fit together to give a homeomorphic copy of the disk, redraw D_1 and D_2 as rectangles and put the three pieces together in the plane. What we get is three rectangles laid end on end, which is homeomorphic to a rectangle and hence to a disk (see Figure 2.9).

Unfortunately, this disk D^2 intersects the Möbius band within the strip R. By perturbing it, we can remove this intersection in $\mathbb{R}^4$. Let $f : D^2 \to \mathbb{R}^3$, $g : M \to \mathbb{R}^3$ be our embeddings into $\mathbb{R}^3$ which agree on the boundary circles. Now define $\bar{f} : D^2 \to \mathbb{R}^4$, $\bar{g} : M \to \mathbb{R}^4$ by $\bar{f}(x) = (f(x), 1 - |x|)$, $\bar{g}(x) = (g(x), 0)$. Here the first entry corresponds to the first three coordinates of $\mathbb{R}^4$, and the last entry corresponds to the fourth coordinate. Now $\bar{f}$ and $\bar{g}$ are embeddings into $\mathbb{R}^4$, which intersect only along the common boundary circle, and hence fit together to give an embedding of P into $\mathbb{R}^4$. We are using the extra dimension to remove the intersection between the original images of f and g. This is analogous to removing the intersection of a figure 8 in the plane by lifting one piece of the 8 slightly into $\mathbb{R}^3$ near the intersection point.

There are useful descriptions of P besides being the union of the Möbius band and a disk along their boundary circles. For instance, it may be described as a quotient space of the 2-sphere S^2 by identifying $x = (x_1, x_2, x_3)$ with $-x = (-x_1, -x_2, -x_3)$; $-x$ is called the *antipodal point* of x. There is a natural identification map $f : S^2 \to P$ and we give P the quotient topology using this map. This description of P is responsible for the name *projective plane* since each pair of points $\{x, -x\}$ which are identified corresponds to a unique line through the origin. Thus P can be thought of the space of lines through the origin in $\mathbb{R}^3$ with the appropriate topology. Another description of P as a quotient space can be derived from the above by noting that only the upper hemisphere is necessary in finding a surjective map onto P. Using the fact that the upper hemisphere is homeomorphic to the disk, P may be described as the quotient space of the disk with opposite points (x_1, x_2) and $(-x_1, -x_2)$ on the boundary circle identified. From this description we can see (as illustrated in Figure 2.10) why P is the union of the Möbius band and the disk. In this figure, B is easily

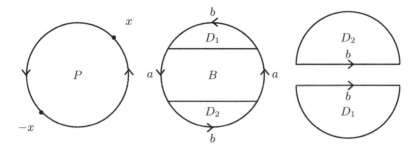

Figure 2.10. Two views of the projective plane.

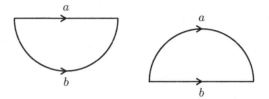

Figure 2.11. Two homeomorphic half disks.

identified to the Möbius band and D_1 and D_2 (with the indicated identifications) together form a disk. Note that we are using the idea of Exercise 2.2.1.

Exercise 2.2.1. Construct a homeomorphism between the lower half of the disk and the upper half of the disk sending the boundary as indicated in Figure 2.11. (Hint: Construct homeomorphisms between each half disk and the whole disk.)

Exercise 2.2.2.

(a) Use the quotient map $f : S^2 \to P$ to show that P is a compact, connected surface.

(b) Use the handle decomposition description of P to show that P is a compact connected surface.

Our final model surface is the Klein bottle, which we denote by K. It is the union of two Möbius bands joined along their boundary circles. We can think of it as being obtained from two projective planes by removing a disk from each and gluing along the boundaries of each disk. We will return to this idea in a more general context of forming connected sums. We can picture K and T at the same time by taking a rectangle and making certain identifications on the boundary. Each starts with the space $D^1 \times D^1$. For the torus and Klein bottle, we first identify $D^1 \times \{-1\}$ with $D^1 \times \{1\}$ by identifying $(x, -1)$ with $(x, 1)$. This forms a cylinder. For the torus we then identify $\{1\} \times D^1$ with $\{1\} \times D^1$ by identifying $(-1, y)$ with $(1, y)$. For the Klein bottle, we identify $\{-1\} \times D^1$ with $\{1\} \times D^1$ by identifying $(-1, y)$ with $(1, -y)$. See Figure 2.12 for an illustration.

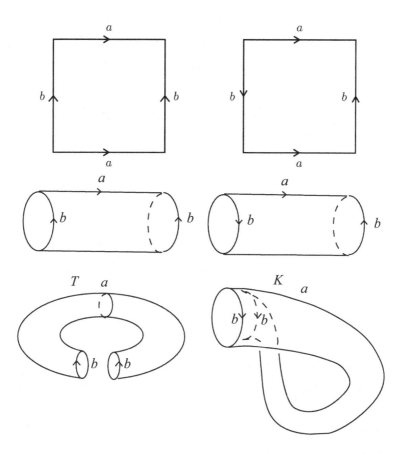

Figure 2.12. Constructing the torus and Klein bottle.

We indicate how the Klein bottle is the union of two Möbius bands glued along their boundary circles in Figure 2.13.

Exercise 2.2.3. Give a description of K as a quotient space of $\mathbb{R}^2$ analogous to our description of T as a quotient space of $\mathbb{R}^2$ in Chapter 1. Use this description to show that K is a surface.

Exercise 2.2.4. Show that K is compact and connected.

We describe two different handle decompositions of K. Each has one 0-handle, two 1-handles, and one 2-handle. They differ in that in one we have a cylinder after the first 1-handle is attached, and in the other we have a Möbius band after the first 1-handle is attached. We will illustrate both decompositions in terms of our picture of the Klein bottle as a rectangle with identifications. We have illustrated in Figure 2.14 the decomposition of K as $h^0 \cup h_1^1 \cup h_2^1 \cup h^2$. But we could have added the two 1-handles in the other order and got a decomposition of K as $h^0 \cup h_2^1 \cup h_1^1 \cup h^2$. Note that in the second decomposition $h^0 \cup h_2^1$ is a cylinder and the attaching map for h_1^1 identifies $\{\pm 1\} \times D^1$ to part of

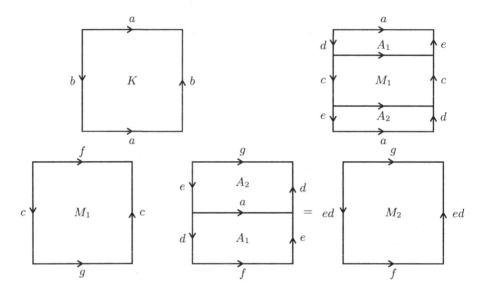

Figure 2.13. The Klein bottle is a union of two Möbius bands.

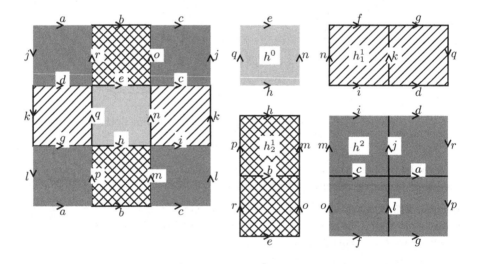

Figure 2.14. A handle decomposition of the Klein bottle.

$\partial(S^1 \times D^1) = S^1 \times \{\pm 1\}$ by identifying $\{-1\} \times D^1$ to an interval in $D^1 \times \{-1\}$ but identifying $\{1\} \times D^1$ to a similar interval in an orientation-reversing fashion. By this we mean that if $f : D^1 \to S^1$ is the embedding of $[-1, 1]$ into S^1, then $\{-1\} \times D^1 \to S^1 \times \{-1\}$ is given by $(-1, x) \to (f(x), -1)$ and $\{1\} \times D^1 \to S^1 \times \{1\}$ is given by $(1, x) \to (f(-x), 1)$.

The four surfaces—S^2 (hereafter just called S), T, P, K—are our model surfaces. We now describe the model surfaces with boundary. The first will be the disk D^2. Note that the exterior of the disk in the plane is homeomorphic via $(x, y) \rightarrow (x/(x^2 + y^2), y/(x^2 + y^2))$ to $D^2 \backslash \{(0, 0)\}$. Hence the exterior of the disk is also a manifold with boundary. By a *disk in a surface* we mean the image $f(D^2)$ of the standard disk $D^2 \subset \mathbb{R}^2$ under a homeomorphism $\tilde{f}$ from $\mathbb{R}^2$ onto an open set in the surface, where $\tilde{f}|D^2 = f$. Since the complement of the interior of the disk in $\mathbb{R}^2$ is a surface with boundary, it follows that the complement of the interior of a disk in a surface is a surface with boundary. The question then arises whether two such complements $M \backslash \text{int } D_1$ and $M \backslash \text{int } D_2$ are homeomorphic if D_1, D_2 are disks in M. This is true when M is connected because of the following important result, which is called the 'disk lemma'. The lemma is a basic one in PL or differential topology for manifolds of any dimension and is based on the idea of a regular neighborhood or tubular neighborhood of a point. Rourke and Sanderson [31] and Hirsch [15] are good sources for treatments from these viewpoints. We will give a more refined version as well as relate it to the Schönflies theorem and the concept of orientation in Section 2.4 and the supplementary exercises at the end of the chapter. The general topological case in higher dimensions is closely connected to recognizing an annular region enclosed between one embedded disk and a larger one, which was a difficult question that was only recently solved by Kirby [16] in dimensions other than 4 and then more recently by Quinn [28] in dimension 4.

Lemma 2.2.1 (Disk lemma, first version). *Let D_1, D_2 be disks in the interior of a connected surface (with boundary) M. Then there is a homeomorphism $h : M \rightarrow M$ (fixed on ∂M) with $h(D_1) = D_2$.*

Exercise 2.2.5. Show by induction that if $D_1, \ldots, D_k$ are k disjoint disks in the interior of a connected surface M, and $D'_1, \ldots, D'_k$ is another collection of k disjoint disks in the interior of M, then there is a homeomorphism $h : M \rightarrow M$ which is the identity on ∂M sending D_i to D'_i.

We define $M_{(p)}$ to be the surface obtained from M by removing p disjoint disks from int M when M is connected. By Exercise 2.2.5 it is well defined up to homeomorphism independent of the particular disks removed. Our model surfaces with boundary will be $S_{(p)}$, $T_{(p)}$, $P_{(p)}$, and $K_{(p)}$. Note that $S_{(1)}$ is a disk and $P_{(1)}$ is a Möbius band.

2.3 Isotopy and attaching handles

In this section we will develop some technical lemmas to understand how the attaching of 1- and 2-handles depends on the particular attaching homeomorphism used. The important thing for the reader to understand from this section is the statements of the lemmas and how they reduce the classification problem to a problem of understanding certain models. The proofs of the lemmas are of less importance, and it may be useful to omit them on the first reading of the section. They are largely an application of ideas of collars, quotient spaces, and properties of self-homeomorphisms of an interval to itself.

Suppose H is a handlebody; we will show that the boundary of H is a disjoint union of a number of circles and that there is a (closed) neighborhood of ∂H of the form $\partial H \times I$. Such a neighborhood is called a *collar* of the boundary, and we will be using the existence of such a collar throughout the section. Any compact manifold with boundary possesses a collar on the boundary. Arguments are given by Rourke and Sanderson [31] and Hirsch [15] in the PL and differentiable cases and by Hatcher [13] for topological manifolds. We consider how $H \cup_f h^1$ depends on the particular embedding of $f : \{\pm 1\} \times D^1 \to \partial H$. In understanding this, we will use the concept of an isotopy.

Definition 2.3.1. We say homeomorphisms $g_0, g_1 : B \to B$ are *isotopic* if there is a homeomorphism $G : B \times I \to B \times I$ with $G(b,t) = (G_t(b), t)$ and $G_0 = g_0, G_1 = g_1$. G is called an *isotopy* between g_0 and g_1. We say embeddings $f_0, f_1 : A \to B$ are *ambient isotopic* if there is an isotopy $G : B \times I \to B \times I$ so that $G_0 = \text{id}$ and $G_1 f_0 = f_1$.

We will be dealing with the notion of ambient isotopy of embeddings of D^1 in S^1 in this section and of embeddings of D^2 in surfaces in the next section. Isotopy of homeomorphisms is an equivalence relation, with transitivity the only difficult property to verify. If F_t gives an isotopy between f and g, and G_t gives an isotopy between g and h, then

$$H_t = \begin{cases} F_{2t} & \text{if } 0 \leq t \leq \frac{1}{2}, \\ G_{2t-1} & \text{if } \frac{1}{2} \leq t \leq 1 \end{cases}$$

gives an isotopy between f and h. Note that the argument is the same one that we used to show that path connectivity is transitive. An isotopy is a path of homeomorphisms connecting two homeomorphisms. In terms of ambient isotopy, there is also an argument in terms of composition. If F, G are ambient isotopies with $F_0 = \text{id}, F_1 f = g$, and $G_0 = \text{id}, G_1 g = h$, then if $H = GF$, we have $H_0 = \text{id}, H_1 f = G_1 F_1 f = G_1 g = h$.

We first review how the circle arises from the interval and from the reals as a quotient space. Let $p : \mathbb{R} \to S^1$ be the map given by $p(t) = (\cos 2\pi t, \sin 2\pi t)$. Then p is a quotient map and we can use it to identify S^1 with $\mathbb{R}/t \sim t+n, n \in \mathbb{Z}$. Any closed interval of length <1 will be mapped homeomorphically onto its image, which will be called an *arc* in the circle. A homeomorphism $f : \mathbb{R} \to \mathbb{R}$ is called *periodic* of period 1 if either $f(x+1) = f(x) + 1$ (for f order preserving), or $f(x+1) = f(x) - 1$ (for f order reversing). Such a homeomorphism f will then induce a homeomorphism $\bar{f}$ by the quotient construction: $\bar{f}([x]) = [f(x)]$. It is also the case that any homeomorphism of S^1 arises as $\bar{f}$ for a periodic homeomorphism f.

Since the homeomorphism f is periodic, it is actually determined by its values on any interval of length 1. To discuss this, we will assume that f is order preserving for simplicity, but the argument is analogous in the order-reversing case. Choosing an interval $[a, a+1]$, f order-preserving periodic implies that $f(a + 1) = f(a) + 1$; thus f gives a homeomorphism by restriction $f' = f|[a, a+1] : [a, a+1] \to [b, b+1]$, where $b = f(a)$. Conversely, whenever we have an order-preserving homeomorphism $f' : [a, a + 1] \to [b, b + 1]$, it determines a periodic homeomorphism $f : \mathbb{R} \to \mathbb{R}$ via noting that any $x \in \mathbb{R}$ can be expressed as $x = t + n, n \in \mathbb{Z}, t \in [a, a + 1)$ and t unique. Then f is defined by $f(x) = f(t) + n$. We leave it as an exercise to check that f is continuous and gives a periodic homeomorphism. Thus order-preserving periodic homeomorphisms of $\mathbb{R}$ may be identified with order-preserving homeomorphisms $f' : [a, a+1] \to [b, b+1]$. In turn, these are identified with orientation-preserving homeomorphisms of the circle. A related fact is that the restriction of p to $[a, a + 1]$ or $[b, b + 1]$ determines a quotient map to the circle, and so $\bar{f}$ can also be regarded as coming from f' via this quotient construction. We indicate the relationship of these maps by the following commutative diagram, where i_a, i_b denote the natural inclusions.

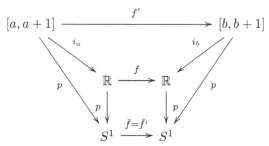

Exercise 2.3.1. Show that the map $f : \mathbb{R} \to \mathbb{R}$ defined above using $f' : [a, a + 1] \to [b, b + 1]$ is a periodic homeomorphism.

Because of these relationships, we may deal with maps from the circle to itself by using periodic maps of $\mathbb{R}$ or maps from the interval $[a, a+1]$ to $[b, b+1]$. This also applies to self-maps of $S^1 \times I$ in a similar manner.

Lemma 2.3.1. *Let I_1, I_2 (resp., I'_1, I'_2) be disjoint arcs in the circle. Then there is a homeomorphism of the circle sending I_1 to I'_1 and I_2 to I'_2. Moreover, this homeomorphism may be chosen to be isotopic to the identity.*

Proof. Note first that any two arcs I, I' in the circle are homeomorphic. Moreover, there is "standard" homeomorphism between them. For consider $p : \mathbb{R} \to S^1$, $p(t) = (\cos 2\pi t, \sin 2\pi t)$. Select intervals $J, J' \subset \mathbb{R}$ so that $p(J) = I$, $p(J') = I'$. Now any two intervals are homeomorphic via the unique order-preserving affine linear homeomorphism from $[a, b]$ to $[a', b']$ sending $ta + (1-t)b$ to $ta' + (1-t)b', 0 \le t \le 1$. This induces (via p) a "standard" homeomorphism from I to I'. We leave it as an exercise to show that this "standard" homeomorphism does not depend on the choice of J, J'.

Regarding S^1 as the quotient space of $\mathbb{R}$, we can find intervals J_1, J_2, J'_1, J'_2 which p sends to I_1, I_2, I'_1, I'_2 homeomorphically. Moreover, these can be chosen so that $v_1 < v_2 < v_3 < v_4 < v_1 + 1$, $v'_1 < v'_2 < v'_3 < v'_4 < v'_1 + 1$, $|v_1 - v'_1| < 1$, and $J_1 = [v_1, v_2], J_2 = [v_3, v_4], J'_1 = [v'_1, v'_2], J'_2 = [v'_3, v'_4]$. We then define a map $f : [v_1, v_1+1] \to [v'_1, v'_i+1]$ by requiring that $f(v_i) = v'_i$ and $f(v_1+1) = v'_1+1$, as

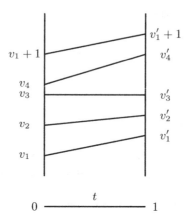

Figure 2.15. Isotoping embeddings.

well as making the map affine linear on each subinterval $[v_i, v_{i+1}]$ and $[v_4, v_1+1]$. This induces the homeomorphism $\bar{f} : S^1 \to S^1$ which sends I_j to I'_j.

To see that it is isotopic to the identity, we construct a map using intervals which induces the isotopy. The map F_t giving the isotopy will be defined at each level t by a map like the one above and will be determined by the images of the vertices. Its graph for a fixed vertex v_j and varying t is just the straight line segment joining v_j and v'_j. We show in Figure 2.15 the images of these vertices as t varies. The subintervals between vertices are then mapped by affine linear maps. These maps then induce the required isotopy of homeomorphisms of the circle. □

Exercise 2.3.2. Show that the "standard" homeomorphism constructed above does not depend on the choice of J, J'.

Note that Lemma 2.3.1 also allows us to move a single interval via an isotopy. We will use it in both situations. Lemma 2.3.1 is used in conjunction with the following lemma to say that we may specify without loss of generality the image arcs without changing the homeomorphism type of $H \cup_f h^1$.

Lemma 2.3.2. *Suppose $H \cup_f h^1$ and $H \cup_g h^1$ are handlebodies, and there is a neighborhood of ∂H in H of the form $\partial H \times I$ where ∂H corresponds to $\partial H \times \{0\}$ and an isotopy $F : \partial H \times I \to \partial H \times I$ with $F_0 f = g$ and $F_1 = id$, the identity. Then there is a homeomorphism from $H \cup_f h^1$ to $H \cup_g h^1$ sending h^1 to h^1 by identification (via the "identity"). The homeomorphism is also the identity outside the collar of the boundary.*

Proof. Define the homeomorphism as follows. First send h^1 to h^1 via the identity. Identifying the neighborhood of ∂H with $\partial H \times I$, send $\partial H \times I \to \partial H \times I$

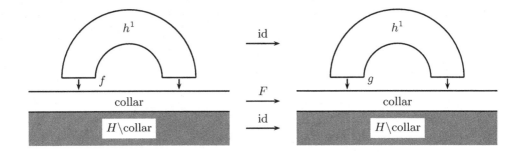

Figure 2.16. Using an isotopy on the collar.

via F, and send $H\backslash(\partial H \times I)$ to $H\backslash(\partial H \times I)$ via the identity. These maps fit together to give a homeomorphism. □

Figure 2.16 gives an illustration of the above proof.

The way to use this lemma with the preceding one is to define g by $F_0 f$, where $F_0(I_1) = I'_1, F_0(I_2) = I'_2$. This allows us to shift our image arcs of the attaching homeomorphism up to isotopy without changing the homeomorphism type of the resulting handlebody.

We now study the dependence of $H \cup_f h^1$ on f, after fixing the image set $f(\{\pm 1\} \times D^1)$. Because of the argument above, we will assume that $f_1(\{-1\} \times D^1) = f_2(\{-1\} \times D^1)$ and $f_1(\{1\} \times D^1) = f_2(\{1\} \times D^1)$ and then compare $H \cup_{f_1} h^1$ and $H \cup_{f_2} h^1$. Consider $f_2^{-1} f_1$; this is a homeomorphism from $\{\pm 1\} \times D^1$ to itself, sending each copy of D^1 to itself. We claim that the only important fact we need to know to compare the homeomorphism type of $H \cup_{f_i} h^1$ is whether both these maps preserve order (or reverse it) or whether one preserves order and the other reverses it. The claim depends on the following lemma, in which we are using I instead of D^1 because we also want to apply it to homeomorphisms of the circle.

Lemma 2.3.3.

(a) *Any order-preserving homeomorphism $f : I \to I$ is isotopic to the identity.*

(b) *Any order-reversing homeomorphism $f : I \to I$ is isotopic to $r'(t) = 1-t$.*

Proof. The idea of the proof of (a) is to graph f and the identity and connect $f(s)$ and s by a straight line. We use the formula $G(s,t) = G_t(s) = (1-t)f(s)+ts$. Then $G_t(-1) = -1, G_t(1) = 1, G_0(s) = f(s), G_1(s) = s$, and G_t is a homeomorphism for each t. We confirm the last fact by noting that if $0 \le s_1 < s_2 \le 1$, then

$$G_t(s_1) = (1 - t)f(s_1) + t(s_1) < (1 - t)f(s_2) + ts_2 = G_t(s_2).$$

We are using here that $0 \le (1-t)$, $t \le 1$ and both $1-t$ and t cannot be 0 since they add to 1. We leave (b) as an exercise. □

Exercise 2.3.3. Deduce part (b) from (a) by composing an order-reversing f with r' to get an orientation preserving $g = fr'$ to which to apply (a).

Here is the version for D^1.

Lemma 2.3.4.

(a) *Any order-preserving homeomorphism $g : D^1 \to D^1$ is isotopic to the identity.*

(b) *Any orientation-reversing homeomorphism $g : D^1 \to D^1$ is isotopic to the reflection $r(t) = -t$.*

Exercise 2.3.4. Use the homeomorphism $h : D^1 \to I, h(s) = \frac{1}{2}(s+1)$ and Lemma 2.3.3 to prove Lemma 2.3.4.

We define an embedding of D^1 into S^1 to be orientation preserving (resp., reversing) if it is the composition of an order-preserving (reversing) embedding of D^1 into $\mathbb{R}$ and the map $p : \mathbb{R} \to S^1$, $p(t) = (\cos 2\pi t, \sin 2\pi t)$. We say that a homeomorphism from S^1 to S^1 is orientation preserving if it comes from a periodic homeomorphism of $\mathbb{R}$ which is order preserving. The composition of an orientation-preserving embedding and an orientation-preserving homeomorphism of S^1 will be orientation preserving. All isotopies of S^1 arise as periodic isotopies of $\mathbb{R}$. This means that each F_t must be orientation preserving when $F_0 = \text{id}$. We use this for the next proposition.

Proposition 2.3.5.

(a) *Suppose $f_1, f_2 : D^1 \to S^1$ are orientation-preserving embeddings. Then they are ambient isotopic.*

(b) *Suppose $f_1, f_2 : D^1 \to S^1$ are orientation-reversing embeddings. Then they are ambient isotopic.*

(c) *Suppose $f_1, f_2 : D^1 \to S^1$ are embeddings so that f_1 preserves orientation and f_2 reverses orientation. Then they are not ambient isotopic.*

Proof.

(a) Let $I_1 = f_1(D^1), I_2 = f_2(D^1)$. By Lemma 2.3.1, there is an ambient isotopy F_t so that $F_0 = \text{id}$ and $F_1(I_1) = I_2$. By construction, F_t arises from a periodic order-preserving homeomorphism of $\mathbb{R}$, so will preserve orientation for all t. Let $g_1 = F_1 f_1$—this is orientation preserving. Hence $g_1^{-1} f_2 : D^1 \to D^1$ is order preserving. By Lemma 2.3.4 there is an ambient isotopy $G_1 : D^1 \times I \to D^1 \times I$ so that $G_0(x) = x, G_1 = g_1^{-1} f_2$. Define $H : I_2 \times I \to I_2 \times I$ by $H_t = g_1 G_t g_1^{-1}$. Then $H_0 = \text{id}, H_1 = g_1 G_1 g_1^{-1} = g_1 g_1^{-1} f_2 g_1^{-1} = f_2 g_1^{-1}$. Then $H_1 g_1 = f_2 g_1^{-1} g_1 = f_2$. Now the isotopy H is the identity on $\{g_1(-1), g_1(1)\}$ since G_t is the identity on the end points. Hence we can extend H_t to all of S^1 by defining it to be the identity on $S^1 \backslash I_2$. Finally, the composition $H_t F_t$ will give an ambient isotopy connecting f_1 and f_2.

(b) The embeddings $f_1 r$, $f_2 r$ are orientation preserving. By (a), there is an isotopy $F_t : S^1 \to S^1$ with $F_0 = \mathrm{id}$, $F_1 f_1 r = f_2 r$. Composing with r gives

$$F_1 f_1 = F_1 f_1 (rr) = (F_1 f_1 r) r = (f_2 r) r = f_2 (rr) = f_2.$$

(c) By the comment above, isotopies of S^1 are orientation preserving at each level F_t so ambient isotopic embeddings are either both orientation preserving or both orientation reversing.

$\square$

There is a result for homeomorphisms of S^1 which is analogous to Lemma 2.3.3 and follows from it.

Lemma 2.3.6.

(a) *Any orientation-preserving homeomorphism $f : S^1 \to S^1$ sending 1 to 1 is isotopic to the identity, where 1 is sent to 1 at each stage of the isotopy.*

(b) *Any orientation-preserving homeomorphism $f : S^1 \to S^1$ is isotopic to the identity.*

(c) *Any orientation-reversing homeomorphism $f : S^1 \to S^1$ is isotopic to $r(x, y) = (x, -y)$.*

Proof.

(a) The orientation-preserving homeomorphism f is induced from an order-preserving homeomorphism $f' : I \to I$, which by Lemma 2.3.3 is isotopic to the identity via an isotopy F_t. Then F_t induces an isotopy $\bar{F}_t : S^1 \to S^1$ between f and the identity.

(b) A rotation $R_\theta (\cos \phi, \sin \phi) = (\cos(\phi + \theta), \sin(\phi + \theta))$ is isotopic to the identity via $F_t = R_{t\theta}$. First isotope f to a map $g = R_\theta f$ which sends 1 to 1. Then apply (a).

(c) This is left as an exercise.

$\square$

Exercise 2.3.5. Prove part (c).

Now consider what happens when we attach a 1-handle to H. The boundary of H consists of a finite number of circles, each of which has a collar neighborhood. Since D^1 is connected, it will be embedded into a single boundary circle. When we form $H \cup_f D^1 \times D^1$, then the images $f(\{-1\} \times D^1) \cup f(\{1\} \times D^1)$ will lie in either one or two boundary circles. We identify each of these boundary circles with S^1 and the collar with $S^1 \times I$. Suppose g is another attaching map which has $f(\{-1\} \times D^1)$ and $g(\{-1\} \times D^1)$ lie in the same boundary circle and $f(\{1\} \times D^1)$ and $g(\{-1\} \times D^1)$ lie in the same boundary circle. Denote by $f_{(-1)}, f_{(1)}$ the restrictions of f to the two attaching intervals, with similar notation for g. Then Lemma 2.3.1 says that there is an ambient isotopy F of the relevant circles so that $F_0 = \text{id}$ and $F_1 f_{(-1)} = f'_{(-1)}, F_1 f_{(1)} = f'_{(1)}$, where $f'_{(-1)}(\{-1\} \times D^1) = g_{(-1)}(\{-1\} \times D^1), f'_{(1)}(\{1\} \times D^1) = g_{(1)}(\{1\} \times D^1)$. Then if $(f'_{(-1)})^{-1} g_{(-1)}$ is order preserving, then there is an ambient isotopy between $f'_{(-1)}$ and $g_{(-1)}$. Composing these ambient isotopies will give an ambient isotopy between $f_{(-1)}$ and $g_{(-1)}$. On the other hand, if $(f'_{(-1)})^{-1} g_{(-1)}$ is order reversing, then there will be no ambient isotopy between $f'_{(-1)}$ and $g_{(-1)}$, and hence no ambient isotopy between $f_{(-1)}$ and $g_{(-1)}$. However, in this case, we have the map $(f'_{(-1)})^{-1} g_{(-1)} r$ order preserving, where $r(x, y) = (x, -y)$ and so we get $f_{(-1)}$ is ambient isotopic to $g_{(-1)} r$. Similarly, we either have $f_{(1)}$ ambient isotopic to $g_{(1)}$ or to $g_{(1)} r$. Note that when $f_{(-1)}, f_{(1)}$ both attach the handle to the same boundary circle, the proof of Lemma 2.3.5 will allow us to construct an ambient isotopy of this circle to achieve these results for both embeddings simultaneously.

Thus there are four possibilities:

(1) $f_{(-1)}$ is ambient isotopic to $g_{(-1)}$ and $f_{(1)}$ is ambient isotopic to $g_{(1)}$;

(2) $f_{(-1)}$ is ambient isotopic to $g_{(-1)} r$ and $f_{(1)}$ is ambient isotopic to $g_{(1)} r$;

(3) $f_{(-1)}$ is ambient isotopic to $g_{(-1)}$ and $f_{(1)}$ is ambient isotopic to $g_{(1)} r$;

(4) $f_{(-1)}$ is ambient isotopic to $g_{(-1)} r$ and $f_{(1)}$ is ambient isotopic to $g_{(1)}$.

In case (1), Lemma 2.3.2 says that there is a homeomorphism between $H \cup_f h^1$ and $H \cup_g h^1$: this homeomorphism will be the standard identification on the handle h^1 and the identity on the complement of a collar on the boundary circle(s) in H, and will use the isotopy on the collar(s). We get similar results in the other three cases. However, there is a homeomorphism between $H \cup_g h^1$ and $H \cup_{gr} h^1$ which is defined by sending H to itself by the identity and sending $h^1 = D^1 \times D^1$ to itself via r. Thus in the first two cases, we get a homeomorphism between $H \cup_f h^1$ and $H \cup_g h^1$. In the third case, we get a homeomorphism between $H \cup_f h^1$ and $H \cup_{g'} h^1$ where $g'_{(-1)} = g_{(-1)}$ and $g'_{(1)} = g_{(1)} r$. Case (4) gives the same space as in case (3) since $H \cup_{g'} h^1 \simeq H \cup_{g'r} h^1$ using the homeomorphism r on h^1. Thus there are at most two different ways to attach a handle up to homeomorphism. Given one way f, the other way which may be different up to homeomorphism is to use the same attaching map on one interval and to compose with r on the other interval. When both intervals are attached to the same circle, these will be different, as we will see below. When they are attached to two different circles, they may or may not be different, depending on connectivity and orientability conditions, which are discussed in the next section.

We next look at how the boundary changes when we attach h^1. Because of the discussion above, we can specify the attaching map on one interval, and the attaching map up to possible composition with a reflection on the other subinterval. The only place the boundary will change is for the circles where the handle is attached. The other circles and their collars will remain after the handle is attached. First consider the case where the handles are attached to a single circle. We draw a collar neighborhood and look at two standard ways of attaching the handle. We can identify the two possibilities in Figure 2.17. In the

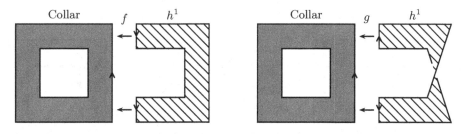

Figure 2.17. Attaching a 1-handle to one boundary circle.

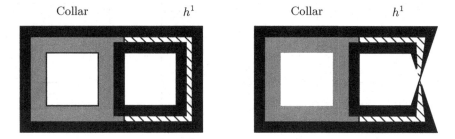

Figure 2.18. New boundary neighborhoods.

one on the left, the new boundary will have two components in place of one. For each component it is easy to find a collar on the boundary that will use part of the old collar which has been adjusted near where the handle is attached to fit together with an adjusted collar on the handle. For the case on the right, the new boundary will still consist of one boundary circle, and there is a Möbius band embedded in the surface using a piece of the collar and the handle h^1. Again, we can piece together a collar using the old collar and a collar on the handle, where we adjust each near where the two pieces meet. In each case, the new boundary is formed from the old by removing the attaching intervals and putting in the new intervals $D^1 \times \{\pm 1\}$. We picture the new collars in Figure 2.18. Since these two surfaces have different numbers of boundary circles, they are not homeomorphic.

Exercise 2.3.6. Find the Möbius band that is referred to above on the right surface in Figure 2.17.

Now consider the case where the handle is attached to two different boundary circles, each of which has a collar. All of the other boundary circles and their collars will remain after this handle is attached. We can again reduce to two cases, which are pictured in Figure 2.19. By examining the two diagrams, we see that after the handle is attached, the two boundary circles have been replaced by a single boundary circle, and we can again find a collar of this new boundary piece which comes from modifying the old collars and the collar on the handle. We again remove the attaching intervals from the boundary and add in $D^1 \times \{\pm 1\}$ to the boundary.

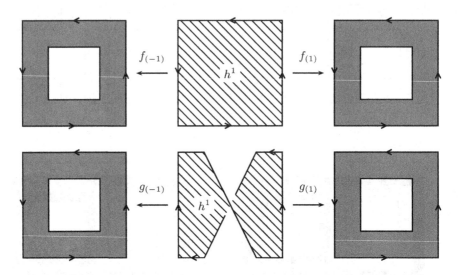

Figure 2.19. Attaching a 1-handle to two boundary circles.

Exercise 2.3.7. Show that the result from attaching a 1-handle to the same boundary circle is never homeomorphic to what you get when it is attached to two different boundary circles. (Hint: Consider the number of boundary circles in each case.)

We summarize the main results in this section. To understand how attaching a 1-handle changes a handlebody, we only need a limited amount of information. First, we need to know the components of the boundary where the handle is attached. Within any individual component, we may specify completely what the image arcs are without changing the homeomorphism type of $H \cup h^1$. Once the image arcs are specified, then the result of attaching the 1-handle is described via one of our standard models. Thus to study the changes in the handlebody induced by adding a 1-handle, it suffices to study what happens in each of our model situations.

Theorem 2.3.7. *Let H be a handlebody with a handle decomposition with 0-handles and 1-handles.*

  (a) *Then ∂H is homeomorphic to the disjoint union of a finite number of circles.*

  (b) *There is a neighborhood $N(\partial H)$ of ∂H which is homeomorphic to $\partial H \times I$, where $\partial H \times \{0\} \subset \partial H \times I$ is sent to $\partial H \subset H$ via the standard identification $(x, 0) \to x$.*

Proof. We outline the argument, leaving the details as Exercise 2.3.8. For 0-handles, the two claims are easily verified. Thus we prove the result by induction on the number of 1-handles, seeing that it holds as each 1-handle is attached. Let $H = H_1 \cup h^1$. We assume by induction hypothesis that (a) and (b) hold for H_1. Hence ∂H_1 is a disjoint union of a number of circles and there is a neighborhood of ∂H_1 of the form $\partial H_1 \times I$. Since we have a collar on the boundary where the 1-handle is attached, our earlier discussion applies. To see what the new boundary will be, we may choose our attaching maps to be one of the two standard ones. Only one or two boundary circles and its neighborhood will be affected by attaching the 1-handle. Thus it suffices to show that the new boundary and neighborhood as claimed in these two cases. Figure 2.18 can be used to show that (a) and (b) still hold after the 1-handle is attached when they are attached to the same boundary component. We leave as an exercise the case where the two arcs are attached to different boundary components, where Figure 2.19 may be used to find the required neighborhood. $\square$

Exercise 2.3.8. Fill in the details in the proof above.

We next consider what happens when we attach a 2-handle h^2 to form $H = H_1 \cup_f h^2$. Here $f : S^1 \to \partial H_1$ is an embedding into the boundary, which consists of a finite number of circles, each of which has a collar. Such an embedding has to be sent into a single circle. Its image is connected and compact. If it missed any point, it would be a closed subarc, which is not homeomorphic to the circle. Thus the image must be the whole circle. If we identify the two circles, then the map would have to be isotopic to either the identity or a reflection by Lemma 2.3.6. But these two give the same result up to homeomorphism, using a homeomorphism which is the identity on H_1 and r on $h^2 = D^2$. The boundary circle to which the handle is attached is removed from the boundary by this operation. This can be seen by identifying the collar neighborhood to the annular region $S^1 \times [1, 2]$ between circles of radius 1 and 2 in the plane and then attaching the handle via the identity to fill in the unit disk. Thus we get the following proposition.

Proposition 2.3.8. *If $f, g : S^1 \to \partial H_1$ are attaching maps for a 2-handle which map to the same boundary circle, then $H_1 \cup_f h^2 \simeq H_1 \cup_g h^2$. This new surface will have one fewer boundary circle than H_1.*

This last proposition says that if we have two handlebodies that are homeomorphic, then if we attach corresponding 2-handles to "fill in" these boundary circles, the resulting surfaces are homeomorphic. There is an alternate proof which does not use the result on isotopic homeomorphisms of the circle. Instead it uses the fact that any homeomorphism of the circle extends to a homeomorphism of the disk by coning; that is, given $h : S^1 \to S^1$, we define $H : D^2 \to D^2$ by $H(\mathbf{0}) = \mathbf{0}$ and $H(r\boldsymbol{x}) = rh(\boldsymbol{x})$, where $|\boldsymbol{x}| = 1$ and $0 \le r \le 1$. The homeomorphism between $H_1 \cup_f h^2$ and $H_1 \cup_g h^2$ is the identity on H_1 and the extension of the homeomorphism $g^{-1}f$ on $D^2 = h^2$. If there are multiple 2-handles, we can use this on each one. In the other direction, the disk lemma and the exercise which follows it says that if two surfaces are homeomorphic, then the surfaces obtained by removing k disks from each of them are also homeomorphic. For it provides a homeomorphism of a surface with itself which sends any connected

k disks in the interior to any other collection. Because of these results, we will largely focus our attention on the 1-handles. However, embedded 2-disks in a surface do play a key role in the concept of orientation, which we will study in the next section.

2.4   Orientation

We now discuss the concept of *orientability* of a surface. Orientability is a property that is frequently easy to grasp intuitively but is relatively difficult to deal with precisely. A more advanced means of handling it is to re-express it in terms of homology conditions and note that the homology computations of orientability agree with one's intuition in the usual examples. In Chapter 6 we will pursue this approach through homology. Here we will deal with it formally through orienting handle decompositions but will discuss other definitions such as embedded Möbius bands and isotopy classes of embedded disks.

Before giving a more formal treatment, let us discuss some of the more common intuitive definitions. The models to think of in each case are the cylinder (orientable) and the Möbius band (nonorientable). A surface (possibly with boundary) embedded in $\mathbb{R}^3$ is said to be orientable if it has two sides, and nonorientable if it has one side. Locally, of course, the surface looks like a plane which cuts space into two halves. Imagine a point on the surface and a vector pointing to one side of the surface. To say that the surface has one side means that we can find a path in M so that if we translate our vector consistently along this path then it will be pointing to the other side of the surface when we return to the original point. See Figure 2.20 for an illustration.

Although this idea may seem fairly understandable, it has many pitfalls. One is that we are implicitly assuming that there is a "normal direction" to the surface towards which our vector can point. This problem can be handled by restricting our attention to differentiable surfaces, where the idea of a normal direction is readily defined. A more serious problem is that we are assuming our surface lies

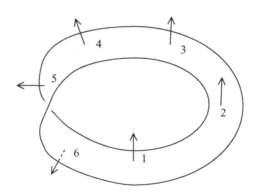

Figure 2.20. Orientation-reversing path via normal vector.

in $\mathbb{R}^3$: no compact nonorientable surface without boundary is embeddable in $\mathbb{R}^3$. One way out of this problem is to imagine the surface to be locally a plane and to orient this plane consistently by choosing a basis $\boldsymbol{v}_1, \boldsymbol{v}_2$ at each point. If the surface sits differentiably in some $\mathbb{R}^n$, there is a tangent plane attached to each point so that projection onto this plane gives a homeomorphism locally. Any other basis for the plane $\boldsymbol{w}_1, \boldsymbol{w}_2$ is related to $\boldsymbol{v}_1, \boldsymbol{v}_2$ via $\begin{pmatrix} \boldsymbol{w}_1 & \boldsymbol{w}_2 \end{pmatrix} = \begin{pmatrix} \boldsymbol{v}_1 & \boldsymbol{v}_2 \end{pmatrix} A$, where A is an invertible 2×2 matrix. To say the basis $\boldsymbol{w}_1, \boldsymbol{w}_2$ determines the same orientation as $\boldsymbol{v}_1, \boldsymbol{v}_2$ means that the determinant of A is positive. If $\det A < 0$, we say $\boldsymbol{w}_1, \boldsymbol{w}_2$ determines the opposite orientation. There are two equivalence classes of orientations at a point, since $\det A$ is either positive or negative. It is common practice to attach a direction of rotation from $\boldsymbol{v}_1$ to $\boldsymbol{v}_2$ at each point of M to indicate the choice of an equivalence class of a basis. Now we say that a surface is orientable if we can do this at each point in a consistent manner. The consistency can be checked locally since M locally looks like $\mathbb{R}^2$ and so each basis can be referred back to $\mathbb{R}^2$ to see whether it is always clockwise or always counterclockwise as a rotation. If M is path connected and there is a global choice of orientation which is locally consistent, then it will be globally consistent in terms of translation along a path in the surface always keeping the chosen orientation. The nonorientability of the Möbius band under this definition is illustrated in Figure 2.21.

The above definition resolves the problem of the previous definition in that the surface need not be embedded in $\mathbb{R}^3$. However, it does depend on a well-defined transition between equivalence classes of bases according to different local descriptions of M as $\mathbb{R}^2$. This can be done if we require our manifold to be differentiable, since a differentiable map from $\mathbb{R}^2$ to $\mathbb{R}^2$ has a linear approximation which can be used to compare bases. In fact, this idea is the basis for the standard definition of orientability for differentiable manifolds in general.

Motivated by the fact that Möbius band epitomizes nonorientablity, we now give the first of three equivalent definitions of orientability for a surface.

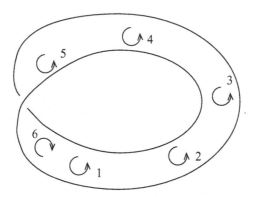

Figure 2.21. Orientation-reversing path via rotation direction.

Definition 2.4.1 (Möbius band version). We call a surface *nonorientable* if it possesses an embedded Möbius band. If it is not nonorientable, then it is said to be *orientable*.

This definition is sometimes useful in establishing that a surface is nonorientable, but it can be difficult to use in seeing that it is orientable. The next two exercises, which use the strength of the Jordan curve theorem and invariance of domain, show that the plane is orientable with this definition.

Exercise 2.4.1. Show that the complement of the center circle in a Möbius band is connected but that the complement of the center circle in an annulus is separated.

Exercise 2.4.2. Use the previous exercise to show that a Möbius band cannot be embedded in the plane. (Hint: Consider the Jordan curve theorem and invariance of domain.)

In the statement of the disk lemma, we are focusing more on the image of the embedding rather than the embedding itself. A refined version of the disk lemma leads to an equivalent definition for orientability that involves the notion of an ambient isotopy. As motivation, suppose that $f : D^2 \to \mathbb{R}^2$ is a standard embedding onto a disk $B(x_0, r_1) \subset B(\mathbf{0}, r)$. Let $R : B(\mathbf{0}, r)$ be a rotation of the plane by angle θ; that is, $R = R_\theta | B(\mathbf{0}, r)$. We can extend R to a homeomorphism of the plane which is the identity outside of $B(\mathbf{0}, r+1)$ by defining it on the annulus $A(r, r+1) = \{x : r \le |x| \le r+1\}$ by $R|(r+s)S^1 = R_{(1-s)\theta}|(r+s)S^1$, $0 \le t \le 1$. This just rotates these circles by smaller angles until we send the circle of radius $r+1$ to itself by the identity. Then we can extend R over the complement of $B(\mathbf{0}, r+1)$ by defining it to be the identity. The homeomorphism R is isotopic to the identity, where the isotopy again just moves the amount of rotation on each circle back to the angle 0. The formula for the isotopy is

$$R_t(x) = \begin{cases} R_{t\theta}(x) & |x| \le r, \\ R_{t(1-s)\theta}(x) & |x| = r+s, \ 0 \le s \le 1, \\ x & |x| \ge r+1. \end{cases}$$

Thus the embedding Rf is ambient isotopic to f via an ambient isotopy which is the identity outside a ball. In particular, if $f = i$ is the standard inclusion, then $Ri = i(R|D^2)$ is ambient isotopic to the identity.

Now consider the reflection $r : D^2 \to D^2$, $r(x, y) = (x, -y)$ and the embedding $f = ir$. It turns out that ir is not ambient isotopic to i. This will be shown in Chapter 6 using homology. We give here an argument that $fr = i$ is not (continuously) differentiably ambient isotopic to f. For suppose there were a diffeomorphism $F : \mathbb{R}^2 \times I \to \mathbb{R}^2 \times I$, $F(x, t) = (F_t(x), t)$ with $F_0 = \mathrm{id}, F_1 i = ir$. The derivative map $DF_t(x)$ will vary continuously and has positive determinant 1 for $t = 0$. Hence it has positive determinant for all t. But $DF_1(\mathbf{0})Di(\mathbf{0}) = Di(\mathbf{0})Dr(\mathbf{0})$ and the left-hand side has positive determinant while the right-hand side has determinant -1, a contradiction. The homology argument is based on the same idea, where we replace the derivative computation with one based on homology. The same type of argument can be used in the differentiable case for embeddings $f : D^2 \to M$ to show f is not isotopic to fr if M is oriented as a differentiable manifold, which will mean that there will be a continuously varying way to compute the sign of the derivative of an embedding by comparing with a choice of basis giving the orientation of the tangent space at each point. This gives a restriction on when embeddings of disks are ambient isotopic and it turns out to be the only restriction in a connected oriented differentiable manifold (see [15]).

We contrast this last example to what happens in a Möbius band, where there is an isotopy which is the identity on the boundary circle and gives an isotopy between f and fr for a standard embedding f. We regard the Möbius band B as a quotient of $\mathbb{R} \times 2D^1 \subset \mathbb{R}^2$ via the equivalence relation $(x, y) \sim (x + 1, -y)$. Now consider the region $B_1 = \mathbb{R} \times D^1 / \sim \subset B$. For $0 \le t \le 1$, the translation $T_t : \mathbb{R} \times D^1 \to \mathbb{R} \times D^1$, $T_t(x, y) = (x + t, y)$ induces a homeomorphism $H_t : B_1 \to B_1$. Moreover, H_1 induces the same map as the reflection $r(x, y) = (x, -y)$ does since $T_1(x, y) = (x + 1, y) \sim (x, -y)$. Thus, if $i : D^2 \to B_1$ is a standard embedded disk about $(0, 0)$, then this shows that $ir = ri$ is ambient isotopic to the identity with isotopy H_t. The boundary circle of B_1 comes from $[-\frac{1}{2}, \frac{1}{2}] \times \{\pm 1\} / \sim$. This is homeomorphic to a circle S^1 using $k(s, -1) = (\cos \pi s, \sin \pi s), k(s, 1) = (\cos(\pi/2 + s), \sin(\pi/2 + s))$. Using k to identify the boundary with S^1, then H_t becomes the rotation $R_{\pi t}$; that is, $R_{\pi t} = k T_t k^{-1}$. In particular, H_1 is rotation by π. We can extend k to a homeomorphism $K : B \backslash \operatorname{int} B_1 \to S^1 \times [1, 2]$ by defining $K(s, t) = (k(s, -1), |t|)$, $t < 0$, and $K(s, t) = (k(s, 1), t)$, $t > 0$. Then we can extend R_π over $S^1 \times [1, 2] \to S^1 \times [1, 2]$ by rotating $R(z, t) = (R_{\pi(2-t)}(z), t)$. This map is isotopic to the identity via $R_u(z, t) = (R_{u\pi(2-t)}(z), t)$. By identifying $B \backslash \operatorname{int} B_1$ with $S^1 \times [1, 2]$; this allows us to extend H_1 to a homeomorphism $H_1 : B \to B$ which is the identity on the boundary and is induced by reflection r on B_1, and extend H_t to an isotopy $H_t : B \to B$ between the identity and H_1. Finally, suppose $g : B \to M$ is an embedding into the interior of a surface M. Then we can define an isotopy in M which is the identity isotopy on $M \backslash g(B)$ and corresponds to our isotopy above on $g(B)$. For the embedded disk $f = gi$, we will have f ambient isotopic to fr.

When we discussed the Schönflies theorem earlier in the case of polygonal curves in the plane in Section 1.8, our main concern was to see that the compact region that was bounded was homeomorphic to a disk. However, the proof that was outlined there actually allows us to find an isotopy which moves the original polygonal curve to a standard triangle, with the isotopy being the identity outside a larger ball. This holds since various homeomorphisms used in the argument can be chosen to be locally based near triangles where they are the identity outside a neighborhood of the triangle and the action within the neighborhood can be shown to be isotopic to the identity. This form generalizes to the general case of simple closed curves in the plane. For a proof, see [3]. There are also other versions which hold in all dimensions. Besides [3], good sources are [22, 5, 6].

Theorem 2.4.1 (Strong form of the Schönflies theorem). *Let $C = f(S^1)$ be a simple closed curve in the plane and $\mathbb{R}^2 \backslash C = A \cup B$ as given by the Jordan curve theorem, with A bounded. Then there is an ambient isotopy $G : \mathbb{R}^2 \times I \rightarrow \mathbb{R}^2 \times I$ with $G_0 = \mathrm{id}$, so that $G_1(C) = S^1$ and $G_1(A) = D^2$. The ambient isotopy can be chosen to be the identity outside a large disk.*

We now state a strong form of the disk lemma for embeddings of a disk into the plane and then in a surface.

Theorem 2.4.2 (Strong form of disk lemma in the plane). *Let $f : D^2 \rightarrow \mathbb{R}^2$ be an embedded disk in the plane. Then there is an isotopy $G : \mathbb{R}^2 \times I \rightarrow \mathbb{R}^2 \times I$ with $G_0 = \mathrm{id}$ and $G_1 f = i$ or $G_1 f = ir$. Here $i : D^2 \rightarrow \mathbb{R}^2$ is the inclusion map of the unit disk and r is the reflection $r(x, y) = (x, -y)$. This isotopy is the identity outside a large disk. Moreover, there is no such isotopy connecting i and ir.*

Theorem 2.4.3 (Strong form of the disk lemma). *Let $f_0, f_1 : D^2 \rightarrow M$ be embedded disks in the connected surface M with boundary ∂M (which may be empty). Then there is an isotopy $G_t : M \rightarrow M$ which is the identity on a collar neighborhood of the boundary and $G_0 = \mathrm{id}, G_1 f_0 = f_1$ or $G_1 f_0 = f_1 r$.*

Note that Theorem 2.4.3 is a two-dimensional analogue of Theorem 2.3.5. In the supplementary exercises, Exercises 2.9.25–2.9.33 derive Theorem 2.4.3 from Theorem 2.4.1.

There are two isotopy equivalence classes of embeddings of disks in $\mathbb{R}^2$ and they are represented by the inclusion and the reflection followed by the inclusion. Theorem 2.4.3 extends this for disks in a connected surface M. The basic idea, which is pursued in the supplementary exercises, is that there is an ambient isotopy so that the image of the disk is contained in a fixed neighborhood homeomorphic to $\mathbb{R}^2$. This allows us to apply Theorem 2.4.2 to compare with a standard embedding. Thus there will be either one or two ambient isotopy equivalence classes of disks in M. The surface will be orientable when there are two classes, and nonorientable when there is just one class. If the surface contains a Möbius band, then we have shown that there is an ambient isotopy between f and fr, so any two embedded disks are ambient isotopic. Moreover, any surface which is nonorientable in terms of possessing an embedded Möbius band will have any two embedded disks ambient isotopic to each other. This leads to the following alternate definition of an orientable surface.

Definition 2.4.2 (Embedded disk version). A connected surface M is orientable iff there are exactly two ambient isotopy equivalence classes of embedded disks in M. If M is nonorientable, then any two embedded disks are ambient isotopic. When a connected surface M is orientable, an *orientation* is a choice of equivalence class of embedded disks in M. An embedded disk in this equivalence class will be called *positively oriented.*

If H_t is an isotopy with $H_0 = \mathrm{id}$, $H_1 g_0 = g_1$, then $h = H_1$ is a homeomorphism with $h g_0 = g_1$. We state this form of the disk lemma.

Theorem 2.4.4 (Alternate form of disk lemma). *Let M be a compact connected surface, possibly with boundary.*

(a) *Suppose M is oriented, and $g_1, g_2 : D^2 \to M$ are positively oriented embedded disks. Then there is a homeomorphism $h : M \to M$ with h the identity on a collar neighborhood of ∂M and $hg_1 = g_2$. Moreover, h is isotopic to the identity.*

(b) *If M is nonorientable, and $g_1, g_2 : D^2 \to M$ are embedded disks, then there is a homeomorphism $h : M \to M$ with h the identity on a collar neighborhood of ∂M and $hg_1 = g_2$. Moreover, h is isotopic to the identity.*

Exercise 2.2.5 can be modified to give the following result.

Corollary 2.4.5. *Let M be a compact connected surface, possibly with boundary. Suppose $g_1, \ldots, g_k : D^2 \to M$ are k disjoint embedded disks in the interior of a connected surface M, and $g'_1, \ldots, g'_k : D^2 \to M$ is another collection of k disjoint embedded disks in the interior of M.*

(a) *If M is oriented and all g_i, g'_i are positively oriented, then there is a homeomorphism $h : M \to M$ which is the identity on a collar neighborhood of ∂M with $hg_i = g'_i$. Moreover, h is isotopic to the identity.*

(b) *If M is nonorientable, then there is a homeomorphism $h : M \to M$ which is the identity on a collar neighborhood of ∂M with $h(g_i) = g'_i$. Moreover, h is isotopic to the identity.*

Now suppose that M is a surface with p boundary circles. If M is oriented, the choice of orientation on M will determine an orientation on each boundary circle. When we identify a collar with $S^1 \times [1, 2]$ with the boundary circle corresponding to $S^1 \times \{1\}$, then if the orientation on the surface corresponds to the usual positive orientation, then the orientation on the boundary corresponds to the usual negative orientation on this circle. This allows us to fill in any boundary circle with the usual positively oriented disk to get an oriented surface N with one fewer boundary circle and a embedded disk g_i for each boundary circle. Applying Corollary 2.4.5 to N, there is a homeomorphism of N which permutes these embedded disks. When we remove the interiors of these disks, this leads to a homeomorphism of M which permutes the boundary components so that the orientation of the boundary circles is preserved. If M is not orientable, then N will also be nonorientable and so we can choose the homeomorphism to achieve any desired result in terms of orientation of the boundary circles. This leads to a homeomorphism of M which may preserve or reverse the orientation on each boundary circle independently.

Corollary 2.4.6. *Let M be a compact connected surface with boundary, $C_1, \ldots, C_k$ a subcollection of boundary circles of M, and σ a permutation of $\{1, \ldots, k\}$. Then there is a homeomorphism $h : M \to M$ which is the identity on a collar neighborhood of $\partial M \backslash \bigcup C_i$ so that $h(C_i) = C_{\sigma(i)}$. If M is oriented, then h will preserve the orientation on the boundary circles. If M is nonorientable, then h can be chosen to preserve or reverse orientation on each boundary circle C_i independently.*

We now return to handle decompositions and give a definition of orientability that will be easy to use and will play a key role in the classification theorem. We give a more direct approach to Corollary 2.4.6 for handlebodies in Section 2.6 and in Exercises 2.9.37 and 2.9.38.

We examine the handle decomposition of the Möbius band in order to motivate the definition we will give for nonorientability of handlebodies. First, note that any handle, being homeomorphic to a disk, has a notion of orientation attached to it. We can think of this as being given a direction of rotation with counterclockwise being thought of as positive and clockwise as negative (since $(1,0),(0,1)$ is taken as the standard basis of $\mathbb{R}^2$ and rotation from $(1,0)$ to $(0,1)$ is counterclockwise). This sense of rotation on the disk induces a preferred orientation on the boundary circle. This is usually indicated by an arrow. On any subset of the circle which excludes at least one point, this preferred direction induces an ordering on an interval in $\mathbb{R}$ using $p : \mathbb{R} \to S^1, p(x) = (\cos 2\pi x, \sin 2\pi x)$. The positive orientation corresponds to the usual ordering of $\mathbb{R}$. Note that the direction of rotation on D^2 is completely determined by the direction of rotation on the circle. Since this may be expressed locally in terms of ordering, we will use it for our definition of orientability of handlebodies.

Definition 2.4.3. An *orientation* for a handle is an orientation of its boundary circle. By this we mean a consistent ordering for any arc (an *arc* is a homeomorphic image of an interval in the circle). By consistent, we mean that if two arcs intersect in an arc, the two orderings agree on the arc of intersection.

Note that this definition allows us to decide on a preferred equivalence class for an embedded disk into the interior of the handle.

We indicate the two possible orientations for our two models of the disk D^2 and $D^1 \times D^1$ by arrows in Figure 2.22. With this in mind, let us look at the handle decomposition for the Möbius band. As Figure 2.23 indicates, a Möbius band has a handle decomposition with one 0-handle and one 1-handle. Note that when the 1-handle is attached the orientation agrees on one of the arcs and disagrees on the other. We leave it as an exercise to check that this phenomenon is not dependent on our choice of orientation for either handle.

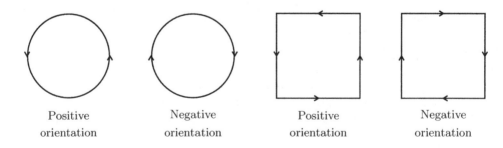

| Positive | Negative | Positive | Negative |
| orientation | orientation | orientation | orientation |

Figure 2.22. Orienting handles.

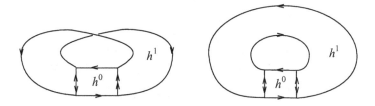

Figure 2.23. Orienting handles on the Möbius band and annulus.

Exercise 2.4.3. Show that no matter how one orients each of the handles in the Möbius band, the 1-handle is attached to the 0-handle so that the orientation agrees on one of the arcs and disagrees on the other.

Let us compare this with the corresponding handle decomposition of the annulus. Note that for the annulus we may choose orientations for each handle so that they disagree on their intersection (i.e. where the 1-handle is attached) (see Figure 2.23).

Definition 2.4.4 (Handlebody version). We say that a handlebody is *orientable* if we may choose an orientation for each handle so that these orientations disagree on identifications; otherwise, it is said to be *nonorientable*. An *orientation* for a handlebody is a consistent choice of orientations for each handle, where consistent means that the orientations disagree on identifications of the boundaries of handles.

The reader may be puzzled by the fact that we want the orientations of the handles to disagree on the arcs that are identified instead of having them agree on both arcs. In the example of the annulus above, we could have easily chosen an orientation of the second handle so that the orientations agree on the identified arcs. The main reason for the condition of making the orientations of arcs disagree as we attach a 1-handle is that this is what is required for small embedded disks with counterclockwise orientations of each handle to be isotopic as we move across the edge where the two handles are joined. Thus the orientation on the edge must change if we are to have a consistent orientation in the adjoining handle. This change allows us to extend an embedded disk in the preferred equivalence class across the edge into the preferred equivalence class in the adjacent disk. It is also needed to consistently orient the new boundary. The boundary of a handlebody will be a disjoint union of circles. If the handlebody is oriented as we have defined it, then the boundary circles will inherit an orientation from the orientations of the individual handles. In Figure 2.23, for example, the boundary of the annulus consists of two circles, each of which is oriented.

Since the structure of a handle decomposition is an inductive one, our definition of orientability is also inductive. In order to use the definition with three or more handles, it has to be the case that the boundary circles have an orientation determined by the orientations of the handles involved at the end of each step. We now indicate why the boundary will inherit a consistent orientation with

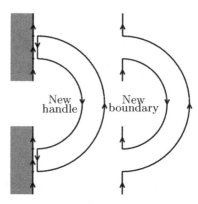

Figure 2.24. Orienting the boundary.

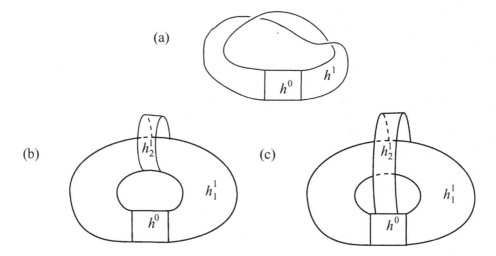

Figure 2.25. Some handlebodies.

our definition when a 1-handle is attached. Suppose that we have oriented the handles consistently so far, so that the boundary has an orientation imposed on it from the orientations of the individual handles. The new handle will be attached along two arcs, which will be identified in some way with two arcs in the boundary so far. If the new handle is attached so that both orientations disagree, then Section 2.3 shows that the change in the boundary is represented by the model in Figure 2.24. But in this model, the new boundary is seen to inherit a consistent orientation.

Exercise 2.4.4. Use the definition to determine whether the handlebodies in Figure 2.25 are orientable or not.

Note that the only problem in orienting a handlebody will occur when a 1-handle is attached, since, when 0-handles are attached, we are taking a disjoint union with a disk, and when 2-handles are attached, they are attached via a homeomorphism of the circle to one of the boundary pieces. One of the two possible orientations on the circle will disagree with the orientation so far on the boundary.

If a handlebody is orientable and connected, then there are precisely two ways to orient the handles consistently (we use the word consistently to mean that orientations always disagree on identified arcs). Note that an individual handle has exactly two choices of orientation. If the handlebody is orientable, then there is at least one way to orient handles in a consistent manner, which we now consider fixed. Note that if we change the orientation of every individual handle from the fixed orientation, then the handles with the new orientations are oriented consistently. We want to see that these are the only two ways to orient the connected handlebody. Suppose we change the orientation on one of the handles h_1 in the handlebody from the fixed orientation. Since the handlebody is connected, this handle must be incident to at least one other handle (where we call two handles incident if they have boundary arcs which are identified). For any handle h_2 incident to h_1, we must change the orientation of h_2 from the fixed orientation in order to be consistent with the new orientation of h_1. Similarly, we would have to change the orientation of any handle h_3 incident to h_2. Continuing this argument inductively, we see that whenever we have a chain $h_1, \ldots, h_k$ of handles with h_i incident to h_{i+i}, then changing the orientation of h_1 from the fixed orientation forces us to change the orientations of all of the handles in the chain in order to be consistent. But a connected handlebody has the property that given any two handles h_a, h_b, there is a chain $h_a = h_1, \ldots, h_k = h_b$ of handles with h_i incident to h_{i+1}. This may be proved inductively on the number of handles in the handlebody and is left as an exercise. Hence, if one of the handles has its orientation changed from the fixed orientation, then we have to change the orientations of all of the handles for the new orientations to be consistent.

Exercise 2.4.5. Show by induction on the number of handles that in a connected handlebody any two handles h_a, h_b may be joined by a chain $h_a = h_1, h_2, \ldots, h_k = h_b$ with h_i incident to h_{i+1}.

The discussion above has consequences for nonorientable handlebodies as well. If a handlebody is nonorientable, then we will be able to orient the first n handles $h_1, \ldots, h_n$ consistently, but the next handle h_{n+1} cannot be oriented consistently with all of these, even if we go back and change orientations on some of the earlier handles. Suppose that we have a handlebody H and that we have succeeded in orienting the handlebody $H_1 = h_1 \cup \cdots \cup h_n$ consisting of the first n handles of H, but that we cannot orient the next handle h_{n+1} consistently with the chosen orientations in H_1. Must H be nonorientable, or can we go back and make better choices of the orientations in H_1 so that we can orient h_{n+1} consistently with the new orientations? If H_1 is connected, then H will be nonorientable. For H_1 is orientable, so the only possibility of changing the orientations in H_1 is to change all of the orientations of the individual handles, in which case h_{n+1} cannot be oriented consistently with all of the new orientations, since it still must agree with one of the new orientations of the boundary of H_1 and disagree with the other in the two intervals where it is attached.

Exercise 2.4.6. Give an example of an oriented handlebody H_1 and a handle h^1 attached to H_1 so that h^1 cannot be oriented consistently with the chosen orientations of the handles in H_1 but that $H_1 \cup h^1$ is orientable.

The condition of orientability of a handlebody is equivalent to the conditions involving embedded disks or embedded Möbius bands. In Chapter 6 we use homology to show that a handle-oriented surface is disk-oriented. We show as one step in the proof of the classification theorem that when a handlebody is nonorientable, there will be an embedded Möbius band and a corresponding isotopy between an embedding f and fr. Using these facts, we outline a proof of the equivalence of the definitions of orientability in Exercise 2.9.39.

2.5   Connected sums

Our classification theorem will be stated in terms of the concept of connected sum. Actually, there are two different definitions involved, that of ordinary connected sum, denoted $\#$, and boundary connected sum, denoted $\amalg$. We work in the context of compact connected surfaces (with or without boundary) for connected sum, and compact connected surfaces with boundary for boundary connected sum. The boundary of such a surface will be a union of a finite number of circles.

We first define *boundary connected sum* of two surfaces M, N with boundary. Choose an embedding f of $\{\pm 1\} \times D^1$ which sends $\{-1\} \times D^1$ into an arc in ∂M and sends $\{1\} \times D^1$ into an arc in ∂N. Then the boundary connected sum of M and N is

$$M \amalg N = M \cup_{f_{(-1)}} D^1 \times D^1 \cup_{f_{(1)}} N = \left(M \bigsqcup N \right) \cup_f D^1 \times D^1.$$

If M, N are handlebodies, then $M \amalg N$ is a handlebody formed from the disjoint union $M \bigsqcup N$ by adding a 1-handle. If M, N are oriented, then there is an additional restriction imposed on the construction that the 1-handle must be attached so that its orientation is consistent with that of M, N, and so $M \amalg N$ will be oriented as well. In terms of handlebodies, this is expressed so the natural orientation of the 1-handle is such that it disagrees on each attaching boundary circle with the orientations of the boundaries of the handlebodies M, N. We say that we are forming the *oriented boundary connected sum* of the two oriented handlebodies. See Figure 2.26 for an illustration of the boundary connected sum of a torus with one hole and a sphere with two holes. Note how the orientations on the boundaries match up so that the result is still oriented. When forming boundary connected sum, all components of the boundary except the two where the 1-handle is being attached are unchanged. For those two components, the

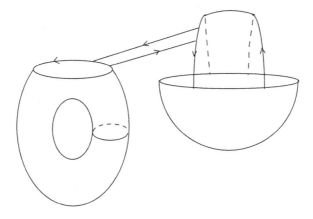

Figure 2.26. Boundary connected sum $T_{(1)} \amalg S_{(2)}$.

effect of the 1-handle is to exchange the two circles that are part of the boundary for a single circle.

We need to know that the construction is well defined, independent of choices made in the embeddings. The first question is dependence on the components of the boundaries of M, N chosen. It turns out that as long as M, N are connected, we can specify this component without changing the homeomorphism type of the result. This depends on Lemma 2.4.6, which says that there are homeomorphisms $h_M : M \to M$, $h_N : N \to N$ sending any one boundary circle to any other, with the restriction that if the surface is oriented, then this homeomorphism will preserve orientations on the boundary circles. Using h_M and a similar h_N on N, we can construct a homeomorphism between $(M \bigsqcup N) \cup_f D^1 \times D^1$ and $(M \bigsqcup N) \cup_{hf} D^1 \times D^1$ which moves the attaching circles as we wish. The homeomorphism will use the identity on $D^1 \times D^1$, h_M on M, and h_N on N. In the expression above, we are using h to denote the restrictions of h_M, h_N to the boundary circles. If $H_1 = M$ and $H_2 = N$ are oriented handlebodies, then our requirements on orientations will allow us to use Lemma 2.3.4 to show that the result of attaching the 1-handle to form the oriented boundary connected sum is well defined up to homeomorphism. If one of H_1, H_2 is not oriented, then the two ways of attaching the 1-handle will lead to the same result up to homeomorphism—hence boundary connected sum is well defined in this case as well. For Lemma 2.4.6 says there is a homeomorphism of a connected nonorientable surface which switches the orientation on a boundary circle. We use such a homeomorphism, say $h : H_2 \to H_2$, to identify $(H_1 \bigsqcup H_2) \cup_f h^1$ and $(H_1 \bigsqcup H_2) \cup_{f'} h^1$, where f' is formed from f by composing with the map h which reverses the orientation on a boundary circle. As an example, we describe such a homeomorphism for $P_{(1)}$, the Möbius band. Think of this space as $D^1 \times D^1 / (1, y) \sim (-1, -y)$. Then the needed homeomorphism is induced as the quotient map from the self-homeomorphism of $D^1 \times D^1$ given

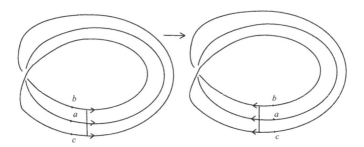

Figure 2.27. Homeomorphism reversing the orientation of the boundary circle.

by $(x, y) \to (-x, y)$. Figure 2.27 shows what happens to the boundary circle for
this map.

Actually, boundary connected sum for surfaces turns out to be well defined
up to homeomorphism even if we do not put on the orientation restrictions of
how we define it. The reason for this is that every oriented surface possesses an
orientation-reversing homeomorphism. For specific surfaces that are involved in
the classification theorem, this can be seen directly by putting them in 3-space
in a symmetric fashion and using a reflection.

We now give a direct argument why there is a homeomorphism which reverses
orientation on every boundary-circle for an oriented handlebody. The argument
is by induction on the number of handles. If there is just one handle, then the
surface is a disk and we can use a reflection. If we add a disjoint disk as a 0-handle
or cap off a boundary circle with a 2-handle, then we can extend an orientation-
reversing homeomorphism. Thus we restrict to the case where the last handle
which is attached is a 1-handle since this is the substantive case. Assume that it
is true for a connected surface with a handle decomposition with n handles and
suppose $H = H_1 \cup_f h^1$ has $n+1$ handles. Whether H_1 is connected or not (it will
either have one or two connected pieces since H is assumed connected), we can
still find by the induction assumption an orientation-reversing homeomorphism h
of H_1 which reverses the orientation on each of its boundary circles. Writing $h^1 = D^1 \times D^1$, consider the homeomorphism $r : D^1 \times D^1 \to D^1 \times D^1$, $r(x, y) = (x, -y)$.
This reverses the orientation of the boundary of h^1. There is a homeomorphism
$k : H_1 \cup_f h^1 \to H_1 \cup_{hfr^{-1}} h^1$ which is defined by using h on H_1 and r on h^1. The
new attaching map hfr^{-1} is chosen to make these fit together. By construction,
k will reverse orientation on the new boundary circles. However, it is not a self-
homeomorphism yet. But then we note that the handles are attached to the
same boundary circles as before and since both r and h reverse orientations,
the map hfr^{-1} attaches the handles with the same orientation convention as f.
Hence there is an isotopy of the boundary circles where the handle is attached
to make these agree. Using this isotopy and a collar within H_1 as in the last
section, we get a homeomorphism $l : H_1 \cup_{hfr^{-1}} h^1 \to H_1 \cup_f h^1$. The composition
$lk : H_1 \cup_f h^1 \to H_1 \cup_f h^1$ now is a homeomorphism which reverses orientation
on each boundary circle.

We summarize this discussion. For a connected oriented surface, there is a homeomorphism which reverses orientation on each boundary circle. If the surface is nonorientable, more is true. For then we could find a homeomorphism which reverses orientation on a single boundary circle and preserves orientation on the other boundary circles. If we have two compact connected surfaces with boundary A, B and we wish attach a handle h^1 via f in order to glue $\{-1\} \times D^1$ to ∂A and glue $\{1\} \times D^1$ to ∂B, we can use self-homeomorphisms of A, B to see that we may specify up to homeomorphism the boundary circle where each is attached. If A, B are both oriented, then this specifies orientations for each boundary circle—if not, then we can just choose an orientation for the boundary circle where each edge is attached. By possibly changing the attaching map using a reflection of h^1, we can assume that $\{-1\} \times D^1$ is mapped into ∂A in an orientation preserving fashion using the standard positive orientation on D^1 (which is the opposite of the orientation it inherits as boundary of $D^1 \times D^1$ oriented in a positive fashion). Thus the only variable up to homeomorphism of the result is how the interval $\{1\} \times D^1$ is attached to ∂B in terms of the orientation. If A and B are oriented and the boundary circles inherit an orientation from this, then we would need $\{1\} \times D^1$ to be attached in an orientation-reversing fashion in order for $A \cup B \cup h^1$ to inherit an orientation consistent with the orientations of A, B. In particular, this is necessary in the case of handlebodies to orient the new handlebody consistently. Nevertheless, as long as B possesses a self-homeomorphism which reverses the orientation of the circle in ∂B where this edge is attached, the resulting surface is well defined up to homeomorphism. This will be true when B is nonorientable using the definition of nonorientability in terms of embedded disks. For orientable surfaces which are given as handlebodies, it is also true by the argument above. However, this result, when oriented, will not necessarily be oriented consistently with both A and B. This justifies the result that $A \amalg B$ is well defined up to homeomorphism, no matter how h^1 is attached, as well as specifying conditions on the attachment for the orientation of h^1 to be consistent with given orientations on A, B in the new surface $A \cup B \cup h^1$.

There is an important special case of boundary connected sum.

Lemma 2.5.1. *The boundary sum of two disks is homeomorphic to a disk:*

$$D_1^2 \amalg D_2^2 \simeq D^2.$$

Thus, if $H = h_1^0 \cup h_2^0 \cup h^1$ is formed from attaching a 1-handle to two 0-handles, then H is homeomorphic to a disk. If H is oriented, the orientation of ∂H imposed from the orientation of the three handles imposes an orientation on the disk.

Proof. We identify $h_1^0 = D_1^2$ with $[-1, 0] \times D^1$, identify $h_2^0 = D_2^2$ with $[1, 2] \times D^1$, and h^1 with $[0, 1] \times D^1$, with the usual orientations. The attaching map of h^1 can be assumed (after isotopy) to map $\{0\} \times D^1$ to h^0 via the identity and the map $\{1\} \times D^1 \to h_2^0$ via either the identity or the reflection $r(1, y) = (1, -y)$. For H to be oriented, we have to have the first case. Then H is homeomorphic to $[-1, 2] \times D^1$ with the standard induced orientation. If we have h_1^0, h_2^0, h^1 given to us as above and the attaching map is r, then these are not consistently oriented. Then we can extend r to a homeomorphism of $[1,2] \times D^1$ to itself via $r(x,y) = (x, -y)$ to give a homeomorphism of $[-1, 0] \times D^1 \cup_{\mathrm{id}} [0, 1] \times D^1 \cup_r [1, 2] \times D^1$ to $[-1, 0] \times D^1 \cup_{\mathrm{id}} [0, 1] \times D^1 \cup_{\mathrm{id}} [1, 2] \times D^1$ which uses the identity on the first two pieces and r on the last piece. Informally, this corresponds to reorienting h_2^0 so that this piece is now oriented. $\square$

We now discuss the connected sum of M, N. Here M, N could be surfaces or surfaces with boundary. The construction will only use the interior of M, N. Intuitively, what we do is remove a disk from M and N and then add a cylinder $D^1 \times S^1$ by identifying $\{-1\} \times S^1$ with the boundary of the disk removed from M and $\{1\} \times S^1$ with the boundary of the disk removed from N. Thus the connected sum

$$M \# N = M \backslash f(\{-1\} \times D^2) \cup_{f \mid \{-1\} \times S^1} D^1 \times S^1 \cup_{f \mid \{1\} \times S^1} N \backslash f(\{1\} \times D^2).$$

If M and N are oriented surfaces, we require our attaching homeomorphisms to preserve the orientation in M and reverse it in N. This allows the orientation on $D^1 \times S^1$ arising as part of the boundary of $D^1 \times D^2$ to fit together with the orientations of M and N to give an orientation on the connected sum. If the disks we remove each represent an oriented 2-handle, then in M the circle $\{-1\} \times S^1$ is identified with the same orientation of the handle and, in N, $\{1\} \times S^1$ is identified with the opposite orientation of the handle.

Another way to phrase connected sum for orientable surfaces is to embed $i_M : D^2 \to M$ in an orientation-preserving fashion and embed a disk $i_N : D^2 \to N$ in an orientation-reversing fashion. This means that the embedding i_M is in the preferred isotopy class of embedded disks for the orientation of M and i_N is not in the preferred isotopy class for embedded disks for its orientation. Then remove the points $i_M(\mathbf{0})$ and $i_N(\mathbf{0})$ from M and N and identify $i_M(\text{int } D^2 \backslash \{\mathbf{0}\})$ with $i_N(\text{int } D^2 \backslash \{\mathbf{0}\})$ via the orientation-preserving homeomorphism $i_N R i_M^{-1}$ where $R : D^2 \backslash \{\mathbf{0}\} \to D^2 \backslash \{\mathbf{0}\}$ is the orientation-reversing homeomorphism $R(re^{i\theta}) = (1 - r)e^{i\theta}$. In terms of the earlier description, the embeddings of $i_M(D^2 \backslash \text{int } \frac{1}{2} D^2)$ and $i_N(D^2 \backslash \text{int } \frac{1}{2} D^2)$ are fitting together to give the cylinder that was used in forming the connected sum. This last description is a bit more difficult to understand, but it is most useful for verifying that connected sum is independent of the choice made up to homeomorphism. For if we choose different embeddings i_M', i_N', then the orientation conventions and Theorem 2.4.2 say that there are isotopies of M, N which connect i_M, i_M' via a path of homeomorphisms of M and connect i_N, i_N' by a similar path. In particular, there are orientation-preserving homeomorphisms $h_M : M \to M$, $h_N : N \to N$ with $h_M i_M = i_M'$, $h_N i_N = i_N'$. Then these induce a homeomorphism between the two ways of forming connected sum by sending $M \backslash \{i_M(\mathbf{0})\}$ to $M \backslash \{i_M'(\mathbf{0})\}$ via h_M and sending $N \backslash \{i_N(\mathbf{0})\}$ to $N \backslash \{i_N'(\mathbf{0})\}$ via h_N. We leave it as an exercise to check that this is consistent with the identifications being made to form the connected sums.

Exercise 2.5.1. Verify that our description of the homeomorphism in pieces does fit together to give a homeomorphism between the two ways of forming connected sum with i_M, i_N and i'_M, i'_N.

If one (or both) of the surfaces is nonorientable, we do not have to make any restrictions on the embeddings into a nonorientable piece by Theorem 2.4.4. Then the same argument shows that the connected sum is well defined in this case. Since any orientable surface given as a handlebody possesses an orientation-reversing homeomorphism, the connected sum will still be well defined for two orientable surfaces without these restrictions as in the boundary connected sum case.

Note that in our description above of the connected sum, we can write $M \# N = M \backslash i_M(\text{int } \frac{1}{2}D^2) \cup_{i_M i_N^{-1} | \frac{1}{2}S^1} N \backslash i_N(\text{int } \frac{1}{2}D^2)$. Thus we are embedding disks into each surface, removing their interiors and gluing the boundaries using the embeddings. Note that since the embeddings i_M, i_N are ambient isotopic to their restrictions to $\frac{1}{2}D^2$ (see Exercise 2.9.30) we could alternatively phrase this as $M \# N = M \backslash i_M(\text{int } D^2) \cup_{i_M i_N^{-1} | S^1} N \backslash i_N(\text{int } D^2)$. This means that we embed two disks, remove their interiors, and glue the boundaries using the embeddings on the boundaries. When both M, N are oriented, we will embed i_M in an orientation-preserving fashion and i_N is an orientation-reversing fashion in order to get the oriented connected sum with an orientation consistent with the orientations on M, N. That this is well defined for different choices i_M, i'_M, i_N, i'_N uses the existence of homeomorphisms h_M, h_N with $h_M i_M = i'_M$, $h_N i_N = i'_N$ so that $h_M : M \backslash i_M(\text{int } D^2) \to M \backslash i'_M(\text{int } D^2)$ and $h_N : N \backslash i_N(\text{int } D^2) \to N \backslash i'_N(\text{int } D^2)$ fit together with identifications to give a homeomorphism. If one of M, N is not oriented, then the existence of a self-homeomorphism of the nonorientable surface which reverses orientation allows us to see that the construction is still well defined, independent of the embeddings chosen. Finally, the fact that an oriented surface possesses a self-homeomorphism which reverses orientation is used to show that the connected sum is well defined generally for surfaces without the orientation restrictions on the embeddings.

We illustrate the connected sum of a torus and a sphere with two holes in Figure 2.28. An important example of a connected sum is $P \# P \simeq K$. This follows since removing a disk from a projective plane gives the Möbius band, and

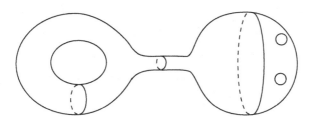

Figure 2.28. The connected sum $T \# S_{(2)}$.

the Klein bottle is formed from two Möbius bands by identifying their boundary circles.

We can reformulate $M \# N$ in terms of the boundary connected sum. Denote by $Q_{(1)}$ the result of removing a disk from Q when Q is a surface without boundary.

Lemma 2.5.2. $(M \# N)_{(1)} \simeq M_{(1)} \amalg N_{(1)}$.

Proof. We first remove a disk from each of M, N, then take the boundary connected sum along arcs in boundary circles, and then fill in with a disk. This is illustrated in Figure 2.29.   □

If M and N are oriented handlebodies and the disks removed are 2-handles, then $M \# N$ will be an oriented handlebody with one more 1-handle and one fewer 2-handle. The cylinder we are adding will give a 1-handle and a 2-handle, as the figure shows.

To illustrate another aspect of handle decompositions, we show how to form the connected sum in the handlebody situation when there are no 2-handles in M, say, to remove. In this case, we can give a new handle decomposition for M so that there will be a 2-handle to use. To do this, we take a small arc in a boundary circle and add a 1-handle to it as illustrated in Figure 2.30. Now attach a 2-handle to fill in the rectangle that was created by the 1-handle. We claim that the result is homeomorphic to what we started with. Use a collar on the boundary circle containing the arc to get a new rectangle and construct a homeomorphism (see Exercise 2.5.2) to push the union of two rectangles into the lower rectangle, leaving the part of the boundary in the original space fixed. Finally, extend

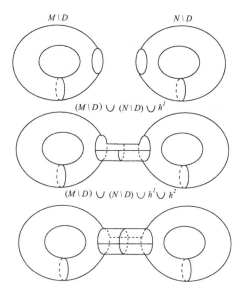

Figure 2.29. Relating the connected sum and the boundary connected sum.

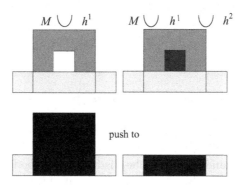

Figure 2.30. Creating an extra 1- and 2-handle.

this homeomorphism by the identity outside a collar neighborhood of our arc. Another way of saying this is that if M, N do not have 2-handles, then we can first attach trivial 1-handles to each to create new boundary circles, then add another 1-handle to form a boundary connected sum, and then finally fill in with a 2-handle. If the surfaces are oriented, then we will be using attaching maps consistent with the orientation at each stage.

Exercise 2.5.2. Construct a homeomorphism from $[-1, 1] \times [-1, 1]$ onto $[-1, 1] \times [-1, 0]$ which fixes pointwise $\{-1\} \times [-1, 0] \cup [-1, 1] \times \{-1\} \cup \{1\} \times [-1, 0]$.

Exercise 2.5.3. Show that $M \amalg N \simeq N \amalg M$.

Exercise 2.5.4. Show that $(M \amalg N) \amalg Q \simeq M \amalg (N \amalg Q)$.

Exercise 2.5.5. Show that $M \# N \simeq N \# M$.

Exercise 2.5.6. Show that $(M \# N) \# Q \simeq M \# (N \# Q)$.

Exercise 2.5.7. Show that $M \amalg D^2 \simeq M$. (Hint: Write $M \amalg D^2$ as $M \cup R_1 \cup R_2$, where R_1, R_2 are rectangles joined along a segment in their boundaries, and use the idea of Exercise 2.5.2.)

Exercise 2.5.8. If N is a surface without boundary, show that $N_{(p)} \amalg S_{(q)} \simeq N_{(p+q-1)}$. (Hint: Consider first the case $q = 1$.)

Exercise 2.5.9. Show that $M \# S^2 \simeq M$. (Hint: Think of what is removed and what is filled in when we form the connected sum with S^2, or use Exercise 2.5.7 and $M \# N = M_{(1)} \amalg N_{(1)} \cup h^2$.)

Exercise 2.5.10. If K is the Klein bottle, use $K \simeq P \# P$ to show that $K_{(1)} \simeq P_{(1)} \amalg P_{(1)}$.

Exercise 2.5.11. Show that $A \# B \# C \simeq A_{(1)} \amalg B_{(1)} \amalg C_{(1)} \cup h^2$. Use this to show that the boundary connected sum of k copies of $A_{(1)}$ is homeomorphic to the connected sum of k copies of A with one disk removed. Denoting the connected sum of k copies of A by $A^{(k)}$, you are to show that $A^{(k)}_{(1)} \simeq A_{(1)} \amalg \cdots \amalg A_{(1)}$

(k copies). Then show that $A^{(k)} \# S_{(p)} \simeq A_{(1)} \amalg \cdots \amalg A_{(1)} \amalg S_{(p)}$. We will use the notation $A^{(k)}_{(p)}$ for this space.

2.6   The classification theorem

In this section we will prove the classification theorem for surfaces. Actually, what we prove is that any connected handlebody is homeomorphic to one of a certain collection of handlebodies. The distinct handlebodies in our collection are in fact not homeomorphic to one another, although the proof of that fact will require results on the fundamental group in the next chapter.

Let $M^{(k)}_{(1)}$ denote the boundary connected sum of k copies of $M_{(1)}$, which, by Exercise 2.5.11, is the same as the connected sum of k copies of M with one disk removed. For a surface M without boundary, we denote by $M^{(k)}$ the connected sum of k copies of M, and by $M^{(k)}_{(p)}$ the result of removing p disks from $M^{(k)}$. By Exercise 2.5.11, this is the same surface as is obtained by taking the $M^{(k)}_{(1)}$ and taking the boundary sum with $S_{(p)}$. The classification theorem says that each compact connected surface (possibly with boundary) is homeomorphic to $S_{(p)}, T^{(k)}_{(p)}$, or $P^{(k)}_{(p)}$, where $p = 0$ corresponds to no boundary. Figure 2.31 illustrates some of these surfaces.

We will show below that we can assume that the 1-handles are attached disjointly and that there is a single 0-handle. Given this, we ignore the 2-handles for the moment (they are only filling in the boundary circles with disks) and think of the surface as given by attaching some 1-handles disjointly to the disk. If we then take the boundary connected sum of two such surfaces and use Lemma 2.5.1, we can get a picture of boundary connected sum in terms of putting together the sum of all of the 1-handles for the two surfaces attached along separate arcs on the boundary of the disk. See Figure 2.32, which illustrates the boundary sum $T_{(1)} \amalg P_{(1)}$. The left-hand side gives the boundary sum using standard handle

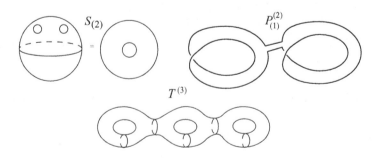

Figure 2.31. Examples of surfaces.

Figure 2.32. Boundary sum with a single 0-handle.

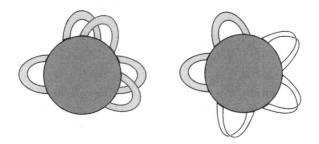

Figure 2.33. Models for $T_{(2)}^{(2)}$ and $P_{(3)}^{(3)}$.

decompositions of $T_{(1)}$ and $P_{(1)}$. The right-hand side gives the handle decomposition of $T_{(1)} \amalg P_{(1)}$ with a single 0-handle with the same 1-handles from the two surfaces attached to a single 0-handle.

We give standard model handle decompositions for $S_{(p)}, T_{(p)}^{(g)}, P_{(p)}^{(h)}$ when $p \geq 1$ as coming from attaching 1-handles to a disk, which is the 0-handle. We will show that there is a homeomorphism of our handlebody to one of these model surfaces. There are three types of subfigures—a trivial handle, a torus pair, and a twisted handle. The trivial handles have the effect of removing a disk from a surface, creating another boundary circle—when attached to the disk it forms an annulus. The torus pair is a pair of linked 1-handles attached which creates $T_{(1)}$ when it is attached to a disk. The twisted handle is a handle attached to a single boundary circle which does not change the number of boundary circles—it creates a Möbius band from the disk. The standard model for $T_{(p)}^{(g)}$ will have g torus pairs and $p-1$ trivial handles, each attached within disjoint intervals along the boundary of the disk. For $S_{(p)}$, there will just be $p-1$ trivial handles. For $P_{(p)}^{(h)}$, there will be h twisted handles and $p-1$ trivial handles, each attached within disjoint intervals along the boundary of the disk. We picture $T_{(2)}^{(2)}$ and $P_{(3)}^{(3)}$ in Figure 2.33.

We will be working with these types of diagrams and manipulating them by isotoping the attaching of 1-handles. We will call this operation "sliding handles". Some of these isotopies will use the existence of another handle to slide one handle over another one. This will usually mean that the isotopy drags an attaching arc of a handle over a boundary arc created by another 1-handle, although sometimes we will slide both attaching arcs over another handle just to

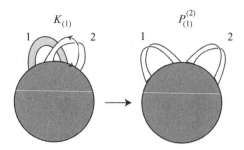

Figure 2.34. Sliding handles to get $P_{(1)}^{(2)} \simeq K_{(1)}$.

put it in a better position. We use this now to illustrate in Figure 2.34 another viewpoint on the homeomorphism $K_{(1)} \simeq P_{(1)} \amalg P_{(1)}$ from Exercise 2.5.10. Here we have slid one attaching arc for handle 1 over a boundary arc for handle 2 in the direction of the arrows to go from a standard picture for $K_{(1)}$ to the standard picture for $P_{(1)}^{(2)}$.

The classification theorem says that each nonorientable surface is homeomorphic to $P_{(p)}^{(k)}$ for some k, p. In particular, $T_{(1)} \amalg P_{(1)}$ is nonorientable. The next result identifies this surface as $P_{(1)}^{(3)}$ and is a critical step in classifying nonorientable surfaces.

Lemma 2.6.1 (Fundamental lemma of surface theory). $T_{(1)} \amalg P_{(1)} \simeq P_{(1)} \amalg P_{(1)} \amalg P_{(1)} \simeq K_{(1)} \amalg P_{(1)}$ *and* $T \# P \simeq P \# P \# P \simeq K \# P$.

Proof. First note that the two statements are equivalent since $T_{(1)} \amalg P_{(1)} \simeq (T \# P)_{(1)}$ and $P_{(1)} \amalg P_{(1)} \amalg P_{(1)} \simeq (P \# P \# P)_{(1)}$. A homeomorphism $T_{(1)} \amalg P_{(1)} \simeq P_{(1)} \amalg P_{(1)} \amalg P_{(1)}$ extends by coning over the disk to a homeomorphism $T \# P \simeq P \# P \# P$. Conversely, if we have a homeomorphism from $T \# P$ to $P \# P \# P$, then by the disk lemma we can specify the image of a disk and get a homeomorphism from $T_{(1)} \amalg P_{(1)}$ to $P_{(1)} \amalg P_{(1)} \amalg P_{(1)}$. The statements involving K just use $K \simeq P \# P$.

We will prove this important result from the viewpoint of each type of connected sum. The first proof comes from sliding handles and is what is used in our proof of the classification theorem. Figure 2.35 illustrates the homeomorphism $T_{(1)} \amalg P_{(1)} \simeq K_{(1)} \amalg P_{(1)} \simeq P_{(1)}^{(3)}$. The first diagram has a pair of 1-handles for $T_{(1)}$ and a twisted 1-handle for $P_{(1)}$. When we slide handle 2 over handle 3, it twists handle 2 and links it with handle 3. We next slide handle 3 over handle 1. This unlinks handles 2 and 3 and leaves the diagram for $K_{(1)} \amalg P_{(1)}$. We then slide handle 1 over handle 2 and get the diagram for $P_{(1)}^{(3)}$.

We now show that $T \# P \simeq P \# P \# P$ from another viewpoint. This viewpoint will be useful in a different proof of the classification theorem and some other exercises in the supplementary exercises at the end of the chapter discussing surgery. Since $P \# P \simeq K$, we have to show that $T \# P \simeq K \# P$. For this, we need a slightly different description of the torus and Klein bottle. The torus can

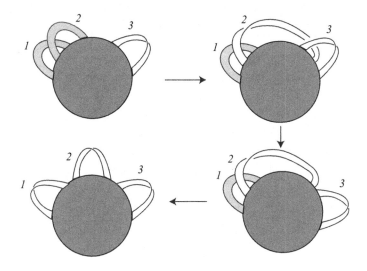

Figure 2.35. Proving the fundamental lemma via handle slides.

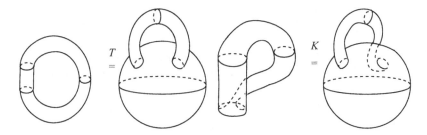

Figure 2.36. Surgery descriptions of T, K.

first be described as the union of two cylinders. Since a cylinder is formed from
a sphere by removing two disks, we can think of a torus as being obtained from
a sphere by removing two disks and gluing back in a cylinder. There is a similar
description for the Klein bottle (the cylinder is put back in a different way).
This construction is a special case of a general construction called surgery (see
Figure 2.36).

If we remove the bottom hemisphere and flatten the upper hemisphere, we
get the view in Figure 2.37 of $T\backslash D^2$ and $K\backslash D^2$. Now taking connected sum of
P with T (resp., K) entails removing a disk from P and gluing back in $T\backslash D^2$
(resp., $K\backslash D^2$). We now remove another disk from P away from the first disk to
get a Möbius band. Figure 2.38 illustrates what we get in the two cases.

That these are homeomorphic can be demonstrated physically with a model:
cut the two models along the line segments in Figure 2.38 and after straightening
each will be a strip with a cylinder attached on one side, and the same identifications of the edges. This indicates how to construct the homeomorphism.

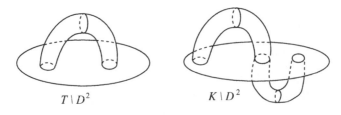

$$T \setminus D^2 \qquad\qquad K \setminus D^2$$

Figure 2.37. $T \setminus D^2$ and $K \setminus D^2$.

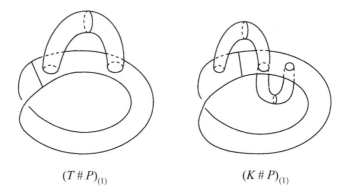

$$(T \# P)_{(1)} \qquad\qquad (K \# P)_{(1)}$$

Figure 2.38. Surgery on a Möbius band.

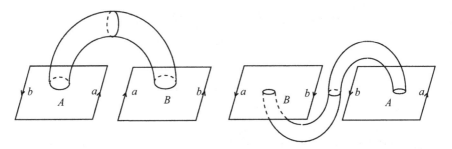

Figure 2.39. Breaking the homeomorphism into pieces.

Alternatively, we can just slide one of the attaching circles for the Klein bottle around the Möbius band so that both circles are attached "on the same side" in a small disk on the Möbius band. The extra intersection we see in 3-space of the cylindrical handle of the Klein bottle with the Möbius band is easily avoided. Figure 2.39 gives another means of getting the homeomorphism. Turn B over and move it to the other side to go from one to the other.

Here is another explanation in terms of isotopy classes of embedded disks. What we are doing when we are forming the connected sum with T or K is taking the boundary of a cylinder $\partial(D^2 \times D^1) = \partial D^2 \times D^1 \cup D^2 \times \{\pm 1\}$, which is oriented so that $D^2 \times \{1\}$ has the standard positive orientation and $D^2 \times \{-1\}$

has the standard negative orientation (draw a picture and check this). We embed
the part of the boundary $D^2 \times \{\pm 1\}$ into the surface, remove it, and add in the
cylinder $\partial D^2 \times D^1$. If the original surface is oriented, the new surface will be
oriented if, for the given orientations (from our orientation of $\partial(D^2 \times D^1)$),
the embeddings both reverse orientations on the disks. Note that another way of
describing these embeddings of the disks is that, if we orient both disks $D^2 \times \{\pm 1\}$
so that they have the standard positive orientation (which is not how they were
oriented above), then for the torus, we are embedding $D^2 \times \{-1\}$ to preserve
orientation and embedding $D^2 \times \{1\}$ to reverse it. In the case of taking the
connected sum of the sphere with the torus, this will be true. But when we
are taking the connected sum of the sphere with the Klein bottle, the disk
$D^2 \times \{1\}$ is embedded with a different orientation. However, when we perform
the same operation within a larger disk in a Möbius band (and hence within
any nonorientable surface), there is no distinction between these two types of
embeddings since they are isotopic, and hence the results of performing these
two constructions are homeomorphic. $\square$

Exercise 2.6.1. Construct a model of $B\#T$ and show that it is $B\#K$. It will
help if the cylinder is attached symmetrically one-fourth and three-fourths of the
way across the strip.

Exercise 2.6.2. Show that $P_{(1)} \sqcup T_{(1)}^{(k)} \simeq P_{(1)}^{(2k+1)}$. Conclude from this that
$P_{(1)}^{(l)} \sqcup T_{(1)}^{(k)} \simeq P_{(1)}^{(2k+l)}$.

We are now ready to state and prove the classification theorem for the
case of handlebodies with 0- and 1-handles. We first make a definition. Let
H be a handlebody with a given handle decomposition. Let n_i be the number of i-handles. Define the *Euler characteristic* $\chi(H)$ of the handlebody H by
$\chi(H) = n_0 - n_1 + n_2$. We will denote by p the number of boundary circles of H.

Theorem 2.6.2. *Let H be a connected handlebody with a handle decomposition
with 0-handles and 1-handles, Euler characteristic $\chi(H)$, and p boundary circles.*

(a) *If H is orientable, then H is homeomorphic to a standard model handlebody for $S_{(p)}$ or $T_{(p)}^{(g)}$, where $\chi(H) = 2 - 2g - p$. The case of $S_{(p)}$
corresponds to $g = 0$ in this formula; that is, $T_{(p)}^{(0)}$ is alternative notation
for $S_{(p)}$.*

(b) *If H is nonorientable, then H is homeomorphic to a standard-model
handlebody for $P_{(p)}^{(h)}$, where $\chi(H) = 2 - h - p$.*

We will simplify the handle decomposition to either remove a copy of $T_{(1)}$ or
$P_{(1)}$ and apply an inductive argument. We use extensively the idea of simplifying
the handle-attaching maps by an isotopy—we will refer to this as *sliding*. We
first note that we may assume that the 1-handles are always attached disjointly
to the boundaries of the 0-handles.

Lemma 2.6.3. *For any handle decomposition H, there is a homeomorphic H'
with corresponding handles so that the 1-handles are attached disjointly to the*

0-handles. The handlebody H' has the same number of handles, and if H is oriented, then H' will have a corresponding orientation.

Proof. The argument is by induction on the number of 1-handles, being trivially true when there is one 1-handle. We suppose that it is true for fewer than k 1-handles and that H has k 1-handles, writing $H = H_1 \cup_f h^1$, where by induction we assume that the 1-handles for H_1 are attached disjointly. The 1-handle h^1 might be attached so that it intersects part of the boundary created from previous 1-handles. However, we can isotope the attaching region off of that part of the boundary and not change the space up to homeomorphism, giving a new handlebody H' where the new 1-handle is disjoint from the previous 1-handles. Since the isotopy preserves orientation of boundary circles, the new handlebody will have an orientation corresponding to the former one.   □

Lemma 2.6.4. *If H is a connected handlebody with 0-handles and disjointly attached 1-handles, then H is homeomorphic to a handlebody H' with a single 0-handle. If H is oriented, then so is H' with corresponding orientations. Moreover, $\chi(H) = \chi(H')$.*

Proof. We work by induction on the number of 0-handles. This is trivially true if there is a single 0-handle. If there is more than one, then H being connected implies that there is a 1-handle h^1 attached to two different 0-handles $h_1^0 \cup h_1^0$. Then $h_1^0 \cup h_2^0 \cup h^1$ is homeomorphic to a disk by Lemma 2.5.1, and if H is oriented, then this disk inherits a consistent orientation with the remaining handles. We can then replace these three handles with a single 0-handle corresponding to this disk. This does not change the Euler characteristic and will have fewer 1-handles. The result then follows by induction.   □

From now on, we will assume that the 1-handles are attached disjointly and there is a single 0-handle.

We first treat the oriented case. We show inductively that H is homeomorphic to the standard model for $T_{(p)}^{(k)}$, where for $k = 0$ this means $S_{(p)}$. For each of the two 1-handle pairs corresponding to $T_{(1)}$, note that the new part of the boundary after attaching plus the three old interior segments enclosed within the attaching interval form an arc of one of the boundary circles. We can use an isotopy to move other handles off of this arc. A picture of what happens to points during this isotopy is given in Figure 2.40. The induction starts with the case of no 1-handles, which is just the disk $S_{(1)}$. Assuming that it is true with k 1-handles, and H has $k + 1$ 1-handles, we express H as $H_1 \cup h^1$, where by induction we can assume that H_1 is already in standard form. If h^1 is attached to the outer boundary circle, we may isotope its attaching intervals into a small interval away from an enclosing interval for the other handles and get a standard diagram where there is one more trivial handle. If it is attached within an inner circle for one of the trivial handles, we can then slide one attaching arc of that handle over it (see Figure 2.41 for this slide). In each case, we get a standard diagram for $T_{(p+1)}^{(g)}$. Note that the $\chi(T_{(p+1)}^{(g)}) = \chi(T_{(p)}^{(g)}) - 1 = 2 - 2g - (p + 1)$, as claimed.

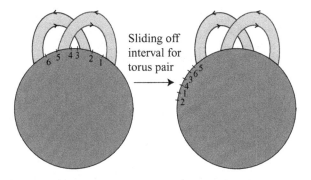

Figure 2.40. Isotoping away from a torus pair.

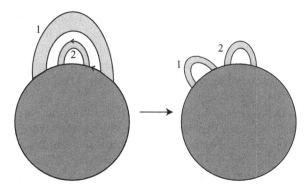

Figure 2.41. Freeing an inner handle by isotopy.

Suppose h^1 is attached to connect the two different circles. This then gives a standard picture for a torus pair using this handle and one of the trivial handles that created a boundary circle to which it was originally attached. We can slide the other handles over this new torus pair to separate it from other trivial handles and use induction to put the torus pairs and trivial handles back in standard position. Thus we have changed the standard model for $T^{(g)}_{(p)}$ to $T^{(g+1)}_{(p-1)}$ here and the new Euler characteristic is $\chi(T^{(g+1)}_{(p-1)}) = \chi(T^{(g)}_{(p)}) - 1 = 2 - 2g - p - 1 = 2 - 2(g+1) - (p-1)$ as claimed. This completes the argument in the orientable case.

We show in Figure 2.42 an illustration of this. There is already one torus pair and two trivial handles labeled 1,2. We attach a new handle 3, which connects two boundary circles, and then handles 1 and 3 form a new torus pair. The attaching arc of handle 2 inside this new torus pair is slid to the right and over three of the boundary arcs in the torus pair (inner 3, inner 1, outer 3) so that it is displayed as a trivial 1-handle attached away from an interval containing all of the attaching subintervals for the torus pair.

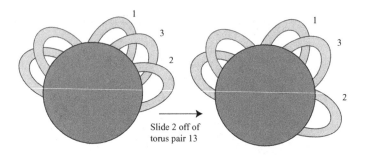

Figure 2.42. Sliding handles to put into normal form.

Now suppose that H is nonorientable, there is a single 0-handle, and the 1-handles are attached disjointly. Then we claim that H is homeomorphic to the standard handlebody with h twisted 1-handles and $p-1$ trivial handles attached to the boundary. Here p is the number of boundary circles and $\chi(H) = 2 - h - p$. Since the handlebody is nonorientable, at least one of the 1-handles is attached to the disk in a twisted fashion.

The argument is again by induction, with the starting case when there is one 1-handle, which is necessarily twisted to give the standard diagram for $P_{(1)}$. We assume that the result is true when there are k 1-handles, and our H has $(k+1)$ 1-handles. We select a 1-handle which is twisted. Since the 1-handles are disjoint, we can consider it the first 1-handle h_1^1. Then if any other 1-handles are attached in the region in the interval used to attach the two attaching regions for h_1^1, we first slide them over the handle h_1^1 to free them from this handle interval. Note that if only one of the subintervals needs to be slid over the region, then this leads to "twisting" or "untwisting". In a picture within 3-space, we will get some double-twisted handles and linked 1-handles, but this is just a feature of the ambient space and not of the surface—the only important thing up to homeomorphism of the surface is whether the intervals are attached to preserve or reverse orientation. After this sliding, the picture will be one twisted handle on one interval in the circle and the other handles in the remaining interval. We now regard the original twisted handle h_1^1 as the last handle. We can then ignore this handle and note by the induction hypothesis that the others can be isotoped to be either in the standard orientable form or in the standard nonorientable form. This rewrites H as $H_1 \cup h_1^1$, where H_1 is standard. In case H_1 is nonorientable, it will be $P_{(p)}^{(h)}$, and there will be h twisted handles and $p - 1$ trivial handles. Here $\chi(H_1) = 2 - h - p$. Then adding h_1^1 to form H just adds one more twisted handle, giving the standard diagram for $P_{(p)}^{(h+1)}$, and $\chi(H) = 2 - (h + 1) - p$.

When H_1 is orientable, then we can assume that H_1 is the standard form for $T_{(p)}^{(g)}$ and $\chi(H_1) = 2 - 2g - p$. Then our diagram for H will have h_1^1 attached as a twisted handle on the outer circle, with $p - 1$ trivial handles and g torus handle pairs. We then use the fundamental lemma of surface theory to do handle

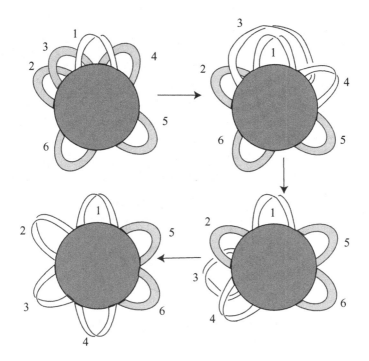

Figure 2.43. Sliding handles to get $P_{(3)}^{(4)}$.

slides to exchange a torus pair plus a twisted handle for three twisted handles, showing $T_{(1)} \amalg P_1 \simeq P_{(1)}^{(3)}$. Repeating this procedure for each torus pair will change our diagram for $T_{(p)}^{(g)} \amalg P_1$ to that for $P_{(p)}^{(2g+1)}$. Moreover, $\chi(H) = \chi(H_1) - 1 = 2 - 2g - p - 1 = 2 - (2g+1) - p$, as required. This completes the inductive step in the nonorientable case.

We consider an example to illustrate our argument (see Figure 2.43). We first identify handle 1 as a twisted handle and move handles 3 and 4 off of an interval containing it by sliding. Note that this has twisted each of these handles and linked them. We then redraw handle 3 and slide the attaching interval for handle 4 over the outside of handles 5 and 6. Now the handle 1 is isolated and the other five handles represent some nonorientable surface. We then use the inductive step to move those into standard position, so that with handle 1 the diagram is a standard diagram for $P_{(3)}^{(4)}$.

In our model surfaces, the outer boundary circle seems different from the inner circles. We show directly without using the equivalence of the different definitions of orientability that there is a homeomorphism of the handlebody which interchanges the outer circle with the inner circle or interchanges any two inner circles. This will be consistent with the orientations in the orientable case. We use the set E which is the union of a collar neighborhood of each circle and a rectangular strip joining them (see Figure 2.44). The figure on the top left

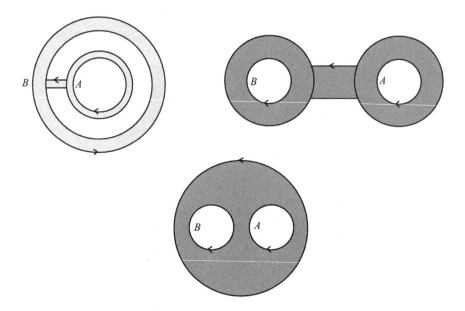

Figure 2.44. Permuting boundary circles of handlebodies.

represents a homeomorphic image of E when we join the outer and inner circles
and the figure on the top right represents a homeomorphic image of E when
we join two inner circles. There is a homeomorphism between them which is
described by being the standard identification of the right annuli and rectangles
and then sending the annulus with boundary circle B on the left to the one on the
right. If we think of each as a standard annulus between circles of radii 1 and 2,
the homeomorphism we want is the composition of an inversion $(r, \theta) \to (2 - r, \theta)$
in polar coordinates, which switches the outer and inner circles with a reflection
$(x, y) \to (-x, y)$. This will exchange the circles but adjust the orientations so
that this composition is consistent with how the rectangle is glued on in each
case. The figure on the top right is homeomorphic to the lower figure. The outer
circle is mapped in a standard fashion to the outer radius of the disk and the two
inner circles are identified. The map on some intermediate circles is pictured in
Figure 2.45. Finally, we can describe the homeomorphism which interchanges the
two circles using the bottom diagram. We basically just want to use a rotation,
but we want the map to be the identity on the boundary. We take a slightly
smaller disk and use a rotation there. On the annular region between the two
disks, we damp out the amount of rotation on these circles from π on the inner
circle to 0 on the outer circle. Using the homeomorphisms of our original regions
E with this figure, this gives us a homeomorphism of E which interchanges the
two circles and is the identity on the other boundary circle of E. This then
extends by the identity outside of E to get the required homeomorphism which
interchanges the two boundary circles. An alternate view of this last step is to

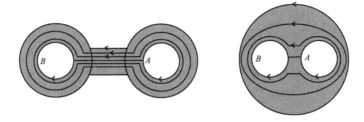

Figure 2.45. Constructing the homeomorphism.

do a slide of one trivial handle over the outer arc of an adjacent one and then slide back to the original position to switch the two boundary circles.

We now deduce the general case from Theorem 2.6.2.

Theorem 2.6.5. *Suppose H is a compact connected surface with a handlebody decomposition having p boundary circles, where $p = 0$ if there is no boundary.*

(a) *If H is orientable, then H is homeomorphic to a standard-model handlebody for $S_{(p)}$ or $T_{(p)}^{(g)}$. Using $g = 0$ for the case of S, g is given by the formula $\chi(H) = 2 - 2g - p$.*

(b) *If H is nonorientable, then H is homeomorphic to a standard-model handlebody for $P_{(p)}^{(h)}$, where $\chi(H) = 2 - h - p$.*

Proof. Suppose h has n_2 2-handles. We remove the 2-handles from H to get H_1, which has no 2-handles but $n_2 + p$ boundary circles and $\chi(H_1) = \chi(H) - n_2$. By Theorem 2.6.2, H_1 is homeomorphic to one of our model handlebodies for: $T_{(n_2+p)}^{(g)}$ in case (a) and $P_{(n_2+p)}^{(h)}$ in case (b), where $\chi(H_1) = 2 - 2g - (n_2 + p)$ in case (a) and $2 - h - (n_2 + p)$ in case (b). Hence $\chi(H) = 2 - 2g - p$ in case (a) and $2 - h - p$ in case (b). Since homeomorphisms of circles extend to homeomorphisms of disks they bound, this homeomorphism extends to the 2-handles we removed to give homeomorphisms from H to models where the 2-handle is attached in a standard manner to one of the boundary circles in the model. When the circle is an inner circle on a trivial handle, then the argument from Exercise 2.5.2 says that filling in the circle with a 2-handle gives a handlebody that is homeomorphic to one without the trivial 1-handle and 2-handle. Thus, as long as we are using n_2 2-handles attached to inner circles, we get standard models for $T_{(p)}^{(2g)}$ (which means $S_{(p)}$ when $g = 0$) or $P_{(p)}^{(h)}$ in the two cases. If a 2-handle is attached to an outer circle, our discussion above says that up to homeomorphism we can regard it as attached to an inner circle. Thus we can use inner circles except for the case when $p = 1$, and attaching it gives the standard handle decomposition for the closed surfaces $S, T^{(g)}, P^{(h)}$ based on those for $S_{(1)}, T_{(1)}^{(g)}, P_{(1)}^{(h)}$. □

Exercise 2.6.3. For each of the orientable surfaces in Figure 2.46, use handle sliding to put it in standard form and identify the surface. Check your answer

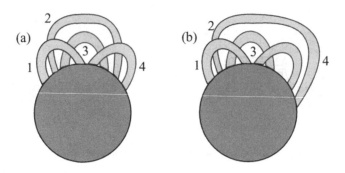

Figure 2.46. Surfaces for Exercise 2.6.3.

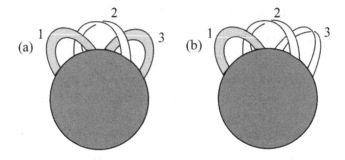

Figure 2.47. Surfaces for Exercise 2.6.4.

by using Theorem 2.6.2. (Hint: First find a torus pair and slide other handles off of it.)

Exercise 2.6.4. For each of the nonorientable surfaces in Figure 2.47, use handle sliding to put it in standard form and identify the surface. Check your answer by using Theorem 2.6.2. (Hint: First find a twisted handle and slide other handles off of it.)

The next six exercises give different arguments for certain steps in the proof of Theorem 2.6.2.

Exercise 2.6.5. Consider the step in the argument for the orientable case where a new torus pair is created by adding a new 1-handle. Show that, if we regard the 0-handle and the torus pair as giving us $T_{(1)}$, then we can remove this from the surface and replace it by a consistently oriented disk with the same boundary to which the other handles are attached. Considering this disk as the new 0-handle and the other 1-handles as attached to its boundary to form a handlebody H_1, show that $H \simeq H_1 \# T$. Then use induction to show that $H \simeq T_{(p)}^{(g+1)}$ when $H_1 \simeq T_{(p)}^{(g)}$.

Exercise 2.6.6. Consider the step in the argument for the orientable case where the new 1-handle is attached to a single boundary circle to form H from H_1. Show that if H_1 is homeomorphic to $T_{(p)}^{(g)}$, then we can add a 2-handle to H, so $H \cup h^2 \simeq H_1$. Thus, removing this 2-handle shows that $H \simeq T_{(p+1)}^{(g)}$.

Exercise 2.6.7. Consider the step in the argument for the nonorientable case where we attach a new 1-handle h^1 to form H from H_1, which, we can assume by induction, is either $T_{(p)}^{(g)}$ or $P_{(p)}^{(h)}$ and is in standard form. Show that we can assume in either case that it is attached away from the torus pairs or twisted handles so we can just consider it as being attached to $S_{(p)}$. In the case where H_1 is orientable, h^1 will necessarily be attached to $S_{(p)}$ so $S_{(p)} \cup h^1$ is nonorientable. However, this is not neccesarily the case when H_1 is nonorientable.

Exercise 2.6.8. Show that if h^1 is attached to a single boundary circle of $H_1 \simeq S_{(p)}$ as a twisted handle to form H, then the union of the 0-handle and h^1 forms a Möbius band $P_{(1)}$. We form a new handlebody H_2 by replacing $h^0 \cup h^1$ by a disk with the same boundary and then attaching the other 1-handles to it. Show that $H_2 \simeq S_{(p)}$ and $H \simeq P \# H_2 \simeq P_{(p)}$.

Exercise 2.6.9. Show that if h^1 is attached to two boundary circles of $S_{(p)}$ in a twisted fashion to form H and we choose another 1-handle h_2^1 so that $h^0 \cup h_2^1$ forms an annulus whose boundary consists of these two circles, then $H_1 = h^0 \cup h_2^1 \cup h^1$ is a Klein bottle with a disk removed $K_{(1)}$. Form a new handlebody H_2 so that H_2 is formed from removing $h^0 \cup h_2^1 \cup h^1$ and replacing it with a disk (the new 0-handle) and then adding in all of the other handles of the original standard handle decomposition of $S_{(p)}$. Show that $H_2 \simeq S_{(p-1)}$ and $H \simeq K \# S_{(p-1)}$.

Exercise 2.6.10. Deduce the general induction step for the nonorientable case by the last three exercises.

2.7 Euler characteristic and the identification of surfaces

Now that we have a list of possible compact connected surfaces up to homeomorphism, how do we identify a given surface in practice? We first need to find a handle decomposition and determine whether the surface is orientable or not. Second, we need to count the number of boundary components. Finally, we need to compute the Euler characteristic. Orientability and the number of boundary circles are topological invariants since homeomorphic surfaces have the same number of boundary circles and are either both orientable or nonorientable (in terms of embedded Möbius bands or isotopy classes of embedded disks). That the Euler characteristic is a topological invariant (i.e. homeomorphic surfaces have the same Euler characteristic) is most easily justified using homology theory and a result from homological algebra. This approach will be used in Chapter 6.

Showing Euler characteristic is an invariant up to homeomorphism is equivalent to showing that no two surfaces in our collection of surfaces $S_{(p)}, T_{(p)}^{(g)}, P_{(p)}^{(h)}$

with the same orientability and number of boundary circles are homeomorphic to each other. For Theorem 2.6.5 can be interpreted as saying that these three quantites determine the surface up to homeomorphism, so if two of them are not homeomorphic and have the same orientability and number of boundary circles, then they would have to have different Euler characteristics. We will relate our definition of Euler characteristic to the fundamental group, which we study in the next chapter, and the fundamental group will provide a proof that the Euler characteristic is a topological invariant.

We now collect our results into the main classification theorem for surfaces.

Theorem 2.7.1 (Classification theorem for surfaces). *A compact connected surface (possibly with boundary) is homeomorphic to either a sphere with p disks removed, $S_{(p)}$, a connected sum of g tori with p disks removed, $T_{(p)}^{(g)}$, or a connected sum of h projective planes with p disks removed, $P_{(p)}^{(h)}$. Orientability, the number p of boundary components, and Euler characteristic χ form a complete set of invariants. If the surface M is orientable with p boundary components and Euler characteristic χ, then $M \simeq T_{(p)}^{(g)}$ with $g = (2 - \chi - p)/2$ (here $S_{(p)} = T_{(p)}^{(0)}$); if the surface is nonorientable with p boundary components and Euler characteristic χ, then $M \simeq P_{(p)}^{(h)}$, with $h = 2 - \chi - p$. If $p = 0$ in either case (a surface with no boundary), then the term (p) may be omitted.*

Exercise 2.7.1. Verify the formulas given for g, h in terms of χ, p in the statement of Theorem 2.7.1.

Once we know that χ is a topological invariant, we can note certain facts about it that can make it easier to calculate. For example, if we get from A to B by filling in p boundary circles of A with disks, then $\chi(B) = \chi(A) + p$. We could start with a handle decomposition of A and add p 2-handles. Conversely, we can get from B to A by removing p disks, and we have the same relation. Now consider taking the boundary sum of two surfaces with boundary. From the handle viewpoint, we are adding a 1-handle, so we get $\chi(A \natural B) = \chi(A) + \chi(B) - 1$. When we form the connected sum of two surfaces, we can write it as $A \# B = (A_{(1)} \amalg B_{(1)}) \cup h^2$, so

$$\chi(A \# B) = \chi(A_{(1)} \amalg B_{(1)}) + 1 = \chi(A_{(1)}) + \chi(B_{(1)}) = \chi(A) + \chi(B) - 2.$$

Exercise 2.7.2. (a) Show that $\chi(A_1 \amalg \cdots \amalg A_n) = \chi(A_1) + \cdots + \chi(A_n) - (n - 1)$.
(b) Show that $\chi(A_1 \# \cdots \# A_n) = \chi(A_1) + \cdots + \chi(A_n) - 2(n - 1)$.

For any orientable surface $T^{(g)}$, the number g is usually called the *genus* of the surface. When this surface is thought of as the boundary of a solid, then the genus is measuring the number of holes in this solid. However, the word "hole" is being used in a different way than we have been using it, which is the number of boundary circles for a surface with boundary. For a nonorientable surface $P^{(h)}$, the number h is sometimes called the number of *crosscaps* in the surface, or the *nonorientable genus*. Interpretations of these numbers will be pursued in the supplementary exercises.

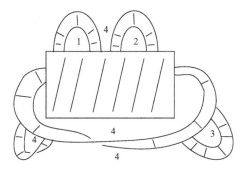

Figure 2.48. Identifying a surface.

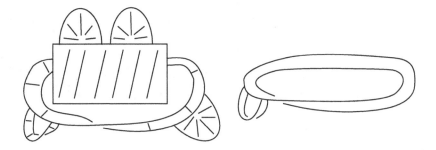

Figure 2.49. New view of a filled-in surface.

We now look at a few examples. Suppose we have a nonorientable surface with one boundary component and Euler characteristic -5. What is it? Since $p = 1$ and $h = 2 - \chi - p = 6$, the surface is $P_{(1)}^{(6)}$.

Consider the surface in Figure 2.48. What is it? First note that it is nonorientable due to the existence of a Möbius band. We count its boundary components. We have numbered them in the figure. Thus $p = 4$. It has a handle decomposition with one 0-handle (the rectangle) and five 1-handles. Hence $\chi = -4$. Thus $h = 2$, and our surface is $P_{(4)}^{(2)}$.

To see this without the computation, first fill in the holes on the boundary circles 1, 2, and 3. Now our surface looks like the left surface in Figure 2.49. Pushing the bumps back into the rectangles (see Exercise 2.5.2), we get the figure on the right. This is a Möbius band with a 1-handle attached (which can be used to get another Möbius band). In Exercise 2.7.3 you are to show that this is $P_{(1)}^{(2)}$. Then our original surface is a Klein bottle with four holes, which is $P_{(4)}^{(2)}$, as we showed above.

Exercise 2.7.3. Show directly that the surface in the right hand side of Figure 2.49 is $P_{(1)}^{(2)} = P_{(1)} \amalg P_{(1)}$ by finding two Möbius bands which are attached together by a 1-handle.

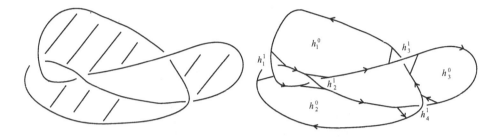

Figure 2.50. Handle decomposition.

Figure 2.51. Surface bounded by a knot.

Consider the shaded surface depicted in Figure 2.50. Its boundary consists of a circle embedded in $\mathbb{R}^3$ (a knot). The figure shows a projection of the knot in the plane with an overcrossing (in terms of this projection) indicated by having one segment pass over another. In order to clarify it, we divide it into seven pieces (which also gives a handle decomposition). Here h_1^0, h_2^0, h_3^0 denote 0-handles (disks) corresponding to the three shaded regions (cut off by line segments joining parts of the knot). Then $h_1^1, h_2^1, h_3^1, h_4^1$ denote the twisted (in $\mathbb{R}^3$) rectangular strips used to join these three disks. Thus, our surface has Euler characteristic $\chi = 3 - 4 = -1$. It has one boundary component. Let us try to orient the handles. We can orient h_1^0 in a counterclockwise fashion and h_2^0, h_3^0 in clockwise fashion, and orient the 1-handles h_1^1, h_2^1, h_3^1 consistently with these. However, it is impossible to orient the last handle h_4^1 consistently with all of the other handles, and so the surface is nonorientable. We can find a Möbius band in the surface as a neighborhood of a circular path running through h_2^1, h_4^1, h_3^1.

By the classification theorem, our surface is homeomorphic to $P_{(1)}^{(2)}$; that is, a Klein bottle with one hole in it or a boundary sum of two Möbius bands. We show this more directly. The 0-handles and 1-handles h_1^1, h_3^1 together give five rectangles laid end on end, which is equivalent to a single rectangle. Adding h_4^1 then gives a Möbius band. Finally, adding h_2^1 gives the Klein bottle with one hole.

We show in Figure 2.51 another surface that has the same knot as boundary. This surface is orientable and is obtained by an algorithm due to Seifert, which produces an orientable surface with boundary a given knot. The bold lines in the second drawing give the boundaries of the three 0-handles. The handle h_1^0 is supposed to lie above the plane that contains most of the knot like the upper hemisphere of the sphere. The three 0-handles are connected by the 1-handles h_1^1 and h_3^1. As in the last example, these five handles fit together to give one rectangle. The handles h_2^1 and h_4^1 are then able to be oriented consistently with these to give an orientable surface. There is one boundary circle given by the knot, and the Euler characteristic is -1. The classification theorem then says that the surface is $T_{(1)}$.

Exercise 2.7.4. By looking at Figure 2.51 as a rectangle with two 1-handles, mimic the argument of the preceding example to show directly that the surface is $T_{(1)}$.

Exercise 2.7.5. Determine which of our models the two surfaces in Figure 2.52 are homeomorphic to. For (b), the exterior region depicts a disk lying over the plane, which can be taken as a 0-handle. The knot which they bound is called the trefoil knot. Use the classification theorem and then give simpler handle decompositions to exhibit these regions more clearly in analogy to our example above. For each surface give handle decompositions with the minimal number of handles.

Consider the surface obtained by taking a hexagon (including the inside) and identifying edges as indicated in Figure 2.53. The identifications on the edges (small letters) induce identifications on the vertices (capital letters) as indicated. We indicate a Möbius band in the surface to show that it is nonorientable.

Although we could find a handle decomposition for this surface, we will identify it by other means. In general, a polygon with some edges identified in pairs will be nonorientable iff there is a pair of identified edges which have orientation (given by arrows) in the same direction (i.e. either both clockwise or both counterclockwise). In such a situation it is easy to find a Möbius band as above; it is harder to show (but true) that, if all identified pairs have opposite orientations, then the surface is orientable. There is one boundary component corresponding to each of the unidentified edges b and d. The question of Euler characteristic

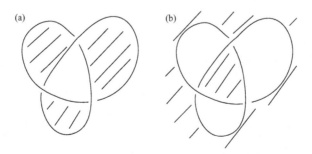

Figure 2.52. Surfaces to identify for Exercise 2.7.5.

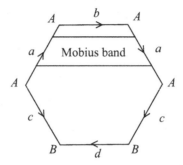

Figure 2.53. Möbius band within identified polygon.

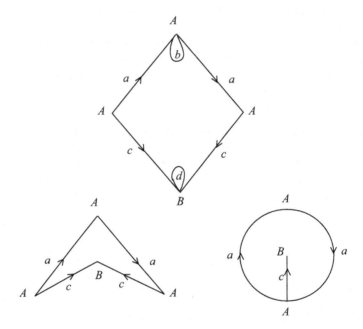

Figure 2.54. Geometrical identification of the surface.

remains. It turns out that there is an easy way to compute the Euler characteristic here. Just count the vertices A, B (2) and the edges a, b, c, d (4) and the
one face given by the hexagon and take the alternating sum $V - E + F = -1$.
This formula $V - E + F$ for a polyhedron divided into polygons is named after
Euler who proved results concerning it but dates back at least to Descartes in
some circumstances. That this computation agrees with our earlier definition in
terms of handle decompositions can be shown using homology theory. Thus we
can identify this surface with $P_{(2)}^{(1)}$.

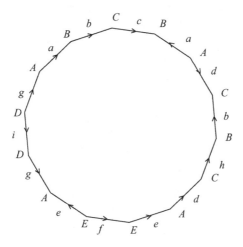

Figure 2.55. Polygon with identifications.

To see this more geometrically, consider Figure 2.54. First note that we can push the boundary circles into the polygon and form a diamond (with identifications) with two holes. Fill in the holes (keeping track of their existence) and then start folding the edges labeled c. As we do so, we can make the edges labeled a deform to halves of a disk as the two copies of c come together to form one interior edge. We are left with a disk with opposite edges identified, which is a description of the projective plane. Recalling the two holes we filled in, our original surface is $P_{(2)}^{(1)}$.

Figure 2.55 gives another example of a polygon with identified edges.

First we find out which vertices are identified. We do this by identifying the initial vertex of a with the initial vertex of a in its other occurrence and the corresponding terminal vertices. We then get the following vertices: $A = $ in $a = $ term $g = $ in $d = $ term e; $B = $ term $a = $ in $b = $ term $c = $ term h; $C = $ in $c = $ term $b = $ term $d = $ in h; $D = $ in $g = $ in $i = $ term i; $E = $ in $e = $ in $f = $ term f.

Thus we get five vertices, nine edges, and one face. The Euler characteristic is -3. The free edges (not identified to another edge) are c, f, h, i. The edges f and i each give a boundary circle, and c and h together give a boundary circle. Hence the number of boundary components is three. The surface is orientable since each pair of edges that are identified occurs with the opposite orientations. Hence the surface is $T_{(3)}$.

The classification theorem can be developed from the point of view of polygons with identifications. For a treatment from this point of view, see the books by Cairns [7] and Massey [18]. We prefer the handlebody approach both because it is of a more geometrical rather than combinatorial flavor and because the idea of a handlebody decomposition is of such importance in the study of manifolds.

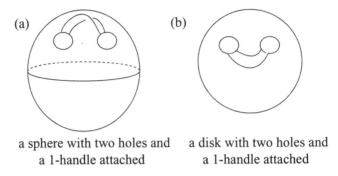

(a)   (b)

a sphere with two holes and   a disk with two holes and
a 1-handle attached   a 1-handle attached

Figure 2.56. Surfaces for Exercise 2.7.7.

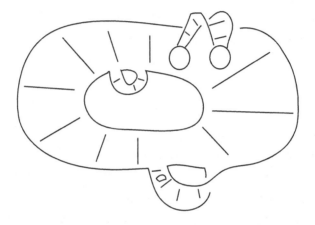

Figure 2.57. Surface for Exercise 2.7.8.

Exercise 2.7.6. Identify the following surfaces:

(a) M has $\chi = -5$, is nonorientable, and has three boundary components.

(b) N is orientable, has $\chi = -2$, and has empty boundary.

Exercise 2.7.7. Identify the surfaces depicted in Figure 2.56.

Exercise 2.7.8. Identify the surface depicted in Figure 2.57. (Hint: First fill in the obvious holes. Then find a handle decomposition of what is left. Also, you might try breaking it into a connected sum of pieces that are easier to identify.)

Exercise 2.7.9. Identify the surfaces in Figure 2.58.

2.8   Simplifying handle decompositions

This section concerns the structure of handle decompositions of a surface or surface with boundary. It may be approached as a project as we will outline

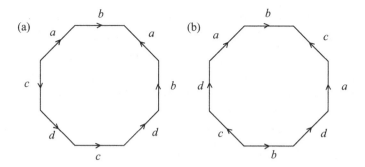

Figure 2.58. Surfaces for Exercise 2.7.9.

proofs of important facets of handle decompositions, including their existence
and how they may be simplified. We also relate the handle decomposition to
matrices and handle operations to matrix manipulations.

We have already made one simplification in our very definition of a handle
decomposition, since we defined a handle decomposition so that the handles
are attached in order of increasing index; that is, we first attached 0-handles,
then 1-handles, and finally the 2-handles. From some methods that generate
handle decompositions, this property may not be satisfied. However, it is easy
to arrange. Show that if we have a handle decomposition of a surface where the
handles are not necessarily attached in order of increasing index, then there is a
directly corresponding handle decomposition where they are attached this way.
The key idea is that a 0-handle does not interfere with any handles that were
attached before it is attached and no handle attached after a 2-handle interferes
with the 2-handle.

This justifies our assumption that a handle decomposition starts with the
0-handles, then the 1-handles, and finally the 2-handles. From this point, we
showed in the proof of the classification theorem that we could find a different
handle decomposition for the surface where the 1-handles were attached disjointly and there was a single 0-handle. For each 1-handle, call the arcs where
the 1-handle is attached to the 0-handle the attaching arcs, and the other two
arcs of the boundary of the 1-handle the transverse arcs. In terms of the structure
of a 1-handle as $D^1 \times D^1$, the attaching arcs are $\{\pm 1\} \times D^1$ and the transverse
arcs are $D^1 \times \{\pm 1\}$. These transverse arcs become part of the boundary at
the stage after the 1-handle is attached, and each 2-handle will intersect some
of them.

Now consider a surface without boundary with a handle decomposition with
a single 0-handle and disjoint 1-handles. As we attach the 2-handles, show that if
the attaching circle of the 2-handle intersects the transverse region of a 1-handle,
then the intersection must be one of the two arcs or both of the arcs. Then show
that when we attach a 2-handle, its boundary circle is divided into subarcs which
alternately run over transverse arcs for 1-handles and arcs on the boundary of the
0-handle. Show that this means that we can form a "dual" handle decomposition

where the original 2-handles are now considered 0-handles, the original 0-handle is now considered a 2-handle, and the 1-handles take the role of 1-handles but with the role of the attaching and transverse arcs reversed. That is, what were previously transverse arcs become attaching arcs and what were attaching arcs are now transverse arcs. By reversing the roles of 0-handles and 2-handles, show that there is a handle decomposition for the surface without boundary with a single 0-handle and a single 2-handle.

Here is another approach to eliminating excess 0-handles and 2-handles without isotoping the 1-handles to be disjoint. Suppose that we have a handle-body H and then we attach a 0-handle h^0 and a 1-handle h^1 so that one attaching arc is attached to a circle in ∂H and the other attaching arc is attached to h^0. Show that $H \cup h^0 \cup h^1$ is homeomorphic to H. Then any handle decomposition which is built up from $H \cup h^0 \cup h^1$ can be simplified to give a handle decomposition without these two handles. Similarly, show that if we have a handle decomposition where $H = H_1 \cup h^1$ and then attach a 2-handle h^2 to H so that it is attached along one transverse arc of h^1 with the remainder of the attaching circle being identified to points in ∂H coming from ∂H_1, then $H \cup h^2 = H_1 \cup h^1 \cup h^2$ is homeomorphic to H_1. Note that the other transverse arc of h^1 remains in the boundary after h^2 is attached, so h^2 is not filling in the last hole in the boundary. For each of these you should think of these homeomorphisms as arising from adding a rectangle to a surface with boundary with an edge of the rectangle being identified with an interval in the boundary. A collar neighborhood of the boundary is useful here, as is the argument in Section 2.5 on how to create an extra 1-handle and 2-handle for a surface with boundary. You should investigate where the other transverse arc is sent under the homeomorphism between $H_1 \cup h^1 \cup h^2$ and H_1. The handles h^0, h^1 in the first case and h^1, h^2 in the second case which we eliminate in the new handle decompositions are called *canceling handles*. Using this idea, give another argument that any handle decomposition for a surface without boundary can be simplified to have a single 0-handle and a single 2-handle.

A connected surface must have at least one 0-handle, and if it has no boundary, it must have at least one 2-handle. Use the Euler characteristic to show that the Euler characteristic and the number of 0- and 2-handles determines how many 1-handles there must be. Use this to show that the minimal number of handles in a handle decomposition of $T^{(g)}$ is $2 + 2g$ and it has $2g$ 1-handles to go with a single 0-handle and single 2-handle. Show that the minimal number of handles for $T^{(g)}_{(p)}$, $p > 0$, is $2k + p$ and give a handle decomposition with exactly that number of handles. Determine the minimal number of handles in handle decompositions of $S_{(p)}$ and $P^{(h)}_{(p)}$, giving handle decompositions with these numbers of handles.

There is an algebraic means of quantifying the twisting and linking of a handle decomposition with 0- and 1-handles with a matrix with entries in the integers mod 2. We consider the handles as being attached disjointly to the boundary of a disk. We say a 1-handle has self-linking 1 or is twisted if when it is attached to the disk, then the result is the Möbius band $P_{(1)}$. If attaching the handle to a disk adds a boundary circle to form $S_{(2)}$, then we say it has self-linking number 0. We say that two 1-handles have linking number 1 if the attaching arcs of one of them lie in two separate components of the complement of the attaching arcs of the other on the boundary of the disk. If they lie in the same component, then the linking number is 0. An example of when this occurs is in forming the handle decomposition of $T_{(1)}$. We then associate a matrix to the handle decomposition as follows: the ii-entry of the matrix is the self-linking number of the i-th handle, and the ij-entry of the matrix will be the linking number of the i-th handle with the j-th handle. For the case of the standard handle decompositions of $P_{(1)}, T_{(1)}, K_{(1)}$, the corresponding matrices are

$$(1), \quad \begin{pmatrix} 0 & 1 \\ 1 & 0 \end{pmatrix}, \quad \begin{pmatrix} 0 & 1 \\ 1 & 1 \end{pmatrix}.$$

The matrix obtained does depend on the ordering of the handles, but changing the order changes the matrix by a congruence using a permutation matrix corresponding to the reordering.

We now want to investigate algebraically what is happening to this matrix for handlebodies formed from attaching disjoint 1-handles to a disk during steps in the proof of the classification theorem.

1. Show that the matrix of a handle decomposition is a symmetric matrix.

2. Show that when we take the boundary connected sum $A \amalg B$ and amalgamate the two original 0-handles and the one handle to form the 0-handle of the sum, then the matrix $M(A \amalg B)$ is

$$\begin{pmatrix} M(A) & 0 \\ 0 & M(B) \end{pmatrix}; \tag{2.8.1}$$

that is, it is the block matrix formed from the individual matrices for A and B.

3. Show that when we change the handle decomposition by sliding one attaching arc of handle h_i^1 over one transverse arc of handle h_j^1, the new matrix entries m'_{pq} change from the old one m_{pq} as follows: (1) if neither p, q is i, then $m'_{pq} = m_{pq}$; (2) if $q \neq i$, then $m'_{iq} = m_{iq} + m_{jq}, m'_{qi} = m_{qi} + m_{qj}$; (3) $m'_{ii} = m_{ii} + m_{jj}$ and $m'_{kk} = m_{kk}, k \neq i$. Note that all of these additions are being done modulo 2. Verify that algebraically we are getting from M to M' by adding row j to row i and then adding column j to column i. This is just the step in the symmetric simplification of a matrix changing M to $M' = EME^t$, where E is the elementary matrix which differs from the identity by a 1 in the ij-position. You should check that this step occurs in the proof of the fundamental lemma via handle sliding and use what happens there as a model for your proof. Verify that when we slide both arcs over a single arc of another handle, it does not change the matrix.

4. Consider the operation of sliding other handles off of an interval for a torus pair that occurs in the proof of the orientable case. Determine how this changes the matrix. For simplicity, consider these two handles for the torus pair as handles 1 and 2 (this just corresponds to conjugation by a permutation

matrix). You should consider how this operation affects any two other handles which have an attaching arc in the interval we are sliding off of. In particular, how does it affect rows and columns 1 and 2 and the diagonal entries.

5. Consider the operation of freeing an inner hole that occurs in the proof. What does that correspond to in terms of the matrix?

6. Consider the operation of sliding an attaching arc off of the interval enclosed by the two arcs of a twisted handle. What does this correspond to in terms of the matrix? In particular, if the twisted handle is labeled as handle 1, then what happens to row and column 1 during this operation.

7. Show that the proof of the classification theorem leads to a proof that the symmetric matrix for a handle decomposition can be reduced to the standard matrix for $S_{(p)}, T_{(p)}^{(g)}, P_{(p)}^{(h)}$ via congruence with elementary matrices and permutation matrices.

8. Show that if the matrix for the handle decomposition is the one for the standard decomposition of $S_{(p)}, T_{(p)}^{(k)}$, or $P_{(p)}^{(k)}$, then the surface is in fact homeomorphic to one of these standard surfaces. One approach to doing this is using induction and splitting off a piece of the surface corresponding to a block in the matrix decomposition.

We now leave the topic of simplifying handle decompositions and address their existence. From various examples relating to figures from knots and links, we have seen how handle decompositions may arise. The proof of the existence of a handle decomposition for a surface typically involves some other structure on the surface such as a triangulation or a smooth structure for the surface. Pursuing either approach involves a number of technicalities. We content ourselves with looking at illustrative examples. The most basic example is that of the torus embedded in 3-space as pictured in Figure 2.59. Here the height function has the property that near each critical point the Hessian matrix of second-order partial derivatives is nondegenerate; that is, it is in one of the following forms in local coordinates given by the first two coordinates: (1) $x^2 + y^2 + p$ (local minimum); (2) $-x^2 + y^2 + p$ or $x^2 - y^2 + p$ (saddle point); (3) $-x^2 - y^2 + p$ (local maximum). The number of minus signs gives the *index* of the critical point. We can think of the surface as being built up from the empty set by looking at the part of it beneath a given height. As we pass each critical point, the surface changes up to homeomorphism by adding a handle whose index is that of the critical point, and then adding a collar on the boundary (which does not change the homeomorphism type). For the torus in Figure 2.59, we indicate the handles that are added. In Morse theory (see [19, 20, 11, 15]) it is shown that each differentiable surface has a function that is qualitatively similar to the height function in the case of the torus, and this may be used to give a handle decomposition for the surface just as the height function was used to give a handle decomposition for the torus.

Use the height function to find handle decompositions for the surfaces in Figure 2.60. Determine which of the model surfaces each surface is homeomorphic to. Then show how to modify the handle decomposition so that it is standard if it is not already.

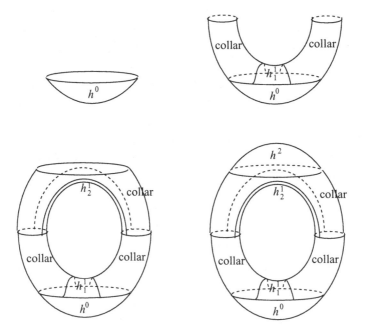

Figure 2.59. Handle decomposition for the torus.

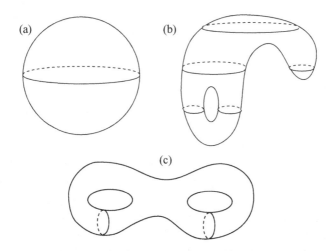

Figure 2.60. Finding handle decompositions for surfaces.

Another way that a handle decomposition may arise is from a triangulation of a surface. Here we decompose the surface into (homeomorphic images of) triangles which have to fit together according to prescribed rules to form the surface. The 0-handles will correspond to the vertices, the 1-handles correspond to the edges, and the 2-handles correspond to the triangles. In order that we can

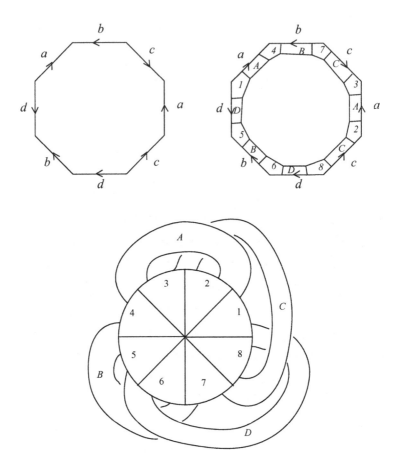

Figure 2.61. Handle decomposition for an identified polygon.

draw this more easily, we will assume that the surface is formed from a polygon with identifications as we discussed before. This will not necessarily give an obvious triangulation, but it does determine a fairly simple handle decomposition. The identifications of edges in pairs will induce identifications of vertices. For each class of identified vertices, their neighborhood will fit together to form a 0-handle. Then the neighborhoods of the identified edges fit together to form a 1-handle. The remainder of the surface will then give a 2-handle. For example, consider Figure 2.61. We pull apart the surface to illustrate the 0- and 1-handles in the corresponding handle decomposition. By sliding C over B we can change the handle decomposition to a standard one for $T^{(2)}$.

Give a picture of the 0- and 1-handles corresponding to the expression of the sphere as a tetrahedron and as a cube. Indicate the boundary circles where the 2-handles will be attached. Give a picture of the 0- and 1-handles corresponding

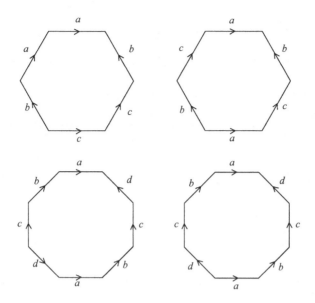

Figure 2.62. Finding handle decompositions for identified polygons.

to the surfaces in Figure 2.62 given as a polygon with identifications. Use this to identify the surfaces in terms of the standard models.

These examples show how to go from a polygon with some pairwise identifications to a handle decomposition. Suppose we have a handle decomposition of a surface which has a single 0-handle and 1-handles which are attached disjointly. Now consider cutting each handle open along $\{0\} \times D^1$ to get some "half-handles" sticking out from the disk. Getting the handles back involves identifying these two central edges. Before doing this, show that we can push the half-handles back into the disk and then think of the result as a disk. Then we are getting the surface from this disk with some arcs on its boundary identified in pairs corresponding to the arcs $\{0\} \times D^1$ for each 1-handle. We picture this process in Figure 2.63. We are starting with a figure (a) from Exercise 2.6.3. Use Exercises 2.6.3 (figure (b)) and 2.6.4 as examples to work out what polygons with identifications they become. Then determine an algorithm which describes the pattern of identifications you will get. Consider the process of handle sliding and determine how this changes the pattern of identifications. Use this to give another proof of the classification theorem, where the goal is to get the pattern of identifications to be a standard pattern.

2.9 Supplementary exercises

Exercises 2.9.1–2.9.8 outline an argument that a compact connected 1-manifold is homeomorphic to S^1.

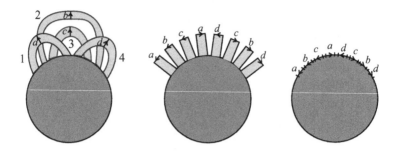

Figure 2.63. Expressing a handlebody as a polygon with identifications.

Exercise 2.9.1. Verify that the circle S^1 is a compact connected 1-manifold. Show that S^1 can be covered by two open sets U, V, each homeomorphic to $\mathbb{R}$, so that $U \cap V$ is homeomorphic to the union of two disjoint open rays.

Exercise 2.9.2. Show that a compact 1-manifold is not homeomorphic to an open set in $\mathbb{R}$.

Exercise 2.9.3. Show that a proper open set in $\mathbb{R}$ may be decomposed (uniquely) as the union of a disjoint (countable) collection of intervals and rays. (Hint: For $x \in U$, find the largest interval $I_x \subset U$ containing x.)

Exercise 2.9.4. Suppose U_i, U_j are two open sets in a 1-manifold M with homeomorphisms $h_i : U_i \to \mathbb{R}$, $h_j : U_j \to \mathbb{R}$, and $h_{ji} = h_j h_i^{-1} : h_i(U_i \cap U_j) \to h_j(U_i \cap U_j)$ is the homeomorphism between the two images of $U_i \cap U_j$. Show that if a finite interval (a, b) is in the decomposition of Exercise 2.9.3 for $h_i(U_i \cap U_j)$, then $h_{ji}(a, b) = \mathbb{R}$. (Hint: Suppose $h_{ji}(a, b) = (c, d)$ and h_{ji} is order preserving. If $\{a_k\}$ is a sequence in (a, b), converging to a, then $\{h_i^{-1}(a_k)\}$ will be a sequence in M converging to both $h_i^{-1}(a)$ and $h_j^{-1}(c)$, which is impossible in a metric space. Similar reasoning will show that the only possibility is $h_{ji}(a, b) = \mathbb{R}$.)

Exercise 2.9.5. Show that one of the three cases must occur:

(a) $h_i(U_i \cap U_j)$ or $h_j(U_i \cap U_j)$ is $\mathbb{R}$;

(b) $h_i(U_i \cap U_j)$ and $h_j(U_i \cap U_j)$ are each a ray;

(c) $h_i(U_i \cap U_j)$ and $h_j(U_i \cap U_j)$ each consists of two rays.

Exercise 2.9.6. Show that in cases (a) and (b) above, $U_i \cup U_j$ is homeomorphic to $\mathbb{R}$.

Exercise 2.9.7. Show that in case (c) above, $U_i \cup U_j$ is homeomorphic to S^1.

Exercise 2.9.8. Use induction on the number of open sets in a covering of M by open sets homeomorphic to $\mathbb{R}$ and the two previous exercises to conclude that a compact connected 1-manifold is homeomorphic to S^1. (Hint: Show that if $U_1 \cup U_2$ already falls into case (c), then any more U_i are superfluous; that is, $U_i \subset U_1 \cup U_2$ by using an argument similar to the one outlined in Exercise 2.9.4.)

Exercise 2.9.9. Show that connected n-manifold is path connected. (Hint: Mimic the proof that a connected open set in $\mathbb{R}^n$ is path connected or use Exercise 1.9.51.)

Exercise 2.9.10. Show that a compact n-manifold has only a finite number of path components, each of which is a compact, connected n-manifold. Conclude that a compact 1-manifold is the disjoint union of a finite number of circles.

Exercise 2.9.11. Show that in a compact connected 1-manifold with nonempty boundary, if $h_i(U_i) = [0, \infty)$ and $h_j(V_j) = \mathbb{R}$ and we do not have $V_j \subset U_i$, and $U_i \cap V_j \neq \emptyset$, then $h_i(U_i \cap V_j)$ and $h_j(U_i \cap V_j)$ must each be rays, and $U_i \cup V_j$ is homeomorphic to $[0, \infty)$.

Exercise 2.9.12. Show that in a compact connected 1-manifold with nonempty boundary, if $h_i(U_i) = [0, \infty)$ and $h_j(U_j) = [0, \infty)$ with $h_i^{-1}(0) \neq h_j^{-1}(0)$, and $U_i \cap U_j \neq \emptyset$, then $h_i(U_i \cap U_j) = (a, \infty), a > 0$ and $h_j(U_i \cap U_j) = (b, \infty)$, with $h_{ji} : (a, \infty) \to (b, \infty)$ an orientation-reversing diffeomorphism, and $U_1 \cup U_2$ is homeomorphic to a closed interval.

Exercise 2.9.13. Show that a compact connected 1-manifold with nonempty boundary is homeomorphic to $[0, 1]$.

Exercise 2.9.14. Show that Exercise 2.9.13 implies Exercise 2.9.8. (Hint: Remove an open interval (a, b) in a neighborhood homeomorphic to $\mathbb{R}$ from the given compact connected 1-manifold.)

Exercise 2.9.15. Show that Figure 2.64 does not represent a 1-manifold. (Hint: Look at a small neighborhood of one of the points where three segments come together. Show that if it were a 1-manifold, there would have to be a standard neighborhood that is homeomorphic to a connected open subset of the line, and derive a contradiction from this.)

Exercise 2.9.16. Show that the map $f : S \to P, f(x) = [x]$, arising from considering P as a quotient space of S by identifying x with $-x$, has the property that given $y \in P$, there is a neighborhood U of y so that $f^{-1}(U) = U_1 \bigsqcup U_2$ and $p|U_i$ is a homeomorphism of U_i onto U.

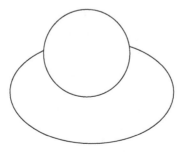

Figure 2.64. Not a 1-manifold.

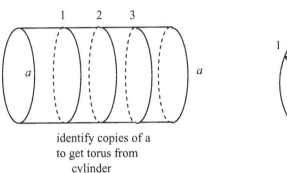

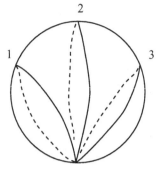

identify copies of a
to get torus from
cylinder

Figure 2.65. Collapsing a wedge in a torus.

Exercise 2.9.17. Construct a continuous map $p : T \to K$ so that, given $x \in K$, there is a neighborhood U of x so that $p^{-1}(U) = U_1 \bigsqcup U_2$ and $p|U_i$ is a homeomorphism of U_i onto U. (Hint: Consider T and K as quotient spaces of $\mathbb{R}^2$, or alternatively, consider T as a quotient space of $[-1,3] \times [-1,1]$ and K as a quotient space of $[-1,1] \times [-1,1]$.)

Exercise 2.9.18. Construct a homeomorphism between the complement of two points in a sphere S and the complement of one circle (properly chosen) in T.

Exercise 2.9.19. Construct a continuous map from the torus $T = S_a^1 \times S_b^1$ onto the sphere S which sends $S_a^1 \times \{p\} \cup \{p\} \times S_b^1$ to the south pole and is a homeomorphism on $T \backslash (S_a^1 \times \{p\} \cup \{p\} \times S_b^1)$. (Hint: Consider a rotating family of planes through the south pole and the intersection with S (see Figure 2.65). Alternatively, think of both S and T as arising from a disk by making certain identifications on the boundary circle.)

Exercise 2.9.20. Remove a small neighborhood of the center circle in the Möbius band which is itself a smaller Möbius band. Show that the resulting space is homeomorphic to a cylinder (see Figure 2.66).

Exercise 2.9.21. Show that there is a circle embedded in $T \backslash D^2$ which, when removed, has connected complement. Use this fact to show that $T_{(p)}^{(k)}$ cannot be embedded in the plane, where $k \geq 1, p \geq 0$.

Exercise 2.9.22. Show that a 2-sphere S cannot be embedded in the plane. (Hint: Use invariance of domain.)

Exercise 2.9.23. Use Exercises 2.9.21, 2.9.22, 2.4.1, and the classification theorem to characterize the compact connected surfaces (with or without boundary) that can be embedded in the plane.

Exercise 2.9.24. Construct an embedding of the Möbius band into the solid torus $S^1 \times D^2$ so that the boundary curve in the Möbius band is embedded on

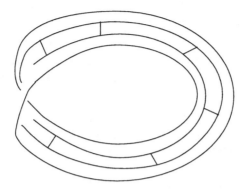

Figure 2.66. Removing a smaller Möbius band.

the boundary torus $S^1 \times S^1$. Describe the composition of the embedding of the boundary circle with the projections of $T = S^1_a \times S^1_b$ onto the circles S^1_a and S^1_b.

The next four exercises use Theorem 2.4.1 to prove Theorem 2.4.2.

Lemma 2.9.1. *Any embedded disk $f : D^2 \to \mathbb{R}^2$ is ambient isotopic to $g : D^2 \to \mathbb{R}^2$ with $g(D^2) = D^2$ and $g(S^1) = S^1$. The ambient isotopy can be chosen to be the identity outside a large disk.*

Exercise 2.9.25. Use Theorem 2.4.1 to prove Lemma 2.9.1.

If $f : D^2 \to D^2$ is a homeomorphism, then we say that f is orientation preserving (reversing) if $f|S^1$ is. We extend Lemma 2.3.6 from circles to disks.

Lemma 2.9.2.

(a) *An orientation-preserving homeomorphism $f : D^2 \to D^2$ is isotopic to the identity.*

(b) *An orientation-reversing homeomorphism $f : D^2 \to D^2$ is isotopic to the reflection $r(x, y) = (x, -y)$.*

Exercise 2.9.26. Prove Lemma 2.9.2. (Hint: (b) will follow from (a) by using a composition of r and the isotopy. The key idea for (a) is to first extend $f|S^1$ to a homeomorphism $F : D^2 \to D^2$ by coning at $\mathbf{0}$, and extend the isotopy F_t between the identity and $f|S^1$ to an isotopy G_t between the identity and this extension. Then $g = G^{-1}f$ will be isotopic to f and will be the identity on S^1. Then show that g is isotopic to the identity via an isotopy that gradually increases the annular region near the boundary which is sent via the identity and compresses the action of g to smaller and smaller disks about the origin.)

Exercise 2.9.27. Suppose $f : D^2 \to D^2$ is isotopic to the identity with the restriction to S^1 given by $F_t : S^1 \to S^1$, with $F_0 = \mathrm{id}$ and $F_1 = f|S^1$. We can extend f to a homeomorphism $\tilde{f} : \mathbb{R}^2 \to \mathbb{R}^2$ as follows. For $1 \le r = |\mathbf{x}| \le 2$, define $\tilde{f}(\mathbf{x}) = rF_{2-r}(\mathbf{x}/r)$. This is just using the isotopy F_t on each circle. Then

$\tilde{f}(x) = x$ for $|x| = 2$. We then define $\tilde{f}(x) = x$ when $|x| \geq 2$. Give an isotopy between $\tilde{f}$ and the identity which extends the isotopy between f and the identity.

Exercise 2.9.28. Deduce all but the last statement of Theorem 2.4.2 from the preceding exercises. That i and ir are not isotopic will be shown in Chapter 6 as an application of homology.

The next five exercises will outline the proof of Theorem 2.4.3 that there are at most two isotopy classes of embedded disks in a surface by reducing it to the strong form of the disk lemma for embedded disks in the plane, Theorem 2.4.2.

Exercise 2.9.29. For $\epsilon < 1$, define $c_\epsilon : [0, \infty) \to [0, \infty)$ to be (i) the identity on $[2, \infty)$, (ii) the affine linear map sending $[0, 1]$ to $[0, \epsilon]$ by multiplying by ϵ, and (iii) the unique affine linear map sending $[1, 2]$ to $[\epsilon, 2]$. Give a formula for c_ϵ and show that c_ϵ is isotopic to the identity with an isotopy k_t which is the identity on $[2, \infty)$.

Exercise 2.9.30. Using c_ϵ of the last exercise, define $C_\epsilon : \mathbb{R}^2 \to \mathbb{R}^2$ by $C_\epsilon(x) = c_\epsilon(|x|)x/|x|$, $x \neq 0$, $C_\epsilon(0) = 0$.

    (a) Show that C_ϵ is isotopic to the identity with identity isotopy outside $2D^2$ via an isotopy K_t.

    (b) For an embedded disk $f : D^2 \to M$, which is the restriction of an embedding $\tilde{f} : \mathbb{R}^2 \to M$, let $\tilde{f}^\epsilon = \tilde{f}C_\epsilon$. Show that $f = \tilde{f}i$ is ambient isotopic to $f^\epsilon = \tilde{f}^\epsilon i$, where the ambient isotopy is the identity off of $f(2D^2)$.

    (c) Show that if $V \subset M$ is an open set about $f(0)$, then f is ambient isotopic to g with $g(D^2) \subset V$.

Exercise 2.9.31. Show that given $x, y \in \text{int } D^2$, there is a homeomorphism $h : D^2 \to D^2$, with $h(x) = y$ and $h(z) = z$ for $z \in S^1$. Show that h is isotopic to the identity where the isotopy $H_t|\partial D = \text{id}$. (Hint: Use coning.)

Exercise 2.9.32. Suppose that M is a connected surface with (possibly empty) boundary and $x, y \in \text{int } M$. Show that there is an isotopy H_t of M which is the identity on ∂M with $H_0 = \text{id}$ and $H_1(x) = y$. If C is a collar of the boundary and $x, y \in M \backslash C$, show that we can assume that H is the identity on C. (Hint: Fix $x \in \text{int } M$. Let $U = \text{int } M$, and consider the set A of those points $y \in U$ so that there is an isotopy H_t of M which is the identity on ∂M with $H_0 = \text{id}$ and $H_1(x) = y$. Show that A is both open and closed in U. Use the connectedness of U to show that $A = U$.)

Exercise 2.9.33. Use the preceding exercises and the strong form of the disk lemma in the plane, Theorem 2.4.2, to show that given any two embedded disks $f_1, f_2 : D^2 \to \text{int } M$, there is an isotopy H_t of M which is the identity on a collar neighborhood of the boundary so that $H_0 = \text{id}$ and $H_1 f_1 = f_2$ or $H_1 f_1 = f_2 r$, which is Theorem 2.4.3. (Hint: Let $U = \tilde{f}_2(\mathbb{R}^2)$ be an open set about $f_2(0)$. Show that there is an ambient isotopy H_1 in M with $H_1(f_1(0)) = f_2((0)$ and $f_1(D^2) \subset U$. Then use Theorem 2.4.2.)

Exercise 2.9.34. A space M is called *homogeneous* if, given $x, y \in M$, there is a homeomorphism $h : M \to M$ with $h(x) = y$. Use Exercise 2.9.32 to show that a connected manifold without boundary is homogeneous.

Exercise 2.9.35. Show that if a homeomorphism $f : \mathbb{R}^2 \to \mathbb{R}^2$ satisfies $f|\mathbb{R}^2 \backslash RD^2 = \mathrm{id}$, then f is isotopic to the identity.

Exercise 2.9.36. Use Lemma 2.9.1 to show that given any two 2-disks D_1, D_2 embedded in the plane, there is an isotopy $H_t : \mathbb{R}^2 \to \mathbb{R}^2$ which is the identity outside of a large disk so that $H_0 = \mathrm{id}$ and $H_1(D_1) = D_2$.

Exercise 2.9.37. From our proof of the classification theorem for handlebodies, a nonorientable surface possesses an embedded Möbius band which we can think of as arising inside a twisted handle and a collar of the 0-handle. For any boundary circle C, we can take a collar of this circle, the Möbius band, and a rectangle joining them to form a subset W which is homeomorphic to the left diagram of Figure 2.67. Explain why this region is homeomorphic to a Möbius band with a disk removed as pictured on the right side of Figure 2.67. Use this homeomorphism to find a homeomorphism of the handlebody which is the identity outside of W and reverses the orientation of C.

Exercise 2.9.38. Use the previous exercise and the discussion preceding Theorem 2.6.5 to give an independent proof of Corollary 2.4.6 for handlebodies.

Exercise 2.9.39. In Chapter 6, it is shown that handle-oriented surfaces are disk-oriented. Use this to show that the three definitions of being orientable for handlebodies are equivalent. (Hint: Call these three definitions handle-orientable, Möbius-orientable, and disk-orientable. Show handle-nonorientable implies Möbius-nonorientable implies disk-nonorientable implies handle-nonorientable.)

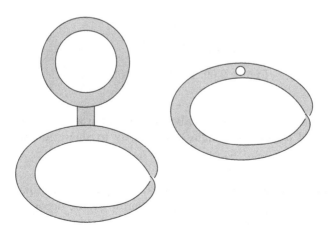

Figure 2.67. Using a Möbius band to reverse orientation on a boundary circle.

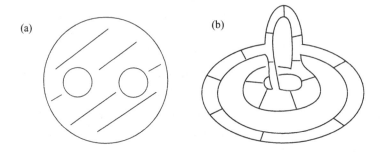

Figure 2.68. Orientable handlebodies.

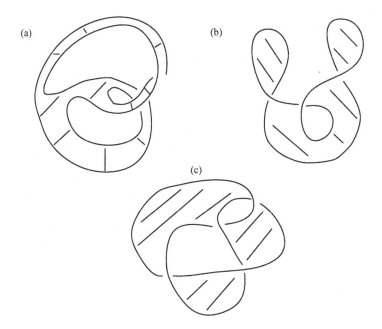

Figure 2.69. Finding a Möbius band.

Exercise 2.9.40. Give a handlebody decomposition for each of the surfaces in Figure 2.68 and orient the handles consistently to show that the surfaces are orientable.

Exercise 2.9.41. Show that the surfaces in Figure 2.69 are nonorientable by finding an embedded Möbius band. Then find a handle decomposition in each case and show that the handles cannot be oriented consistently.

Exercise 2.9.42. Show that a handlebody with only one 0-handle is connected.

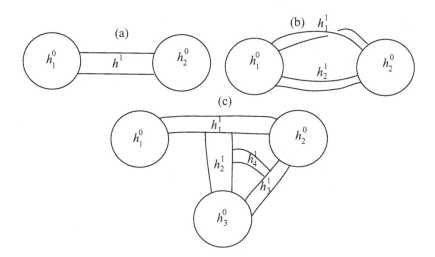

Figure 2.70. Decompositions with a single 0-handle.

Exercise 2.9.43. In the connected handlebodies of Figure 2.70, find a new decomposition with exactly one 0-handle.

Exercise 2.9.44. Find a new handle decomposition for Figure 2.70(c) so that the 1-handles are all attached to the boundaries of the 0-handles.

Exercise 2.9.45. What is the minimal number of 1-handles required for a handle decomposition of $T\#T$. Construct a handle decomposition of $T\#T$ with this minimal number of 1-handles. (Hint: Consider the Euler characteristic.)

Exercise 2.9.46. Construct a handle decomposition of $(T\#T)_{(1)}$ with three 0-handles, seven 1-handles, and one 2-handle.

Exercise 2.9.47. Write the surfaces in Figure 2.71 as connected sums, or boundary connected sums, of other surfaces, none of which is a disk or a sphere. Indicate in the figure what the two pieces are and which of the standard surfaces (with boundary) each is homeomorphic to.

Exercise 2.9.48. In each of the surfaces with boundary in Figure 2.72, find an arc A joining two points on the boundary so that if we cut the surface along the arc and open it up (equivalently, remove a neighborhood $A \times (-1,1)$ of the arc), we get two surfaces, neither of which is homeomorphic to a disk. Relate this to a boundary connected-sum decomposition.

Exercise 2.9.49. In each of the surfaces in Figure 2.73, find an embedded circle C which separates the surface so that if we cut the surface along the circle (equivalently, remove a neighborhood $C \times (-1,1)$ of the circle), we get two surfaces, neither of which is homeomorphic to $S_{(p)}$. Relate this to a connected-sum decomposition.

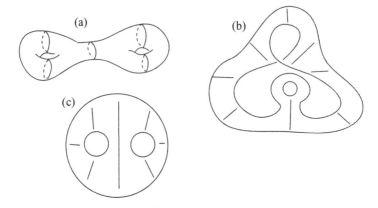

Figure 2.71. Connected sums.

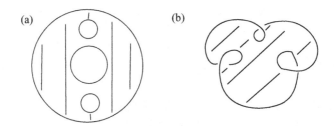

Figure 2.72. Separating arcs and boundary sums.

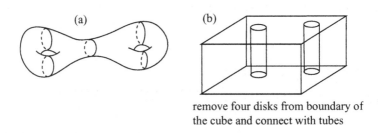

remove four disks from boundary of
the cube and connect with tubes

Figure 2.73. Separating circles and connected sums.

Exercise 2.9.50. Show that the disk with the upper semicircle identified to the lower semicircle as indicated in Figure 2.74(a) is homeomorphic to S. (Hint: Send the arc a to an arc in the xz-plane joining $(-1, 0, 0)$ to $(1, 0, 0)$ going through the south pole and send vertical lines to the intersection of S with planes parallel to the yz-plane.)

Exercise 2.9.51. Show that the disk with the upper semicircle identified to the lower semicircle as indicated in Figure 2.74(b) is homeomorphic to P.

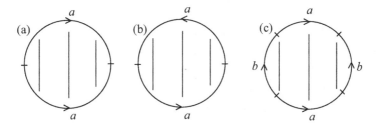

Figure 2.74. Quotients of the disk.

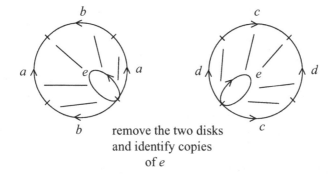

remove the two disks
and identify copies
of e

Figure 2.75. Connected sum and words.

Exercise 2.9.52. Show that the disk with identifications on the boundary circle as indicated in Figure 2.74(c) is homeomorphic to T.

Exercise 2.9.53. If the boundary of a disk is divided into $2n$ edges and the edges are identified in pairs, show that the quotient space represents a compact connected surface.

Exercise 2.9.54. Suppose a surface S comes from a disk by dividing its boundary into $2n$ edges and identifying the edges in pairs. Associate to S the word given by starting at a vertex and reading counterclockwise around the circle (and reading a^{-1} whenever the arrow on a is clockwise). For example, we get words aa^{-1}, aa, and $aba^{-1}b^{-1}$ for Figure 2.74(a), (b), and (c), respectively. Show that if $w(X)$ represents the word associated to X in this fashion and similarly $w(Y)$ (using different letters) is associated to Y, then the juxtaposition $w(X)w(Y)$ is associated to $X\#Y$. (e.g. $bbcdc^{-1}d^{-1}$ is associated to $P\#T$). (Hint: Form the connected sum as indicated in Figure 2.75).

Exercise 2.9.55. Show that $T^{(k)}$ may be represented as a disk with identifications on the boundary associated to the word $a_1b_1a_1^{-1}b_1^{-1}\ldots a_kb_ka_k^{-1}b_k^{-1}$. Show that $P^{(k)}$ may be represented as a disk with identifications on the boundary associated to the word $a_1a_1\cdots a_ka_k$.

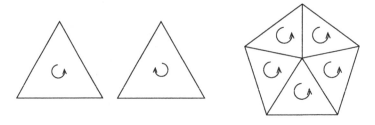

Figure 2.76. Orienting a triangulation.

Exercise 2.9.56. Show that the words $a_1a_1a_2a_2a_3a_3, a_1a_1a_2a_3a_2^{-1}a_3, a_1a_1a_2a_3$ $a_2^{-1}a_3^{-1}$ are associated to homeomorphic surfaces.

Exercise 2.9.57. Suppose $T_1, \ldots, T_n$ are triangles in $\mathbb{R}^N$ so that: (i) any two triangles are disjoint or they intersect in a common edge or vertex; (ii) every edge is the edge of precisely two triangles; (iii) for any vertex v, the triangles containing v may be ordered cyclically $T_{v,1}, T_{v,2}, \ldots, T_{v,k}, T_{v,1}$ so that subsequent triangles intersect along an edge with only the vertex v in common as in right figure in Figure 2.76. Show that $X = \cup_{i=1}^n T_i$ is a surface (called a *triangulated surface*).

Exercise 2.9.58. Find homeomorphic images of S and T in $\mathbb{R}^3$ that are triangulated surfaces.

Exercise 2.9.59. A triangulated surface is *orientable* if we can orient each triangle (by giving a clockwise or counterclockwise direction to its edges) so that for any edge, the two triangles containing the edge impose opposite orientations on the edge (see Figure 2.76). Show how to orient your triangulations of S and T in this manner.

Exercise 2.9.60. Indicate how a triangulated surface which cannot be oriented (in terms of orienting the triangles consistently) will contain a Möbius band.

In the problems below, you are asked to identify surfaces; that is, tell which one of the surfaces $S_{(p)}, T_{(p)}^{(g)}, P_{(p)}^{(h)}$ the given surface is homeomorphic to.

Exercise 2.9.61. Which surface is the surface of a coffee cup homeomorphic to?

Exercise 2.9.62. Identify the following surface. Take an empty box (with thickness) and poke a small hole in each face. The surface is the surface of the box after the holes are poked.

Exercise 2.9.63. Identify the surfaces in Figure 2.68.

Exercise 2.9.64. Identify the surfaces in Figure 2.69.

Exercise 2.9.65. Identify the surfaces in Figure 2.70.

Exercise 2.9.66. Identify the surfaces in Figure 2.71.

Exercise 2.9.67. Identify the surfaces in Figure 2.72.

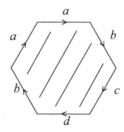

Figure 2.77. Quotient of a hexagon.

Figure 2.78. Surface for Exercise 2.9.70.

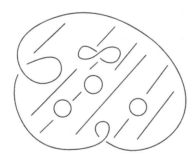

Figure 2.79. Surface for Exercise 2.9.71.

Exercise 2.9.68. Identify the surfaces in Figure 2.73.

Exercise 2.9.69. Identify the surface formed from a hexagon by identifying edges as indicated in Figure 2.77.

Exercise 2.9.70. Identify the surface in Figure 2.78.

Exercise 2.9.71. Identify the surface in Figure 2.79.

Exercise 2.9.72. Identify the surface in Figure 2.80.

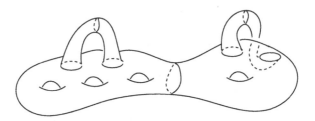

Figure 2.80. Surface for Exercise 2.9.72.

Figure 2.81. Surface for Exercise 2.9.73.

Exercise 2.9.73. Identify the surface in Figure 2.81.

Exercise 2.9.74. Identify the surface obtained by taking the connected sum of two disks. Identify the surface obtained by taking the boundary connected sum of two disks.

Exercise 2.9.75. Using a model of T in $\mathbb{R}^3$ which is symmetric with respect to the yz-plane and reflection through that plane, construct a homeomorphism from $T_{(1)}$ to $T_{(1)}$ which reverses orientation on the boundary circle.

Exercise 2.9.76. Using a model for general orientable surface $M = T^{(g)}_{(p)}$ in $\mathbb{R}^3$, construct a homeomorphism of M which reverses orientation on each boundary circle.

Exercise 2.9.77. Construct a homeomorphism between $T\backslash\{p\}$ and the surface in Figure 2.82 (without the boundary circle).

Exercise 2.9.78. Describe a homeomorphism between the two surfaces in Figure 2.83.

Exercise 2.9.79. Recall that an isotopy $F : A \times I \to B \times I$ between embeddings $f_0, f_1 : A \to B$ is called ambient if there is an isotopy $H : B \times I \to B \times I$ with $F_t = H_t f_0$. Show that if $A = I$ and $B = S^1$, then every isotopy of embeddings of I into S^1 is ambient.

Exercise 2.9.80. A *homotopy* between continuous maps $f_0, f_1 : A \to B$ is a continuous map $F : A \times I \to B, F(x,t) = F_t(x)$ with $F_0 = f_0 F_1 = f_1$ (f_0 and f_1

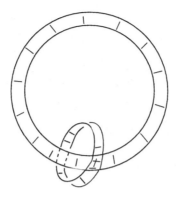

Figure 2.82. $T\backslash\{p\}$.

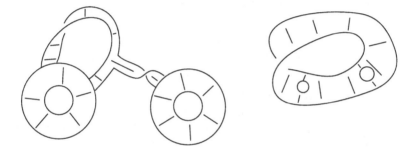

Figure 2.83. Constructing a homeomorphism.

are said to be *homotopic*). If f_0, f_1 are embeddings and each F_t is an embedding, then $F(x,t) = (F_t(x), t)$ will be an isotopy. Thus a homotopy is a generalization of an isotopy. Show that $f_0(x) = x$ and $f_1(x) = -x$ are homotopic embeddings from D^1 to D^1 but they are not isotopic. (Hint: Construct a homotopy F so that F maps $\{x\} \times I$ to the straight line between $(x,0)$ and $(-x,1)$. Show that no isotopy exists because f_0 preserves order and f_1 reverses order.)

Exercise 2.9.81. Give a homeomorphism between $D^1 \times D^1 \cup_f h^1$ and $D^1 \times D^1 \cup_g h^1$ where $f(-1,x) = (-1,x)$, $f(1,x) = (1,x)$, and $g(-1,x) = (1,x)$, $g(1,x) = (-1,x)$.

Exercise 2.9.82. Give a homeomorphism h between $D^1 \times D^1 \cup_f h^1$ and $D^1 \times D^1 \cup_g h^1$ where $f(-1,x) = (-1,x)$, $f(1,x) = (1,x)$, $g(-1,x) = (x,1)$, $g(1,x) = (x,-1)$, and, moreover, h restricts to the identity on a small disk in the interior of $D^1 \times D^1$ as well as on h^1.

In the next group of exercises, we give an outline of another proof of the classification theorem. It is modeled on a proof given by Hirsch [15]. We will be working with a surface without boundary here, which is given to us with a handle decomposition. The argument is based on an inductive argument which uses induction on the number of 1-handles. The idea is to simplify the surface by finding within it a torus or projective plane and then writing the given surface M as $N\#T$ or $N\#P$, where N has fewer 1-handles and is known by induction; or to write M as $N_1\#N_2$ where N_1, N_2 each have fewer 1-handles and are known by induction; or to just show that M has a handle decomposition with fewer 1-handles and so is known already by induction.

The result we wish to prove by induction on the number of 1-handles is the following: If M is a connected handlebody without boundary and is oriented, then M is homeomorphic to $T^{(2g)}$, where $\chi(M) = 2 - 2g$. If M is a connected handlebody without boundary and is nonorientable, then M is homeomorphic to $P^{(h)}$, where $\chi(M) = 2 - h$.

Exercise 2.9.83. Show that, if M has no 1-handles, then $M \simeq S^2$. (Hint: First show that it has a single 0-handle and a single 2-handle by connectivity.)

Exercise 2.9.84. Show that, if M is nonorientable and there is exactly one 1-handle, then M has exactly three handles and is homeomorphic to P.

We assume that the result is known when M has fewer than k 1-handles and M has a handle decomposition with k 1-handles. Let h^1 denote the first 1-handle which is attached. Then it is attached to either one or two 0-handles.

Exercise 2.9.85. Show that, if h^1 is attached to two 0-handles $h_1^0 \cup h_2^0$ via f, then $h_1^0 \cup h_2^0 \cup_f h^1$ is homeomorphic to a disk. Thinking of this disk as a new 0-handle to replace h_1^0, h_2^0, h^1, show that M has a handle decomposition with fewer 1-handles and use induction to prove the result.

Suppose that h^1 is attached to a single 0-handle h^0. Then either $h^0 \cup h^1$ is a cylinder $S^1 \times I$ or is a Möbius band B. Suppose first that the original handle decomposition is oriented. Then $h^0 \cup h^1 = C$ is a cylinder. Now remove the interior of the cylinder and glue in two disks D_1^2, D_2^2 to $N' = M \backslash \text{int } C$ along the boundary circles, oriented so that they form a new oriented handle decomposition of $N = N' \cup D_1^2 \cup D_2^2$ where we now think of D_1^2, D_2^2 as 0-handles to which the remaining handles of M are attached to form N.

Exercise 2.9.86. Show that, if N' is connected, then $M \simeq T\#N$, where N is an oriented surface and we are forming oriented connected sum. Conclude from this that $M \simeq T^{(g)}$ with $\chi(M) = 2 - 2g$.

Exercise 2.9.87. Show that, if N' is not connected, then it is the union of two connected oriented surfaces N_1, N_2 and M is the oriented connected sum of $N_1\#N_2$. Conclude from this that $M \simeq S$ and $\chi(S) = 2$ or $M \simeq T^{(g)}$ with $\chi(M) = 2 - 2g$.

Next suppose that the handle decomposition for M is not orientable, but $h^0 \cup h^1$ is a cylinder. We remove $\text{int } C$ as before and attach the two disks which impose the same orientations on their boundaries as C to form N from N'. Suppose first that N is connected.

Exercise 2.9.88. Show that, if the handle decomposition for N is orientable, then $N \simeq T^{(k)}$ for some k and $M \simeq T^{(k)}\#K \simeq P^{(h)}$. Moreover $\chi(M) = 2 - h$.

Exercise 2.9.89. Show that, if the handle decomposition for N is nonorientable, then $N \simeq P^{(k)}$ for some k and $M \simeq P^{(k)} \# K \simeq P^{(h)}$ with $\chi(M) = 2 - h$.

Exercise 2.9.90. Next suppose that N is not connected and write $N = N_1 \cup N_2$, where these are each connected. Then show that one of N_1, N_2 is nonorientable and so is homeomorphic to $P^{(k)}$. Show that the other piece is either $T^{(p)}$ or $P^{(l)}$. Show that $M \simeq P^{(h)}$ with $\chi(M) = 2 - h$.

Exercise 2.9.91. The remaining case is when $h^0 \cup h^1 \simeq B$. Then replacing it by a disk with the same boundary gives N which has one fewer 1-handle, and so by induction is homeomorphic to either $T^{(k)}$ or $P^{(l)}$, depending on whether the handle decomposition is orientable or not. Show that this implies that M is homeomorphic to $P^{(h)}$ with $\chi(M) = 2 - h$.

Consider the following operation on a surface, called a *surgery of index 1*. Embed $\{-1, 1\} \times D^2$ into M via f, remove $f(\{-1, 1\} \times \text{int} \, D^2)$, and replace it by $D^1 \times S^1$ (gluing via $f|\{-1, 1\} \times S^1$). We write the result as $\chi(M, f) = (M \backslash f(\{-1, 1\} \times \text{int} \, D^2) \cup_f D^1 \times S^1$. This is the operation performed in getting T and K from S and in forming a connected sum when $M = M_1 \bigsqcup M_2$. For an oriented surface, we can guarantee that the result of the surgery is oriented by choosing the embeddings of the disks to have one preserve orientation and the other reverses it. Basically, we are using a fixed orientation on $\partial(D^1 \times D^2) = \partial D^1 \cup D^2 \cup D^1 \times \partial D^2$ and arranging that the orientation on $D^1 \times \partial D^2$ fits together with the orientation on the complement of $f(\partial D^1 \times D^2)$ in M. That the two disks should be embedded with opposite orientations just comes from the fact that they inherit opposite orientations as part of $\partial(D^1 \times D^2)$.

A related operation is a surgery of index 2 where we embed $D^1 \times S^1$ into N via g, remove $g(\text{int} \, D^1 \times S^1)$, and replace it by $\{-1, 1\} \times D^2$ (gluing via $g|\{-1, 1\} \times S^1$). Here $\chi(N, g) = (M \backslash \text{int} \, D^1 \times S^1) \cup_g \{-1, 1\} \times D^2$. We do this in a fashion consistent with the orientation in the case of an oriented surface. Note that these operations (properly done) are inverses of one another: $\chi(\chi(M, f), g) = M$ when g embeds $D^1 \times S^1$ into the second factor of $\chi(M, f) = M \backslash f(\{-1, 1\} \times \text{int} \, D^2) \cup_f D^1 \times S^1$. Similarly, $\chi(\chi(N, g), f) = N$ for properly chosen f. In a connected manifold, any two embedded disks lie in a larger disk.

Exercise 2.9.92. Suppose $H : M \times I \to M \times I$ is an isotopy in a connected surface with H_0 the identity and $H_1 f = f'$. Show that $\chi(M, f)$ is homeomorphic to $\chi(M, f')$.

Exercise 2.9.93. Using the fact that $\chi(S, f)$ is homeomorphic to either T or K depending on the embedding f, show that for any connected $M, \chi(M, f)$ is homeomorphic to either $M \# T$ or $M \# K$.

Exercise 2.9.94. Use the preceding two exercises to give another proof that $P \# T$ is homeomorphic to $P \# K$.

Exercise 2.9.95. Figure 2.84 shows $\chi(T, g)$ for one $g : D^1 \times S^1 \to T$. What surfaces do we get if we perform surgeries using the three embeddings of $D^1 \times S^1$ indicated in Figure 2.85.

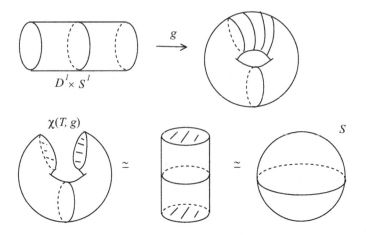

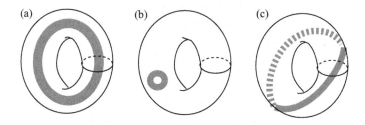

Figure 2.84. Surgery on the torus to get a sphere.

Figure 2.85. Other surgeries on the torus.

Exercise 2.9.96.

(a) Construct a homeomorphism from T to T which sends the center curve $g(\{0\} \times S^1)$ in Figure 2.84 to the center curve $g(\{0\} \times S^1)$ in Figure 2.85(a).

(b) Show that there does not exist a homeomorphism from T to T sending $g(\{0\} \times S^1)$ in Figure 2.84 to $g(\{0\} \times S^1)$ in Figure 2.85(b).

(c) Construct a homeomorphism from T to T which sends the center curve $g(\{0\} \times S^1)$ in Figure 2.84 to $g(\{0\} \times S^1)$ in Figure 2.85(c).

Exercise 2.9.97. Show that if $g : D^1 \times S^1 \to T$ is an embedding so that $g(\{0\} \times S^1)$ does not separate T (i.e. removing it leaves a path connected surface), then $\chi(T, g)$ is homeomorphic to S. (Hint: Note that $\chi(T, g) = M$, for some compact connected surface and the inverse surgery will give T back. But it also gives us either $M \# K$ or $M \# T$ by the discussion of 1-surgery. Use the classification theorem.)

Exercise 2.9.98. Suppose $g_1, g_2 : D^1 \times S^1 \to T$ are two embeddings, $g_1(\{0\} \times S^1) = C_1, g_2(\{0\} \times S^1) = C_2$, and C_1, C_2 do not separate T. Show there is a homeomorphism $h : T \to T$ with $h(C_1) = C_2$. (Hint: Use the preceding exercise and the disk lemma.)

Define the *genus* $g(M)$ of a connected surface M to be the maximal number k of disjoint circles $C_1, \ldots, C_k$ that can be embedded in M so that $M \backslash (C_1 \cup \cdots \cup C_k)$ is connected.

Exercise 2.9.99. Show that the genus is a topological invariant of the surface; that is, homeomorphic surfaces have the same genus.

The following facts are useful in the next exercises. If $C = f(\{0\} \times S^1)$ is an embedded circle in an orientable surface M, then f extends to an embedding of $D^1 \times S^1$ into M, and $M \backslash C$ is connected iff $M \backslash f(D^1 \times S^1)$ is connected. If $C = f(\{0\} \times S^1)$ is an embedded circle in a nonorientable surface, then either there is an extension of f to an embedding of $D^1 \times S^1$ into M as above or there is an embedding of the Möbius band B into M with C corresponding to the center circle. In either case, the connectivity of $M \backslash C$ is equivalent to the connectivity of the complement of the neighborhood of C (either $f(D^1 \times S^1)$ or $f(B)$). If $C_1, \ldots, C_k$ are disjoint embedded circles, then we can choose the extensions so that the neighborhoods $f_i(D^1 \times S^1)$ or $f_i(B)$ are disjoint.

In the following group of exercises we will use the classification theorem to find the genus of all surfaces. A different approach to the classification theorem is to use the idea of nonseparating curves, surgeries, and the Euler characteristic to give a proof of the classification theorem. This is usually done within the context of triangulated surfaces and uses the relation of surgery operations and the Euler characteristic. This is the approach taken in the book by Armstrong [1], for example.

Exercise 2.9.100. Show that the genus of S is 0 and the genus of T is 1. (Hint: To see that 2 is not the genus of T, note what happens when we do a surgery on the first circle.)

Exercise 2.9.101. Show that the genus of $T^{(k)}$ is $\geq k$ by exhibiting k disjoint circles in $T^{(k)}$ which do not separate $T^{(k)}$.

Exercise 2.9.102. Show that the result of doing surgery in $T^{(k)}$ via $g : D^1 \times S^1 \subset T^{(k)}$ such that $g(\{0\} \times S^1)$ does not separate $T^{(k)}$ is $T^{(k-1)}$.

Exercise 2.9.103. Show that the genus of $T^{(k)}$ is k. (Hint: Use induction on k, and the previous exercise.)

Exercise 2.9.104. Show that, for any nonorientable M, there is a curve C which does not separate so that there does not exist $g : D^1 \times S^1 \to M$ with $g(\{0\} \times S^1) = C$.

Exercise 2.9.105. Show that the genus of P is ≥ 1.

Consider the following operation on a surface M. Embed D^2 in M via f, remove $f(\mathrm{int}\, D^2)$, and sew in a Möbius band B via f on the boundary circle. Let $\eta(M, f) = M \backslash f(D^2) \cup_f B$. Note that in a nonorientable surface there is an inverse of this operation. First embed a Möbius band via $g : B \to N$, remove $g(\mathrm{int}\, B)$, and sew in a disk, $\eta(N,g) = (N \backslash g(\text{int } B)) \cup_g D^2$. Done properly, $\eta(\eta(N,g),f) = N, \eta(\eta(M,f),g) = M$. These two operations are sometimes called nonorientable surgeries of index 1 and 2.

Exercise 2.9.106. Show that $\eta(M,f)$ is homeomorphic to $M \# P$.

Exercise 2.9.107. Show that the genus of $P^{(k)}$ is $\geq k$.

Exercise 2.9.108. Show that the genus of P is 1. (Hint: If a curve does not separate P, show that its neighborhood is B and not $D^1 \times S^1$.)

Exercise 2.9.109. Show that the genus of $P^{(k)}$ is k.

3

The fundamental group and its applications

3.1 The main idea of algebraic topology

In this chapter we study, through the example of the fundamental group, the general method of algebraic topology. We give a means of associating to a geometric problem a (hopefully easier) algebraic problem to solve. Consider the problem of distinguishing between two surfaces. How can we tell, for example, that there is no homeomorphism between the sphere and the torus? According to algebraic topology, to solve this problem we should transform it into an algebraic problem that is readily solved. To each topological space X, associate to it some algebraic object, say a group $g(X)$. Do this in such a way that homeomorphic spaces have isomorphic groups associated to them. Thus one way of telling that X is not homeomorphic to Y is by showing that $g(X)$ is not isomorphic to $g(Y)$. Of course, this works only when we can readily compute $g(X)$ and $g(Y)$ and decide whether or not they are isomorphic. This method is successful in distinguishing between surfaces.

We now formalize the informal discussion above. In this section, we start by discussing briefly the concept of a group. Readers with a previous course in abstract algebra should just skim over the group theory material in this section to become familiar with our notation and viewpoint. We then apply these ideas to discuss how algebraic topology uses group theory to answer topological questions. In later sections and chapters, we will introduce more sophisticated results from group theory as it is needed.

Definition 3.1.1. A *group* $(G, \cdot)$ is a set G together with a binary operation $G \times G \to G, (a, b) \to a \cdot b$, which we will call multiplication, satisfying:

(1) there is an element $e \in G$, called the *identity* of the group, so that

$$g \cdot e = e \cdot g = g \quad \text{for all } g \in G;$$

(2) for each $g \in G$, there is an element $g' \in G$ with $g \cdot g' = g' \cdot g = e$. The element g' is called the *inverse* of g and is denoted g^{-1};

(3) given $g_1, g_2, g_3 \in G$, then $(g_1 \cdot g_2) \cdot g_3 = g_1 \cdot (g_2 \cdot g_3)$. This property is called *associativity*.

Example 3.1.1. Let us look at some examples.

- The simplest possible group is the group with only one element $\{e\}$. This is called the *trivial group*.

- The integers $(\mathbb{Z}, +)$ with the operation of addition form a group. The identity is 0 and the inverse of a is $-a$. Similarly, the rationals $\mathbb{Q}$, the reals $\mathbb{R}$, and the complex numbers $\mathbb{C}$ also form groups under addition.

- Note that the set of natural numbers $\mathbb{N}$ does not form a group under addition since there is no identity element. Even after adding $0, \mathbb{N} \cup \{0\}$ still does not form a group under addition since there are no inverses. The integers do not form a group under multiplication since there are no inverses (1 does act as an identity). The rationals almost form a group under multiplication except for the fact that 0 has no multiplicative inverse.

- The rationals with 0 deleted, $\mathbb{Q}\backslash\{0\}$, form a group under multiplication— the identity is 1 and the inverse of p/q is q/p. Similarly, $\mathbb{R}\backslash\{0\}$ and $\mathbb{C}\backslash\{0\}$ form groups under multiplication.

- An example of a finite group is $\mathbb{Z}_p$, the integers modulo p. The elements of the group are $0, 1, \ldots, p-1$. We add two elements as if they were integers; if the sum is greater than or equal to p, then we subtract off p from the sum; that is, $a \cdot b = a+b$ if $a+b < p$, and $a \cdot b = a+b-p$ if $a+b \geq p$. The identity is 0 and the inverse of a is $p - a$ if $a \neq 0$. For example, $\mathbb{Z}_3 = \{0, 1, 2\}$, with multiplication table as follows:

$\cdot$	0	1	2
0	0	1	2
1	1	2	0
2	2	0	1

We can also regard $\mathbb{Z}_p$ as equivalence classes of integers where $a \sim b$ if p divides $a - b$. Then the group operation on $\mathbb{Z}_p$ is induced from addition in the integers. Of course, we have to check that it is well defined; that is, if $a \sim b, c \sim d$, then $a+b \sim c+d$. But $a \sim b$ implies $a - b = mp$, and $c \sim d$ implies $c - d = np$; hence $(a + c) - (b + d) = (m + n)p$, so $a + c \sim b + d$.

- A somewhat more complicated example is given by the permutation group S_n. An element of S_n is a permutation of $\{1, 2, \ldots, n\}$; that is, is a bijection of $\{1, 2, \ldots, n\}$ onto itself. The identity for the group is the identity permutation. Since each permutation is a bijection, it has an inverse, which will be the inverse in the group. The multiplication involved is just the composition of functions. For example, S_3 is the group of permutations of

{1, 2, 3}. If our permutation $f : \{1, 2, 3\} \to \{1, 2, 3\}$ is denoted by $[abc]$, where $f(1) = a, f(2) = b, f(3) = c$, then S_3 has six elements: $[123]$, $[132]$, $[213]$, $[231]$, $[312]$, and $[321]$. Some sample products are $[213] \cdot [132] = [231]$, $[132] \cdot [213] = [312]$, $[213] \cdot [213] = [123]$. We are using functional notation in computing these compositions, working from right to left. Note that it is not always the case that $a \cdot b = b \cdot a$. A group where $a \cdot b = b \cdot a$ for all a, b is called *abelian*. Thus S_3 is not abelian.

Exercise 3.1.1. Find the inverse of each element of S_3.

The circle S^1 can be regarded as a subset of the complex numbers $\mathbb{C}$ by identifying the complex number $a + ib$ with the point (a, b) in the plane. The length $|z| = |a + ib|$ is defined by $|z|^2 = a^2 + b^2$; it agrees with the length of $(a, b) \in \mathbb{R}^2$. The circle then represents the complex numbers of length 1. Since $|z_1 \cdot z_2| = |z_1 \cdot z_2|$, complex multiplication restricts to give a group operation on S^1. Multiplication satisfies $(\cos\theta + i\sin\theta)(\cos\phi + i\sin\phi) = \cos(\theta + \phi) + i\sin(\theta + \phi)$, so that it corresponds to addition of angles. The identity is $1 = 1 + i0$ and the inverse of $a + ib$ is $a - ib$.

If $(G, \cdot)$ is a group and $H \subset G$ is a subset of G so that the multiplication in G restricts to give a multiplication in H (i.e., $h_1, h_2 \in H$ for $h_1, h_2 \in H$), the identity e of G is in H, and the inverse of any element of H is in H, then $(H, \cdot)$ will form a group, called a *subgroup* of G. In the example above, the circle $(S^1, \cdot)$ is a subgroup of $(\mathbb{C}\backslash\{0\}, \cdot)$, where the dot denotes complex multiplication. $(\mathbb{Z}, +)$ is a subgroup of $(\mathbb{Q}, +)$ and $(\mathbb{R}, +)$. If we regard $\mathbb{R}$ as a subset of $\mathbb{C}$, then each of these is a subgroup of $(\mathbb{C}, +)$.

The set of all homeomorphisms of a topological space X forms a group with operation the composition of homeomorphisms, denoted Homeo(X). The identity element is given by the identity homeomorphism, and f^{-1} is the inverse of f in this group. Subgroups of Homeo(X) are important to both topology and other forms of geometry. For example, the set of rigid motions of the plane studied in Chapter 1 forms a subgroup $m(\mathbb{R}^2)$ of Homeo($\mathbb{R}^2$). Recall that a rigid motion is a homeomorphism $f : \mathbb{R}^2 \to \mathbb{R}^2$ so that $d(f(a), f(b)) = d(a, b)$, where d denotes Euclidean distance. Any rigid motion can be written as a composition of a translation, a rotation, and a reflection. The translations $\mathcal{T}(\mathbb{R}^2) = \{T_a : a \in \mathbb{R}^2\}, T_a(x) = x + a\}$, form a subgroup of $m(R^2)$ (and Homeo(R^2)), as do the rotations $\mathcal{R}(\mathbb{R}^2) = \{R_\theta : \theta \in \mathbb{R}\}, R_\theta(r\cos\phi, r\sin\phi) = (r\cos(\theta + \phi), r\sin(\theta + \phi))$. The reflections do not form a subgroup since they do not contain the identity homeomorphism. Both the translations $\mathcal{T}(\mathbb{R}^2)$ and the rotations $\mathcal{R}(\mathbb{R}^2)$ form abelian groups, but $m(\mathbb{R}^2)$ is not abelian.

In the late nineteenth century Felix Klein advocated studying the geometry of a space X in terms of subgroups of Homeo(X). Traditional Euclidean geometry of the plane can be studied in terms of $m(\mathbb{R}^2)$, whereas the topology of the plane is expressible in terms of Homeo($\mathbb{R}^2$). In Euclidean geometry, we are interested in subsets of the plane which are transformed to each other via rigid motions of the plane. For example, two triangles T_1, T_2 are congruent if there is a rigid motion f with $f(T_1) = T_2$. The Schönflies theorem says that any two simple closed curves in the plane are equivalent under the group of homeomorphisms of the plane; that is, if C_1, C_2 are simple closed curves in the plane, then there is a homeomorphism f of the plane with $f(C_1) = C_2$. Part of Klein's viewpoint was that the algebraic structure of these groups of homeomorphisms is useful in studying the geometry. In particular, any general fact known about groups (or special classes of groups) could then be applied to these particular groups of homeomorphisms.

Exercise 3.1.2. Verify that $m(\mathbb{R}^2)$ is not abelian. (Hint: Compute $T_a R_\theta(\mathbf{0})$ and $R_\theta T_a(\mathbf{0})$).

In order to familiarize ourselves somewhat with the definition of a group, let us verify that the identity element is unique and that inverses are unique. Suppose that there are two elements e_1, e_2 with $g \cdot e_i = e_i \cdot g = g$ for all $g \in G, i = 1, 2$. Taking $g = e_1$, we get $e_1 \cdot e_2 = e_1$; taking $g = e_2$, we get $e_1 \cdot e_2 = e_2$. Hence $e_1 = e_2$. We now show that if g_1 is a left inverse for g (i.e. $g_1 \cdot g = e$), and if g_2 is a right inverse for g (i.e. $g \cdot g_2 = e$), then $g_1 = g_2$ and so is an inverse for g $[g_1 = g_1 \cdot e = g_1 \cdot (g \cdot g_2) = (g_1 \cdot g) \cdot g_2 = e \cdot g_2 = g_2]$. This same argument shows that inverses are unique; we leave the details as an exercise.

Exercise 3.1.3. Show that inverses are unique.

We want to put an equivalence relation on groups, called *isomorphism*. We will regard two groups as being the same if there is an isomorphism between them; they are then called *isomorphic*. Isomorphism is the group theoretic analogue of homeomorphism for topological spaces. The analogue of a continuous map is called a *homomorphism*. As a continuous map is consistent with the topologies of the two spaces involved in the sense that inverse images of open sets are open, a homomorphism is consistent with the multiplications of the two groups in that the image of a product of two elements is the product of the images of the two elements.

Definition 3.1.2. Let $(G_1, \cdot)$ and $(G_2, \times)$ be groups. Then a function $f : G_1 \to G_2$ is called a *homomorphism* if $f(a \cdot b) = f(a) \times f(b)$.

Example 3.1.2. Here are some examples.

- $f : (\mathbb{Z}, +) \to (\mathbb{Z}, +), f(x) = 2x$, is a homomorphism $[f(a + b) = 2(a + b) = 2a + 2b = f(a) + f(b)]$.
- Another example of a homomorphism is given by $f : (\mathbb{Z}, +) \to (\mathbb{Z}_p, +), f(x) = [x]$. For $f(a + b) = [a + b] = [a] + [b] = f(a) + f(b)$, since addition in $\mathbb{Z}_p$ is defined in terms of addition in $\mathbb{Z}$.

A property that any homomorphism must have is that it must send the identity to the identity. Let e_1 denote the identity of G_1 and e_2 that of G_2. Then, if $f : (G_1, \cdot) \to (G_2, \times)$ is a homomorphism, $f(e_1) = f(e_1 \cdot e_1) = f(e_1) \times f(e_1)$. Multiply both sides $f(e_1)^{-1}$ to get $e_2 = f(e_1)$.

Exercise 3.1.4. Decide whether the following maps are homomorphisms:

(a) $f : \mathbb{Z}_p \to \mathbb{Z}, f([n]) = n$;

(b) $f : \mathbb{Z}_4 \to \mathbb{Z}_2, f([n]) = [n]$;

(c) $f : \mathbb{Z}_2 \to \mathbb{Z}_4, f([n]) = [2n]$;

(d) $f : \mathbb{Z}_2 \to S_3, f(0) = [123], f(1) = [213]$;

(e) $f : (\mathbb{C}, +) \to \mathcal{T}(\mathbb{R}^2), f(a) = T_a$;

(f) $f : S^1 \to \mathcal{R}(\mathbb{R}^2), f((\cos \theta, \sin \theta)) = R_\theta$;

(g) $p : (\mathbb{R}, +) \to (S^1, \cdot), p(t) = (\cos 2\pi t, \sin 2\pi t)$;

(h) $f_a : (\mathbb{R}, +) \to (\mathbb{R}, +), f_a(b) = ab$.

Definition 3.1.3. $f : G_1 \to G_2$ is an *isomorphism* if f is a bijective homomorphism and the set inverse f^{-1} is a homomorphism as well; equivalently, a homomorphism $f : G_1 \to G_2$ is an isomorphism if there is a homomorphism $g : G_2 \to G_1$ with $fg = 1_{G_2}, gf = 1_{G_1}$, where 1_X denotes the identity function of X. We will denote an isomorphism by $\simeq$: $G_1 \simeq G_2$ means that there is an isomorphism between G_1 and G_2; we say that G_1 is *isomorphic* to G_2.

Example 3.1.3. Here are some examples.

- $f : (\mathbb{Z}, +) \to (\mathbb{Z}, +), f(x) = 2x$, is not an isomorphism since it is not surjective. However, if we denote by $(2\mathbb{Z}, +)$ all multiples of 2 with the usual addition, then $\bar{f} : (\mathbb{Z}, +) \to (2\mathbb{Z}, +), \bar{f}(x) = 2x$ is a 1–1, surjective homomorphism. Moreover, $\bar{f}^{-1}(y) = y/2$ is also a homomorphism, so $\bar{f}$ is an isomorphism.

- $\mathbb{Z}$ is not isomorphic to $\mathbb{Z}_p$ since they have a different number of elements, and an isomorphism is a bijection.

- S_3 is not isomorphic to $\mathbb{Z}_6$ even though they each have six elements (for $\mathbb{Z}_6$ is abelian and S_3 is not). If we choose $a, b \in S_3$ with $a \cdot b \neq b \cdot a$, and $f : S_3 \to \mathbb{Z}_6$ were a homomorphism, then $f(a \cdot b) = f(a) + f(b)$, which equals $f(b) + f(a) = f(b \cdot a)$ since $\mathbb{Z}_6$ is abelian. But this means that f is not 1–1 and hence is not an isomorphism.

- Consider the set $\{-1, 1\}$ with operation the usual multiplication of integers. This group is isomorphic to $(\mathbb{Z}_2, +)$. An isomorphism from $(\mathbb{Z}_2, +)$ to $(\{-1, 1\}, \cdot)$ is given by sending 0 to 1 and 1 to -1. We leave the details as an exercise.

Exercise 3.1.5. Show that the map from $(\mathbb{Z}_2, +)$ to $(\{-1, 1\}, \cdot)$ defined above is in fact an isomorphism. In the future we will denote $\{-1, 1\}$ with the operation of multiplication by $\mathbb{Z}_2$ as well because of this isomorphism.

Exercise 3.1.6. Decide whether the following groups are isomorphic. Either construct an isomorphism and prove it is an isomorphism or show why there is no isomorphism.

(a) $S_2 = $ permutations of $\{1, 2\}$ and $\mathbb{Z}_2$,

(b) $(\mathbb{Z}, +)$ and $(\mathbb{Q}, +)$,

(c) $(\mathbb{C}, +)$ and $\mathcal{T}(\mathbb{R}^2)$,

(d) $(S^1, \cdot)$ and $\mathcal{R}(\mathbb{R}^2)$,

(e) $(\mathbb{R}, +)$ and $(S^1, \cdot)$.

In Chapter 1 we showed that sometimes (but not always) a bijective continuous map is a homeomorphism. The following proposition is an analogue of this.

Proposition 3.1.1. *A bijective homomorphism $f : (G_1, \cdot) \rightarrow (G_2, \times)$ is an isomorphism.*

Proof. We have to show that f^{-1}, which exists since f is bijective, is a homomorphism. Thus we must show that $f^{-1}(a \times b) = f^{-1}(a) \cdot f^{-1}(b)$. But $f(f^{-1}(a) \cdot f^{-1}(b)) = f(f^{-1}(a)) \times f(f^{-1}(b)) = a \times b$ and $f(f^{-1}(a \times b)) = a \times b$. Since f is a bijection, the result follows. $\qquad\square$

Another property of homomorphisms that is useful in conjunction with the above proposition is the following.

Proposition 3.1.2. *A homomorphism $f : (G_1, \cdot) \rightarrow (G_2, \times)$ is 1–1 iff $f(a) = e_2$ implies $a = e_1$. (Here e_i denotes the identity of G_i.)*

Proof. Suppose f is 1–1 and $f(a) = e_2$. Since $f(e_1) = e_2$ as well, then f 1–1 implies that $a = e_1$. Conversely, suppose that $f(a) = e_2$ implies that $a = e_1$. Let a, b be given with $f(a) = f(b)$. Then $f(a \cdot b^{-1}) = f(a) \times f(b)^{-1} = e_2$. Hence $a \cdot b^{-1} = e_1$ and thus $a = b$. $\qquad\square$

There are two important subgroups related to a homomorphism $f : G_1 \rightarrow G_2$. First, the *kernel* of f is the subgroup $\ker f = \{g \in G_1 : f(g) = e_2\}$. The last proposition says that f is 1–1 precisely when $\ker f$ is the trivial group. Another subgroup related to f is the *image* of f, $\operatorname{im} f = \{h \in G_2 : h = f(g)$ for some $g \in G_1\}$. By definition, f is surjective precisely when $\operatorname{im} f = G_2$.

Exercise 3.1.7. Verify that $\ker f$ and $\operatorname{im} f$ are subgroups.

Here is another useful definition and group theoretical result. A subgroup $H \subset G$ is called *normal* if whenever $g \in G, h \in H$, then $ghg^{-1} \in H$. Note that subgroups of abelian groups are automatically normal. When H is normal, we can form a new group, called the quotient group G/H of right cosets of H. The elements of this group are the sets $Hg = \{hg : h \in H\}$. They are multiplied by $(Hg_1)(Hg_2) = Hg_1g_2$. The condition of normality is what is needed to see that this makes sense and forms a group. There is naturally a surjective homomorphism $Q : G \rightarrow G/H$. The kernel of Q is just H. Thus a normal subgroup is the kernel of a homomorphism. We leave it as an exercise to check that the kernel of a homomorphism is always normal. A basic isomorphism theorem in group theory states that whenever $f : G_1 \rightarrow G_2$ is a homomorphism, then there is an induced isomorphism $\bar{f} : G_1/\ker f \rightarrow \operatorname{im} f$.

Exercise 3.1.8. Show that if $f : G_1 \rightarrow G_2$ is a homomorphism, then $\ker f$ is a normal subgroup.

Exercise 3.1.9. Show that if S_3 denotes the permutations of $\{1, 2, 3\}$, then the subgroup consisting of the two permuations $[123], [213]$ is not normal.

Here is a general construction by which we can form a new group from two groups A and B. We will denote all multiplications by a centered dot. The *direct sum* of A and B, denoted $A \oplus B$, is the set of pairs of (a, b), with $a \in A, b \in B$. The multiplication is defined componentwise using the multiplications in A and B. That is, $(a, b) \cdot (c, d) = (a \cdot c, b \cdot d)$. For example, we could form $\mathbb{Z} \oplus \mathbb{Z}$ where $(a, b) \cdot (c, d) = (a + c, b + d)$. Similarly, we could form $\mathbb{Z} \oplus \mathbb{Z}_p$. A group G is *finitely generated* if there exist $g_1, \ldots, g_n \in G$ so that every element of G is expressible as products of these elements. It is an important theorem in group theory that all finitely generated abelian groups are formed (up to isomorphism) by taking direct sums of copies of $\mathbb{Z}$ and $\mathbb{Z}_p$ for various p; thus a general finitely generated abelian group "looks like" $\mathbb{Z} \oplus \cdots \oplus \mathbb{Z} \oplus \mathbb{Z}_{p_1} \oplus \cdots \oplus \mathbb{Z}_{p_k}$. The number of copies of $\mathbb{Z}$ (called the *rank* of the group) and the various p_i (called the *torsion coefficients* of the group) distinguish these groups up to isomorphism. For example, $\mathbb{Z}$ is not isomorphic to $\mathbb{Z} \oplus \mathbb{Z}$, even though they each have the same number of elements; that is, they are not isomorphic even though there is a bijection between them. For a homomorphism, $f : \mathbb{Z} \to \mathbb{Z} \oplus \mathbb{Z}$ is determined completely by $f(1)$. Suppose $f(1) = (a, b)$. Then $f(n) = (na, nb) = n(a, b)$ and cannot be surjective. An important subclass of finitely generated groups are those which are isomorphic to the direct sum of n copies of $\mathbb{Z}$; these groups are called finitely generated *free abelian* groups.

With this bit of group theory as a background, we describe more precisely the method of algebraic topology. We wish to assign to each topological space a group in a "consistent" manner. The precise name for the correspondence is a *(covariant) functor*. We denote the functor by F. If we denote topological spaces by $\mathcal{T}$ and groups by $\mathcal{G}$, then F is a special type of function from $\mathcal{T}$ to $\mathcal{G}$. For each topological space $X, F(X)$ will be a group. Next, suppose that $f : X \to Y$ is a continuous map between topological spaces X and Y. Then to f we wish to associate a homomorphism $F(f) : F(X) \to F(Y)$. Moreover, we wish this correspondence to obey two rules:

(1) $F(1_X) = 1_{F(X)}$, where 1_A denotes the identity on A;

(2) $F(fg) = F(f)F(g)$.

Together these imply that if $f : X \to Y$ is a homeomorphism, then $F(f) : F(X) \to F(Y)$ is an isomorphism. For f being a homeomorphism implies that there is a continuous function $g : Y \to X$ with $fg = 1_Y$ and $gf = 1_X$. Then $1_{F(Y)} = F(1_Y) = F(fg) = F(f)F(g)$ and $1_{F(X)} = F_{1_X} = F(gf) = F(g)F(f)$. Hence $F(f)$ is an isomorphism, and so homeomorphic spaces have isomorphic groups associated to them.

Let us return to the classification of surfaces to see how this works in practice. We showed in Chapter 2 that each compact connected surface (without boundary) is homeomorphic to either a sphere, a connected sum of g tori, or a connected sum of h projective planes. We quoted a result involving Euler characteristic to say that these possibilities were distinct. Another proof involves algebraic topology as indicated above. That is, to each surface $S, T^{(g)}, P^{(h)}$, we associate via a functor F a group $F(S), F(T^{(g)}), F(P^{(h)})$ and note that no two of the groups obtained are isomorphic. Hence no two of the surfaces are homeomorphic. We will use the fundamental group in Section 3.4 to show that S, T, and P are not homeomorphic. The fundamental group will be used to distinguish completely between compact connected surfaces (i.e. nonhomeomorphic surfaces have nonisomorphic fundamental groups), once we have proved the Seifert–van Kampen theorem so that we can compute the fundamental group of each surface.

Exercise 3.1.10. Suppose F is a functor from topological spaces and continuous maps to finitely generated abelian groups and homomorphisms so that for compact connected surfaces (without boundary) A, B, $F(A \# B) = F(A) \oplus F(B)$. For finitely generated abelian groups, it is true that $G_1 \oplus G_2$ is isomorphic to G_1 iff $G_2 = \{e\}$. Show that this implies that $F(S)$ is the trivial group. Show that if $F(T)$ is not $\{e\}$, then F distinguishes $T^{(g)}, g \geq 0$; that is, $F(T^{(p)})$ is not isomorphic to $F(T^{(q)})$, for $p \neq q, p, q \geq 0$.

3.2　The fundamental group

In this section we define a group, called the *fundamental group*, which we associate in a functorial manner to a topological space. Its definition requires not only a topological space X but also a point $x \in X$, called the *base point*. If X is path connected, the group obtained does not depend up to isomorphism on the base point x. We will denote the fundamental group by $\pi_1(X, x)$. Its elements will be equivalence classes of paths in X which run from x to x (*loops at x*).

Intuitively, two loops are equivalent if we can continuously deform one loop to the other. The precise definition involves the notion of a homotopy. What the fundamental group measures is the "distinct" (up to homotopy) loops in a space. Any space has the constant loop that stays at x. The question is: are there any loops in the space that cannot be continuously deformed to the constant loop, and, if so, how many distinct ones up to homotopy? To give some feel for this, here are some answers that we will derive. For the sphere S^2, the answer is that there are none, so $\pi_1(S^2, x) \simeq \{e\}$. For the circle, there are an infinite number and $\pi_1(S^1, x) \simeq \mathbb{Z}$. A loop corresponding to the integer n is $p_n(t) = (\cos 2\pi n t, \sin 2\pi n t)$; it wraps around the circle n times. For the torus $T, \pi_1(T, x) \simeq \mathbb{Z} \oplus \mathbb{Z}$. This comes from regarding the torus as $S^1 \times S^1$. A loop representing (m, n) is $(p_m(t), p_n(t))$. It wraps around the first circle m times and the second circle n times. In particular, these calculations furnish a proof that the sphere is not homeomorphic to the torus since their fundamental groups are not isomorphic.

We now define $\pi_1(X, x)$. First, its representatives are loops at $x, f :$ $(I; \{0, 1\}) \to (X, x)$. By this notation we mean that f is a continuous function from $I = [0, 1]$ to X with $f(0) = f(1) = x$. Two loops f_0, f_1 are called *homotopic* if there is a continuous function $F : (I \times I, \{0, 1\} \times I) \to (X, x)$ with $F(s, 0) = f_0(s), F(s, 1) = f_1(s), F(0, t) = F(1, t) = x$. F is called a *homotopy* (of loops at x) between f_0 and f_1. To denote that f_0 and f_1 are homotopic with a homotopy F, we write $f_0 \sim_F f_1$. If the particular homotopy is unimportant, we write $f_0 \sim f_1$. Sometimes $F(s,t)$ is denoted by $F_t(s)$; note that F_t is a loop at x. Intuitively, f_0 and f_1 are homotopic if there is a path F_t of loops at x connecting f_0 and f_1. Figure 3.1 illustrates a homotopy between two loops in the plane.

We show that homotopy gives an equivalence relation on loops at x. First, note that $f \sim f$. Define $F : I \times I \to X$ by $F(s,t) = f(s)$. We are just taking the constant path of loops, each loop being f. Second, $f \sim g$ implies $g \sim f$. For if F is a homotopy between f and g, then $G(s,t) = F(s,1-t)$ is a homotopy between g and f. Here we are just traversing the path of loops connecting f and g in the opposite direction. Next, suppose $f \sim_F g$ and $g \sim_G h$. Then we must see why $f \sim_H h$. The idea is pictured in Figure 3.2. What we want to do is just put the homotopies together one on top of the other. To get the appropriate domain space, we have to reparametrize.

Instead of having F defined on $I \times [0,1]$, we would like to redefine it on $I \times [0,\frac{1}{2}]$. This is easily done by using the unique linear order-preserving

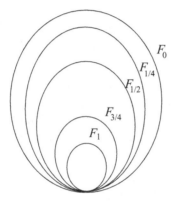

Figure 3.1. Homotopic loops.

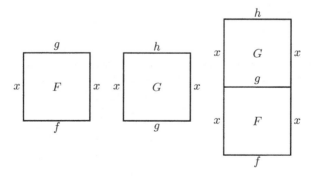

Figure 3.2. Transitivity of homotopy.

homeomorphism $[0, \frac{1}{2}] \rightarrow [0,1], t \rightarrow 2t$, and then taking the composition $I \times [0, \frac{1}{2}] \xrightarrow{(s,2t)} I \times I \xrightarrow{F} X$. We also want G defined on $I \times [\frac{1}{2}, 1]$ instead of $I \times I$. Thus we take the unique affine linear order-preserving homeomorphism $[\frac{1}{2}, 1] \rightarrow [0,1], t \rightarrow 2t - 1$, and then take the composition $I \times [1/2, 1] \xrightarrow{(s,2t-1)} I \times I \xrightarrow{G} X$. Piecing these together at $I \times \{\frac{1}{2}\}$ gives the required homotopy H. Formally,

$$H(s,t) = \begin{cases} F(s, 2t) & \text{if } 0 \le t \le \frac{1}{2}, \\ G(s, 2t - 1) & \text{if } \frac{1}{2} \le t \le 1. \end{cases}$$

H is well defined since $F(s, 1) = g(s) = G(s, 0)$ and these are giving $H(s, \frac{1}{2})$ in the two parts of the definition. Thus homotopy of loops $\sim$ is an equivalence relation; we will denote the equivalence class of f by $[f]$. The equivalence classes are the elements of the group $\pi_1(X, x)$.

Next, we need to define a multiplication on equivalence classes. We first define a multiplication on loops at x, denoted by $*$. We then define a multiplication on equivalence classes of loops from this, which we denote by $\bar{*}$, using the formula $[f]\bar{*}[g] = [f * g]$. We have to show that if $f_0 \sim f_1, g_0 \sim g_1$, then $f_0 * g_0 \sim f_1 * g_1$ to see that the definition of the multiplication $\bar{*}$ does not depend on the representative chosen from an equivalence class.

Intuitively, $f * g$ is defined by first going along the loop f and then going along the loop g. The problem again is the domain of definition. We first reparametrize f to be defined on $[0, \frac{1}{2}]$ and reparametrize g to be defined on $[\frac{1}{2}, 1]$ and then put them together. Formally,

$$f * g(s) = \begin{cases} f(2s) & \text{if } 0 \le s \le \frac{1}{2}, \\ g(2s - 1) & \text{if } \frac{1}{2} \le s \le 1. \end{cases}$$

This is the composition $[0, \frac{1}{2}] \xrightarrow{2s} [0,1] \xrightarrow{f} X$ for $0 \le s \le \frac{1}{2}$, and is the composition $[\frac{1}{2}, 1] \xrightarrow{2s-1} [0,1] \xrightarrow{g} X$ for $\frac{1}{2} \le s \le 1$. The continuity of $f * g$ follows from the piecing lemma. To see that $*$ induces a well-defined operation on equivalence classes, we need to see that $f_0 \sim f_1, g_0 \sim g_1$ implies that $f_0 * g_0 \sim f_1 * g_1$. The idea is that if F is the homotopy between f_0 and f_1 and G is the homotopy between g_0 and g_1, then we can form $F * G$ by composing homotopies as we have composed loops to get a homotopy between $f_0 * g_0$ and $f_1 * g_1$. The idea of the argument for independence of representatives for addition up to homotopy is illustrated in Figure 3.3. We leave the details as an exercise.

Exercise 3.2.1. Fill in the details of the above argument to show that $\bar{*}$ is well defined.

To show that $\pi_1(X, x)$ is a group, we have to show that f is associative and that there is an identity and inverses. The identity equivalence class will be

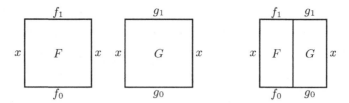

Figure 3.3. Addition of homotopies.

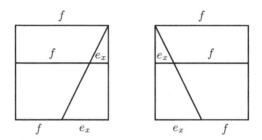

Figure 3.4. $f * e_x \sim f \sim e_x * f$.

represented by $e_x : I \to X, e_x(s) = x$; that is, the constant loop at X. Thus we need to see why $f * e_x \sim f \sim e_x * f$. Basically, $f * e_x$ goes along the loop f from 0 to $\frac{1}{2}$ and then stays at x from $\frac{1}{2}$ to 1. The idea of the homotopy is to gradually increase the time it takes to traverse the path f until we go along f from 0 to 1. More formally, what we need is a one-parameter family of maps $p_t : [0, s_t] \to [0, 1]$ so that $s_0 = \frac{1}{2}$ and $s_1 = 1$, and $p_0(s) = 2s, p_1(s) = s$. The simplest maps to take come from making $t \to s_t$ affine linear and making p_t affine linear. That $t \to s_t$ is affine linear and $0 \to \frac{1}{2}, 1 \to 1$ means that $s_t = \frac{1}{2} + \frac{1}{2}t$ since the affine linear map sending a_1 to b_1 and a_2 to b_2 is given by $t \to b_1 + (b_2 - b_1)(t - a_1)/(a_2 - a_1)$. Then the order-preserving affine linear map $p_t : [0, \frac{1}{2} + \frac{1}{2}t] \to [0, 1]$ is given by using the same formula, $p_t(s) = 2s/(1 + t)$. Thus the homotopy is given by

$$F(s, t) = \begin{cases} f(2s/(1+t)) & \text{if } 0 \le s \le (1+t)/2, \\ x & \text{if } (1+t)/2 \le s \le 1. \end{cases}$$

A picture of this homotopy (actually the reparametrization of f at various levels of t) is depicted in Figure 3.4. Since we are always using affine linear maps determined by their values on the end points, we will tend to emphasize the pictures that lead to the formulas, rather than the formulas themselves, which follow directly, though sometimes tediously, from the pictures.

Note that F is continuous by the piecing lemma since it is defined on the union of two closed sets, and restricts to a continuous function on each, with the definitions agreeing on the intersection. Note also that $F(s, 0) = f * e_x(s), F(s, 1) = f(s)$.

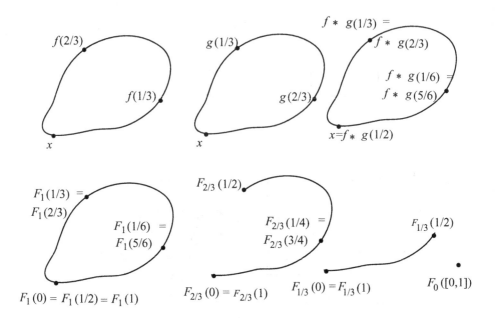

Figure 3.5. The inverse of a loop.

We leave it as an exercise to use the right half of Figure 3.4 and our method above to write down a homotopy between $e_x * f$ and f.

Exercise 3.2.2. Write down a formula for a homotopy between $e_x * f$ and f.

We next attack the problem of finding an inverse for $[f] \in \pi_1(X, x)$. The idea is to just go along f in the opposite direction; that is, $\bar{f}(s) = f(1 - s)$ will represent the inverse. The homotopy between $f * \bar{f}$ and e_x is found by going less and less along f and then retracing our steps, changing the parametrization so that at the end we are just staying at x. Figure 3.5 depicts this for a loop in the plane.

The main problem in writing down a formula for this homotopy is just getting the reparametrization correct. During the first half of the time interval, we want F_t to go through the first tth part of f, and during the last half we want it to go through the last tth part of $\bar{f}$. We use the affine linear functions $[0, \frac{1}{2}] \rightarrow [0, t], s \rightarrow 2st$, and $[\frac{1}{2}, 1] \rightarrow [1 - t, 1], s \rightarrow (1 - t) + 2t(s - \frac{1}{2}) = 2t(s - 1) + 1$. The formula is

$$F(s,t) = \begin{cases} f(2st) & \text{if } 0 \le s \le \frac{1}{2}, \\ \bar{f}(2t(s - 1) + 1) & \text{if } \frac{1}{2} \le s \le 1. \end{cases}$$

F will be continuous since $f(2st)$ and $\bar{f}(2t(s - 1) + 1)$ are continuous on their domains and when $s = \frac{1}{2}, f(t) = \bar{f}(1 - t)$. Note that $F(s, 0) = x = e_x(s)$ and $F(s, 1) = f * \bar{f}(s)$.

We leave it as an exercise to give a homotopy between $\bar{f} * f$ and e_x.

Exercise 3.2.3. Write down a homotopy between $\bar{f} * f$ and e_x.

Finally, to show that the group operation is associative, we have to show that $(f * g) * h$ is homotopic to $f * (g * h)$. Here are the formulas for each composition:

$$(f * g) * h(s) = \begin{cases} (f * g)(2s) & \text{if } 0 \le s \le \frac{1}{2} \\ h(2s - 1) & \text{if } \frac{1}{2} \le s \le 1 \end{cases} = \begin{cases} f(4s) & \text{if } 0 \le s \le \frac{1}{4}, \\ g(4s - 1) & \text{if } \frac{1}{4} \le s \le \frac{1}{2}, \\ h(2s - 1) & \text{if } \frac{1}{2} \le s \le 1, \end{cases}$$

$$f * (g * h)(s) = \begin{cases} f(2s) & \text{if } 0 \le s \le \frac{1}{2} \\ (g * h)(2s - 1) & \text{if } \frac{1}{2} \le s \le 1 \end{cases} = \begin{cases} f(2s) & \text{if } 0 \le s \le \frac{1}{2}, \\ g(4s - 2) & \text{if } \frac{1}{2} \le s \le \frac{3}{4}, \\ h(4s - 3) & \text{if } \frac{3}{4} \le s \le 1. \end{cases}$$

Getting a homotopy is just a matter of homotoping the parametrization. Figure 3.6 is supposed to be suggestive of how to get the formula. Here is another way to see that the maps are homotopic. Each is a composition of $k : ([0, 3], \{0, 3\}) \to (X, x)$ with an order-preserving linear homeomorphism $m : [0, 1] \to [0, 3]$. Here

$$k(s) = \begin{cases} f(s) & \text{if } 0 \le s \le 1, \\ g(s - 1) & \text{if } 1 \le s \le 2, \\ h(s - 2) & \text{if } 2 \le s \le 3. \end{cases}$$

Now $f * (g * h) = km_1$ where the map m_1 is the piecewise affine linear map sending $[0, \frac{1}{2}]$ to $[0, 1]$, $[\frac{1}{2}, \frac{3}{4}]$ to $[1, 2]$, and $[\frac{3}{4}, 1]$ to $[2, 3]$. It is given by the formula

$$m_1(s) = \begin{cases} 2s & \text{if } 0 \le s \le \frac{1}{2}, \\ 4s - 1 & \text{if } \frac{1}{2} \le s \le 1. \end{cases}$$

Similarly, $(f * g) * h$ is given by km_2 where

$$m_2(s) = \begin{cases} 4s & \text{if } 0 \le s \le \frac{1}{2}, \\ 2s + 1 & \text{if } \frac{1}{2} \le s \le 1. \end{cases}$$

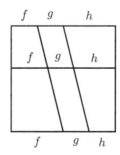

Figure 3.6. Associativity of $*$ up to homotopy.

The argument used in Chapter 2 to show that self-homeomorphisms of the interval that preserve order are isotopic can be used here to show that any two order-preserving homeomorphisms from $[0, 1]$ to $[0, 3]$ are homotopic, fixing the image of the end points during the homotopy. The homotopy between m_1 and m_2 is given by the formula $M(s, t) = (1 - t)m_1(s) + tm_2(s)$. This is called a *straight line homotopy* since for fixed s, we have $M_t(s) = M(s, t)$ moving along the straight line joining $m_1(s)$ to $m_2(s)$ in a linear fashion. Now a homotopy between $f * (g * h) = km_1$ and $(f * g) * h = km_2$ is given by $H(s, t) = kM_t(s)$: we just compose the homotopy corresponding to the two parametrizations with k. Because of this fact, we could divide up the interval any way we want in forming $f * g * h$ and get the same result up to homotopy; in particular, we could use f on $[0, \frac{1}{3}]$, g on $[\frac{1}{3}, \frac{2}{3}]$, and h on $[\frac{2}{3}, 1]$ if we wanted to.

This completes the verification that $\pi_1(X, x)$ is a group. We now look at the functorial properties of the correspondence $(X, x) \to \pi_1(X, x)$. Suppose $f : (X, x) \to (Y, y)$ is a continuous map. Let $p : (I, \{0, 1\}) \to (X, x)$ represent $[p] \in \pi_1(X, x)$. Then define the homomorphism $F(f) = f_* : \pi_1(X, x) \to \pi_1(Y, y)$ by $f_*([p]) = [fp] \in \pi_1(Y, y)$. To see that this is well defined, we must show that if $p \sim q$, then $fp \sim fq$. But, if $P : I \times I \to X$ is a homotopy between p and q with $P(s, 0) = p(s), P(s, 1) = q(s), P(0, t) = x = P(1, t)$, then we can define a homotopy between fp and fq by fP. For $fP(s, 0) = fp(s), fP(s, 1) = fq(s), fp(1, t) = fp(0, t) = f(x) = y$. Note that f_* is a homomorphism. For

$$f_*([p] \bar{*} [q]) = [f(p * q)] = [fp * fq] = [fp] \bar{*} [fq] = f_*[p] \bar{*} f_*[q].$$

We next check that $F(1_X) = 1_{F(X)} = 1_Y$ and $F(gf) = F(g)F(f)$. First, $(1_X)_*[f] = [1_X f] = [f]$. Also, $(gf)_*[p] = [(gf)p] = [g(fp)] = g_*[fp] = g_* f_*[p]$. Thus the association of $\pi_1(X, x)$ to (X, x) and $F(f) = f_*$ to f is a functor between (pointed) topological spaces with continuous maps and groups with homomorphisms. Thus homeomorphic spaces will have isomorphic fundamental groups (with corresponding base points). We depict the functorial properties of F by the following diagram:

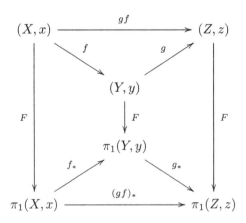

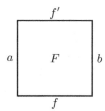

Figure 3.7. $f \sim f'$ rel 0,1.

The role of the base point will be pursued later as a project. Informally, the main result is that the fundamental group of a path-connected space is independent of the choice of base point up to isomorphism, and the isomorphism itself between $\pi_1(X, x_1)$ and $\pi_1(X, x_2)$ can be chosen to depend on a path connecting x_1 to x_2.

We end this section with a calculation. A path-connected space X with $\pi_1(X, x) \simeq \{e\}$ is called *simply connected*. The next proposition says that $\mathbb{R}^n$ and D^n are simply connected.

Proposition 3.2.1. $\pi_1(\mathbb{R}^n, \; \boldsymbol{x}) \simeq \{e\}$. *If D^n denotes the unit disk in $\mathbb{R}^n$, then* $\pi_1(D^n, \boldsymbol{x}) \simeq \{e\}$.

Proof. In either case, we have to show that any loop at $\boldsymbol{x}$ in the space is homotopic within the space to the constant loop at $\boldsymbol{x}$. The idea is just to contract the loop via radial lines to $\boldsymbol{x}$ using a straight line homotopy. The homotopy is given by $F(s,t) = (1-t)f(s) + t\boldsymbol{x}$. Then $F(s,0) = f(s), F(s,1) = \boldsymbol{x} = e_{\boldsymbol{x}}(s)$, and $F(0,t) = \boldsymbol{x} = F(1,t)$ since $f(0) = f(1) = \boldsymbol{x}$. Note that if $f(s) \in D^n$, then $F(s,t) \in D^n$ as well. $\qquad\square$

We close this section by noting that many of the constructions involved in forming the fundamental group apply to paths as well as loops. For example, if f, g are paths with $f(0) = a, f(1) = b, g(0) = b, g(1) = c$, then we can form a product $f * g$ using the same formula as before. For paths $f, f' : (I, 0, 1) \to (X, a, b)$, we say that f is homotopic to f' rel 0,1, written as $f \sim f'$ rel 0,1, if there is a continuous map $F : I \times I \to X$ so that $F(x, 0) = f(s)$, $F(s, 1) = f'(s)$, $F(0, t) = a$, $F(1, t) = b$. We illustrate such a homotopy with Figure 3.7. Then $f \sim f'$ rel 0,1, $g \sim g'$ rel 0,1 imply $f * g \sim f' * g'$ rel 0,1. There are also corresponding statements about associativity, $(f * g) * h \sim f * (g * h)$ rel 0,1 when these are defined.

3.3   The fundamental group of the circle

In this section we will compute $\pi_1(S^1, \mathbf{1})$, where $\mathbf{1} = (1, 0)$. Fundamental for this computation will be the map $p : \mathbb{R} \to S^1$, $p(x) = (\cos 2\pi x, \sin 2\pi x)$. Recall from Section 1.7 that p is locally a homeomorphism; that is, given $x \in \mathbb{R}$, there is an interval I_x containing x (any interval of length less than 1 will do) so

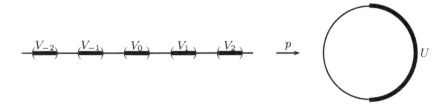

Figure 3.8. Covering of neighborhood for $p : \mathbb{R} \to S^1$.

that $p \mid I_x$ is a homeomorphism onto its image. Moreover, if $U = p(I_x)$, then $p^{-1}(U) = \bigsqcup_{n \in \mathbb{Z}} V_n$, where $V_0 = I_x$ and V_n comes from V_0 by translating it by n. Each V_i is mapped homeomorphically to U by p and they are permuted by the family of translations $T_n(x) = x + n$ (see Figure 3.8).

Recall also that, if we put the equivalence relation $\sim$ on $\mathbb{R}$, $a \sim b$ iff $a - b \in \mathbb{Z}$, then p induces a homeomorphism $\bar{p}$ between $\mathbb{R}/\!\sim$ and S^1. The point $\mathbf{1}$ corresponds to $[0]$ under this homeomorphism. We divide the circle into two open sets A and B, where $A = S^1\backslash\{-1\}, B = S^1\backslash\{\mathbf{1}\}$. Then $p^{-1}(A) = \mathbb{R}\backslash\{n + \frac{1}{2} : n \in \mathbb{Z}\} = \bigcup_{n \in \mathbb{Z}}(n - \frac{1}{2}, n + \frac{1}{2})$ and $p^{-1}(B) = \mathbb{R}\backslash\{n : n \in \mathbb{Z}\} = \bigcup_{n \in \mathbb{Z}}(n - 1, n)$. Note that $p \mid (n - \frac{1}{2}, n + \frac{1}{2})$ is a homeomorphism onto A for each n, and $p \mid (n - 1, n)$ is a homeomorphism onto B for each n. We call these homeomorphisms $p_{n,1/2}$ and p_n and denote their inverses by $q_{n,1/2}$ and q_n. Note that if $f : X \to S^1$ is a continuous map and $f(X) \subset A$, then $q_{n,1/2}f$ has the property that $pq_{n,1/2}f = p_{n,1/2}q_{n,1/2}f = f$. A continuous map $\tilde{f} : X \to \mathbb{R}$ so that $p\tilde{f} = f$ is called a *lifting* of f. We depict the lifting $\tilde{f}$ of f via the following diagram:

We have seen that if $f(X) \subset A$, then f has a lifting. Of course, f has many liftings, one for each integer n. If, however, we specify that $\tilde{f}(x_0) \in (n_0 - \frac{1}{2}, n_0 + \frac{1}{2})$ and X is connected, then we claim that there is a unique lifting $\tilde{f}$ for f. For $\tilde{f}$ continuous and X connected we have that $\tilde{f}(X)$ is connected. Since $f(X) \subset A, \tilde{f}(X) \subset \bigcup_{n \in \mathbb{Z}}(n - \frac{1}{2}, n + \frac{1}{2})$. Since $\tilde{f}(X)$ is connected and $\tilde{f}(x_0) \in (n_0 - \frac{1}{2}, n_0 + \frac{1}{2})$, this implies that $\tilde{f}(X) \subset (n_0 - \frac{1}{2}, n_0 + \frac{1}{2})$; the details are left as an exercise. Now $q_{n_0,1/2}f = \tilde{f}$ gives a lifting of f. If $\tilde{f}'$ were another lifting with $f'(x_0) \in (n_0 - \frac{1}{2}, n_0 + \frac{1}{2})$, then $p\tilde{f}'(x) = p\tilde{f}(x)$ and $p \mid (n_0 - \frac{1}{2}, n_0 + \frac{1}{2})$ a homeomorphism imply $\tilde{f}'(x) = \tilde{f}(x)$. Analogously, if $f(X) \subset B$, then $q_n f$ gives a lifting for each n, and there is a unique lifting $\tilde{f} = q_{n_0}f$ with $\tilde{f}(x_0) \in (n_0 - 1, n_0)$ when X is connected.

We can find liftings of continuous maps into A or B, and the lifting is unique once the image of a single point is specified. We now discuss the general problem of finding a lifting of $f : (I, \{0, 1\}) \to (S^1, 1)$. We note that I is a compact metric space and $\{A, B\}$ gives an open cover for S^1. Hence $\{f^{-1}(A), f^{-1}(B)\}$ gives an open cover for I. Since I is compact, this cover has a Lebesgue number $\delta > 0$. Thus any set of diameter less than δ has its image under f contained in either A or B. Choose an integer k with $1/k < \delta$, and divide the interval $I = [0, 1]$ into k subintervals of length $1/k, I = I_1 \cup \cdots \cup I_k, I_j = [(j - 1)/k, j/k]$. By our choice of k, $f(I_j) \subset A$ or $f(I_j) \subset B$. Thus we can find a lifting of $f \mid I_j$; moreover, this lifting is unique if we specify the image of one point, $\tilde{f}((j - 1)/k)$. We now claim that $f : I \to S^1$ has a unique lifting $\tilde{f}$ with $\tilde{f}(0) = 0$. We first show that how to get $\tilde{f}$. Look at $f \mid I_1$. Since $f(0) \notin B$, we must have $f(I_1) \subset A$; thus there is a unique lifting $\widetilde{(f \mid I_1)} : I_1 \to \mathbb{R}$ with $\widetilde{(f \mid I_1)}(0) = 0$. We define $\tilde{f} \mid I_1$ as $\widetilde{(f \mid I_1)}$. In particular, this defines $\tilde{f}$ on $1/k$. Now look at $f \mid I_2$. Since $f(I_2) \subset A$ or $f(I_2) \subset B$, there is a unique lifting $\widetilde{(f \mid I_2)}$ with $\widetilde{(f \mid I_2)}(1/k) = \tilde{f}(1/k)$. We then define $\tilde{f} \mid I_2 = \widetilde{(f \mid I_2)}$. We can continue in this fashion to define a unique lifting of f over successive subintervals, yielding the unique lifting $\tilde{f}$. Uniqueness follows inductively from the uniqueness properties of each of the liftings $\widetilde{(f \mid I_j)}$. We leave the details as an exercise.

Exercise 3.3.1. Let $g : X \to \cup J_n$ be continuous, where $\{J_n\}$ are pairwise disjoint open intervals, $n \in \mathbb{Z}$, and X is connected. Show that if $g(x_0) \in J_{n_0}$, then $g(X) \subset J_{n_0}$.

Exercise 3.3.2. Modify the argument outlined above to prove the unique path lifting property, Theorem 3.3.1.

Theorem 3.3.1 (Unique path lifting property). *If $f : I \to S^1$ is a continuous map with $f(0) = x_0$ and $p(\tilde{x}_0) = x_0$, then there is a unique lifting $\tilde{f} : I \to \mathbb{R}$ (i.e. $\tilde{f}$ is continuous and $p\tilde{f} = f$) with $\tilde{f}(0) = \tilde{x}_0$.*

We now want to use the unique path lifting property to define a homomorphism $\bar{h} : \pi_1(S^1, 1) \to \mathbb{Z}$, which we will show is an isomorphism. Since $\pi_1(S^1, 1)$ consists of equivalence classes of loops, we will actually define a function h from loops at 1 to $\mathbb{Z}$ and show that h gives the same value to homotopic loops, hence determining a function $\bar{h} : \pi_1(S^1, 1) \to \mathbb{Z}$. We then show that $\bar{h}$ is a homomorphism, is 1–1, and is onto, and so is an isomorphism.

Suppose $f : (I, \{0, 1\}) \to (S^1, 1)$ is continuous. By Theorem 3.3.1 there is a unique lifting $\tilde{f} : I \to \mathbb{R}$ with $\tilde{f}(0) = 0$. We define $h(f) = \tilde{f}(1) \in \mathbb{Z}$. $\tilde{f}(1)$ is an integer since $p\tilde{f}(1) = f(1) = 1$ and $p^{-1}(\{1\}) = \mathbb{Z} \subset \mathbb{R}$. We then define $\bar{h} : \pi_1(S^1, 1) \to \mathbb{Z}$ by $\bar{h}([f]) = h(f)$.

Well defined: To see that $\bar{h}$ is well defined, we have to show that if $f \sim f'$, then $h(f) = h(f')$; that is, $f \sim f'$ implies $\tilde{f}(1) = \tilde{f}'(1)$. Denote a homotopy between f and f' by $F : I \times I \to S^1$ with $F(s, 0) = f$, $F(s, 1) = f'$, $F(0, t) = 1 = F(1, t)$. We claim that there is a unique lifting $\tilde{F}$ of F with $\tilde{F}(0, 0) = 0$, for $\{F^{-1}(A), F^{-1}(B)\}$ is an open cover of the compact metric space $I \times I$, and hence has a Lebesgue number $\delta > 0$. Choose n so that any square of side length $1/n$ has diameter less than δ. Now subdivide $I \times I$ into n^2 squares $I_j \times I_k$ of side length $1/n, 1 \leq j, k \leq n$. Hence $F(I_j \times I_k) \subset A$ or $F(I_j \times I_k) \subset B$. Now consider

$F \mid I_1 \times I_1$. Since $F(0,0) \notin B$, we must have $F(I_1 \times I_1) \subset A$, and so $q_{0,1/2}(F \mid I_1 \times I_1) = \widetilde{F \mid I_1 \times I_1} = \tilde{F}_{11}$ gives a lifting of $F_{11} = F \mid I_1 \times I_1$ with $\tilde{F}_{11}(0,0) = 0$. Moreover, Exercise 3.3.1 may be used as in the discussion preceding it to show that $\tilde{F}_{11}$ is the unique lifting of F_{11} sending $(0,0)$ to 0. Next consider $F_{21} = F \mid I_2 \times I_1$ (in general, let $F_{jk} = F \mid I_j \times I_k$). Since $F_{21}(I_2 \times I_1) \subset A$ or $F_{21}(I_2 \times I_1) \subset B$, there is a unique lifting $\tilde{F}_{21} : I_2 \times I_1 \to \mathbb{R}$ with $\tilde{F}_{21}(1/n,0) = \tilde{F}_{11}(1/n,0)$. We may inductively define $\tilde{F}_{11}, \tilde{F}_{21}, \ldots, \tilde{F}_{n1}, \tilde{F}_{12}, \ldots, \tilde{F}_{n2}, \ldots, \tilde{F}_{1n}, \ldots, \tilde{F}_{nn}$ with $\tilde{F}_{jk}(j/n, (k-1)/n) = \tilde{F}_{(j+1)k}(j/n, (k-1)/n)$ and $\tilde{F}_{1k}(0, k/n) = \tilde{F}_{1(k+1)}(0, k/n)$. That any two of these agree on their common interval of intersection follows from the fact that they agree at one point and the unique path lifting property. Call the map that they define $\tilde{F}$; that is, $\tilde{F} \mid I_j \times I_k = \tilde{F}_{jk}$. It is continuous by the piecing lemma. $F \mid \{0\} \times I$ maps $0 \times I$ to 1 and $\tilde{F}(0,0) = 0$; hence unique path lifting shows that $\tilde{F}(0,t) = 0$ since this is one lifting of $F \mid 0 \times I$. Unique path lifting also implies $F \mid I \times \{0\} = \tilde{f}$ and $F \mid I \times \{1\} = \tilde{f}'$. Finally, $\tilde{F} \mid \{1\} \times I$ is a lifting of $F \mid \{1\} \times I$, which sends $1 \times I$ to 1. Since $\tilde{F}(1,0) = \tilde{f}(1)$, unique path lifting implies that $\tilde{F}(1,t) = \tilde{f}(1)$. But these statements imply that $\tilde{F}(1,1) = \tilde{f}'(1) = \tilde{f}(1)$. Thus h is well defined. We illustrate how the map $\tilde{F}$ is defined when $n = 3$ in Figure 3.9.

Exercise 3.3.3. Use the argument given above to prove the homotopy lifting theorem 3.3.2.

Theorem 3.3.2 (Homotopy lifting theorem). *Suppose* $F : I \times I \to S^1$ *is a continuous map. Then there is a continuous lifting* $\tilde{F} : I \times I \to \mathbb{R}$ *satisfying* $p\tilde{F} = F$. *Moreover, this is unique if we also require* $\tilde{F}(0,0) = \tilde{x}_0$, *where* $p(\tilde{x}_0) = F(0,0)$.

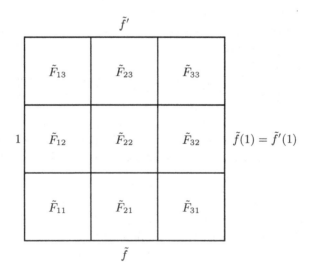

Figure 3.9. Lifting a homotopy.

Homomorphism: We now show that $\bar{h}$ is a homomorphism. Suppose $\bar{h}([f]) = m, \bar{h}([g]) = n$. Then we have to show that $\bar{h}([f]\bar{*}[g]) = m + n$. But $\bar{h}([f]\bar{*}[g]) = \bar{h}([f * g]) = h(f * g)$. Thus we need to find a lifting $\widetilde{f * g}$ of $f * g$ with $\widetilde{f * g}(0) = 0$ and evaluate $\widetilde{f * g}(1)$. But $\tilde{f}$ and $\tilde{g}$ (after we translate $\tilde{g}$ to begin at $\tilde{f}(1) = m$) together give a lifting. Define $\tilde{g}_m(t) = \tilde{g}(t) + m$. Then

$$\widetilde{f * g}(t) = \begin{cases} \tilde{f}(2t) & \text{if } 0 \le t \le \frac{1}{2}, \\ \tilde{g}_m(2t - 1) & \text{if } \frac{1}{2} \le t \le 1 \end{cases}$$

is the required lifting of $f * g$. Note that $\widetilde{f * g}(1) = \tilde{g}_m(1) = \tilde{g}(1) + m = m + n$.

Exercise 3.3.4. Check that $\widetilde{f * g}$ as defined above is a lifting of $f * g$ with $\widetilde{f * g}(0) = 0$.

Onto: Next we have to show that $\bar{h}$ is onto. We do this by exhibiting $[f] \in \pi_1(S^1, 1)$ with $\bar{h}([f]) = n$. We first define $\tilde{f}$ as $\tilde{f}(t) = nt$. Then we define $f = p\tilde{f}$. Since $\tilde{f}$ is a lifting of f and $\tilde{f}(0) = 0, \tilde{f}(1) = n$, then $\bar{h}([f]) = \tilde{f}(1) = n$ as required. Note that f just wraps around the circle $|n|$ times (in the counterclockwise direction for positive n, clockwise for negative n). It is given by the formula $f(t) = (\cos 2\pi nt, \sin 2\pi nt)$.

1–1: By Proposition 3.1.2 we only have to show that if $\bar{h}([f]) = 0$, then $[f] = [e_1]$. But $\bar{h}([f]) = 0$ means that $\tilde{f}(1) = 0$, where $\tilde{f}$ is the unique path lifting of f with $\tilde{f}(0) = 0$. Hence $\tilde{f} : (I, \{0, 1\}) \to (\mathbb{R}, 0)$ and thus represents an element of $\pi_1(\mathbb{R}, 0) \simeq \{e\}$. Thus $[\tilde{f}] = [e_0]$ and $p_*([\tilde{f}]) = p_*[e_0] = [e_1]$ since a homomorphism must send the identity to the identity. Thus $[f] = [p\tilde{f}] = p_*([f]) = [e_1]$.

We have proved the following theorem.

Theorem 3.3.3. $\pi_1(S^1, 1) \simeq \mathbb{Z}$. *This isomorphism is given by the function* $\bar{h} : \pi_1(S^1, 1) \to \mathbb{Z}$ *which sends* $[f]$ *to* $\tilde{f}(1)$ *where* $\tilde{f}$ *is the unique lifting of* f *with* $\tilde{f}(0) = 0$.

Much of the remainder of this chapter will consist of applications of this computation and its underlying ideas.

Exercise 3.3.5. Show that if f, f' are homotopic loops (rel 0,1) at 1, and $\tilde{f}, \tilde{f}'$ are the unique lifts with $\tilde{f}(0) = \tilde{f}'(0) = 0$, then there is $G : I \times I \to \mathbb{R}, G(s, t) = G_t(s)$, with $G_0 = \tilde{f}, G_1 = \tilde{f}', G(0, t) = 0, G(1, t) = \tilde{f}(1) = \tilde{f}'(1)$. (Hint: Examine the proof of Theorem 3.3.3.)

Exercise 3.3.6. Define a new multiplication on loops at 1 in S^1 by $f \circ g(s) = p(\tilde{f}(s) + \tilde{g}(s))$ where $\tilde{f}, \tilde{g}$ are the unique lifts of f, g with $\tilde{f}(0) = 0 = \tilde{g}(0)$. Show that this multiplication is well defined on homotopy classes, that is, $f \sim f', g \sim g'$ implies $f \circ g \sim f' \circ g'$. (Hint: Use Exercise 3.3.5.)

Exercise 3.3.7. With the definition of the last exercise, show that the operation $[f] \bar{\circ} [g] = [f \circ g]$ makes the set of homotopy classes of loops at 1 in S^1 into a group, which we will denote by $\pi'_1(S^1, 1)$. (Hint: Utilize the group structure in $\mathbb{R}$: for example, to define the inverse of $[f]$, we want a $[g]$ with $p(\tilde{f}(s) + \tilde{g}(s))$ homotopic to the constant map at $\mathbf{1}$. But, if $\tilde{g}(s) = -\tilde{f}(s)$, then $p(\tilde{f}(s) + \tilde{g}(s)) = p(0) = \mathbf{1}$. So choose $g(s) = p(-\tilde{f}(s))$.)

Exercise 3.3.8. Using the notation of the preceding exercises, show that $\pi_1(S^1, \mathbf{1}) \simeq \pi'_1(S^1, \mathbf{1})$. (Hint: As sets, they are the same, so try to use the identity map and show that it is a homomorphism using the two group structures. It is useful, although not necessary, to use Theorem 3.3.3.)

3.4   Applications to surfaces

In Chapter 2 we used the Euler characteristic to give an argument that the surfaces $S_{(p)}, T^{(k)}_{(p)}$, and $P^{(k)}_{(p)}$ are pairwise nonhomeomorphic for distinct (k, p). However, the justification of the invariance of Euler characteristic for homeomorphic surfaces was not given there. Distinguishing these surfaces can be based instead on the computations of their fundamental groups. The general case is deferred to later in the chapter when we discuss the Seifert–van Kampen theorem which facilitates more efficient calculation. However, we will do some of the easier calculations in this section to at least show that S, T, and P have distinct fundamental groups.

The easiest case to handle is the computation of the fundamental group of the torus. For this, we regard T as $S^1 \times S^1$ and we choose $(\mathbf{1}, \mathbf{1})$ as base point. The result $\pi_1(T, (\mathbf{1}, \mathbf{1})) \simeq \mathbb{Z} \oplus \mathbb{Z}$ follows from the following proposition on the fundamental group of a product space.

Theorem 3.4.1. $\pi_1(X \times Y, (x, y)) \simeq \pi_1(X, x) \oplus \pi_1(Y, y)$.

Proof. Let $p_1 : X \times Y \to X$, $p_2 : X \times Y \to Y$ be the projections. Then note that $(p_1)_* : \pi_1(X \times Y, (x, y)) \to \pi_1(X, x)$ and $(p_2)_* : \pi_1(X \times Y, (x, y)) \to \pi_1(Y, y)$ are homomorphisms. Define a homomorphism $P : \pi_1(X \times Y, (x, y)) \to \pi_1(X, x) \oplus \pi_1(Y, y)$ by $P(\alpha) = ((p_1)_*(\alpha), (p_2)_*(\alpha))$. It is a homomorphism since $(p_1)_*$ and $(p_2)_*$ are and multiplication in a direct sum is done coordinatewise. Define a homomorphism $Q : \pi_1(X, x) \oplus \pi_1(Y, y) \to \pi_1(X \times Y, (x, y))$ by $Q([f], [g]) = [h]$, where $h(t) = (f(t), g(t))$. We leave it to the reader to check that Q is a homomorphism and Q is an inverse to P. $\qquad\square$

Exercise 3.4.1. Show that Q is a homomorphism and is an inverse for P.

We now compute $\pi_1(S^2, \mathbf{1})$, where $\mathbf{1}$ denotes $(1, 0, 0)$ here. We show that $\pi_1(S^2, \mathbf{1})$ is the trivial group; that is, each loop at $\mathbf{1}$ is homotopic to the constant loop. The idea of the proof is that $S^2 \backslash \{-\mathbf{1}\} \simeq \mathbb{R}^2$, and this allows us to homotope any loop that misses $\{-\mathbf{1}\}$ (via the image of a straight line homotopy in $\mathbb{R}^2$) to the constant loop. The argument reduces to this situation by first homotoping the loop in S^2, which may hit $-\mathbf{1} = (-1, 0, 0)$, to a loop which misses this point.

Our argument will be similar to the argument computing $\pi_1(S^1, \mathbf{1})$. We first write S^2 as $A \cup B$, where $A = S^2 \backslash \{\mathbf{1}\}$ and $B = S^2 \backslash \{-\mathbf{1}\}$. We saw in Chapter 2 that $A, B \simeq \mathbb{R}^2$ (via rotation and stereographic projection). Suppose $f : (I, \{0, 1\}) \to (S^2, \mathbf{1})$ is a loop at $\mathbf{1}$. Our goal is to homotope f to f' so that $f'(I) \subset B$. Since B is homeomorphic to $\mathbb{R}^2$ and $\pi_1(\mathbb{R}^2, c) \simeq \{e\}$, then f' is homotopic to the constant loop. First note that if $g_0, g_1 : [a, b] \to \mathbb{R}^2$ are paths with $g_i(a) = x, g_i(b) = y$, then g_0 is homotopic to g_1 *relative to the end points*; that is, there is a homotopy $F : [a, b] \times I \to \mathbb{R}^2$ with $F(a, t) = x, F(b, t) = y, F(s, 0) = g_0(s), F(s, 1) = g_1(s)$. Just use the straight line homotopy; the details are left as an exercise. In particular, any g is homotopic relative to the end points to the straight line path joining $g(a)$ to $g(b)$. Next note that if we take a loop $g : [0, 1] \to X$ with $[a, b] \subset [0, 1]$ and if $g \mid [a, b]$ is homotopic relative to the end points to $g' \mid [a, b]$, then g is homotopic to g'' with $g'' \mid [a, b] = g' \mid [a, b]$ and $g'' \mid [0, 1]\backslash(a, b) = g \mid [0, 1]\backslash(a, b)$. The idea is to use the constant homotopy on $[0, 1]\backslash(a, b)$ and the homotopy between $g \mid [a, b]$ and $g' \mid [a, b]$ on $[a, b]$; the details are left as an exercise.

Now consider the cover $f^{-1}(A), f^{-1}(B)$ of I. Let δ be a Lebesgue number of this cover and choose n so that $1/n < \delta$. Subdivide $[0, 1]$ into n equal subintervals with $1/n < \delta$, so that $f([k/n, (k + 1)/n])$ is a subset of either A or B. Let $m(f)$ be the number of subintervals containing a point x with $f(x) = -1$. If $m(f) = 0$, then $f(I) \subset B$ as desired, and so f is homotopic to the constant loop. Our proof is by induction on the number $m(f)$. Since we know the result for $m(f) = 0$, we have started our inductive proof. We have to give a means of homotoping the given f to f' so that $m(f') < m(f)$, for then knowing our statement for $m(f) \le p$ implies it for $m(f) \le p + 1$. Note that the first and last subintervals are sent to B (since $f(0) = f(1) = 1$), so they contain no points sent to -1. Note also that any subinterval containing a point sent to -1 is sent entirely to A. Now, moving from left to right, select the first subinterval $[a, a + 1/n]$ containing a point sent to -1, and let $[a, b]$ denote the interval formed from $[a, a + 1/n]$ together with all the consecutive subintervals containing points sent to -1. Note that $f(a) \ne -1$, $f(b) \ne -1$: if $f(a) = -1$, then $[a, a + 1/n]$ would not be the first subinterval containing a point sent to -1; if $f(b) = -1$, then we could have included at least one more subinterval in our consecutive subintervals containing points sent to -1. Note that $f[a, b] \subset A$. Using the fact that A is homeomorphic to $\mathbb{R}^2$, we can show that $f \mid [a, b]$ is homotopic to $f'[a, b]$ relative to the end points, where $f(a) = f'(a), f(b) = f'(b)$, and $f'([a, b]) \subset B$; that is, the image of this interval misses the point -1. Then f is homotopic to f'' with $f'' \mid [a, b] = f' \mid [a, b]$ and $f'' \mid [0, 1]\backslash(a, b) = f \mid [0, 1]\backslash(a, b)$. But $m(f'') < m(f)$ and so by induction f'' (hence f) is homotopic to a loop in B as required.

Exercise 3.4.2. Show that if $g_0, g_1 : [a, b] \to \mathbb{R}^2$ are maps with $g_0(a) = g_1(a), g_0(b) = g_1(b)$, then g_0 is homotopic to g_1 relative to the end points.

Exercise 3.4.3. Using the notation of the previous proof, show that if $g : [a, b] \to A$ has $g(a), g(b) \ne -1$, then g is homotopic to g' relative to the end points, where $g'([a, b]) \subset B$. (Hint: Just choose g' with $g'([a, b]) \subset A \cap B$ and show that there is a homotopy since $A \simeq \mathbb{R}^2$.)

Exercise 3.4.4. Show that if $g \mid [a, b]$ is homotopic to $g' \mid [a, b]$ relative to the end points (where $g : [0, 1] \to X$ is a loop at x), then g is homotopic (as loops) to g'' where $g'' \mid [a, b] = g' \mid [a, b]$ and $g'' \mid [0, 1]\backslash(a, b) = g \mid [0, 1]\backslash(a, b)$.

We give another approach to this calculation of $\pi_1(S^2, \mathbf{1})$. As before, start with $f : (I, \{0, 1\}) \to (S^2, \mathbf{1})$ and find a Lebesgue number δ for the cover $\{f^{-1}(A), f^{-1}(B)\}$. Partition the interval into subintervals of length $1/n < \delta$. If we consider the path $f_i = f \mid [i-1, i]$, then we can consider f as the product of the paths $f = f_1 * f_2 * \cdots * f_n$. This is a product of paths, not of loops. Now amalgamate the subintervals which are sent to A and the subintervals which are sent to B so that we can regard $f = g_1 * h_1 * g_2 * h_2 * \cdots * g_{k+1}$, where g_i is sent to B and h_i is sent to A. The first and last subintervals are sent to B since the end points are sent to $\mathbf{1} \in B$. We alternately name the interior vertices $v_1, w_1, v_2, w_2, \ldots$. Now each interior vertex v_i of the subdivision of I will be sent to a point v_i' of $A \cap B$ and the interior vertex w_i is sent to w_i' of $A \cap B$. Since $A \cap B$ is path connected, we can choose a path p_i in $A \cap B$ from v_i' to $\mathbf{1}$ and a path q_i in $A \cap B$ from w_i' to $\mathbf{1}$. Note that the paths $p_i * \bar{p}_i$ and $q_i * \bar{q}_i$ are homotopic relative to the end points to the constant paths c_i, d_i that stay at v_i', w_i'. Also, the compositions $\alpha * c_i, \alpha * d_i$ are homotopic to α for any path α where this composition makes sense. We can use these homotopies to homotope f to

$$g_1 * c_1 * h_1 * d_1 * \cdots * h_k * d_k * g_{k+1} \sim g_1 * p_1 * \bar{p}_1 * h_1 * q_1 * \cdots * \bar{q}_k * g_{k+1}.$$

Now let $g_1' = g_1 * p_1$, $h_i' = \bar{p}_i * h_i * q_i$, $g_i' = \bar{q}_{i-1} * g_i * p_i$, $i \neq 1, k+1$, $g_{k+1}' = \bar{q}_k * g_{k+1}$. Then f is homotopic relative to the end points to the composition $g_1' * h_1' * \cdots * h_k' * g_{k+1}'$. The advantage of this new composition is that each of g_i', h_i' is a loop at $\mathbf{1}$. The loops g_i' are loops in B, and the loops h_i' are loops in A. Since each of A, B is homeomorphic to $\mathbb{R}^2$, then these loops are homotopic relative to the end points to constant loops at $\mathbf{1}$. Thus f is homotopic to the constant loop at $\mathbf{1}$. Since f was an arbitrary loop at $\mathbf{1}$, this shows that $\pi_1(S^2, \mathbf{1})$ is the trivial group.

The argument above can be used to prove the following theorem. We leave the details as an exercise.

Theorem 3.4.2. *Suppose that A, B, and $A \cap B$ are path connected open sets in $X = A \cup B$ and $c \in A \cap B$. Then if $\pi_1(A, c)$ and $\pi_1(B, c)$ are trivial groups, then $\pi_1(X, c)$ is also the trivial group.*

Exercise 3.4.5. Prove Theorem 3.4.2.

We now show that $\pi_1(P, [\mathbf{1}]) \simeq \mathbb{Z}_2$. We will just outline the argument and leave the details as an exercise. We use the description of $\mathbb{Z}_2$ as $\{-1, 1\}$ with the usual multiplication in $\mathbb{R}$. Recall the map $p : S^2 \to P$ given by regarding P as the quotient space of S^2, where $\boldsymbol{x} \sim -\boldsymbol{x}$. For a small open set V about $\boldsymbol{x} \in S^2$, we have $V \cap T(V) = \phi$, where $T(\boldsymbol{x}) = -\boldsymbol{x}$. Note that $pT = p$. Then the two sets $V, T(V)$ will each map homeomorphically via p to an open set U about $[\boldsymbol{x}] \in P$, and the map T interchanges these two preimages $p^{-1}(U) = V \cup T(V)$. All that is necessary for U is that it has small enough diameter that it is contained in an open hemisphere. We will choose U so that it is path connected. We can then use compactness of P to get a cover $U_1, \ldots, U_n$ of P of open sets with $p^{-1}(U_i) = U_{i1} \cup U_{i2}$, where $p \mid U_{ij} : U_{ij} \to U_i$ is a homeomorphism, $T(U_{i1}) = U_{i2}$. Now suppose $[f] \in \pi_1(P, [\mathbf{1}])$. Then $\{(f^{-1}(U_i)\}$ gives an open cover of I, and so has a Lebesgue number $\delta > 0$. Choose k so that $1/k < \delta$ and subdivide I into k equal subintervals of length $1/k$, $I = I_1 \cup \cdots \cup I_k$. Then $f(I_m)$ always lies in some U_i. Thus $f \mid I_m$ lifts to S^2; that is, there is a map $(p \mid U_{ij})^{-1} f = q_{ij} f = \tilde{f}$ so that $p\tilde{f} = f$. Of course, there are two possibilities for j, but we can argue as in the computation of the fundamental group of the circle that if we specify $\tilde{f}$ on one point, then the lifting is unique. We can then piece together the liftings as in the proof that $\pi_1(S^1, 1) \simeq \mathbb{Z}$ to show that there is a unique lifting $\tilde{f}$ of f with $\tilde{f}(0) = 1$. Now define $\bar{h} : \pi_1(P, [1]) \to \mathbb{Z}_2 = \{-1, 1\}$ by

$$\bar{h}([f]) = \begin{cases} -1 & \text{if } \tilde{f}(1) = -1, \\ 1 & \text{if } \tilde{f}(1) = 1. \end{cases}$$

Seeing that $\bar{h}$ is well defined requires an argument similar to the one used in showing $\pi_1(S^1, 1) \simeq \mathbb{Z}$. If $f \sim f'$, then let $F : I \times I \to P$ be a homotopy. Then F can be lifted to $\tilde{F} : I \times I \to S^2$ with $\tilde{F}_0 = \tilde{f}, \tilde{F}_1 = \tilde{f}', \tilde{F}_0(0, t) = 1$. Then the unique path-lifting property (as applied to a lifting to S^2 using the unique liftings of maps to U_i once one point is specified) is applied to show that $\tilde{F}(1, t) = \tilde{f}(1) = \tilde{f}'(1)$.

To see that $\bar{h}$ is a homomorphism, we need to piece together lifts $\tilde{f}, \tilde{g}$ of f, g to get a lift of $f * g$. If $\tilde{f}(1) = 1$, then $\tilde{f} * \tilde{g}$ will be such a lift. However if $f(1) = -1$, then $\tilde{f}$ and $\tilde{g}$ do not fit together since $\tilde{f}$ ends at -1 and $\tilde{g}$ begins at 1. In our proof that $\pi_1(S^1, 1) \simeq \mathbb{Z}$, a similar problem was surmounted by translating $\tilde{g}$; that is, replacing it by $\tilde{g}_n(t) = \tilde{g}(t) + n$. The important fact about $\tilde{g}_n$ used was that $\tilde{g}_n(0) = n$ and $p\tilde{g}_n(t) = p\tilde{g}(t) = g(t)$. The map $\tilde{g}_n$ is formed from $\tilde{g}$ by composing with $T_n : \mathbb{R} \to \mathbb{R}, T_n(s) = s + n$. Since $pT_n = p$, $p(T_n\tilde{g}) = (pT_n)\tilde{g} = p\tilde{g} = g$. For $p : S^2 \to P$, the map $T : S^2 \to S^2, T(\boldsymbol{x}) = -\boldsymbol{x}$, plays the same role as T_n above. For $pT = p$, $p(T\tilde{g}) = (pT)\tilde{g} = p\tilde{g} = g$. Thus if $\tilde{f}(1) = -1$, we may lift $f * g$ to $\tilde{f} * (T\tilde{g})$, and use this lifting to verify that $\bar{h}$ is a homomorphism.

We can show that $\bar{h}$ is onto by explicitly constructing a path $\tilde{f} : I \to S^2$ which connects 1 to -1 and then getting f as $p\tilde{f}$. To see that $\bar{h}$ is 1–1, we use the calculation $\pi_1(S^2, 1) = \{e\}$ and an analogous argument to the proof used for computing $\pi_1(S^1, 1)$.

Exercise 3.4.6. Fill in the details to show that $\pi_1(P, [1]) \simeq \mathbb{Z}_2$.

As the reader may suspect, it is not just a coincidence that there are parallel proofs to show that $\pi_1(S^1, 1) \simeq \mathbb{Z}$ and $\pi_1(P, [1]) \simeq \mathbb{Z}_2$. For S^1, our proof used the map $p : \mathbb{R} \to S^1$, and for P it used the map $p : S^2 \to P$. Both of these maps are examples of covering maps and $(\mathbb{R}, p, S^1), (S, p, P)$ are called *covering spaces*.

Definition 3.4.1. We will assume that in our treatment of covering spaces in this book that each of A, B are path-connected, locally path-connected Hausdorff spaces. If $p : A \to B$ is a continuous surjective map, we say p is a *covering map* and (A, p, B) (or just A if p, B are clear by the context) is a *covering space* if, for each $b \in B$, there is a path-connected open set U containing x so that $p^{-1}(U)$ is the disjoint union $\bigsqcup_{k \in \mathcal{K}} U_k$ of open sets and $p \mid U_k : U_k \to U$ is a homeomorphism. The set U is said to be *evenly covered*. Thus the requirement can be rephrased as saying that there is a covering of B by path-connected, evenly covered open sets.

Definition 3.4.2. Two covering spaces (A_1, p_1, B) and (A_2, p_2, B) are said to be *equivalent* (or *isomorphic*) if there is a homeomorphism $h : A_1 \to A_2$ so that $p_2 h = p_1$.

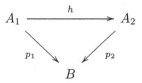

Exercise 3.4.7. Verify that $(\mathbb{R}, p, S^1), (S^2, p, P)$ are covering spaces.

In the two examples in the last exercise, the base spaces S^1, P arise as quotients of the spaces $\mathbb{R}, S^2$ via actions of groups. In the case of $p : \mathbb{R} \to S^1$, we get $S^1 \simeq \mathbb{R}/x \simeq T_n x$ where $T_n x = x + n$. For the second example, $P \simeq S^2/x \sim Ax$, where $Ax = -x$ generates a group isomorphic to $\mathbb{Z}_2 = (\pm 1, \cdot)$ consisting of A and the identity. There is a similarly constructed covering space $p : T \to K$. If we regard $T = S^1 \times S^1$, then there is a homeomorphism $h : T \to T$ which is given by $T(z, w) = (-z, \bar{w})$, where we regard z, w as complex numbers and $\bar{w}$ is the complex conjugate of w. The quotient space is the same as formed from $D^1 \times S^1$ by identifying $(-1, w)$ to $(1, \bar{w})$. When we look at a small disk neighborhood of a point, then h will rotate it half way around the first circle but reverse its orientation as it maps it to a disjoint disk. In the quotient $T/(z, w) \sim h(z, w)$, these two disks get identified to a single disk. This image disk in the quotient is then evenly covered by the two original disks in T and so the map $p : T \to T/(z, w) \sim h(z, w) \simeq K$ is a covering map.

Exercise 3.4.8. Verify that $T/(z, w) \sim h(z, w)$ is homeomorphic to K where we use our original definition of K as a quotient of $D^1 \times D^1$.

Just as we are interested in topological spaces up to homeomorphism or groups up to isomorphism, we are interested in covering spaces up to equivalence, regarding equivalent covering spaces as being essentially the same. One of the most important facts about covering spaces is the unique path lifting property. That is, if we are given a path $f : I \to B$ with $f(0) = x$ and $p(\tilde{x}) = x$, then there is a unique lifting $\tilde{f} : I \to A$ with $\tilde{f}(0) = \tilde{x}$; that is, $\tilde{f}$ is a path in A with $p\tilde{f} = f$. This is proved by a Lebesgue number argument together with the fact that it is easy to lift paths whose images lie in an evenly covered open set in the definition.

It turns out that there is an intimate connection between covering spaces and the fundamental group. In particular, if we have a covering space (A, p, B) with $\pi_1(A, a) \simeq \{e\}$ (such covering spaces do exist for "nice" B) such as $p : \mathbb{R} \to S^1$ and $p : S^2 \to P$, then we can use A to compute $\pi_1(B, b)$ by a path-lifting scheme analogous to the ones used for S^1 and P. Much more is true, however. It turns out that equivalence classes of covering spaces may be classified in terms of subgroups of $\pi_1(B, b)$. This interplay between topology and algebra is one

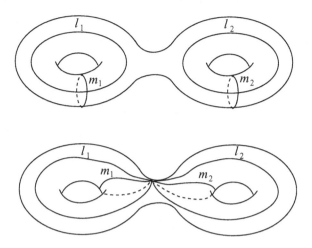

Figure 3.10. Generating loops for $\pi_1(T\#T, x)$.

of the most beautiful in mathematics. Theorems about covering spaces may be deduced from purely algebraic results, and, conversely, covering space theory may be used to give topological proofs of theorems in algebra.

In Chapter 4, we give a development through a set of exercises of some of the most important features of covering spaces. For excellent fuller treatments of covering space theory (as well as the fundamental group), we recommend [18] and [13].

We indicate why T is not homeomorphic to $T\#T$ without giving the full computation of the fundamental groups which would distinguish them. An argument showing $T^{(k)} \not\simeq T^{(p)}$ for $k \neq p$ could be given by generalizing this argument. Basically, the idea is that in T, there are two generating loops for the fundamental group, but in $T\#T$, at least four loops are required to generate it. Figure 3.10 shows these generating loops, first without taking base points into account and then with a common base point.

There is a continuous function $g : T\backslash D^2 \to T$ arising as follows. Take a disk $D_1^2 \subset T$ and a smaller disk $D_2^2 \subset D_1^2$. We can regard the pair (D_1^2, D_2^2) as being homeomorphic to the pair $(D^2, (\frac{1}{2})D^2)$ of standard disks of radii 1 and $\frac{1}{2}$, respectively. Then there is a map $(D^2, (\frac{1}{2})D^2) \to (D^2, \mathbf{0})$ which takes each radial line from $\boldsymbol{x}/2$ to $\boldsymbol{x} \in S^1$ and sends it to the radial line from $\mathbf{0}$ to $\boldsymbol{x}$. Note that this map is the identity on S^1. Using the homeomorphism between (D_1^2, D_2^2) and $(D^2, (\frac{1}{2})D^2)$, there is a corresponding map $(D_1^2, D_2^2) \to (D_1^2, x_0)$ where x_0 corresponds to the center of the disk. The map from $T\backslash D_1^2$ onto T which we want is the identity outside of D_1^2 and then uses the map $(D_1^2, D_2^2) \to (D_1^2, x_0)$. This has the property that there are loops in $T\backslash D_1^2$ which get mapped to the generating loops in $\pi_1(T, \mathbf{1})$ as in Figure 3.11.

This map can be extended to give a continuous map $f_1 : T\#T \to T$ by regarding $T\#T$ as $(T\backslash D^2)_1 \cup (T\backslash D^2)_2$ and just sending $(T\backslash D^2)_1$ to T as above and sending $(T\backslash D^2)_2$ to $\mathbf{1}$. There is an analogous map f_2 which collapses

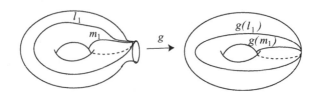

Figure 3.11. Collapsing $T \backslash D$ to T.

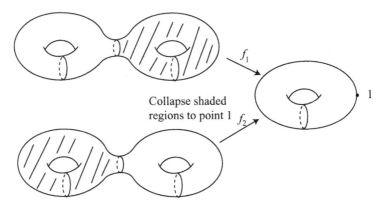

Collapse shaded
regions to point 1

Figure 3.12. Two collapses of $T \# T$ to T.

$(T \backslash D^2)_1$. Figure 3.12 illustrates these maps. Now $((f_1)_*, (f_2)_*) : \pi_1(T \# T) \rightarrow \pi_1(T) \oplus \pi_1(T)$ (with appropriate base points) is surjective since there are loops in $T \# T$ which are sent to each of the generating loops in each factor of T. Hence $\pi_1(T \# T)$ maps onto $\mathbb{Z} \oplus \mathbb{Z} \oplus \mathbb{Z} \oplus \mathbb{Z}$. But it is a fact from group theory that there is no surjective homomorphism from $\mathbb{Z} \oplus \mathbb{Z}$ onto $\mathbb{Z} \oplus \mathbb{Z} \oplus \mathbb{Z} \oplus \mathbb{Z}$. Basically, this follows from the fact that $\mathbb{Z} \oplus \mathbb{Z}$ has two generators and $\mathbb{Z} \oplus \mathbb{Z} \oplus \mathbb{Z} \oplus \mathbb{Z}$ requires four generators.

Exercise 3.4.9. Give a formula for the map $(D^2, \frac{1}{2}D^2) \rightarrow (D^2, \mathbf{0})$ sending S^1 to S^1 via the identity and the circle of radius $\frac{1}{2}$ to $\mathbf{0}$ as described above which sends radial lines to radial lines. (Hint: Use polar coordinates.)

Exercise 3.4.10. In linear algebra, it is shown that there is no surjective vector space homomorphism from $\mathbb{R}^2$ onto $\mathbb{R}^4$ (since the dimension of the image space of a homomorphism is always less or equal to the dimension of the domain space). Use this to show that there is no group homomorphism from $\mathbb{Z} \oplus \mathbb{Z}$ onto $\mathbb{Z} \oplus \mathbb{Z} \oplus \mathbb{Z} \oplus \mathbb{Z}$. (Hint: A group homomorphism from $\mathbb{Z} \oplus \mathbb{Z}$ is completely determined by its values on $(1, 0)$ and $(0, 1)$. A homomorphism to $\mathbb{Z} \oplus \mathbb{Z} \oplus \mathbb{Z} \oplus \mathbb{Z}$ is surjective iff $(1, 0, 0, 0), (0, 1, 0, 0), (0, 0, 1, 0), (0, 0, 0, 1)$ are in the image of the homomorphism. Go from an assumed group homomorphism to a corresponding vector space homomorphism.)

3.5 Applications of the fundamental group

In this section we will prove a number of results which use our calculation of $\pi_1(S^1, 1) \simeq \mathbb{Z}$. Our first application will be to show that any continuous map $f : D^2 \to D^2$ has a fixed point; that is, there is $x \in D^2$ with $f(x) = x$. An analogous result is true for maps from D^n to D^n. For $n = 1$, it can be proved using the intermediate value theorem and the map $g(x) = f(x) - x$. For $n > 2$, there is a proof which utilizes homology theory in a role analogous to the way we will use the fundamental group. First we need a couple of results which will be useful in other applications as well.

Lemma 3.5.1. *If $f : S^1 \to A$ extends to a continuous map $F : D^2 \to A$, then $f_* : \pi_1(S^1, 1) \to \pi_1(A, f(1))$ is the trivial homomorphism, that is, $f_*(\alpha) = [e_{f(1)}]$.*

Proof. We have $f = Fi$ where i is the inclusion of S^1 into D^2. Hence $f_* = F_* i_*$, and i_* is the trivial homomorphism since $\pi_1(D^2, 1) \simeq \{[e_1]\}$. Thus $f_*(\alpha) = F_* i_*(\alpha) = F_*([e_1]) = [e_{f(1)}]$. $\qquad\square$

Proposition 3.5.2. *There does not exist a continuous map $f : D^2 \to S^1$ with $fi = 1_{S^1}$.*

Proof. If there were, then $(fi)_* = 1_{\pi_1(S^1, 1)}$. But $1_{S^1}(\alpha) = \alpha$ and $(fi)_*(\alpha) = f_*(i_*(\alpha)) = [e_{f(1)}]$. $\qquad\square$

Definition 3.5.1. *If $A \subset X$ and $i : A \to X$ is the inclusion, then a continuous map $r : X \to A$ with $ri = 1_A$ is called a* retraction.

Proposition 3.5.2 is saying that there is no retraction from D^2 onto S^1. The following result generalizes Proposition 3.5.2.

Theorem 3.5.3 (No-retraction theorem).

(a) *If there is a retraction $r : X \to A$, then the homomorphism $r_* : \pi_1(X, a) \to \pi_1(A, a)$ is surjective.*

(b) *If $\pi_1(A, a)$ is not the trivial group and $\pi_1(X, a)$ is the trivial group, then there is no retraction from X onto A.*

Exercise 3.5.1. Prove Theorem 3.5.3.

We now apply Proposition 3.5.2 to prove that any continuous map of the disk to itself must fix at least one point.

Theorem 3.5.4 (Fixed point theorem for D^2). *Let $f : D^2 \to D^2$ be continuous. Then there exists $x \in D^2$ with $f(x) = x$.*

Proof. Suppose not. Then $x - f(x)$ is never zero. Consider the ray going from $f(x)$ in the direction of x. There is a unique point different from $f(x)$ on this ray that is on the unit circle; call this point $g(x)$. Since $g(x)$ lies on the line through x and $f(x)$, the point $g(x)$ has the form $x + t(x - f(x))$, for some $t \geq 0$. See Figure 3.13 for an illustration of the construction of g. Calculation shows that t

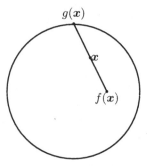

Figure 3.13. Constructing $g : D^2 \to S^1$.

depends continuously on $\boldsymbol{x}$ (see Exercise 3.5.2) and hence $g(\boldsymbol{x})$ is a continuous function from D^2 to S^1. But, if $\boldsymbol{x}$ is already on the unit circle, we have $g(\boldsymbol{x}) = \boldsymbol{x}$. Thus the existence of g contradicts Proposition 3.5.2.   □

Exercise 3.5.2. Show that the correspondence from $\boldsymbol{x}$ to t in the last proof is continuous.

Exercise 3.5.3. Show that a continuous map $f : D^1 \to D^1$ must have a fixed point.

For our next applications, we need to introduce the notion of the degree of a map from S^1 to S^1. Although this could be approached using the fundamental group, this involves technical difficulties involving the role of the base point, so we will just introduce degree using the ideas behind the computation of $\pi_1(S^1, \mathbf{1})$ rather than the fundamental group itself.

First we need to define a homotopy of maps from A to B.

Definition 3.5.2. We say that $f : A \to B$ is *homotopic to* $g : A \to B$ if there is a continuous map $F : A \times I \to B$ with $F(x, 0) = f(x), F(x, 1) = g(x)$.

This is essentially the same definition that we gave for a homotopy of loops in defining the fundamental group, with the principal difference being that in that case we had requirements on where the points $0, 1$ go during the homotopy. The definition is also reminiscent of the concept of an isotopy, which is a special type of homotopy where all of the maps involved are embeddings or homeomorphisms.

Let $f : S^1 \to S^1$ be a continuous map, and consider the composition $f' = fp : I \to S^1$, where $p(t) = e^{2\pi i t}$. Then using the unique path-lifting property, there is a lift $\tilde{f}' : I \to \mathbb{R}$ with $f' = p\tilde{f}'$ once we choose the lift $\tilde{f}'(0)$. Uniqueness implies that any two lifts $\tilde{f}_1', \tilde{f}_2'$ of f' differ by $\tilde{f}_2'(s) = \tilde{f}_1'(s) + n$ for the integer $n = \tilde{f}_2'(0) - \tilde{f}_1'(0)$. For both $\tilde{f}_2'(s), \tilde{g}_2'(s) = \tilde{f}_1'(s) + n$ are lifts of f' which satisfy $\tilde{f}_2'(0) = \tilde{g}_2'(0)$. Since $\tilde{f}'$ is a lift of f' and $f'(0) = f'(1)$, we have $p\tilde{f}'(1) = p\tilde{f}'(0)$, and so $\tilde{f}'(1) - \tilde{f}'(0)$ is an integer. Our argument above that two lifts differ by addition of an integer then implies that the difference $\tilde{f}'(1) - \tilde{f}'(0)$ is an integer which is independent of the lift chosen. This leads to the definition of the degree of the map f.

Definition 3.5.3. Let $f : S^1 \to S^1$ be a continuous map, with $f' = fp : I \to S^1$. Choose a lift $\tilde{f}' : I \to \mathbb{R}$ with $p\tilde{f}' = f'$, and define the *degree* of f by

$$\deg f = \tilde{f}'(1) - \tilde{f}'(0).$$

The discussion before the definition shows that $\deg f$ is well defined, independent of the lift chosen. We next show that homotopic maps have the same degree.

Proposition 3.5.5. *If* $f, g : S^1 \to S^1$ *are homotopic, then* $\deg f = \deg g$.

Proof. Let $F : S^1 \times I \to S^1$ be the homotopy. Define $F' : I \times I \to S^1$ by $F's, t) = F(p(s), t)$. Note that $F'(s, 0) = f'(s), F'(s, 1) = g'(s)$. Theorem 3.3.2 then says that there is a unique lift $\tilde{F}' : I \times I \to \mathbb{R}$ of F' once we specify $\tilde{F}'(0, 0)$ lying above $F'(0, 0) = f(0)$. Then $\tilde{F}'(s, 0) = \tilde{f}'$ is a lift of f' and $\tilde{F}'(s, 1) = \tilde{g}'$ is a lift of g'. Moreover, $\tilde{F}' \mid \{0\} \times I$ and $\tilde{F}' \mid \{1\} \times I$ are lifts of $F' \mid \{0\} \times I, F' \mid \{1\} \times I$, respectively, and these maps satisfy $F'(0, t) = F(1, t) = F'(1, t)$. Thus they are lifts of the same map from I to S^1, and so they differ by an integer constant (which is $\tilde{F}'(1, 0) - \tilde{F}'(0, 0)$). This means that $\tilde{F}'(1, 1) - \tilde{F}'(1, 0) = \tilde{F}'(0, 1) - \tilde{F}'(0, 0)$; call the common value N. We get

$$\tilde{F}'(1, 1) - \tilde{F}'(0, 0) = (\tilde{F}'(1, 1) - \tilde{F}(0, 1) + (\tilde{F}(0, 1) - \tilde{F}(0, 0))$$
$$= (\tilde{g}'(1) - \tilde{g}'(0)) + N = \deg(g) + N,$$
$$\tilde{F}'(1, 1) - \tilde{F}'(0, 0) = (\tilde{F}'(1, 1) - \tilde{F}(1, 0)) + (\tilde{F}(1, 0) - \tilde{F}(0, 0))$$
$$= N + (\tilde{f}'(1) - \tilde{f}'(0)) = N + \deg(f).$$

Hence $\deg f = \deg g$. $\qquad\qquad\square$

The constant map $f : S^1 \to S^1, f(z) = c \in S^1$, has f' being constant and so lifts to a constant map to $\mathbb{R}$. This gives the following lemma.

Lemma 3.5.6. *A constant map* $f : S^1 \to S^1$ *has degree* 0.

The following lemma is the analog of Lemma 3.5.1.

Lemma 3.5.7. *If* $f : S^1 \to S^1$ *extends to a map* $F : D^2 \to S^1$, *then the degree of* f *is zero.*

Proof. Let $G : S^1 \times I \to D^2$ be the continuous map $G(z, t) = tz$. This sends $S^1 \times \{1\}$ to the circle by $(z, 1) \to z$ and sends $S^1 \times \{0\}$ to the center point of the disk, sending $\{z\} \times I$ to the radial line joining the origin to z. The composition $FG : S^1 \times I \to S^1$ is a homotopy between f and the constant map that sends the circle to $F(\mathbf{0})$. Hence the degree of f is zero. $\qquad\square$

It is useful to use the language of complex numbers in dealing with this. To a point (x, y) in the plane, we associate the complex number $x + iy$, thereby identifying $\mathbb{R}^2$ with $\mathbb{C}$. Then S^1 corresponds to the complex numbers of length 1. The pair $(\cos 2\pi t, \sin 2\pi t)$ is associated to the complex number $e^{2\pi it} = \cos 2\pi t + i \sin 2\pi t$. Our covering map $p : \mathbb{R} \to S^1$ becomes $p(t) = e^{2\pi it}$

in this notation. The notation is convenient because of the key property of the exponential function, $e^{z+w} = e^z e^w$. This gives

$$p(t + s) = e^{2\pi i(t+s)} = e^{2\pi it}e^{2\pi is} = p(t)p(s).$$

To compute $\deg k$, we first form $k(p(t)) = k(e^{2\pi it})$. Lifting this to $\mathbb{R}$ means finding $\tilde{k} : I \to \mathbb{R}$ with $e^{2\pi i\tilde{k}(t)} = k(e^{2\pi it})$. We then take $\tilde{k}(1) - \tilde{k}(0) = \deg k$.

Exercise 3.5.4. Suppose $f : S^1 \to S^1$ factors as $f(z) = f_1(z) \cdot f_2(z)$, where the center dot denotes complex multiplication and $f_i : S^1 \to S^1$. Show that $\deg f = \deg f_1 + \deg f_2$. (Hint: To compute $\deg f_i$, lift $f_i' = f_i p$ to $\tilde{f}_i'$ so that $p\tilde{f}_i' = f_i p$. Use $p(x+y) = p(x) \cdot p(y)$ to show that $\tilde{f}'(s) = \tilde{f}_1'(s) + \tilde{f}_2'(s)$ is a lifting of $f' = fp$.)

For our next results, we need a couple of lemmas concerning the degrees of certain maps from S^1 to S^1.

Lemma 3.5.8. *The degree of the map $f(z) = z^n$, where z denotes a complex number of length 1 and z^n denotes z multiplied by itself n times, is n.*

Proof. The map $f' : I \to S^1$ is $f'(t) = e^{2\pi int}$, which lifts to $\tilde{f}'(t) = nt$. Then

$$\deg f = \tilde{f}'(1) - \tilde{f}'(0) = n.$$

$\square$

Lemma 3.5.9. *Let $f : S^1 \to S^1$ be a continuous map with $f(-x) = -f(x)$. Then f has odd degree.*

Proof. We need to find a lifting of $fp : [0, 1] \to S^1$. Suppose $f(1) = a$. Let f_1 denote the restriction of f to the upper half of the unit circle. Note that this determines f completely since $f(-x) = -f(x)$. Let p_1 denote the restriction of p to the first half of the unit interval. Let h denote a lifting of $f_1 p_1$ to a map into $\mathbb{R}$ and denote by $\tilde{a}$ the image of 0. Since $f(-1) = -f(1)$, $h(\frac{1}{2})$ will be of the form $\tilde{a} + k + \frac{1}{2}$. Denote by f_2, p_2 the restrictions of f, p to the last half of the unit interval. Since $f(-x) = -f(x)$, we have $f_2 p_2(t) = -f_1 p_1(t - \frac{1}{2})$, which is equal to

$$e^{\pi i}f_1 p_1\left(t - \tfrac{1}{2}\right) = e^{\pi i}e^{2\pi ih(t-1/2)} = e^{2\pi i(1/2+h(t-1/2))} = e^{2\pi i(h(t-1/2)+k+1/2)}.$$

This means that $b(t) = h(t - \frac{1}{2}) + k + \frac{1}{2}$ will be a lifting of $f_2 p_2$. Then $h(t)$ and $b(t)$ fit together to give a lifting of c of fp with $c(0) = \tilde{a}$ and $c(1) = \tilde{a} + 2k + 1$. This means that the degree of f is $2k + 1$, which is odd. $\square$

We next note that whenever we have a map $f : rS^1 \to \mathbb{R}^2 \backslash \{\mathbf{0}\}$, there is a standard way to get a corresponding map $\bar{f} : S^1 \to S^1$. Here rS^1 denotes the circle of radius r about the origin in the plane. We just define $\bar{f} = ufm_r$, where $m_r(z) = rz, u(z) = z/|z|$. This allows us to define $\deg f$ as $\deg \bar{f}$. Here are some properties of $\deg f$ which follow from the properties of $\deg \bar{f}$ by taking compositions of homotopies with u, m_r.

Proposition 3.5.10. *If $f, g : rS^1 \to \mathbb{R}^2 \backslash \{0\}$ are continuous, then*

(1) *if f is homotopic to g, then $\deg f = \deg g$;*

(2) *if f extends to a continuous map $F : rD^2 \to \mathbb{R}^2 \backslash \{0\}$, then $\deg f = 0$;*

(3) *if $A(r_1, r_2) = \{z : r_1 \leq z \leq r_2\}$, and $F : A(r_1, r_2) \to \mathbb{R}^2 \backslash \{0\}$ is continuous, then*
$$\deg F \mid r_2 S^1 = \deg F \mid r_1 S^1;$$

(4) *when restricted to any circle $rS^1, r > 0$, the map $f(z) = z^n$ has degree n as a map of rS^1 to $\mathbb{R}^2 \backslash \{0\}$.*

Exercise 3.5.5. Prove Proposition 3.5.10.

We now use the degree to prove the fundamental theorem of algebra, which says that a polynomial must have a complex root.

Theorem 3.5.11 (Fundamental theorem of algebra). *Let $P(z) = z^n + a_{n-1}z^{n-1} + \cdots + a_1 z + a_0$ be a polynomial of the complex variable z with complex coefficients a_i. Then P has a root; that is, there is a complex number z_0 with $P(z_0) = 0$.*

Proof. Suppose not. Let $M = \max(|a_0|, \ldots, |a_{n-1}|)$. Choose $k \geq 1, 2nM$. Then on the circle kS^1 about the origin of radius k we have $P(z) = z^n(1 + a_{n-1}/z + \cdots + a_0/z^n) = z^n(1 + b(z))$, where $|b(z)| \leq \frac{1}{2}$ by our choice of k. Since $P(z)$ is never zero, P maps the plane into the set $\mathbb{R}^2 \backslash \{0\}$. $P(z)$ is homotopic to z^n as maps on kS^1 via the homotopy $F(z, t) = z^n(1 + tb(z))$, and so has degree n. But this contradicts the fact that P must have degree 0 since it extends to a map of kD^2. □

We now state a slightly more refined version of this result.

Corollary 3.5.12. *A polynomial $P(z) = z^n + a_{n-1}z^{n-1} + \cdots + a_1 z + a_0$ with complex coefficient factors as $P(z) = (z - r_1) \cdots (z - r_n)$.*

Proof. By the theorem, we can find one root, which we will call r_1. Then $P(r_1) = 0$ means that P factors as $(z - r_1)Q(z)$, where the degree of $Q(z)$ is $n - 1$. The result now follows by induction. □

Exercise 3.5.6. Suppose that $P(z) = (z - r_1) \cdots (z - r_n)$ and $R_1 < R_2$ are radii of circles on which $P(z) \neq 0$.

(a) Show that if $\deg P \mid R_1 S^1 \neq \deg P \mid R_2 S^1$, then there is a root r of P with $R_1 < |r| < R_2$.

(b) Show that if $|r_i| < R$ for all i, then $\deg P \mid RS^1 = n$.

Exercise 3.5.7. Suppose $P(z) = P_1(z)P_2(z)$ is a polynomial with no roots on the circle of radius R. Show that
$$\deg P | RS^1 = \deg P_1 | RS^1 + \deg P_2 | RS^1.$$

(Hint: Use Exercise 3.5.4.)

Exercise 3.5.8. Suppose $P(z) = (z - r_1) \cdots (z - r_n)$, where $|r_i| < R$ for $i \leq k, |r_i| > R$ for $i > k$. Show that deg $P \mid RS^1 = k$. (Hint: Let $P_1(z) = (z - r_1) \cdots (z - r_k), P_2(z) = (z - r_{k+1}) \cdots (z - r_n)$.)

Exercise 3.5.9. Show that deg $P \mid R_2 S^1 - $ deg $P \mid R_1 S^1$ counts the number of roots (counting multiplicity) of the polynomial P in the annulus $A(R_1, R_2)$. Note that we are assuming there are no roots on the boundary of the annulus so this will be defined.

Our next result is known as the Borsuk–Ulam theorem. It has a generalization in higher dimensions. Two points on a sphere of the form $\boldsymbol{x} = (x_1, \ldots, x_n)$ and $-\boldsymbol{x} = (-x_1, \ldots, -x_n)$ are called *antipodal points*. Lemma 3.5.9 says that a continuous map from S^1 to S^1 which sends antipodal points to antipodal points must have odd degree. We use it to prove a version of the Borsuk–Ulam theorem.

Theorem 3.5.13 (Borsuk–Ulam theorem). *There does not exist a continuous map $f : S^2 \to S^1$ which sends antipodal points to antipodal points.*

Proof. Suppose there were such a map. Consider its restriction to the equator of the sphere, which is S^1. Since the map extends over the upper hemisphere (which is homeomorphic to the disk), Lemma 3.5.1 implies that it has degree 0. However, Lemma 3.5.9 implies that it has odd degree; hence we are led to a contradiction.   □

Here is another version of this result.

Theorem 3.5.14 (Borsuk–Ulam theorem). *A continuous map $f : S^2 \to \mathbb{R}^2$ must send some pair of antipodal points to the same point.*

Exercise 3.5.10. Prove Theorem 3.5.14 by using

$$g(\boldsymbol{x}) = \frac{f(\boldsymbol{x}) - f(-\boldsymbol{x})}{|f\boldsymbol{x}) - f(-\boldsymbol{x})|}$$

to get a map to S^1 where we can apply Theorem 3.5.13 if f did not send any pair of antipodal points to the same point.

Here is an amusing corollary, where we are regarding the earth as being a sphere, and our function f as measuring temperature and pressure.

Corollary 3.5.15. *At any time there will be two antipodal points on the surface of the earth with the same temperature and the same barometric pressure.*

Our next result is another popular corollary of the Borsuk–Ulam theorem, although it is somewhat more complicated to set up as such. It is known as the ham sandwich theorem, where the three regions referred to are imagined to be two pieces of bread and a piece of ham.

Theorem 3.5.16 (Ham sandwich theorem). *Let R_1, R_2, R_3 be three connected open regions in $\mathbb{R}^3$, each of which is bounded, with finite volume. Then there is a plane which cuts each of them in half by volume.*

Proof. We will only give a sketch of the proof, leaving most of the details to the reader. First choose a large ball about the origin which encloses the three regions. For each point x on the boundary sphere, consider the family of planes perpendicular to the line joining x to the origin. Since our regions are connected and open, there will be a unique one of these planes that cuts the region R_i in half by volume. Let $d_i(x)$ denote the directed distance of this plane from the origin, calling it positive if the plane is on the same side of the origin as x and negative if it is on the opposite side. It can be shown that the functions d_i are continuous and that they satisfy $d_i(-x) = -d_i(x)$. Let $f : S^2 \to \mathbb{R}^2$ be given by $f(x) = (d_1(x) - d_2(x), d_1(x) - d_3(x))$. By Theorem 3.5.14 there is a point x with $f(-x) = f(x)$. But this implies that $d_1(x) = d_2(x) = d_3(x)$. This then gives a plane which cuts all three regions in half.   □

Exercise 3.5.11. Fill in the details to prove the 'ham sandwich theorem'.

Exercise 3.5.12. Give an example to show that the analog of the 'ham sandwich theorem' is not true if there are four regions; that is, we cannot always slice a ham and cheese sandwich in half by volume without positioning the ingredients.

3.6   Vector fields in the plane

In this section, we will study vector fields in the plane. In the next section, we will extend our results to vector fields on surfaces, and prove a theorem of Poincaré (later generalized by Hopf) relating the index of the vector field to the Euler characteristic of the surface. These sections provide nice applications of the ideas developed in this chapter to surfaces as well as introducing some ideas of differential topology. They are independent of the remainder of the chapter following them, and so may be studied before or after the remaining sections.

We first consider the case of a vector field defined on a subset X of the plane.

Definition 3.6.1. A *vector field* on $X \subset \mathbb{R}^2$ is a continuous function $v : X \to \mathbb{R}^2$.

We will think of a vector at x as emanating from x; that is, it will be thought of as a vector with initial point x and terminal point $x + v(x)$. We give some examples with corresponding pictures.

Let $X = \mathbb{R}^2$ and $v(x) = x$. Then each vector points out from the origin in the direction of the initial point. Its length increases in terms of the length of the initial point. Figure 3.14(a) illustrates this vector field. For our next example, we again let $X = \mathbb{R}^2$ but now $v(x) = (1, 0)$. This is a constant vector field pointing to the right. See Figure 3.14(b). Our third example again uses $X = \mathbb{R}^2$, and we consider each $z \in \mathbb{R}^2$ as a complex number; that is, we identify $\mathbb{R}^2$ with $\mathbb{C}$. We now define $v(z) = z^3$, where we are using complex multiplication. We depict this vector field in Figure 3.14(c). Our final example is a vector field defined on $X = \mathbb{R}^2 \backslash \{0\}$; this vector field blows up at the origin and does not extend over the origin. Using the same convention as the last example, it is $v(z) = 1/z$. It is illustrated in Figure 3.14(d).

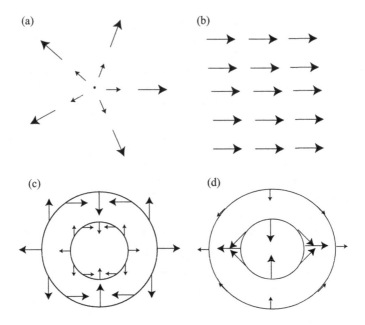

Figure 3.14. Examples of planar vector fields.

We now restrict each of the vector fields in our four examples to the unit circle and describe how the vector moves as we move once around the circle in the counterclockwise direction. In the first example, the vector moves once around the circle since $v(\boldsymbol{x}) = \boldsymbol{x}$. In the second example, the vector field is constant at $(1,0)$. In the third example, the vector field moves around the circle three times in the counterclockwise direction. In the last example, it moves once around the circle in the clockwise direction. In each of these examples, the vector field restricted to the circle is a function from S^1 to itself. The degrees of these four functions are, respectively, $1, 0, 3, -1$. The degrees tell how many times the vector field moves around the circle as we move around the circle once. Of course, these examples are rather special in that each has unit length on the unit circle. However, we could still define the notion of degree of a vector field on the unit circle as long as the vector field does not vanish (i.e. it is not equal to $\boldsymbol{0}$) on the circle. Recall that we defined the degree of a map $v : S^1 \to \mathbb{R}^2 \backslash \{\boldsymbol{0}\}$ by using the composition uv, where $u(z) = z/|z|$ maps $\mathbb{R}^2 \backslash \{\boldsymbol{0}\}$ by radial projection. This is equivalent to considering the normalized vector field $\underline{v}(z) = v(z)/|v(z)|$ and its degree. We can also consider the degrees for vector fields defined on other circles about the origin besides the unit circle. We can then use Proposition 3.5.10 in Section 5 to show that as long as the vector field v does not vanish in the region between the two circles, the degree defined through the composition $\underline{v}m_r$ will be the same on any circle of radius r about the origin. In the examples we gave, it is the case that $\underline{v}m_r$ is the same map as $v|S^1$ instead of just homotopic to it as we would expect. This reflects the symmetry of these vector fields.

Suppose we are interested in the behavior of the vector field on some other circle besides one centered at the origin, say a circle about a point x. Then we could get a function from S^1 to S^1 by first using $m_{x,r}(z) = x + rz$ from S^1 to the circle about x of radius r and then composing with v. To do this, we only need to know that the vector field does not vanish on the circle of radius r about x. The degree that we get does not depend on the particular radius chosen if the vector field v does not vanish on the annular region between the two circles.

If we do this at an x where $v(x) \neq 0$ and choose a disk containing x over which $v(x) \neq 0$, then properties of the degree imply that the restriction to the boundary of the disk has degree 0. Thus the only situation where the computation of the degree of $\underline{v}m_{x,r}$ for small r is of interest is where the vector field vanishes inside the disk $B(x, r)$. In this case, suppose that for a small disk about x, the only point where the vector field vanishes is at x itself. Then if we take any circle about x in this disk and compute the degree of the map $\underline{v}m_{x,r}$ for this circle, then the answer does not depend on the particular circle chosen.

Exercise 3.6.1. Give the details for the claims made in the the the preceding argument that the degree of $\underline{v}m_{x,r}$ does not depend on the choice of radius r as long we are inside a disk where v only vanishes at x.

Definition 3.6.2. A point x where a vector field v vanishes is called a *singularity* of the vector field. A singularity is called *isolated* if there is a deleted neighborhood of the singularity where the vector field in nonzero. For an isolated singularity x, we define the *index* of the singularity by first choosing a small disk about x of radius r so that the only zero inside this disk occurs at x, and then define $i(x)$ as the degree of $\underline{v}m_{x,r}$. If the vector field has only a finite number of singularities, all of which are isolated, then the *index* of the vector field is defined to be the sum of the indices of the singularities; that is

$$I(v) = \sum i(x),$$

where the sum is taken over all of the singularities of v. If a vector field has no singularities, then we define the index of the vector field to be zero.

We look at some examples. For our first three examples, each has only one singularity, and the indices are 1,0,3. The last example had a different type of singularity, one where the vector field blew up rather than vanished, and so the definition does not strictly apply. We now consider an example with two isolated singularities. Suppose our vector field is given by $v(z) = z(z - 2)$. Then the vector field vanishes only at 0 and $2 = (2, 0)$, and each of these singularities is isolated. We first choose a circle of radius 1 about 0. Note that $z - 2$ extends to a map of the unit disk into $\mathbb{R}^2 \backslash \{0\}$. Let $H(z, t) = tz - 2$. Then H is a homotopy into $\mathbb{C} \backslash \{0\}$ between $z - 2$ and the constant map -2. Consider $V(z, t) = zH(z, t)$ and $\underline{V}(z, t) = V(z, t)/|V(z, t)|$. Then $\underline{V}(z, t)$ gives a homotopy between $\underline{v}(z)$, whose degree we wish to compute, and $-z$, as maps from S^1 to S^1. But $-z$ is homotopic to z as maps of S^1 to S^1, via the homotopy $k(z, t) = e^{\pi i t}z, 0 \leq t \leq 1$. Hence the degree of the map is 1, and thus the index of the singularity at 0 is 1. We now consider the index of the singularity at 2. We restrict to the circle of radius 1 about 2 in order to compute it. We first look at $vm_{2,1}(z) = (2+z)z$. As before, we can homotope the $(2+z)$ term to 2 and so get a homotopy of this to $2z$. When we normalize, this gives a homotopy of $\underline{v}m_2, 1(z)$ to z, so the index of the singularity at 2 is also 1. Then the index of this vector field is 2.

Exercise 3.6.2. Give an explicit construction to the homotopy of $\underline{v}m_{2,1}$ to z that is described in the preceding argument.

In our next example, we will not write down an explicit formula, but will give a description of it in pieces, referring to Figure 3.15. The vector field has isolated singularities at -2 and 2. On the circle of radius 1 about -2, the vector field spins once about the circle in the counterclockwise direction as we traverse the left half of the circle and is constantly vertical on the right half of the circle. On smaller concentric circles to this one, the vector field behaves similarly except that the length of the vectors goes to 0 as we approach the singularity at -2. On the left half of the circle about 2, the vector field points upward. On the right half, it moves once around the circle in the clockwise direction as we move from the bottom to the top. Again, the vector field is defined similarly on concentric circles with the lengths of the vectors approaching 0 as we go to the singularity at 2. The index of the vector field at -2 is 1, and the index of the vector field at 2 is -1. Hence the index of this vector field is 0. In the region between the two circles indicated in Figure 3.15, the vector field is vertical. On concentric oblong curves going to infinity, the vector field repeats its behavior on the oblong curve joining the two circles. This example shows how a vector field can have two singularities which cancel each other off as far as their contribution to the index.

We now draw another example with canceling singularities in Figure 3.16. This vector field will be constantly $(1,0)$ outside an oblong region where all of the action is taking place. The index of the two singularites is 1 and -1, and the total index is 0. This last vector field is useful in modifying a given vector field so that the number of singularities is increased but the index remains unchanged. What we will do is first modify the vector field near a nonsingular point to be constant there. Then we remove the constant vector field and replace it by a

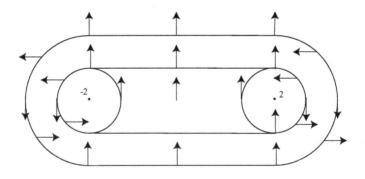

Figure 3.15. Example of canceling singularities.

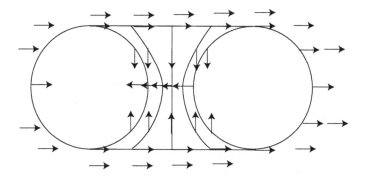

Figure 3.16. Another example of canceling singularities.

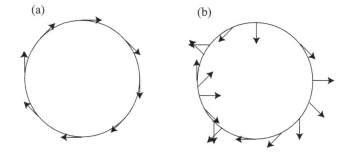

Figure 3.17. Vector fields for Exercise 3.6.3.

copy of the one in Figure 3.16 near the selected point. By doing this, we can get examples of vector fields with arbitrarily high numbers of singularities but with a prescribed index.

Exercise 3.6.3. Each part of Figure 3.17 depicts the vector field on a circle about an isolated singularity. Determine the index of each of the singularities.

Exercise 3.6.4. Give an example of a vector field in the plane which has two singularities of index 1 and one singularity of index -1. (Hint: Start with a vector field with one singularity of index 1 and then introduce two new singularities that look like Figure 3.16.)

Exercise 3.6.5. Show that $f : S^1 \to S^1, f(z) = c$, is homotopic to $g(z) = 1$.

Exercise 3.6.6. Show that, if $f : S^1 \to \mathbb{C}\backslash\{0\}$ satisfies $f(z) = f_1(z)f_2(z)$ (complex multiplication of f_1 and f_2) and f_2 is homotopic to f_3, then f is homotopic to $g(z) = f_1(z)f_3(z)$.

Exercise 3.6.7. Show that, if $f(z) : S^1 \to \mathbb{C}\backslash\{0\}$ extends over the unit disk to $F : D^2 \to \mathbb{C}\backslash\{0\}$, then $f(z)$ is homotopic to $g(z) = 1$. (Hint: Use the map $G : S^1 \times I \to D^2, G(z,t) = tz$ composed with F to get a homotopy to $F(\mathbf{0})$. Then use path connectivity in $\mathbb{C}\backslash\{0\}$ to move $F(\mathbf{0})$ to 1.)

Exercise 3.6.8. Give an explicit homotopy between $f(z) = 2z - 5$ and $g(z) = 1$ as maps from S^1 to $\mathbb{C}\backslash\{0\}$.

Exercise 3.6.9. Suppose $v(z) = v_1(z)v_2(z)$ (complex multiplication) is a vector field with an isolated singularity at $\boldsymbol{x}$. Suppose $v_1(\boldsymbol{x}) = \boldsymbol{0}, v_2(\boldsymbol{x}) \neq \boldsymbol{0}$, and the only singularity of v inside $B(\boldsymbol{x}, r)$ is at $\boldsymbol{x}$. Show that the index of v at $\boldsymbol{x}$ as computed on the circle of radius r about $\boldsymbol{x}$ is the same as the index of v_1 as computed on the same circle.

Exercise 3.6.10. Suppose $v(z) = (z - \boldsymbol{x}_1) \cdots (z - \boldsymbol{x}_n)$ is given as a factored polynomial. Show that v has isolated singularities $\boldsymbol{x}_1, \ldots, \boldsymbol{x}_n$ and that each has index 1, so the index of the vector field is n. (Hint: Use the preceding exercise.)

We have already seen that if a nonzero vector field v extends over a disk, then the degree of the map used in defining the index on the boundary circle is 0. We prove a partial converse; that is, when a vector field v has an isolated singularity at $\boldsymbol{x}$ and the index of this singularity is 0, then there is a vector field w with $w = v$ outside of a disk about $\boldsymbol{x}$, and w has no singularity within the disk. This is equivalent to showing that a vector field which is defined on a circle of radius r about $\boldsymbol{x}$ and has $\underline{v}m_{\boldsymbol{x},r}$ with degree 0 may be extended to a vector field which is nonzero over the disk of radius r about $\boldsymbol{x}$.

Lemma 3.6.1. *Let $f : S^1 \to \mathbb{R}^2\backslash\{0\}$ be a continuous map with the composition of a map of degree 0. Then f extends to a continuous map from D^2 to $\mathbb{R}^2\backslash\{0\}$.*

Proof. Note that ui is the identity on S^1. The map iu is homotopic to the identity on $\mathbb{R}^2\backslash\{0\}$, via the homotopy $H(z, t) = (1-t)z + tz/|z|$. This homotopy pushes radially back to the boundary circle. That the degree of uf is zero means that $ufp, p(t) = e^{2\pi i t}$ is homotopic to the constant loop at $uf(1)$. Hence $(iu)fp$ is homotopic to a constant map, and using H, we see that fp is homotopic to a constant map. Call this homotopy $K : I \times I \to (\mathbb{R}^2\backslash\{0\}) \times I$. It will send $\{0\} \times I$ and $\{1\} \times I$ to a path $\alpha(t)$ joining $uf(1)$ to $f(1)$. Then K induces (regarding S^1 as a quotient space of I via p) a homotopy $\bar{K} : S^1 \times I \to \mathbb{R}^2\backslash\{0\}$ with $\bar{K}|1 \times I = \alpha$ and $\bar{K}(z, 0) = uf(1), \bar{K}(z, 1) = f$,

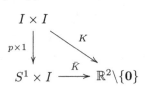

But D^2 is a quotient space of $S^1 \times I$ via the map $r : S^1 \times I \to D^2, r(z, t) = tz$, so $\bar{K}$ induces an extension of f to the disk.

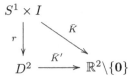

Exercise 3.6.11. Verify that $r : S^1 \times I \to D^2$ is a quotient map and that $\bar{K}'$ gives an extension of f.

By using the standard identification of the unit disk with a disk of radius r about $\boldsymbol{x}$, this lemma impies that if $\underline{v}m_{\boldsymbol{x},r}$ has degree 0, then it will extend over the disk of radius r about $\boldsymbol{x}$ to a map of $\mathbb{R}^2 \backslash \{\boldsymbol{0}\}$. When we piece this together with v, we get the vector field w as claimed.

We next study the situation where we have a vector field v which has all of its singularities inside a disk D_1. We first look at the case where v has only singularities at z_1 and z_2. Consider a straight line joining z_1 and z_2, and an oblong neighborhood about this line formed by taking small disks of the same radius about z_1 and z_2 together with lines connecting the right side of the one of the disks and the left side of the other, as depicted in Figure 3.18(a). Now replace the vector field with one which agrees with v outside of the oblong neighborhood and is defined inside the neighborhood by "radial damping"; that is, on smaller concentric oblong curves, the vectors will be in the same direction as at the corresponding point on the boundary curve but with decreasing length tending to 0 at the center point. See Figure 3.18(b) for an illustration.

We compare the degree of the restriction of the normalized vector field on the large circle boundary of D_1 to the indices of the vector field at z_1 and z_2. We claim that this degree is just the sum of the indices. First note that there is a homotopy between the standard map of S^1 to the boundary of D_1 and the homeomorphism onto the boundary of the oblong region. This homotopy can be thought of as arising from a description of both the oblong region and the disk D_1 as unions of line segments from their boundaries to a center point $\boldsymbol{x}_0$ which lies inside the oblong region. During the homotopy the line segment going to the boundary of the oblong region is stretched to the larger segment going from $\boldsymbol{x}_0$ to the boundary circle of the disk. See Figure 3.19 for an illustration of the image of the boundary circle during the homotopy.

We need a way to compare the degree of the composition of $\underline{v}$ with the standard map of the circle to the boundary of the oblong region (in terms of the construction above) and the sum of indices of the singularities at the two points.

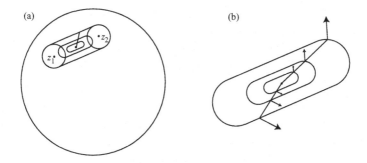

Figure 3.18. Merging two singularities.

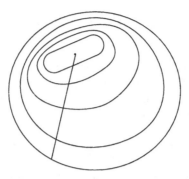

Figure 3.19. Homotoping the boundary circle.

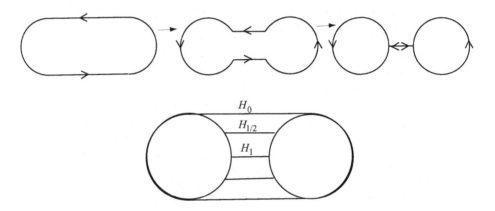

Figure 3.20. Computing the degree on the boundary.

We do this by deforming, by a homotopy, the map of S^1 to the boundary of the oblong region to the map which traverses each of the disks about z_1 and z_2 once in the counterclockwise direction and also goes back and forth along the line joining the centers of these disks. See Figure 3.20. The final map can be thought of as sending the circle to the union of the two circles and the line segment by sending the first quarter of the circle once around the circle C_1 about z_1, sending the second quarter of the circle to the line segment from C_1 to C_2, sending the third quarter once around C_2, and then sending the last quarter back along the line segment to our starting point. To compute the degree of this map, we first wrap the interval around the circle via p, then compose this map with the map described above. We compose with v to get a map of the interval into the circle. We then lift the map to a map of I into $\mathbb{R}$ and compare the lifting of 0 with that of 1. The difference gives the degree. We denote the composition we have formed by f and the lifting by $\tilde{f}$, with $\tilde{f}(0) = a$. Then $\tilde{f}(\frac{1}{4}) = a + n_1$, where n_1 is the index of the singularity at z_1. Now suppose $\tilde{f}(\frac{1}{2}) = a + n_1 + b$. The number b tells us how the image of the path joining the two circles lifts. Next suppose $\tilde{f}(\frac{3}{4}) = a + n_1 + b + n_2$. Since we went around C_2 once as went from $\frac{1}{2}$ to $\frac{3}{4}$, the number $n_2 = i(z_2)$. From $\frac{3}{4}$ to 1, we are going backwards along the same path that we used from $\frac{1}{4}$ to $\frac{1}{2}$, and thus the lifting will be a reparametrization of that lifting (going backwards) shifted along the reals by some integer. This means that as we go from $\frac{3}{4}$ to 1, the image points will be moved backwards by b units. Thus $\tilde{f}(1) = \tilde{f}(\frac{3}{4}) - b = a + n_1 + n_2 = a + i(z_1) + i(z_2)$, and the degree of the map is $i(z_1) + i(z_2)$.

We are now ready to prove our main result concerning indices of vector fields in the plane.

Proposition 3.6.2. *Suppose v is a vector field defined on the disk $B(\boldsymbol{x}, r)$ centered at $\boldsymbol{x}$ which has isolated singularities at $z_1, \ldots, z_k$. The degree of the map $\underline{v}m_{\boldsymbol{x}, r}$ is the sum of the indices of the singularities; that is, the index of the vector field may be computed by computing the degree using the vector field on the boundary circle.*

Proof. We prove this by induction on the number k of singularities. If $k = 1$, then we may use a homotopy between the map $m_{\boldsymbol{x}, r}$ and the map m_{z_1, r_1} onto a small circle around the singularity analogous to that in Figure 3.19 to show that the degree of $\underline{v}m_{\boldsymbol{x}, r}$ is the same as the degree of $\underline{v}m_{z_1, r_1}$. The details are left as an exercise.

Now suppose that we know the result when there are k isolated singularities inside D and our vector field has $k + 1$ singularities inside D. Choose two singularities at z_1 and z_2, so that the line joining z_1 and z_2 does not intersect the other singularities. Replace the vector field v by a vector field w which agrees with v outside of the oblong neighborhood but is changed inside by radial damping as in Figure 3.18(b). This replaces two singularities by one singularity at the center point of this oblong region, and the argument just preceding this proposition shows that the index of this new singularity will be the sum of the indices of v at z_1 and z_2. The vector field w only has k singularities, and the induction hypothesis shows that the sum of the singularities for w is equal to the degree of the map $\underline{v}m_{\boldsymbol{x}, r}$. Since the sum of the indices for w is the same as the sum of the indices for v, the result follows. $\square$

Exercise 3.6.12. Verify the claim made in the first paragraph of the proof that the degree of $\underline{v}m_{\boldsymbol{x}, r}$ is the same as the degree of $\underline{v}m_{z_1, r_1}$.

We will need a slight refinement of this proposition. Consider the region R formed from the disk $B(\boldsymbol{x}, r)$ by removing k small disks about points $z_1, \ldots, z_k$ in D. Suppose we have a vector field v defined in R. Then we claim that $I(v)$ is given by the difference between the degree of the map $\underline{v}m_{\boldsymbol{x}, r}$ and the sum of the degrees of the map on the small circles about z_i. We could define a vector field on D by making it equal to v in R and extending it over the small disks about the points z_i by radial damping. This new vector field will have all of the old singularities of v as well as new singularities at z_i. The sum of the indices of the new singularities will be the sum of the degrees of the maps $\underline{v}m_{z_i, r_i}$. Hence the difference of the degree on the outer circle and the sum of the degrees on the small circles will be the sum of the indices of the original vector field. We state this result for future reference.

Proposition 3.6.3. *Let v be a vector field on a region R in the plane formed from a disk D by removing k small disks about points of z_i of radii r_i. Then the index of v may be calculated by taking the difference of the degree of $\underline{v}m_{x,r}$ and the sum of the degrees of the maps $\underline{v}m_{z_i,r_i}$.*

3.7   Vector fields on surfaces

We now discuss vector fields on surfaces and their indices. We first need to discuss the notion of a differential structure on a surface. Recall that a surface can by covered by open sets $\{U_i\}$ so that there are homeomorphisms $h_i : U_i \to V_i$, where V_i is an open set in $\mathbb{R}^2$. We impose a differential structure on the surface by finding a cover so that if $U_i \cap U_j \neq \emptyset$, then the homeomorphism $h_i h_j^{-1} : h_j(U_i \cap U_j) \to h_i(U_i \cap U_j)$ is a diffeomorphism. This means that $h_i h_j^{-1}$ is required to be differentiable with differentiable inverse. By differentiable, we will mean infinitely differentiable here for simplicity. We will say that $\{(U_i, h_i)\}$ gives a *differential structure* on the surface. A surface M together with a differential structure $\mathcal{S}$ will be called a *smooth surface*, and will be denoted by $(M, \mathcal{S})$ whenever we want to emphasize the differential structure or just by M alone when the differential structure is implicitly clear. We say that two differential structures $\mathcal{S}_1 = \{(U_i, h_i)\}, \mathcal{S}_2 = \{(V_j, g_j)\}$ on M are equivalent if, whenever $U_i \cap V_j \neq \emptyset$, the composition $h_i g_j^{-1} : g_j(U_i \cap V_j) \to h_i(U_i \cap V_j)$ is a diffeomorphism. We are only interested in equivalence classes of differential structures.

Suppose we have selected equivalence class differential structures on the surfaces M, N with structures coming from $\{(U_i, h_i)\}$ on M and $\{(V_j, g_j)\}$ on N. We say that a map $f : M \to N$ is a diffeomorphism if f is a homeomorphism and the maps $g_j f h_i^{-1}$ (where they are defined) are diffeomorphisms. This does not depend on the particular representative of the differential structure since corresponding local descriptions of the surface are related by diffeomorphisms. Basically, we are determining whether a map is differentiable by referring it back to a local description of the domain and range as being open sets in $\mathbb{R}^2$. We get the same answer independent of the particular local description we take because the two local descriptions are related by a diffeomorphism.

Although our definition of a differential structure is rather abstract, differential structures frequently arise in rather concrete ways, such as having the surface embedded smoothly in some $\mathbb{R}^n$ with well-defined tangent planes. In this case the local descriptions can come from projections onto these tangent planes, together with identifying the tangent plane with $\mathbb{R}^2$.

Exercise 3.7.1. Suppose $\mathcal{S}_1, \mathcal{S}_2$ are equivalent differential structures on M. Show that $f : M \to \mathbb{R}$ is differentiable in terms of $\mathcal{S}_1$ iff it is differentiable in terms of $\mathcal{S}_2$. Thus equivalent differential structures have exactly the same differentiable maps to $\mathbb{R}$.

There are two important questions to be asked about differential structures on surfaces. The first is whether every surface possesses a differential structure. The answer is yes, and this is easy to see for the compact, connected surfaces that we have classified. For the orientable ones, we can use an embedding of the surface into $\mathbb{R}^3$ to get our differential structure. For the nonorientable ones, we could either use an embedding into $\mathbb{R}^4$ or break it up into pieces and impose consistent differential structures on the pieces. We could get a differential structure on the Möbius band (e.g. by using an embedding into $\mathbb{R}^3$) and then impose a consistent structure on a disk which overlaps the Möbius band slightly so that their union is the projective plane. Our answer is relatively easy for compact connected surfaces since we know that each one is homeomorphic to one of our model surfaces and so we can just work with the concrete models.

The next question is whether $(M, \mathcal{S})$ and $(N, \mathcal{T})$ being homeomorphic implies that they are in fact diffeomorphic. The answer to this question turns out to be yes as well. That is, if M and N are considered as smooth surfaces, then M is homeomorphic to N iff M is diffeomorphic to N. The subject of differential topology studies smooth manifolds up to diffeomorphism. Some references for an introduction to differential topology at a beginning graduate level are [5,15,21,25]. One of the big breakthroughs in topology happened when it was shown that the answers to the corresponding questions of existence and uniqueness of differential structures above were sometimes no in higher dimensions. That is, there exist manifolds which do not possess differential structures as well as differentiable manifolds which are homeomorphic but not diffeomorphic. This last result was discovered first and is due to John Milnor in 1956. Shortly after Milnor's discovery, Michael Kervaire produced an example of a manifold which has no differential structure. These two results led to a great flourishing in differential topology in the late 1950s and 1960s. Later, Robion Kirby and Laurence Siebenmann produced similar results for PL structures on a manifold which arise from nice triangulations of the manifold. A PL structure can be thought of as lying somewhere between the structure as a manifold and a differential structure. There has been a great deal of work in comparing the notions of a topological manifold structure, a PL structure, and a differential structure. Much of the analysis has relied on constructions in algebraic topology. Most of the results obtained early on concerned manifolds of dimension ≥ 5. More recently, in the 1980s and 1990s, there has been another explosion of results in studying similar questions about four-dimensional manifolds. Here it turns out that "most" compact, connected 4-manifolds do not have differential structures, and, of those that do, there are frequently infinitely many distinct differential structures. A surprising result, completely different from what happens in any other dimension, is that $\mathbb{R}^4$ itself has uncountably many distinct differential structures. These last results arise out of gauge theory, which is a part of differential topology which was motivated by ideas in physics.

For surfaces, all three types of structures turn out to lead to the same results. This is in itself useful, however, since it means that we are allowed to use techniques in differential topology and PL topology to study questions about surfaces, and our results do not depend on the particular structure used. This can lead to an exciting interplay between ostensibly very different ideas. We will see this interplay as we discuss vector fields on surfaces.

We indicate how differential structures on M and N can be used to get a differential structure on $M\#N$. We earlier defined the connected sum in terms of removing disks from M and N and sewing in a cylinder. Suppose D_M, D_N are the disks whose interiors to be removed, and $h_M : D \to M, h_M(\mathbf{0}) = p, h_N : D \to D_N, h_N(0) = q$ are diffeomorphisms (in terms of the given differential structures). Now int $D\backslash\{\mathbf{0}\}$ is homeomorphic to the interior $S^1 \times (0,1)$ of a cylinder $S^1 \times I$, so sewing in the cylinder is equivalent to identifying int $D_M\backslash\{p\}$ and int $D_N\backslash\{q\}$ to what will be the sewed-in cylinder. Note that $S^1 \times \{0\}$ is supposed to correspond to ∂D_M and $S^1 \times \{1\}$ is supposed to correspond to ∂D_N. Hence the identification required is to identify int $D_M\backslash\{p\}$ to int $D\backslash\{\mathbf{0}\}$ via h_M^{-1}, compose with a diffeomorphism $r :$ int $D\backslash\{\mathbf{0}\} \to$ int $D\backslash\{\mathbf{0}\}, r(t\boldsymbol{x}) = (1 - t)\boldsymbol{x}, 0 < t < 1$, and then identify int $D\backslash\{\mathbf{0}\}$ to int $D_N\backslash\{q\}$ via h_N. Note that $h_N r h_M^{-1}$ is a diffeomorphism in terms of the differential structures on $M\backslash\{p\}$ and $N\backslash\{q\}$ coming from those on M, N. Hence these differential structures piece together to give a differential structure on $M\#N = (M\backslash\{p\}) \cup (N\backslash\{q\})/x \sim h_N r h_M^{-1}(x)$, where $x \in$ int $D_M\backslash\{p\}$. A picture of this construction is given in Figure 3.21.

Note that a radial line segment (in terms of h_M) running from ∂D_M toward p is being identified with a radial line segment (in terms of h_N) running from

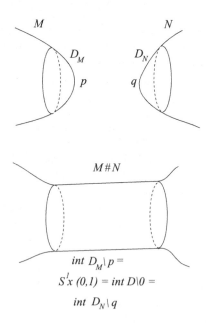

Figure 3.21. Forming connected sum differentiably.

q out to ∂D_N. This is illustrated in Figure 3.22. The circle $h_M(\frac{1}{2}S^1)$ is identified to $h_N(\frac{1}{2}S^1)$ by identifying $h_M(x)$ to $h_N(x)$. We can think of $M\#N$ as
$M\backslash h_M(\mathrm{int}\frac{1}{2}D)\cup_{h_M h_N^{-1}} N\backslash h_N(\mathrm{int}\frac{1}{2}D)$. The curve ℓ (which is differentiable) then
becomes the union of two curves ℓ_1, ℓ_2 as indicated in Figure 3.23. Thus we can
think of $M\#N$ as being formed by removing the disks $h_M(\mathrm{int}\frac{1}{2}D)$, $h_N(\mathrm{int}\frac{1}{2}D)$
and gluing their boundary circles.

   We comment briefly on orientation conventions. To form oriented connected
sums consistent with given orientations on M and N, it is necessary for one of the
embeddings h_M, h_N to preserve orientation and the other to reverse orientation.
Then we can extend the orientations on $M\backslash\frac{1}{2}D$ and $N\backslash\frac{1}{2}D$ to $M\#N$ as the
orientations on the circles $h_M(\frac{1}{2}S^1)$ and $h_N(\frac{1}{2}S^1)$ will disagree as required. From

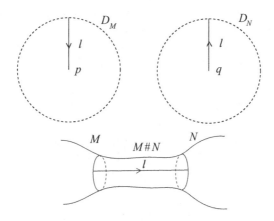

Figure 3.22. Identified radial lines.

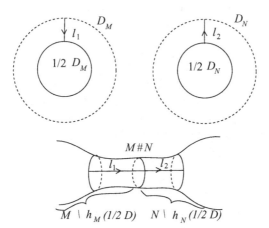

Figure 3.23. Connected sum via gluing along a circle.

the point of view of r, it reverses orientation and so the map $h_N r h_M$ will preserve orientations as the two manifolds are glued together.

We now define the notion of a *tangent vector field* on a surface. If the surface already sits differentiably in some $\mathbb{R}^n$ so that there is a well-defined notion of a tangent plane to the surface at each point on it, then a tangent vector field is just a continuous selection of a tangent vector at x for each point x on the surface. By a tangent vector at x, we mean a vector in $\mathbb{R}^n$ (thought of as emanating from x) which is tangent to a differentiable curve in the surface which passes through x; that is, it is $f'(0)$, where $f : \mathbb{R} \to M \subset \mathbb{R}^n$ is differentiable and $f(0) = x$. The set of all tangent vectors at x (usually thought of as elements of the plane of points $x + tf'(0)$ for f as above and $t \in \mathbb{R}$) is called the tangent plane at x. Suppose we have a tangent vector field in this sense, and that U_i is an open set in M and h_i is a homeomorphism from U_1 onto an open set in $\mathbb{R}^2$ that is part of the differential structure imposed on M via its embedding into $\mathbb{R}^n$ from projection onto the tangent plane. This last condition means that h_i extends to a differentiable function from an open set in $\mathbb{R}^n$ to $h_i(U_i)$. The differential of this function will then map tangent vectors at $x \in U_i$ linearly to vectors in $\mathbb{R}^2$ at $h_i(x)$. Thus a vector field on M determines through the differential a related vector field on $h_i(U_i)$ for all i. These vector fields are related as follows: if $\boldsymbol{v}_k$ represents the vector field on $h_k(U_k)$, then $\boldsymbol{v}_j(h_j(x)) = D(h_j h_i^{-1})(h_i(x))\boldsymbol{v}_i(h_i(x))$. Here $Df(a)$ is the differential of the differentiable map f evaluated at a. It is a linear map from $\mathbb{R}^2$ to $\mathbb{R}^2$ and is represented by the matrix of partial derivatives. Since $h_j h_i^{-1}$ is a diffeomorphism, the differential $D(h_j h_i^{-1})(a)$ is an invertible linear transformation for all values of a where it is defined. Thus we may regard a vector field on M as being given by consistent vector fields in each of the open sets $h_i(U_i)$, where consistent means that the equation $\boldsymbol{v}_j(h_j(x)) = D(h_j h_i^{-1})(h_i(x))\boldsymbol{v}_i(h_i(x))$ holds. Given a vector field in this latter sense, we can use the differential structure maps to reconstruct a vector field in the sense of an embedded surface. We will work with the local definition since it will allow us to refer our questions back to questions about vector fields in $\mathbb{R}^2$ where we can apply the results of the last section.

We need an alternate description of the induced differential. Suppose $f : U \subset R^2 \to V \subset \mathbb{R}^2$ is a differentiable map and $\boldsymbol{v}$ is a tangent vector to U at x. Then there is a differentiable curve $g : \mathbb{R} \to U$ with $g(0) = x, g'(0) = \boldsymbol{v}$. Then $Df(x)\boldsymbol{v} = \boldsymbol{w}$ means that $\boldsymbol{w} = (fg)'(0)$. Thus, to determine the differential, we just have to see what f does to certain curves through a point. Moreover, the fact that the differential is a linear map means that we only have to determine what is happening to the two curves g_1, g_2 with $g_i(0) = x$ and $g_i'(0) = \boldsymbol{v}_i$, with $\boldsymbol{v}_1, \boldsymbol{v}_2$ linearly independent.

We look at what this means for $M \# N$ where we glue the two circles together. Then the curve ℓ is differentiable. In terms of h_M^{-1}, it is sent to a curve that cuts $\frac{1}{2}S^1$ perpendicularly going inward. In terms of h_N^{-1}, it is sent to a curve that cuts $\frac{1}{2}S^1$ perpendicularly going outward. Thus the differential will identify an exterior normal vector to $\partial(M \backslash h_M(\frac{1}{2}\text{int } D))$ to an interior normal to $\partial(N \backslash h_N(\frac{1}{2}\text{int } D))$. Since the circle $\frac{1}{2}S^1$ is identified to itself via the identity, a tangent vector

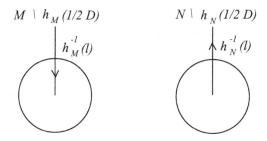

Figure 3.24. Identifying vectors for a connected sum.

to $\partial(M\backslash h_M(\frac{1}{2}\text{int } D))$ will be identified to the corresponding tangent vector to $\partial(N\backslash h_N(\frac{1}{2}\text{int } D))$. See Figure 3.24.

We now apply these ideas to study vector fields on the sphere. We will take our covering of the sphere to be given by $S^2\backslash\{S\}, S^2\backslash\{N\}$, where S, N denote the south and north poles. We then take the homeomorphism h_S, h_N to be given by stereographic projection from the appropriate poles. Using our earlier computations of h_S, h_N, we may compute $h_N h_S^{-1}(a, b) = (1/s^2)(a, b)$, where $s^2 = a^2 + b^2$. The map sends the circle of radius r to the circle of radius $1/r$ via $rz \rightarrow (1/r)z$, where $z \in S^1$. In particular, it sends the unit circle identically to itself. Qualitative features of this calculation can be found geometrically using the projection maps. For example, if C denotes a circle formed from the sphere by intersecting it with a plane parallel to the xy-plane, then stereographic projection from either pole will send C to a circle in the plane. If C lies in the southern hemisphere, then projection from the south pole sends C to a circle of radius greater than 1 and projection from the north pole sends C to a circle of radius less than 1. Moreover, starting from a point in the plane, $h_N h_S^{-1}$ will send the point x in the same plane containing $x, N, \mathbf{0}$, so it will preserve radial lines. What our calculation does is make precise how the radii of these circles correspond.

We compute the differential of the map $h_N h_S^{-1}$ by computing the matrix of partial derivatives, giving

$$A = \frac{1}{s^4}\begin{pmatrix} b^2 - a^2 & -2ab \\ -2ab & a^2 - b^2 \end{pmatrix}.$$

It is easiest to see how this matrix works on a vector by choosing an appropriate basis. The most useful basis is $v_1 = (a, b), v_2 = (-b, a)$. The first vector is an exterior normal to the circle of radius s, and the second is a tangent vector to the circle. Computation gives $Av_1 = -(1/s^2)v_1, Av_2 = (1/s^2)v_2$. Again, we could have seen this geometrically except for the exact eigenvalues. Our map sends a circle to a circle along radial lines, so a tangent vector to the circle must be sent to a tangent vector to the image circle and hence to a multiple of itself. The fact that a ray perpendicular to the circle is sent to a ray perpendicular to the image circle with its orientation reversed implies that an exterior normal to a circle is sent to an interior normal to the image circle. A vector field on S^2

corresponds to two related vector fields on $\mathbb{R}^2$, where they are related by matrix A at corresponding points. Note that we can retain all of the information just by looking at the vector fields on the two unit disks, and their relationship via A restricted to the unit circle, which $h_N h_S^{-1}$ sends identically to itself. For each $z \in S^1$, what $A(z)$ does is reflect each vector through the line perpendicular to the line from the origin to z. A useful way to describe this reflection is that it is the composition of first rotating the circle to send z back to 1 (which is achieved by multiplying by $\bar{z} = z^{-1}$ where $\bar{z}$ denotes complex conjugation), then reflecting in the vertical line (which sends z to $-\bar{z}$), and then rotating back to z (which is achieved by multiplying by z). The composition, which is $A(z)v(z)$, sends the vector field $v(z)$ to the vector field $-z^2\overline{v(z)}$.

We will assume that our vector field has already been normalized to have length 1 where we are doing our computations since we will be doing this normalization as part of computing the index. Suppose that our initial field has degree n. Then $v(z)$, considered as a map from S^1 to S^1, is homotopic to the map z^n. Hence the map $-z^2\bar{v}(z)$ is homotopic to the map $-z^2 z^{-n} = -z^{2-n}$, which is homotopic to z^{2-n}, and hence has degree $2 - n$. Thus multiplication by $A(z)$ converts a unit vector field of degree n on S^1 to a vector field of degree $2 - n$. Thus a vector field on S^2 with no singularities on the equator corresponds to two different vector fields on the unit disk, say v_1, v_2, where v_1 and v_2 are related on the unit circle in such a way that if n is the degree of the normalized vector field $\underline{v}_1$ on the circle, then $2 - n$ is the degree of the normalized vector field $\underline{v}_2$.

We now describe what we mean by the index of a (tangent) vector field on a surface. We first define the *index of a singularity*; that is, the index at a point where the vector field vanishes. The way that this is done is to refer the vector field back to a vector field in an open set of the plane via one of the defining homeomorphisms giving the differential structure, and compute the index of the singularity there. For this to be well defined, we have to show that the result is independent of the particular homeomorphism h_i chosen. This turns out to be true, and the proof is an interesting exercise in advanced calculus using the differential as a linear approximation to $h_j h_i^{-1}$. It works by showing that up to homotopy (which is all right for the purpose of calculating the degree) we can assume that the map $h_j h_i^{-1}$ is linear and orthogonal (so that it maps circles to circles) in a neighborhood of the singularity. This leads to comparing the degrees of $\underline{v}(z)$ and the composition $A\underline{v}A^{-1}(z)$. The fact that the degrees of compositions multiply and the degrees of A, A^{-1} are ± 1 since they are orthogonal shows that the two degree computations give the same result. Thus we can compute the index of the singularity by referring the problem back to the plane. We then define the *index of the vector field on the surface* with a finite number of isolated singularities to be the sum of the indices of those singularities. We will show by expressing the vector field in terms of related planar vector fields that the index of a vector field on a compact, connected surface without boundary is just the Euler characteristic of the surface.

We now consider the index of a vector field on the sphere which has only a finite number of isolated singularities. After a small perturbation of the vector field that does not change the index, we can assume that the vector field has no singularities on the equator. Then we can regard the vector field as two vector fields v_1, v_2 on the unit disk related as we described above. But we showed in the last section that for a vector field on the unit disk, the degree of the normalized vector field on the boundary circle gives the sum of the indices of the singularities inside the disk. Thus the index $I(v)$ of the vector field v on S^2 will be the sum of the indices $I(v_1)$ and $I(v_2)$ of the two vector fields on the disk. Our discussion above shows that these are related by $I(v_2) = 2 - I(v_1)$. Hence $I(v) = I(v_1) + (2 - I(v_1)) = 2$, independent of the particular vector field chosen. In particular, this means that there does not exist a vector field on S^2 which never vanishes. We state this as a theorem.

Theorem 3.7.1. *Let v be a (tangent) vector field on S^2 with a finite number of isolated singularities. Then the index $I(v)$ is 2. In particular, this means that there does not exist a nonvanishing tangent vector field on S^2.*

Exercise 3.7.2. Draw a picture to show the effect on the vector field $v(z) = z^2$ on the unit circle when we reflect $v(z)$ through the line perpendicular to the line from the origin to z. Your picture should show the images of the vectors pictured in Figure 3.25. Verify from your picture that the degree of the new vector field is 0.

Exercise 3.7.3. An orthogonal linear transformation from S^1 to S^1 is the composition of the reflection $r(z) = \bar{z}$ with multiplication by a number in S^1 (i.e. $g(z) = az$, where $a \in S^1$) or a map of the form of g itself. Show that the degree of r is -1. Show that the degree of an orthogonal linear transformation is ± 1. Determine the inverses of g and r, and show that they have the same degree as the original map g or r.

Exercise 3.7.4. Prove that any continuous map $f : S^1 \to S^1$ is homotopic to z^n for some n. (Hint: Use the information that it has degree n for some n and reinterpret what that means in terms of certain maps from the interval being homotopic to show that f and z^n are homotopic.)

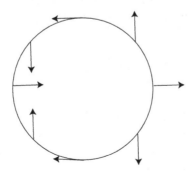

Figure 3.25. The vector field $v(z) = z^2$.

Exercise 3.7.5.

(a) Give an example of a vector field on the sphere with one singularity.

(b) Give an example of a vector field on S^2 with one singularity at the north pole and one singularity at the south pole.

(c) Give an example of a vector field on S^2 with four isolated singularities of indices $1, 1, 1, -1$.

Exercise 3.7.6. Show that, if $f : S^2 \to \mathbb{R}^3$ is continuous, then there must be some $x \in S^2$ so that $f(x) = kx$, where $k \in \mathbb{R}$ and x is thought of as a point in $\mathbb{R}^3$. (Hint: Reinterpret f as a (not necessarily tangent) vector field on S^2 and get a tangent vector field from it by projection onto the tangent plane at x.)

We now consider vector fields on the torus. This time we will start with a description of the torus and its differential structure that is akin to the description of the sphere as the union of two disks. This comes from thinking of the torus as sitting in $\mathbb{R}^3$ symmetrically with respect to the xz-plane and then slicing it into two halves by that plane. The two halves are each homeomorphic to annuli and the two pieces are glued together by identifying points on the boundary circles. We need to understand how tangent vectors on the circles in the two pieces correspond. Since the circles are sent identically to each other, vectors tangent to the circles must be sent to the corresponding tangent vector on the other circle. The more interesting phenomenon occurs for vectors that are normal to each of the circles. To see what happens there, we must consider a curve on the torus that goes from the front half to the back half and cuts the equator circles perpendicularly. In terms of our description of the torus as the union of two annuli, this curve is represented by the union of two radial lines. As the curve leaves one of the annuli, it enters the other one at the corresponding point. Hence an exterior normal to one of the annuli must get identified to an interior normal at the corresponding point of the other annulus. We picture this identification in Figure 3.26.

The way these vectors are being identified on corresponding circles is exactly the same way they were identified for the equator of S^2. Our work there tells us the relationship between the degrees of the vector fields on the circles which get identified: "the sum of the degrees must be equal to 2".

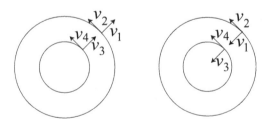

Figure 3.26. Corresponding vectors in the torus.

Suppose we have a vector field on the torus with no singularities on the two equator circles coming from cutting the torus in half as described. This can always be arranged without changing the index by a slight change in the vector field if there are only a finite number of singularities. This vector field can be thought of as being given by two related vector fields on two annuli, where the vector fields on corresponding boundary circles are related so that the sum of their degrees is 2. Suppose that our vector field on the torus has only a finite number of isolated singularities so that we can define the index. We call the vector field v and denote the vector fields on the two annuli by v_1, v_2. We call the degrees of the vector fields on the annuli as d_{jo}, d_{ji}, where o, i are used to denote the outer and inner circles and $j = 1, 2$ denotes the copy of the annulus. Then we have $I(v) = I(v_1) + I(v_2)$ and $I(v_j) = d_{jo} - d_{ji}$. We also have $d_{1o} + d_{2o} = 2 = d_{1i} + d_{2i}$. Putting these together gives

$$I(v) = I(v_1) + I(v_2) = (d_{1o} - d_{1i}) + (d_{2o} - d_{2i}) = (d_{1o} + d_{2o}) - (d_{1i} + d_{2i}) = 0.$$

Theorem 3.7.2. *Let v be a tangent vector field on the torus T with a finite number of isolated singularities. Then the index $I(v) = 0$.*

Note that by this theorem it is possible to have a vector field on T with no singularities. We leave it as an exercise to construct such a vector field.

Exercise 3.7.7. Give an example of a vector field on T with no singularities. (Hint: First find a nonvanishing vector field on a circle.)

We next consider vector fields on the projective plane P. We use the map $p : S^2 \to P$ to impose a differential structure on P. We saw earlier that there was a cover $\{U_i\}$ of P so that $p^{-1}(U_i) = U_{i1} \cup U_{i2}$, and $p|U_{ij} : U_{ij} \to U_i$ is a homeomorphism with inverse q_{ij}. Also, $T(\boldsymbol{x}) = -\boldsymbol{x}$ is a homeomorphism interchanging U_{i1} with U_{i2}. In terms of the differential structure on S^2, T is in fact a diffeomorphism since it is the restriction of a diffeomorphism of $\mathbb{R}^3$. We could impose the differential structure on S^2 by taking diffeomorphisms $h_{i1} : U_{i1} \to \bar{U}_i$ onto open subsets of $\mathbb{R}^2$ (diffeomorphisms in terms of the given differential structure on S^2) together with diffeomorphisms $h_{i2} : U_{i2} \to \bar{U}_i$ given by $h_{i2} = h_{i1}T$. We then use this to impose a differential structure on P by using the maps $\bar{h}_i : U_i \to \bar{U}_i, \bar{h}_i = h_{i1}q_{i1} = h_{i2}q_{i2}$. Because of this correspondence of differential structures on S^2 and P, a vector field on P can be thought of as a vector field on S^2 which is *equivariant* with respect to T. Equivariance means $v(T(\boldsymbol{x})) = DT_{\boldsymbol{x}}(v(\boldsymbol{x}))$.

Suppose we have a vector field on P with a finite number of isolated singularities at $x_1, \ldots, x_k$. Then the vector field determined on S^2 will have isolated singularities at $x_{11}, \ldots, x_{k1}, x_{12}, \ldots, x_{k2}$, where $p(x_{ij}) = x_i$. To compute the index at x_i and the index at x_{ij}, we refer both vector fields to the *same* vector field in $\bar{U}_i$ and so $i(x_i) = i(x_{ij}), j = 1, 2$. But this implies that $2 = \sum i(x_{ij}) = 2(\sum i(x_i))$, and hence the index of our vector field is 1.

We have thus shown the following theorem.

Theorem 3.7.3. *Let v be a tangent vector field on the projective plane P with a finite number of isolated singularities. Then the index $I(v) = 1$.*

Exercise 3.7.8.

(a) Show that the map $p : S^2 \to P$ is differentiable in terms of the differential structures on S^2 and P as defined above.

(b) Show that the map $p : S^2 \to P$ is locally a diffeomorphism.

(c) Show that a map $f : P \to \mathbb{R}$ is differentiable iff the composition $fp : S^2 \to \mathbb{R}$ is differentiable.

Note that in the last three theorems, we have shown that for S, T, P, the index of a vector field is equal to the Euler characteristic of the surface on which it lies. We might conjecture that this is true for any compact connected surface. In fact, this is true and was first proved by Henri Poincaré in 1885 and extended to n-manifolds by Heinz Hopf in 1926. We will prove this by comparing the index of a vector field on the connected sum $M\#N$ with the indices on M and N. This approach is fruitful since all compact connected surfaces are homeomorphic to connected sums of S, T, P. Suppose that we know that for any vector field on M, N, its index is given by the Euler characteristic. We want to show that the same is true for $M\#N$. The connected sum $M\#N$ is formed from M, N by removing a disk from each and gluing together corresponding boundary points. There are two important curves we need to examine on $M\#N$. The first is just the curve that runs around the circle that is formed when the circle in $M\backslash\operatorname{int} D^2$ is identified to the boundary circle in $N\backslash\operatorname{int} D^2$. If we consider a vector field on $M\#N$ as determining vector fields on $M\backslash\operatorname{int} D^2$ and $N\backslash\operatorname{int} D^2$, then the tangent vectors to this circle in $M\backslash\operatorname{int} D^2$ will be identified with corresponding tangent vectors to the boundary circle in $N\backslash\operatorname{int} D^2$. The second curve we need to consider is one which runs from $M\backslash\operatorname{int} D^2$ into $N\backslash\operatorname{int} D^2$ and cuts the identified circles perpendicularly. This curve will have its tangent vector on the boundary circle of $M\backslash\operatorname{int} D^2$ an exterior normal and it will be an interior normal for the boundary circle of $N\backslash\operatorname{int} D^2$ at the corresponding point. The vector field on $D_M\backslash\operatorname{int} D^2$ extends to a vector field on M by radial damping of the vector field over the disk, so that this new vector field will have all of the old singularities as well as a singularity at the center point of the disk. We may similarly extend the vector field on $N\backslash\operatorname{int} D^2$ to a vector field on N. Denote the original vector field on $M\#N$ by v and the restrictions of the extended vector fields on M, N to $M\backslash\operatorname{int} D^2, N\backslash\operatorname{int} D^2$ by v_M, v_N. We assume as before that v has only a finite number of isolated singularities and that none of them occur on the circle where the connected sum is formed. Let d_M, d_N denote the degrees of the vector fields v_M, v_N on their boundary circles. Then by our initial assumption we have $I(v_M) + d_M = \chi(M), I(v_N) + d_N = \chi(N)$. The vector fields on the two disks are related on their boundaries on the same way that they were for the sphere. Hence $d_M + d_N = 2$. This implies that

$$I(v) = I(v_M) + I(v_N) = \chi(M) - d_M + \chi(N) - d_N = \chi(M) + \chi(N) - 2 = \chi(M\#N).$$

This proves the following lemma.

Lemma 3.7.4. *Suppose M, N are surfaces so that the index of any vector field on each is given by the Euler characteristic. Then the same is true of $M\#N$.*

We may use this lemma together with our results for S, T, P to show by induction that the index of a vector field on any compact connected surface is given by the Euler characteristic.

Theorem 3.7.5. *The index of a vector field on a compact connected surface is given by the Euler characteristic.*

Exercise 3.7.9. Prove Theorem 3.7.5 as outlined above.

Exercise 3.7.10. Give an alternate proof of Theorem 3.7.5 for the surface $T^{(k)}$ as being formed from two regions in the plane with k holes in it, by identifying points on the boundary circles. Such a description arises from thinking of $T^{(k)}$ as sitting in $\mathbb{R}^3$ symmetrically with respect to the xz-plane and then slicing it by that plane and flattening out the two halves. Give an analysis of how tangent vectors on identified circles correspond similar to our analysis for T to relate the indices of the vector fields on the two halves. See Figure 3.27 for a picture of this decomposition in the case of $T^{(3)}$.

Our result above is indicative of the interplay between topological ideas and ideas from analysis provided by a differential structure on a surface. The existence of a vector field requires a differential structure. Somehow the index still manages to measure something topological on the surface, independent of what vector field is chosen. Thus we can use topological ideas to say something about vector fields on the surface from the Euler characteristic, and we can use differential notions to actually compute the Euler characteristic by imposing a vector field on the surface and computing the index of that vector field. We could use this last method to identify an unknown surface. Interplay of this type leads to many exciting areas of research in differential topology. A very nice introduction to differential topology can be found in [21].

Exercise 3.7.11. Suppose M is a compact, connected oriented surface with a vector field of index -2. Identify M.

Exercise 3.7.12. Show that there are vector fields on D^2 with a prescribed integer as index and so Theorem 3.7.5 does not apply to surfaces with boundary without some additional hypothesis.

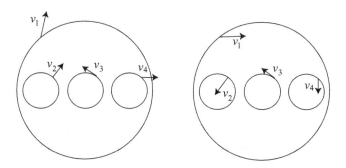

Figure 3.27. Corresponding vector fields from $T^{(3)}$.

Exercise 3.7.13. By adding disks to a surface with boundary, find a formula for the index of a vector field on a surface with boundary where the vector field is required to be an exterior normal vector field on each boundary circle.

3.8   Homotopy equivalences and π_1

We showed earlier that the fundamental group was invariant under homeomorphisms. It has the stronger property that it is invariant under homotopy equivalences as well. We first need to define some terminology.

Definition 3.8.1. Two continuous maps $f_0, f_1 : X \to Y$ are *homotopic* if there is a continuous map $F : X \times I \to Y, F(x,t) = F_t(x)$, so that $F_0(x) = f_0(x), F_1(x) = f_1(x)$. If f_0, f_1 both send $X' \subset X$ to $Y' \subset Y$, then we say that they are continuous maps of the pair $f_i : (X, X') \to (Y, Y')$. They are called homotopic as maps of pairs if $F_t : X' \to Y'$. If, in addition, both of them send X' by the same map (i.e. $f_0|X' = f_1|X'$), we say that they are homotopic rel X' if the additional condition $F_t|X' = f_0|X' = f_1|X_0$ is satisfied.

Note that our definition of homotopy of loops in forming the fundamental group is using homotopy rel $\{0, 1\}$ of maps of pairs $(I, \{0, 1\}) \to (X, x_0)$. An important special case of a homotopy of maps of pairs is when $X' = \{x_0\}$ and $Y' = \{y_0\}$. Homotopy of maps of pairs behaves nicely under composition.

Proposition 3.8.1. *Let* $f_0, f_1 : (X, X') \to (Y, Y')$ *be homotopic maps with homotopy* F, *and* $g_0, g_1 : (Y, Y') \to (Z, Z')$ *be homotopic maps with homotopy* G_t. *Then* $g_0 f_0, g_1 f_1 : (X, X') \to (Z, Z')$ *are homotopic maps.*

Proof. The composition $G_t F_t$ gives the homotopy as maps of pairs. $\square$

We specialize the last proposition to get a result on the induced map on the fundamental group.

Proposition 3.8.2. *If* $g_0, g_1 : (X, x_0) \to (Y, y_0)$ *are homotopic continuous maps of pairs, then* $(g_0)_* = (g_1)_* : \pi_1(X, x_0) \to \pi_1(Y, y_0)$.

Proof. Equality means that, if $f : (I, \{0, 1\}) \to (X, x_0)$ represents an element of $\pi_1(X, x_0)$, then $g_0 f$ is homotopic to $g_1 f$ as maps of pairs. This follows from the last proposition using the constant homotopy for f. $\square$

We now introduce the concept of a homotopy equivalence, which is a generalization of a homeomorphism.

Definition 3.8.2. The continuous map $g : X \to Y$ is called a *homotopy equivalence* if there is a continuous map $h : Y \to X$ with hg homotopic to the identity 1_X on X and gh homotopic to the identity 1_Y on Y. The map h is called the *homotopy inverse* of g. If $g : (X, X_0) \to (Y, Y_0)$ and $h : (Y, Y_0) \to (X, X_0)$ and the homotopies are homotopies of maps of pairs, then we say that g is a *homotopy equivalence of pairs*. A special case of importance is where $g : (X, X_0) \to (X_0, X_0)$

and $g|X_0$ is the identity, $h : (X_0, X_0) \to (X, X_0)$ is the inclusion, and the homotopies are the identity on X_0. Then we say that g is a *deformation retraction* of X onto X_0, or that X *deformation-retracts* onto X_0 with deformation retraction g. In the case when $X_0 = \{x_0\}$ is a point in X and X deformation-retracts to x_0, we say that X is *strongly contractible* to x_0. A deformation retraction to a point is a homotopy between the identity map and the constant map to the point. When X is homotopy equivalent to a point (but with no condition on what happens to the point during the homotopy to the identity), then we say that X is *contractible*. When two spaces are homotopy equivalent, we say that they have the same *homotopy type*. We can also talk of the homotopy type of a pair (X, X_0).

We give some examples.

Any convex set in $\mathbb{R}^n$ deformation-retracts to a point x_0 in it. The deformation retraction uses the straight line homotopy $F_t(x) = (1 - t)x + tx_0$. This homotopy also works if the set $X \in \mathbb{R}^n$ is *star shaped* at x_0, which means that line segments joining x_0 to any $x \in X$ lie in the set.

Sets can be contractible but not deformation-retract to every point in them. A standard example is the comb space $C = \{(x, y) \in \mathbb{R}^2 : x = 0 \text{ or } 1/n, n \in \mathbb{Z}, y \in [0, 1]\}$. This is depicted in Figure 3.28. The figure is deceptive, however, as the collection of vertical line segments approaching $x = 0$ which are getting closer and closer together appear as a black area there due to the resolution. If we zoomed in, they would look more like the line segments to their right. The comb space deformation-retracts to any point on the interval $[0, 1]$ on the x-axis by first deforming to the interval via $F_t(x, y) = (x, (1 - 2t)y)$ on the first half of the time interval, $0 \leq t \leq \frac{1}{2}$, and then contracting via a straight line homotopy on the last half of the time interval via $F_t(x, y) = ((2 - 2t)x + (2t - 1)x_0, 0), \frac{1}{2} \leq t \leq 1$. However, there is no deformation retraction to the point $(0, 1)$. For this point would have to stay fixed during the homotopy, requiring points $(1/n, 1)$ converging to it to stay nearby by a continuity argument. The fact that C is not locally path connected at $(0, 1)$ can be used to show that this is not possible. The details are left as an exercise. Hatcher [13], p. 18 gives an example of a space X which is contractible but does not deformation-retract to any of its points.

Figure 3.28. Comb space.

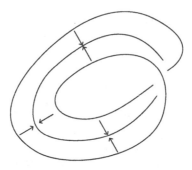

Figure 3.29. Deformation retraction of Möbius band onto the center circle.

Exercise 3.8.1. Show that the comb space C is contractible but does not deformation retract to the point $(0, 1)$.

An important example of a space that deformation-retracts to a subset is where $\mathbb{C}\backslash\{0\}$ deformation-retracts to S^1. The deformation retraction is given by $F_t(\boldsymbol{x}) = (1 - t)\boldsymbol{x} + t(\boldsymbol{x}/|\boldsymbol{x}|)$. Another example is where the Möbius band deformation-retracts to its center circle (see Figure 3.29). If we think of the Möbius band as a quotient space $M = D^1 \times D^1/(-1, y) \sim (1, -y)$, then the deformation retraction is the map $\bar{F}_t$ induced from the map $F_t : D^1 \times D^1 \to D^1 \times D^1$ given by $F_t(x, y) = (x, (1 - t)y)$.

We give an example of a homotopy equivalence of pairs, where the full space is the same and the subspace changes. Let $X = \mathbb{R}, Y = \{x \in \mathbb{R} : |x| \geq \frac{1}{2}\}, Z = \{x \in \mathbb{R} : |x| \geq 1\}$. Let $f : (X, Y) \to (X, Z)$ be given by $f(x) = 2x$ and $g : (X, Z) \to (X, Y)$ by $g(x) = x$. Then f is a homotopy equivalence of pairs since $F_t(x) = (1 + t)x$ is a homotopy between the identity on (X, Y) and $gf(x) = 2x$ and also a homotopy between the identity on (X, Z) and $fg(x) = 2x$. Now form the quotient spaces of all of these spaces by identifying the points of Z to a single point. Then the map $h : X \to S^1$,

$$p(t) = \begin{cases} e^{\pi i t} & \text{if } |t| \leq 1, \\ (-1, 0) & \text{if } |t| \geq 1, \end{cases}$$

induces homeomorphisms of $(X/Z, Y/Z, Z/Z)$ to $(S^1, A, (-1, 0))$ where A is the subset of S^1 with nonpositive first coordinate. The homotopy equivalence of pairs above induces a homotopy equivalence of pairs (S^1, A) to $(S^1, (-1, 0))$. Note that A deformation-retracts to $(-1, 0)$ and our homotopy equivalence is an extension of this to a homotopy equivalence of pairs.

Exercise 3.8.2. By using a rotation of the circle, use the example above to show that there is a homotopy equivalence of pairs $(S^1, B) \to (S^1, (1, 0))$, where B is the subset of points in S^1 with nonnegative x-coordinate.

For another example involving surfaces, consider the torus with one point removed. We think of the torus as a quotient space of $D^1 \times D^1$, where we identify

$(-1, y) \sim (1, y), (x, -1) \sim (x, 1)$. We remove the point $p = (0, 0)$ and then push points radially to the boundary of $D^1 \times D^1$. This uses the map $\bar{F}_t$ induced by $F_t : D^1 \times D^1 \to D^1 \times D^1$ given by $F_t(x, y) = ((1 - t)x + tx/|x|, (1 - t)y + ty/|y|)$. This deformation-retracts the torus minus a point onto the one point union of two circles. In general, the one point union of two spaces is called the *wedge product*, and is denoted by $\vee$. Thus we have a deformation retraction of $T\backslash\{p\}$ onto $S^1 \vee S^1$. Using the exact same map but different identifications to give the Klein bottle results in a deformation retraction of $K\backslash\{p\}$ onto $S^1 \vee S^1$ as well. Instead of removing a point, we could also remove a disk from the middle and use the same maps to give deformation retractions from $T_{(1)}$ and $K_{(1)}$ onto $S^1 \vee S^1$.

Any surface can be expressed as a disk with identifications on the boundary (see Exercise 2.9.57). The surface $T^{(k)}$ results from a disk with the boundary divided into $4k$ equal edges and identified by the pattern $a_1 b_1 a_1^{-1} b_1^{-1} \ldots a_k b_k a_k^{-1} b_k^{-1}$. We depict this in Figure 3.30 for $k = 2$. Then the radial deformation retract of a disk minus a point (or a disk minus a smaller disk) radially onto its boundary circle induces a deformation retraction of $T^{(k)}\backslash\{p\}$ or $T_{(1)}^{(k)}$ onto $S_1^1 \vee \cdots \vee S_{2k}^1$, the one point union of $2k$ copies of the circle, which is what we get as a quotient space of the boundary circle when we make these identifications there. Similarly, $P^{(k)}$ is expressed as the quotient space of the disk with the boundary circle divided into $2k$ equal edges and identified in the pattern $a_1 a_1 \ldots a_k a_k = a_1^2 \ldots a_k^2$. Then $P^{(k)}\backslash\{p\}$ or $P_{(1)}^{(k)}$ deformation retracts onto $S_1^1 \vee \cdots \vee S_k^1$.

If we look at $S_{(p)}$, then this deformation-retracts onto the one-point union (wedge) W_{p-1} of $p - 1$ circles when $p > 1$. For $p = 2$, this is just pushing an annulus out onto a boundary circle radially. For $p > 2$, it is a little more difficult to write down a formula, but we will give a description in terms of a handle decomposition with one 0-handle and then $(p - 1)$ 1-handles attached symmetrically about the boundary. In Figure 3.31 we show this for $p = 3$. For each 1-handle, we extend the attaching arcs radially into the center of the 0-handle. The union of the core arcs $D^1 \times \{0\}$ with this extension from the two attaching points give us

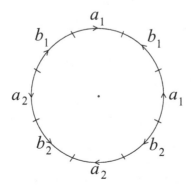

Figure 3.30. The surface $T^{(2)}\backslash\{p\}$ as a quotient space.

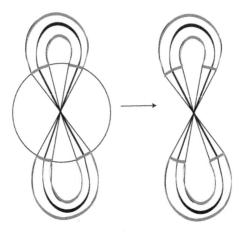

Figure 3.31. Deformation-retracting $S_{(3)}$ onto $S^1 \vee S^1$.

an embedding of $W_2 = S^1 \vee S^1$ into $S_{(3)}$ here. Then we can deformation-retract this as illustrated in the figure onto a thickened version of W_2. We then deformation retract this to W_2 by collapsing the orthogonal line segments that extend into the coning regions the deformation retraction of $D^1 \times D^1$ onto $D^1 \times \{0\}$. Putting together these deformation retractions (and reparametrizing) gives a deformation retraction from $S_{(3)}$ onto the wedge W_2 of two circles. The general case is analogous and gives a deformation retraction of $S_{(p)}$ onto the one-point union W_{p-1} of $p-1$ circles.

The idea of the last example can be modified to show that whenever we attach k 1-handles disjointly to the boundary of a 0-handle, then the result deformation retracts onto a homeomorphic copy of the wedge product W_k of k circles. Note that $T_{(p)}^{(k)}$ can be described in this way with $2k$ 1-handles used to form $T_{(1)}^{(k)}$ and $p-1$ more 1-handles used to form $T_{(p)}^{(k)}$. Similarly, we can form $P_{(p)}^{(k)}$ using a single 0-handle and $k+p-1$ 1-handles. Thus we get the following result.

Proposition 3.8.3. *If $p \geq 1$, then there is a deformation retraction of:*

(a) $S_{(p)}$ *onto a subset which is homeomorphic to the wedge W_{p-1} of $p-1$ copies of the circle;*

(b) $T_{(p)}^{(g)}$ *onto a subset which is homeomorphic to the wedge W_{2g+p-1} of $2g+p-1$ copies of the circle;*

(c) $P_{(p)}^{(h)}$ *onto a subset which is homeomorphic to the wedge W_{h+p-1} of $h+p-1$ copies of the circle.*

Here is another approach; it gives a homotopy equivalence of pairs instead of a deformation retraction, but has the advantage that the maps are easier to understand. We take X to be a surface with boundary with a handle decomposition with a single 0-handle and k 1-handles attached disjointly. We look at the pair (X, p), where p is the center of the 0-handle. The other pair is (Y, p), where

Y is the subset of X which consists of radial lines from p out to the attaching spheres of the 1-handles together with the cores $D^1 \times \{0\}$ of the 1-handles. We first take the deformation retraction of the disk to its center and let the attaching regions follow along. At the end of this retraction, the image of the 1-handles is stretched into cone-shaped regions emanating out from the center together with the 1-handles. Then use a deformation retraction from the 1-handles to their cores extended over the cone-shaped regions to get to Y. Note that p stays fixed over the homotopy but Y moves within itself with the part within the disk moved to p and the rest of Y stretched to fill up Y.

Note that the number of copies of the circle that we have can be rephrased as $1 - \chi$, where χ is the Euler characteristic of the handle decomposition. If we started with another handle decomposition of the surface, then the methods of Section 2.10 can be used to show that we can deform the handle decomposition to another one so that the Euler characteristic does not change (e.g. by merging a pair of 0-handles and a connecting 1-handle to get a new 0-handle) so that the new handle decomposition is of the type used above and has the same Euler characteristic. Thus the Euler characteristic of a surface codes the homotopy type of the surface in terms of the wedge product of $1 - \chi$ copies of the circle. To see that the Euler characteristic is an invariant that can be used to distinguish surfaces, we need to see that the wedge product of k copies of S^1 is not homotopy equivalent to the wedge product of l copies of S^1 when $k \neq l$. To do this, we will use the fundamental group. First we will need to see that how a homotopy equivalence of pairs relates to the fundamental group.

Proposition 3.8.4. *Let* $g : (X, x_0) \to (Y, y_0)$ *be a homotopy equivalence of pairs with homotopy inverse* h. *Then* g_* *induces an isomorphism on fundamental groups.*

Proof. We apply Proposition 3.8.1 to gf, which is homotopic to the identity. It says that $g_* f_* = \mathrm{id}$. Similarly, $f_* g_* = \mathrm{id}$. Thus f_* is an isomorphism with inverse g_*. $\square$

In the next section we will compute the fundamental group of the wedge product of k copies of the circle, showing that it is the free group F_k on k letters. We will also show in Section 3.10 that the fundamental group of a path-connected space does not depend on the choice of base point and the conclusion of the proposition holds whether the homotopy equivalence is one of pairs or not.

Exercise 3.8.3.

(a) Show that the relation of homotopy equivalence is an equivalence relation.

(b) Show that the relation of homotopy equivalence of pairs is an equivalence relation.

Exercise 3.8.4. Show that there is a homotopy equivalence between the letter θ and the symbol 8. (Hint: The map g from θ to 8 collapses the middle line to the center point of 8. The map h from 8 to θ sends the center point of 8 to the midpoint of the central segment. Then nearby parts of 8 are mapped to the rest

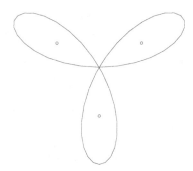

Figure 3.32. $\mathbb{R}^2\backslash\{x_1 \cup x_2 \cup x_3\}$ deformation-retracts to W_3.

of the central segment and the remainder of the 8 is mapped to the top and bottom parts of θ. You need to show that hg is homotopic to the identity on θ and gh is homotopic to the identity on 8. Just give a clear description of your homotopies—exact formulas are not required.)

Exercise 3.8.5. Show that the complement of p points in the plane is homotopy equivalent to the wedge W_p of p copies of S^1. (Hint: Draw p teardrop-shaped circles joined at the center point of the plane and use a radial homotopy from the outside and inside of these circles to describe a deformation retraction onto this set. Here the p points removed are in the inside of the teardrop-shaped circles. See Figure 3.32 where $p = 3$. Again, a geometric description of the homotopy in terms of radial homotopies is being sought, not a formula.)

Exercise 3.8.6. For the letters of the alphabet as typed below, classify each letter as homotopy equivalent to either a point, a circle, or the wedge of two circles:

$$\text{ABCDEFGHIJKLMNOPQRSTUVWXYZ}$$

We next look at the union X of the unit circle Y and the segment s from $x = (\frac{1}{2}, 0)$ to $y = (1, 0)$. We claim that there is a homotopy equivalence of pairs $f : (X, x) \to (Y, y)$. The map f sends the segment s to y and is the identity on Y. The homotopy inverse $g : Y \to X$ takes a small arc $A = A_+ \cup A_-$ onto $s \cup A$. Here A_+ is the arc from 1 to $e^{2i\epsilon}$ and A_- is its reflection in the x-axis. Now A_+ can be identified as the union of two arcs, each of which is homeomorphic to a line segment and parametrized by the angle $A_+ \simeq [0, \epsilon] \cup [\epsilon, 2\epsilon]$. The union $s \cup A_+$ can be identified as the union of the two intervals $s \cup [0, 2\epsilon]$. The map g from A_+ to $s \cup A_+$ then comes from sending $[0, \epsilon]$ homeomorphically onto s using an affine linear map from $[0, \epsilon]$ to $[\frac{1}{2}, 1]$ and sending $[\epsilon, 2\epsilon]$ to $[0, 2\epsilon]$. Putting all of these together we get a homeomorphism from A_+ to $s \cup A_+$. The map from A_- to $s \cup A_-$ is described similarly; it is just the conjugate of this map using reflection. The definition of g on $Y\backslash\text{int } A$ is the identity. Now look at the map $gf : (X, x) \to (X, x)$. This sends $X\backslash(s \cup A)$ via the identity. It sends

$s \cup A_+ \to s \cup A_+$ to itself. This set is homeomorphic to an interval $[0,1]$ where $s \simeq [0, \frac{1}{2}]$ and $A_+ \simeq [\frac{1}{2}, 1]$. The self-map is equivalent under this homeomorphism to the map that sends $[0, \frac{1}{2}]$ to 0, sends $[\frac{1}{2}, \frac{3}{4}]$ to $[0, \frac{1}{2}]$, and sends $[\frac{3}{4}, 1]$ to $[\frac{1}{2}, 1]$. But any self-map $[0,1]$ to $[0,1]$ which sends 0 to 0 and 1 to 1 is homotopic to the identity. Going back to $gf : s \cup A \to s \cup A$, this implies that this is homotopic to the identity. This then leads to a homotopy between the identity and $gf : X \to X$. This homotopy preserves x so is a homotopy of pairs (X, x). Now consider $fg : Y \to Y$. This is the identity on $Y \backslash \text{int } A$. If we identify A_+ with $[0, 2\epsilon] \simeq [0, 1]$, then this map preserves 0 and 2ϵ, so is homotopic to the identity. Hence $fg : (Y, y) \to (Y, y)$ is homotopic to the identity. We record this result for future use.

Proposition 3.8.5. *Let X denote the union of the unit circle and the line segment s from $(\frac{1}{2}, 0)$ to $(1, 0)$. Then the map $f : (X, (\frac{1}{2}, 0)) \to (S^1, (1, 0))$ that collapses s to $(1, 0)$ and is the identity on S^1 is a homotopy equivalence of pairs. In particular, it induces an isomorphism $f_* : \pi_1(X, (\frac{1}{2}, 0)) \to \pi_1(S^1, (1, 0))$.*

A loop which corresponds to the generator is the composition $\alpha * \gamma * \bar{\alpha}$, where $\gamma(s) = e^{2\pi i s}$ is a standard generator for $\pi_1(S^1, (1, 0))$ and α is the path running along s from $(\frac{1}{2}, 0)$ to $(1, 0)$. This path is sent to the constant path in Y at $(1, 0)$ and so $\alpha * \gamma * \bar{\alpha}$ is sent to $c * \gamma * c$ which is homotopic to γ.

Exercise 3.8.7. Show that the proposition can be extended to prove that if we take the wedge W_k of k circles and add an interval I joined by identifying 1 to the wedge point, the pair $(X, 0) = (I \cup_1 W_k, 0)$ is homotopy equivalent to (W_k, w) where w is the wedge point.

Now consider $D^2 \backslash \{0\}$. This space deformation-retracts to $Z = D^2 \backslash \text{int } \frac{1}{4} D^2$. Then the space Z deformation-retracts to $X = s \cup S^1$ from Proposition 3.8.5. To see the latter deformation retraction, consider Figure 3.33. It shows that there is an arc on $\frac{1}{4} S^1$ and each point on the arc can be joined via a straight line to a point on s. The deformation retraction maps these line segments along themselves to points on s. The rest of Z then is described via line segments from

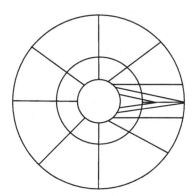

Figure 3.33. A deformation retraction.

the remaining arc in $\frac{1}{4}S^1$ to S^1 as illustrated in Figure 3.33. This part then deformation retracts onto the circle along these line segments. Putting these together, we get the following result.

Proposition 3.8.6. *If $X = s \cup S^1$ is the union of the unit circle and the segment from $(\frac{1}{2}, 0)$ to $(1, 0)$, then there is a deformation retraction from $D^2 \backslash \{\mathbf{0}\}$ onto X.*

When we look at the circle at radius $\frac{1}{2}$, then the image of a loop running around it once under the deformation retraction is a loop of the form $\alpha * \gamma * \bar{\alpha}$ which gives a generator of $\pi_1(X, (\frac{1}{2}, 0))$.

Now we look at the case of a surface without boundary again. We regard it as coming from a disk with identifications on the boundary, such as forming $T^{(k)}$ by identifying edges via the pattern $a_1 b_1 a_1^{-1} b_1^{-1} \cdots a_k b_k a_k^{-1} b_k^{-1}$. Earlier in this section we saw how this led to a deformation retraction from $D^2 \backslash \{\mathbf{0}\}$ onto the quotient space of the boundary, which we identified to the wedge W_{2k}. We want to point out here that there is another deformation retraction of this quotient space minus the center point to the space $X = s \cup W_{2k}$ where we add to the boundary an edge s connecting $(\frac{1}{2}, 0)$ to $(1, 0)$, which we choose as one of the identified vertices. We just take the argument in Proposition 3.8.6 and use it for this quotient space of the disk. This also works for $P^{(k)}$ and gives the following result.

Proposition 3.8.7. (a) *Consider the surface $T^{(k)}$ as given by the quotient of the disk under identifications on the boundary circle with the pattern $a_1 b_1 a_1^{-1} b_1^{-1} \ldots a_k b_k a_k^{-1} b_k^{-1}$. Let p be the center point in the disk and $q = (1, 0)$ a vertex on the boundary where a_1 begins, and $r = (\frac{1}{2}, 0)$. We denote the identified boundary as W_{2k}. Let s be the linear edge joining r to q and $X = s \cup W_{2k}$. Then $T^{(k)} \backslash \{p\}$ deformation retracts onto X.*

(b) *Consider $P^{(k)}$ as given by the quotient of the disk under identification on the boundary circle with the pattern $a_1 a_1 \ldots a_k a_k$. Let p be the center point in the disk and $q = (1, 0)$ a vertex on the boundary where a_1 begins, and $r = (\frac{1}{2}, 0)$. We denote the identified boundary as W_k. Let s be the linear edge joining r to q and $X = s \cup W_k$. Then $P^{(k)} \backslash \{p\}$ deformation-retracts onto X.*

Although the first deformation retraction we gave onto the quotient of the boundary is much simpler, it has a disadvantage in computations of the fundamental group involving the Seifert–van Kampen theorem of the next section which the above deformation retraction avoids. If we look at the interior of the disk, it is embedded into this quotient space. When we delete the center p, this deformation-retracts onto the circle $\frac{1}{2}S^1$. The loop $\beta(s) = \frac{1}{2}e^{2\pi i s}$ which runs around this circle once generates the fundamental group of the circle and hence the fundamental group of the interior of the disk with p deleted, which deformation retracts to it. When we look at its image under the deformation retraction onto $s \cup W_{2k}$, this loop β is mapped to a loop $\alpha * \gamma * \bar{\alpha}$ where α runs along s from r to q and γ gives a loop in the wedge W which is described by $a_1 b_1 a_1^{-1} b_1^{-1} \ldots a_k b_k a_k^{-1} b_k^{-1}$ in the case of $T^{(k)}$ and by $a_1 a_1 \ldots a_k a_k$ in the case of $P^{(k)}$.

Exercise 3.8.8. (a) Show that the boundary connected sum of two surfaces with boundary is homotopy equivalent to the union of the two surfaces with a line segment joining their boundaries. (Hint: The line segment is the core of the 1-handle forming the boundary connected sum. Reduce the problem to finding a deformation retraction from $D^1 \times D^1$ onto $\{\pm 1\} \times D^1 \cup D^1 \times \{0\}$.)

(b) Show that the boundary connected sum of two surfaces with boundary is homotopy equivalent to the wedge product of two surfaces.

3.9 Seifert–van Kampen theorem and its application to surfaces

We return to calculating the fundamental group. The most powerful technique besides using covering spaces is the Seifert–van Kampen theorem, which computes the fundamental group of a path-connected space X which is expressed as the union of two path-connected open sets A, B with path-connected intersection. A great deal of difficulty in stating the theorem is understanding the algebraic construction of a free product with amalgamation used in describing the result. If G_1, G_2 are groups, then the *free product* $G_1 * G_2$ is the group which is formed from words in the elements of G_1, G_2, where the only relations involved are relations in G_1, relations in G_2, and identifying the identity of G_1 with the identity of G_2. That is, an element of $G_1 * G_2$ can be written as a product $x_1 x_2 \cdots x_n$ where $x_i \in G_1$ or $x_i \in G_2$. When elements of G_i are adjacent, they may be replaced by the product in G_i. In this way, we can always represent an element as an alternating product of elements in G_1 and G_2: this expression is called a *reduced word*. The only relation between elements of the two groups is that we identify the identity element of G_1 with the identity element of G_2 and may interchange these to change the word. There are natural injective homomorphisms i_1, i_2 from G_1, G_2 to $G_1 * G_2$. The free product is characterized algebraically up to isomorphism by the universal property that whenever there are homomorphisms ϕ_i from G_1, G_2 to another group H, then there is a homomorphism ϕ from $G_1 * G_2$ to H satisfying $\phi i_j = \phi_j$.

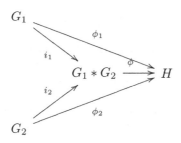

If we write an element of $G_1 * G_2$ as a reduced word $w = x_1 y_1 \cdots x_k y_k$, then $\phi(w) = \phi_1(x_1)\phi_2(y_1)\cdots\phi_1(x_k)\phi_2(y_k)$. Here $x_i \in G_1, y_i \in G_2$, and the first and last elements could be missing.

For our purposes, one of the most important examples of a free product is formed when $G_1 = G_2 = \mathbb{Z}$. Here $G_1 * G_2$ is the free group F_2 on two letters. By iterating the construction, we can take the free product of k copies of $\mathbb{Z}$, giving the free group F_k on k letters. Elements are just words in these k letters, with the group operation being juxtaposition. There are some delicate algebraic issues in the construction of free products. We will treat this somewhat informally and refer the reader to more advanced texts such as [5,13] for details.

A more refined algebraic construction uses a pair of homomorphisms $\psi_1 : K \to G_1, \psi_2 : K \to G_2$ to form the free product with amalgamation $G_1 *_K G_2$. This can be defined as the quotient group of the free product where we take the quotient by the normal subgroup generated by elements of the form $\psi_1(k)\psi_2(k)^{-1}$. There is a natural homomorphism $\bar{i}_j : G_j \to G_1 *_K G_2$ which is induced by the composition $i_j : G_j \to G_1 * G_2$ with the quotient map $G_1 * G_2 \to G_1 *_K G_2$. These homomorphisms no longer are injections in general. More informally, we are allowed to change a word by replacing the element $\psi_1(k) \in G_1$ with the element $\psi_2(k) \in G_2$ as well as the earlier operations of replacing an element of G_i by an equivalent expression of the element as a product and identifying the two identity elements. Note that the free product itself is a special case of the free product with amalgamation where K is the trivial group with one element. A special case of importance is when G_2 is the trivial group. If N denotes the normal subgroup generated by the image of $\psi_1(K)$, then $G_1 *_K G_2 \simeq G_1/N$. The amalgamated free product is characterized by the following universal property. Suppose $\phi_1 : G_1 \to H, \phi_2 \to H$ are homomorphisms so that $\phi_1\psi_1 = \phi_2\psi_2$. Then there is a unique homomorphism $\phi : G_1 *_K G_2 \to H$ with $\phi\bar{i}_j = \phi_i$.

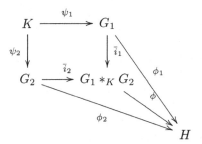

We now state the Seifert–van Kampen theorem.

Theorem 3.9.1 (Seifert–van Kampen theorem). *Let X be a path-connected space with base point x_0. Suppose $X = A \cup B$, where A, B are path-connected open sets in X with path-connected intersection $A \cap B$ containing x_0. Let $\psi_A : \pi_1(A \cap B, x_0) \to \pi(A, x_0), \psi_B : \pi(A \cap B, x_0) \to \pi_1(B, x_0)$ be the homomorphisms induced by the inclusions. Then $\pi_1(X, x_0)$ is isomorphic to the amalgamated free product $\pi_1(A, x_0) *_{\pi_1(A\cap B, x_0)} \pi_1(B, x_0)$.*

Corollary 3.9.2.

(a) *If $\pi_1(A \cap B, x)$ is the trivial group, then $\pi_1(X, x)$ is the free product of $\pi_1(A, x) * \pi_1(B, x)$.*

(b) If $\pi_1(B,x)$ is the trivial group, then $\pi_1(X,x) \simeq \pi_1(A,x)/N$ where N is the normal subgroup of $\pi_1(A,x)$ which is generated by the image of $\pi_1(A \cap B, x) \to \pi_1(A,x)$.

We want to note some important special cases of this result. Before doing so, however, we need to introduce some more algebraic terminology. If we start with a free product F_k, with generators $x_1, \ldots, x_k$, we will write $F_k = \langle x_1, \ldots, x_k \rangle$. Suppose we pick a finite number of elements $r_1, \ldots, r_m \in F_k$. They are expressible as words in the symbols $x_1, \ldots, x_k$. Let $N(r_1, \ldots, r_m)$ denote the smallest normal subgroup of F_k which contains $r_1, \ldots, r_m$. Elements of $N(r_1, \ldots, r_m)$ are expressible as finite products of conjugates $g r_i g^{-1}$ of the elements $r_1, \ldots, r_m$. Then we denote by $\langle x_1, \ldots, x_k | r_1, \ldots, r_m \rangle$ the quotient $F_k/R(r_1, \ldots, r_m)$. We call this the finitely presented group with generators $x_1, \ldots, x_k$ and relations $r_1, \ldots, r_m$.

We first look at a wedge of circles W_2. When $k = 2$ and $W_2 = S_1^1 \vee S_2^1$, let x be the wedge point. We can decompose this into two open sets $A \cup B$, where A is the union of S_1^1 and a small arc about x in S_2^1. Similarly, let B be the union of S_2^1 and the union of a small arc about x in S_1^1. Then $A \cap B$ deformation-retracts to x, so case (a) of Corollary 3.9.2 applies. Also, A deformation-retracts to S_1^1 and B deformation-retracts to S_2^1. This means that $\pi_1(W_2, x) \simeq \pi_1(A,x) * \pi_1(B,x) \simeq \pi_1(S_1^1, x) * \pi_1(S_2^1, x) \simeq F_1 * F_1 = F_2$. This argument can then be adapted to prove the following proposition.

Proposition 3.9.3. $\pi_1(W_k, x) \simeq F_k$.

Exercise 3.9.1. Prove Proposition 3.9.3.

Applying Proposition 3.8.3, which says that surfaces with boundary deformation-retract to wedges of circles, and Proposition 3.8.4, which says that a deformation retraction induces an isomorphism on the fundamental group, we get the following theorem.

Theorem 3.9.4.

(a) $\pi_1(S_{(p)}, x) \simeq F_{p-1}$, $p \geq 1$.

(b) $\pi_1(T_{(p)}^{(g)}, x) \simeq F_{2g+p-1}$, $p \geq 1$.

(c) $\pi_1(P_{(p)}^{(h)}, x) \simeq F_{h+p-1}$, $p \geq 1$.

Strictly speaking, this only applies as indicated to some base point in the wedge where we are deformation-retracting the surface. However, we will show in Section 3.10 that the fundamental group of a path-connected surface is independent of the base point up to isomorphism.

We have already shown that $\pi_1(S, x) \simeq \{e\}$. To deal with another surface without boundary, we regard it as the quotient of a disk with identifications. For $T^{(g)}$, the identifications on the boundary are given by the word $a_1 b_1 a_1^{-1} b_1^{-1} \ldots a_g b_g a_g^{-1} b_g^{-1}$. For $P^{(h)}$, the identifications on the boundary are given by the word $a_1 a_1 \ldots a_h a_h$. The Seifert–van Kampen theorem then leads to the following result.

Theorem 3.9.5.

(a) $\pi_1(T^{(g)}, x) \simeq \langle a_1, b_1, \ldots, a_g, b_g | a_1 b_1 a_1^{-1} b_1^{-1} \ldots a_g b_g a_g^{-1} b_g^{-1} \rangle$.

(b) $\pi_1(P^{(h)}, x) \simeq \langle a_1, \ldots, a_h | a_1^2 \ldots a_h^2 \rangle$.

Proof. We prove part (a). The argument for (b) is essentially the same. In writing $T^{(g)}$ as a quotient of a disk, let A be the complement of the center point p of the disk, and let B be the interior of the disk. Let q be the point $(1,0)$ which is taken as the vertex of a_1 and $r = (\frac{1}{2}, 0)$. We take the point r as our base point for the calculation. Note that since B is contractible to $r, \pi_1(B, r)$ is trivial. Then $A \cap B$ deformation retracts onto the circle at radius $\frac{1}{2}$, and so we can identify $\pi_1(A \cap B, r)$ to $\pi_1(\frac{1}{2}S^1, r) \simeq \mathbb{Z}$. The generator is represented by the loop $\beta(s) = \frac{1}{2}e^{2\pi i s}$. The set A deformation retracts onto the union of the boundary of the disk, which is a wedge W_{2g} of circles, with the segment s running from r to q by Proposition 3.8.7. By Proposition 3.8.7, there is a homotopy equivalence of pairs between $(s \cup W_{2g}, r)$ and (W_{2g}, q). This induces an isomorphism of fundamental groups and so this fundamental group can be identified with F_{2g}. By part (b) of the corollary to the Seifert–van Kampen theorem, $\pi_1(T^{(2g)}, r)$ is the quotient of $\pi_1(A)$ by the normal subgroup generated by the image of $\pi_1(A \cap B, r)$. This last group is generated by β. When we look at the image of β under the deformation retraction of A onto $s \cup W_{2g}$, it is sent to $\alpha * \gamma * \bar{\alpha}$. Here α runs along s from r to q and γ runs once around the unit circle and considers the image in the quotient space. But the pattern of identifications means that γ represents the word $a_1 b_1 a_1^{-1} b_1^{-1} \ldots a_g b_g a_g^{-1} b_g^{-1} \in \pi_1(W_{2g}, q) \simeq F_{2g} = \langle a_1, b_1, \ldots, a_g, b_g \rangle$. Note that under the isomorphism $\pi_1(A, r) \simeq \pi_1(s \cup W_{2g}, r) \simeq \pi_1(W_{2g}, q)$, the class $[\beta]$ maps to the class $[\gamma] = a_1 b_1 a_1^{-1} b_1^{-1} \ldots a_g b_g a_g^{-1} b_g^{-1}$, so $\pi_1(A, r)/N(\operatorname{im} \pi_1(A \cap B, r)) \simeq \langle a_1, b_1, \ldots, a_g, b_g | a_1 b_1 a_1^{-1} b_1^{-1} \ldots a_g b_g a_g^{-1} b_g^{-1} \rangle$. $\square$

The above proof is complicated by having to deal with the base point r in the intersection. In the next section we will show that if X is path connected and $a, b \in X$ with α a path from a to b, then there is an isomorphism from $\pi_1(X, b)$ to $\pi_1(X, a)$ given by sending $[\gamma] \in \pi(X, b)$ to $[\alpha * \gamma * \bar{\alpha}]$. This is just the inverse of the isomorphism that we were using in the argument above, which we found existed because there was a homotopy equivalence of pairs. Let us use this isomorphism instead in the argument and use the standard deformation retraction of A to the boundary W_{2g}. We start with $\pi_1(T^{(2g)}, r) \simeq \pi_1(A, r)/N(\operatorname{im} \pi_1(A \cap B, r))$. But now we use the isomorphism of $\pi_1(A, r)$ with $\pi_1(A, q)$ and see where the class $[\beta]$ maps to under this isomorphism. It is sent to the class $[\bar{\alpha} * \beta * \alpha]$. Thus we can identify the fundamental group as $\pi_1(A, q)/N([\bar{\alpha} * \beta * \alpha])$. We then use the radial deformation retraction of A onto the quotient W_{2g} of the boundary circle. The path α just maps to a constant path at the base point q here and β maps to the loop γ which goes once around the boundary circle, which in the quotient space represents $a_1 b_1 a_1^{-1} b_1^{-1} \ldots a_g b_g a_g^{-1} b_g^{-1} \in F_{2g} = \pi_1(W_{2g}, q)$.

In general, this last argument proves the following result.

Theorem 3.9.6. *Suppose X is a quotient space of D^2 where we divide the boundary up into p edges and make identifications of edges in a pattern so that all* *vertices are identified. The quotient of the boundary will be a wedge product W_k. Call this common vertex q and write $\pi_1(W_k, q) = F(a_1, \ldots, a_k)$, where we use the same notation for generators as edges after identifications. Let w be the word in these generators which is the image of the standard generator $[\gamma] \in \pi_1(S^1, q)$ under the quotient map $\pi_1(S^1, q) \to \pi_1(W_k, q)$. Then $\pi_1(X, q) \simeq \langle a_1, \ldots, a_k | w \rangle$.*

Proof. From the proof, the Seifert–van Kampen theorem applies to compute $\pi_1(X, r) \simeq \pi_1(A, r)/N([\beta])$. We then use the isomorphisms from $\pi_1(X, r) \simeq \pi_1(X, q), \pi_1(A, r) \simeq \pi_1(A, q)$ to reduce the problem of computing $\pi_1(A, q)$ and the image of $\bar{\alpha} * \beta * \alpha$ within it. Using the radial deformation retraction of A onto the boundary, we can identify $\pi_1(A, q)$ with $\pi_1(W_k, q)$, and so need to see where what element $\bar{\alpha} * \beta * \alpha$ represents. As before, this represents the image of a generating loop γ of $\pi_1(S^1, q)$ which we have defined as the word w. □

We used above the result from Section 3.10 that the fundamental group of a path-connected space does not depend on the base point chosen up to isomorphism as well as the specific description of the isomorphism. It is also shown in Section 3.10 that a homotopy equivalence $f : X \to Y$ induces an isomorphism $f_* : \pi_1(X, x) \to \pi_1(Y, f(x))$ whether it is a homotopy equivalence of pairs or not.

The fundamental groups that are occurring are complicated since they are nonabelian in general. In particular, it is a nontrivial problem to distinguish such groups up to isomorphism. One way of dealing with this is to abelianize the fundamental groups by taking their quotients π_1^{ab} by the *commutator subgroup*, which is the smallest normal subgroup which contains each commutator $ghg^{-1}h^{-1}$ of elements g, h of the group. For $T^{(g)}$, this abelianization is $2g\mathbb{Z}$, the direct sum of $2g = 2 - \chi$ copies of $\mathbb{Z}$. The Euler characteristic is detected as $2 - 2g$ in this abelianization. For $P^{(h)}$, the abelianization is $(h-1)\mathbb{Z} \oplus \mathbb{Z}_2$. To see this, rechoose the generators to be $g_1 = a_1 \ldots a_h, a_2, \ldots, a_h$, and the abelianized relation to be $g_1^2 = 1$. Thus nonorientability can be detected by the presence of $\mathbb{Z}_2$ in the abelianized fundamental group, and the Euler characteristic $\chi = 2 - h$ is detected through the occurrence of $h - 1$ in the number of copies of $\mathbb{Z}$ in the fundamental group calculation. Since the fundamental group is an invariant of homeomorphism type (in fact, of homotopy type), this can be used to show that the Euler characteristic is also an invariant under homeomorphism and is independent of the handle decomposition as claimed earlier. Alternatively, we can just use the abelianization of the fundamental group in the same manner that we used the Euler characteristic to prove that the abelianized π_1 and the number of boundary circles will distinguish a surface up to homeomorphism.

Exercise 3.9.2. Apply the Seifert–van Kampen theorem to compute the fundamental group of the following spaces:

(a) $S^1 \vee S^2$;

(b) $S^n, n \geq 3$;

(c) $S^1 \vee S^n, n \geq 3$.

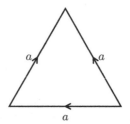

Figure 3.34. Dunce hat.

Exercise 3.9.3. Consider the *dunce hat* D, which is made from a triangle by making identifications of its three edges via the pattern $aa^{-1}a^{-1}$ as indicated in Figure 3.34.

(a) For a point x in the interior of the triangle, compute $\pi_1(D\backslash\{x\}, y)$, where y is also in the interior of the triangle.

(b) Using the decomposition of D with $A = D\backslash\{x\}$, $B = $ interior of triangle, compute $\pi_1(D, y)$.

(c) For a standard neighborhood N of a point x on the edge of the triangle, compute $\pi_1(N\backslash\{x\}, y)$. Note that N is the union of three standard half-disk neighborhoods of the three points that are being identified where the small edge segments near those points are also being identified.

(d) For a vertex x of the triangle (note all three vertices are being identified to one point) and a standard neighborhood N, compute $\pi_1(N\backslash\{x\}, y)$. Here N is formed from three wedges of disk neighborhoods with certain identifications.

Exercise 3.9.4. The projective plane can be considered as the quotient of the disk, where we identify $\boldsymbol{x} \sim -\boldsymbol{x}$ for $\boldsymbol{x} \in S^1$. The *pseudoprojective plane* P_k is the quotient of the disk, where we identify $\boldsymbol{x} \sim e^{2\pi i/k}\boldsymbol{x}$ for $\boldsymbol{x} \in S^1$.

(a) Show that $S^1/\boldsymbol{x} \sim e^{2\pi i/k}\boldsymbol{x}$ is homeomorphic to S^1.

(b) Use the Seifert–van Kampen theorem to compute $\pi_1(P_k, x)$. (Hint: Use A to be the complement of the center of the disk in P_k and B to be the interior of the disk.)

Exercise 3.9.5. Consider the torus $T = S^1 \times S^1 = A \cup B$, where $A = S^1 \times C \cup \{1\} \times S^1$ and $B = S^1 \times D \cup \{1\} \times S^1$. Here $C = \{\boldsymbol{x} \in S^1 : x_1 \leq \frac{1}{2}\}$, $D = \{\boldsymbol{x} \in S^1 : x_1 \geq -\frac{1}{2}\}$. Use the Seifert–van Kampen theorem to compute $\pi_1(T, (1, 0))$. (Hint: In computing the induced maps, see where generators are sent geometrically.)

Exercise 3.9.6. Compute $\pi_1^{ab}(M)$ for the surfaces in Figure 3.35, and use your results and the number of boundary components, to classify the surfaces.

Exercise 3.9.7. Compute $\pi_1^{ab}(M)$ for the surfaces in Figure 3.36, and use your results and the number of boundary components, to classify the surfaces.

attach 2-handle as indicated

(a)   (b)

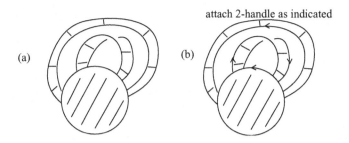

Figure 3.35. Surfaces for Exercise 3.9.6.

attach 2-handle as indicated

(a)   (b)

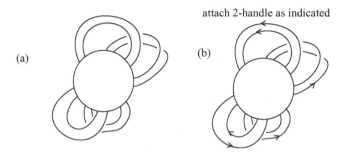

Figure 3.36. Surfaces for Exercise 3.9.7.

In order to prove the Seifert–van Kampen theorem, we will need some preliminaries.

Lemma 3.9.7. *Suppose that f is a loop in $X = A \cup B$, where A, B open in X and $A, B, A \cap B$ are path-connected sets containing x_0. If the interval I is subdivided into subintervals, each of which is mapped to A or to B in order to express $f = f_1 * \cdots * f_n$ as a product of paths in A or B, then we can homotope f relative to the end points to re-express it as $f = f_1' * \cdots * f_n'$, where f_i' is a loop in A or B at x_0.*

Proof. Let $v_j = f(j/n), j = 1, \ldots, n-1$. If v_j is in the intersection $A \cap B$, then we can find a path p_i in $A \cap B$ which runs from v_j to x_0. If it is in A (resp., B) but not in the intersection, we can choose such a path in A (resp., B). We first compose $f : (I, \{0,1\}) \to (X, x_0)$ with a map $I \to I$ which sends a small subinterval about j/n to j/n and stretches out the intervening subintervals via affine linear maps to cover $[(j-1)/n, j/n]$. The composition of this map with f will give $f_1 * c_1 * f_2 * \cdots * c_{n-1} * f_n$. Here c_j denotes the constant map at v_j. This is homotopic to f relative to the end points by the argument we used in proving the constant map serves as the identity in π_1. Up to homotopy, we can replace the maps on the small subintervals mapped to v_j by the composition $p_i * \bar{p}_i$. Then f is homotopic to $f_1' * \cdots * f_n'$, with $f_1' = f_1 * p_1$, $f_j' = \bar{p}_{j-1} * f_j * p_j$, $j = 2, \ldots, n-1$, $f_n' = \bar{p}_{n-1} * f_n$. $\qquad\square$

Lemma 3.9.8. *Let $X = A \cup B$ as in the statement of the Seifert–van Kampen theorem. Then the map $\phi_{AB} : \pi_1(A, x_0) * \pi_1(B, x_0) \to \pi_1(X, x_0)$ which is determined from induced maps from the inclusions $\phi_A : \pi_1(A, x_0) \to \pi_1(X, x_0)$ and $\phi_B : \pi_1(B, x_0) \to \pi_1(X, x_0)$ is surjective.*

Proof. Let f represent an element of $\pi_1(X, x_0)$. Then $\{f^{-1}(A), f^{-1}(B)\}$ is an open cover of I. Since I is a compact metric space, there is a Lebesgue number $\delta > 0$ for this cover. Choose n so that $1/n < \delta$ and subdivide the interval into n subintervals of length $1/n$. Each subinterval is mapped to A or B, so Lemma 3.9.7 says that f is homotopic relative to the end points to a product $f_1' * \cdots * f_n'$, where each f_j' is a loop at x_0 on the jth subinterval. Then $[f_1'] \bar{*} \cdots \bar{*} [f_n']$ is an element of $\pi_1(A, x_0) * \pi_1(B, x_0)$ which maps to $[f]$.   □

Since $\phi_A \psi_A = \phi_B \psi_B$, the map ϕ_{AB} induces a surjective map

$$\bar{\phi}_{AB} : \pi_1(A, x_0) *_{\pi_1(A \cap B, x_0)} \pi_1(B, x_0) \to \pi_1(X, x_0).$$

To show that this map is an isomorphism, we show that any element $[f_1'] \bar{*} \cdots \bar{*} [f_n']$ which maps to the identity element is the identity in the amalgamated free product. What this means is that we can reduce the product $[f_1'] \bar{*} \cdots * \bar{*} [f_n']$ to the identity element by using relations in $\pi_1(A, x_0)$, relations in $\pi_1(B, x_0)$, and identifying $\psi_A(\alpha)$ with $\psi_B(\alpha)$ to transfer an element from $\pi_1(A, x_0)$ to $\pi_1(B, x_0)$. The condition that $[f_1'] \bar{*} \cdots * \bar{*} [f_n']$ represents the identity of $\pi_1(X, x_0)$ means that there is a homotopy $F : I \times I \to (X, x_0)$ which satisfies $F(s, 0) = f_1' * \cdots * f_n'(s)$, $F(s, 1) = F(0, t) = F(1, t) = x_0$. We need to use F to get the required equivalence of $[f_1'] \bar{*} \cdots \bar{*} [f_n']$ to $[e_{x_0}]$ in the amalgamated free product. We first use F to pull back the cover $\{A, B\}$ to $I \times I$. Using the fact that $I \times I$ is a compact metric space, we find its Lebesgue number $\delta > 0$ and choose m so that $\sqrt{2}/mn < \delta$. Letting $k = mn$, we then subdivide $I \times I$ into k^2 subrectangles of side length $1/k$. Each of these subrectangles maps to A or to B. When adjacent rectangles map to A and B, this means that the edge joining them maps to $A \cap B$. In the next lemma, we prove the analogue of Lemma 3.9.7 for maps from the square.

Lemma 3.9.9. *Suppose that $F : I \times I \to X = A \cup B$, where A, B are open in X and $A, B, A \cap B$ are path-connected sets containing x_0, and if $K = \{0, 1\} \times I \cup I \times \{1\}$, then $F|K(s, t) = x_0$. If the rectangle $I \times I$ is subdivided into k^2 equal subrectangles, each of which is mapped to A or to B, then there is another map $F' : I \times I \to X$ which agrees with F except in small disk neighborhoods of the vertices where the rectangles come together, changing the map on those disks where the vertex is not sent to x_0, and $F'|K = F|K$. On $I \times \{0\}$, the new map F_0' represents the same map in the free product as the map F_0.*

To prove this lemma, we need another technical construction. What we first do is compose F with a map $G : I \times I$, which is the identity outside small disks about those vertices that are not sent to x_0. On these disks, we use a map which sends a smaller disk to the center and then stretches the annular region between the smaller disk and the whole disk onto the whole disk. These maps are modeled on the map from D^2 to itself which in polar coordinates sends $re^{i\theta} \to 0$ if $r \leq \frac{1}{2}$ and sends $re^{i\theta} \to 2(r - \frac{1}{2})e^{i\theta}$ if $\frac{1}{2} \leq r \leq 1$. After composing F with G, disks about each vertex $(i/k, j/k)$ are sent the single point v_{ij} where $(i/k, j/k)$ was originally sent. If $v_{ij} \in A \cap B$, then choose a path p_{ij} in $A \cap B$ from v_{ij} to x_0. If $v_{ij} \notin A \cap B$, then select the path to lie in A or in B, depending whether $v_{ij} \in A$ or $v_{ij} \in B$. Now replace the composition GF on the small disks about the vertices with a map on the disk about the ij-vertex so that on radial line segments it is the path p_{ij}. This new map will be F'. Now it will send each vertex to x_0. Note also that on the bottom edge of the rectangle, we will have replaced each f_i by a product of loops in the same set A or B as f_i which represents the same element in $\pi_1(A, x_0)$ or $\pi_1(B, x_0)$. Hence it represents the same element in the free product.

On each level $I \times \{j/k\}$, our subdivision expresses the map restricted to that level as $f_{1j} * f_{2j} * \cdots * f_{kj}$, which determines an element in the free product, with $f_{10} * f_{20} * \cdots * f_{k0}$ determining our original element and each f_{ik} always the map sending the edge to x_0. We look at the vertical edge of the rectangle connecting (i, j) to $(i, j - 1)$. The restriction of F' to this edge (directed downward) determines a loop which we denote as h_{ij}. Note that h_{0j} and h_{kj} each send the edge to x_0. We start replacing the element $f_{10} * f_{20} * \cdots * f_{k0}$ by $h_{01} * f_{10} * f_{20} * \cdots * f_{k0}$ using the fact that h_{01} represents the identity and so the new product is homotopic to the old one, using the equivalence in $\pi_1(A, x_0)$ or $\pi_1(B, x_0)$, depending on whether the lower left rectangle R_{11} is sent to A or to B. In general, we denote by R_{ij} the rectangle in the jth row and ith column when $I \times I$ is subdivided; this notation uses the second index to index the height of the rectangle, with the height changing from 1 to k as we move upward.

We illustrate in Figure 3.37 the subdivision and component loops when $k = 3$.

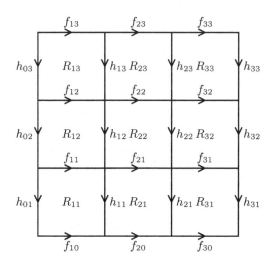

Figure 3.37. Subdivision when $k = 3$.

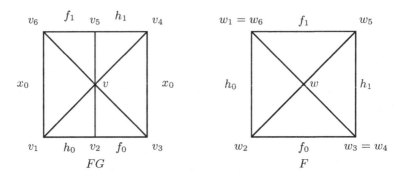

Figure 3.38. Reparametrizing a homotopy.

Let us suppose R_{11} is sent to A. We want to replace $h_{01} * f_{10}$ by $h_{11} * f_{11}$ in this product. Since the first rectangle is sent to A, we can do this in the free product as long as these represent the same element of $\pi_1(A, x_0)$. We now prove a lemma that says that this is true.

Lemma 3.9.10. *Consider a map F from a rectangle $I \times I \to C$ which sends the corner points to x_0. Let h_0, h_1 denote the restriction to the vertical edges (directed downward) and f_0, f_1 denote the restriction to the horizontal edges (directed to the right). Then $f_0 * h_0$ and $h_1 * f_1$ represent the same element of $\pi_1(C, x_0)$.*

Proof. We need to compose F with an appropriate map G from $I \times I$ to itself. The map we choose will use a subdivision of $I \times I$ into six triangles, and we will map each triangle affine linearly to either a triangle or edge in $I \times I$. The map G is determined by where the vertices are mapped. The vertices in the domain $I \times I$ are $v_1 = (0, 0), v_2 = (\frac{1}{2}, 0), v_3 = (1, 0), v_4 = (1, 1), v_5 = (\frac{1}{2}, 1), v_6 = (0, 1), v = (\frac{1}{2}, \frac{1}{2})$. Their image vertices under G are given by $w_1 = w_6 = (0, 1), w_2 = (0, 0), w_3 = w_4 = (1, 0), w_5 = (1, 1), w = (\frac{1}{2}, \frac{1}{2})$. The map G sends v_j to w_j and v to w. Figure 3.38 depicts the image of vertices under G and the images of edges under F and the composition FG. □

Using the above lemma, we can then say that $h_{01} * f_{10}$ represents the same element of $\pi_1(A, x_0)$ as does $f_{11} * h_{11}$. This means that $h_{01} * f_{10} * f_{20} * \cdots * f_{k0}$ and $f_{11} * h_{11} * f_{20} * \cdots * f_{k0}$ represent the same elements of the free product. We then look at the next rectangle R_{21}. If R_{21} is still sent to the same set A as R_{11}, then Lemma 3.9.10 can be used to replace the term $h_{11} * f_{20}$ with $f_{21} * h_{21}$ as elements of $\pi_1(A, x_0)$. This allows us to say that $f_{11} * h_{11} * f_{20} * \cdots * f_{k0}$ and $f_{11} * f_{21} * h_{21} * \cdots * f_{k0}$ represent the same element in the free product. The other possibility is that R_{21} is sent to B. This means that the common edge of the two rectangles is sent to $A \cap B$. Hence h_{11} represents an element of $\pi_1(A \cap B, x_0)$. In R_{11}, we need to consider this as an element of $\pi_1(A, x_0)$, whereas in rectangle R_{21} we need to consider this as an element of $\pi_1(B, x_0)$ in order to homotope $h_{11} * f_{20}$ to $f_{21} * h_{21}$ in $\pi_1(B, x_0)$. Although this is not allowable in the free product, this identification of $\psi_A([h_{11}])$ with $\psi_B([h_{11}])$ is precisely what the amalgamated free product allows.

We then proceed across the bottom row of rectangles, using the rectangle R_{j1} that is sent to A (resp., B) to replace the product of terms $h_{(j-1)1} * f_{j0}$ with $f_{j1} * h_{j1}$ in $\pi_1(A, x_0)$ (resp., $\pi_1(B, x_0)$). Whenever there are adjacent rectangles $R_{(j-1)1}$ and R_{j1} sent to distinct sets A and B, then we have to work in the amalgamated free product to identify $\psi_A([h_{(j-1)1}])$ with $\psi_B([h_{(j-1)1}])$. At the end of working our way down the bottom row of rectangles, we have an equivalence within the amalgamated free product of the elements represented by $h_{01} * f_{10} * f_{20} * \cdots * f_{k0}$ and $f_{11} * f_{21} * \cdots * f_{k1} * h_{k1}$. Since h_{k1} is the map to the base point x_0, we may omit this term in the amalgamated free product. Thus we conclude that in the amalgamated free product the loops $f_{10} * f_{20} * \cdots * f_{k0}$ and $f_{11} * f_{21} * \cdots * f_{k1}$ represent the same element. We then use the same argument on the jth row of rectangles to show that $f_{1(j-1)} * f_{2(j-1)} * \cdots * f_{k(j-1)}$ and $f_{1j} * f_{2j} * \cdots * f_{kj}$ represent the same elements of the amalgamated free product. In the move from one row to another, there may be some vertically adjacent rectangles $R_{j(p-1)}$ and R_{jp} which are sent to different sets A and B. In this case, the loop $f_{j(p-1)}$ is in $A \cap B$, and we will have to use the identification of $\psi_A([f_{j(p-1)}])$ with $\psi_B([f_{j(p-1)}])$ in the amalgamated free product. After moving over all rows of rectangles, we get an equivalence in the amalgamated free product between $[f_{10}] \mp \cdots \mp [f_{k0}]$ and $[f_{1k}] \mp \cdots \mp [f_{kk}] = [e_{x_0}]$. This says that $\bar{\phi}_{AB}$ is an isomorphism between the amalgamated free product and $\pi_1(X, x_0)$.

As an example to clarify and illustrate the proof, we suppose that $k = 3$ and the rectangles $R_{11}, R_{13}, R_{21}, R_{23}, R_{33}$ are sent to A and the others are sent to B. We then indicate the steps used in getting the equivalence of the classes in the amalgamated free product represented by the bottom map $f_{10} * f_{20} * f_{30}$ and the top map $f_{13} * f_{23} * f_{33}$. We use subscripts A, B to indicate equivalences in $\pi_1(A, x_0), \pi_1(B, x_0)$, the symbol id to indicate insertion or deletion of the identity element, and AB to indicate an identification using ψ_A, ψ_B in the amalgamated free product. Even in this simple case, the complete details become rather complicated; however, they consist of applying the same basic steps over each square of the subdivision.

$$[f_{10}]_A \mp [f_{20}]_B \mp [f_{30}]_A \sim_{\text{id}} [h_{01}]_A \mp [f_{10}]_A \mp [f_{20}]_B \mp [f_{30}]_A$$

$$\sim_A [f_{11}]_A \mp [h_{11}]_A \mp [f_{20}]_B \mp [f_{30}]_A$$

$$\sim_{AB} [f_{11}]_A \mp [h_{11}]_B \mp [f_{20}]_B \mp [f_{30}]_A$$

$$\sim_B [f_{11}]_A \mp [f_{21}]_B \mp [h_{21}]_B \mp [f_{30}]_A$$

$$\sim_{AB} [f_{11}]_A \mp [f_{21}]_B \mp [h_{21}]_A \mp [f_{30}]_A$$

$$\sim_A [f_{11}]_A \mp [f_{21}]_B \mp [f_{31}]_A \mp [h_{31}]_A$$

$$\sim_{\text{id}} [f_{11}]_A \mp [f_{21}]_B \mp [f_{31}]_A$$

$$\sim_{\text{id}} [h_{02}]_A \mp [f_{11}]_A \mp [f_{21}]_B \mp [f_{31}]_A$$

$$\sim_A [f_{12}]_A \mp [h_{12}]_A \mp [f_{21}]_B \mp [f_{31}]_A$$

$$\sim_{AB} [f_{12}]_A \bar{*} [h_{12}]_B \bar{*} [f_{21}]_B \bar{*} [f_{31}]_A$$

$$\sim_B [f_{12}]_A \bar{*} [f_{22}]_B \bar{*} [h_{22}]_B \bar{*} [f_{31}]_A$$

$$\sim_{AB} [f_{12}]_A \bar{*} [f_{22}]_B \bar{*} [h_{22}]_A \bar{*} [f_{31}]_A$$

$$\sim_A [f_{12}]_A \bar{*} [f_{22}]_B \bar{*} [f_{32}]_A \bar{*} [h_{32}]_A$$

$$\sim_{\mathrm{id}} [f_{12}]_A \bar{*} [f_{22}]_B \bar{*} [f_{32}]_A$$

$$\sim_{AB} [f_{12}]_B \bar{*} [f_{22}]_B \bar{*} [f_{32}]_A$$

$$\sim_{\mathrm{id}} [h_{03}]_B \bar{*} [f_{12}]_B \bar{*} [f_{22}]_B \bar{*} [f_{32}]_A$$

$$\sim_B [f_{13}]_B \bar{*} [h_{13}]_B \bar{*} [f_{22}]_B \bar{*} [f_{32}]_A$$

$$\sim_B [f_{13}]_B \bar{*} [f_{23}]_B \bar{*} [h_{23}]_B \bar{*} [f_{32}]_A$$

$$\sim_{AB} [f_{13}]_B \bar{*} [f_{23}]_B \bar{*} [h_{23}]_A \bar{*} [f_{32}]_A$$

$$\sim_A [f_{13}]_B \bar{*} [f_{23}]_B \bar{*} [f_{33}]_A \bar{*} [h_{33}]_A$$

$$\sim_{\mathrm{id}} [f_{13}]_B \bar{*} [f_{23}]_B \bar{*} [f_{33}]_A.$$

This last product is the product of three representatives of the identity, and so represents the identity in the amalgamated free product.

3.10   Dependence on the base point

This section is a project which explores the way the fundamental group depends on the base point. You should verify all of the claims being made during our discussion. We will have a standard assumption here that our space is path connected. If a space is not path connected, it can be written as the disjoint union of its path components, which are path connected. The path component containing x is just the set of all $y \in X$ so that there is a path joining y to x. Then the fundamental group $\pi_1(X, x)$ will depend only on the path component containing the base point x and will be unaffected by other path components.

Let α be a path in X which connects $\alpha(0)$ and $\alpha(1)$. This will induce an isomorphism $\alpha_* : \pi_1(X, \alpha(1)) \to \pi_1(X, \alpha(0))$, defined by

$$\alpha_*([f]) = [\alpha * f * \bar{\alpha}].$$

Here $\bar{\alpha}(t) = \alpha(1 - t)$ just gives the path from $\alpha(1)$ to $\alpha(0)$ formed by traversing α backwards. Because of our argument proving associativity, we will adopt the convention that $\alpha * f * \bar{\alpha}$ uses α on $[0, \frac{1}{3}], f$ on $[\frac{1}{3}, \frac{2}{3}], \bar{\alpha}$ on $[\frac{2}{3}, 1]$. See Figure 3.39 for an illustration of the path $\alpha * f * \bar{\alpha}$. Verify that the homotopy class of the result is well defined, independent of the choice of f within $[f]$ since if $f \sim_F f'$, then $\alpha * f * \bar{\alpha} \sim \alpha * f' * \bar{\alpha}$. Just use the constant homotopies on the part mapped via $\alpha, \bar{\alpha}$, and the homotopy F in middle. This is illustrated in Figure 3.40. You should also verify that the result depends only on the homotopy class of the path α rel $0, 1$. That α_* is an isomorphism comes from the fact that $\bar{\alpha}_*$ gives its inverse. One step of verifying this is the equation

$$\bar{\alpha}_* \alpha_* [f] = [\bar{\alpha} * \alpha * f * \bar{\alpha} * \alpha] = [\bar{\alpha} * \alpha][f][\bar{\alpha} * \alpha]^{-1} = [e_{\alpha(0)}][f][e_{\alpha(0)}].$$

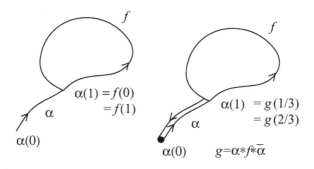

Figure 3.39. Isomorphism $\alpha_* : \pi_1(X, \alpha(1)) \to \pi_1(X, \alpha(0))$.

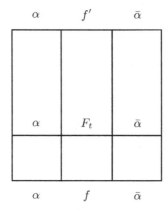

Figure 3.40. $f \sim f'$ implies $\alpha * f * \bar{\alpha} \sim \alpha * f' * \bar{\alpha}$.

The key idea here is that the parametrization does not affect the result up to homotopy and running along α and then following that by going back along $\bar{\alpha}$ is homotopic relative to the end points to the constant map at $\alpha(0)$. The details are analogs of arguments in Section 3.2 and are left as an exercise.

Now suppose that α is a path from x to y and β is a path from y to z. Then the path $\alpha * \beta$ from x to z satisfies the functorial property $(\alpha * \beta)_* = \alpha_* \beta_*$. To see that use the representative $\alpha * \beta * f * \bar{\beta} * \bar{\alpha}$, and note that, up to homotopy, we may choose to spread the parametrization within the five maps in whatever proportion we wish.

Now suppose $f, g : A \to B$ are homotopic, and we choose a as the base point of A. Suppose that $F : A \times I \to B$ gives a homotopy between f and g. Let $\alpha(t) = F(a, t)$. Then $\alpha(0) = f(a)$ and $\alpha(1) = g(a)$. The map α induces an isomorphism $\alpha_* : \pi_1(B, g(a)) \to \pi_1(B, f(a))$.

Proposition 3.10.1. *Let $f, g : A \to B$ be homotopic maps with homotopy F and $\alpha(t) = F(a, t)$. Then we have the following commutative diagram, which means that $\alpha_* g_* = f_*$.*

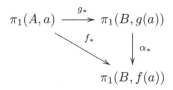

Give a proof of the proposition by following the outline below. Let $p : I \to A$ represent an element $[p] \in \pi_1(A, a)$. Then $f_*([p])$ is represented by fp and $g_*([p])$ is represented by gp. Now $G = F(p \times 1) : I \times I \to B$ has $G(s, 0) = f(s), G(s, 1) = g(s), G(0, t) = \alpha(t) = G(1, t)$. There is a continuous map $Q : I \times I \to I \times I$ which sends $\{0\} \times I$ to $(0, 0)$ and $\{1\} \times I$ to $(1, 0)$, sends $I \times \{1\}$ to $\{0\} \times I \cup I \times \{1\} \cup \{1\} \times I$, and sends $I \times \{0\}$ via the identity to itself. Figure 3.41 illustrates how to construct Q. To construct Q you should divide the region up into triangles and map the triangles affine linearly so that the maps on triangles are determined by the values on the vertices. The images of the two "corner triangles" will be intervals—there is collapsing occurring in Q. Then show that $H = GQ : I \times I \to B$ provides the required homotopy to show that the diagram commutes.

Now consider the situation where α, β are paths in S^1 connecting a to b. Then we want to show that $\pi_1(S^1, a) \simeq \mathbb{Z}$ being abelian implies that $\alpha_* = \beta_*$. The key is to understand what happens when we look at $\gamma_* : \pi_1(S^1, a) \to \pi_1(S^1, a)$ when γ is a loop at a. Then

$$\gamma_*([f]) = [\gamma * f * \overline{\gamma}] = [\gamma] \,\overline{*}\, [f] \,\overline{*}\, [\overline{\gamma}] = [\gamma] \,\overline{*}\, [f] \,\overline{*}\, [\gamma]^{-1}.$$

Show that $\pi_1(S^1, a)$ being abelian implies $\gamma_*([f]) = [f]$. Applying this to $\gamma = \alpha * \bar{\beta}$, show that this implies $\beta_* = \alpha_*$. This means that we can identify $\pi_1(S^1, a)$ with $\pi_1(S^1, 1)$ using any path from 1 to a. After making this identification, we will just write $\pi_1(S^1)$ for the fundamental group, ignoring the choice of base point since this standard identification exists between the fundamental groups using two different base points. Using this identification, we can consider an

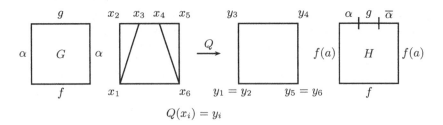

$$Q(x_i) = y_i$$

Figure 3.41. Reparametrizing the homotopy.

induced map $f_* : \pi_1(S^1) \to \pi_1(S^1)$ as being a homomorphism from the integers to themselves. Any such homomorphism from the integers to the integers is completely determined by its value on $1 \in \mathbb{Z}$. If we identify $\pi_1(S^1)$ with $\mathbb{Z}$, then $f_*(1)$ will be some integer, which we will define to be the *induced degree* of the homomorphism, or, more informally, the induced degree of the map. You should formulate this homomorphism as a composition of homomorphisms involving paths and induced maps. With all of these identifications, then show how Proposition 3.10.1 says that homotopic maps from the circle to the circle have the same induced degree. You should then prove that the induced degree agrees with the degree of the map using our earlier definition of the degree of a map $f : S^1 \to S^1$.

We now look at the case of general X. Use the argument above to show that if α, β are paths connecting x to y, then $\alpha_*([f])$ is conjugate to $\beta_*([f])$. This means that there is an element $\eta \in \pi_1(X, x)$ so that $\beta_*([f]) = \eta \bar{*} \alpha_*([f]) \bar{*} \eta^{-1}$. Form a new group $\pi_1^{ab}(X, x)$ by abelianizing the fundamental group $\pi_1(X, x)$. This may be expressed informally as identifying γ with $\eta \bar{*} \gamma \bar{*} \eta^{-1}$ for any η, or, more formally, by taking the quotient of $\pi_1(X, x)$ by the commutator subgroup of $\pi_1(X, x)$, which is the smallest normal subgroup generated by the commutators $\eta \bar{*} \gamma \bar{*} \eta^{-1} \bar{*} \gamma^{-1}$. Then show that after going to the abelianization, the induced map α_* induces a map $\alpha'_* : \pi_1^{ab}(X, y) \to \pi_1^{ab}(X, x)$. Show that this map is independent of the path α; that is, $\alpha'_* = \beta'_*$. This allows us to make a standard identification of $\pi_1^{ab}(X, x)$ with $\pi_1^{ab}(X, y)$ for any $x, y \in X$.

With this identification, denote this common group $\pi_1^{ab}(X)$. Then show that any continuous map $f : X \to Y$ induces a map $f_* : \pi_1^{ab}(X) \to \pi_1^{ab}(Y)$. Show that if f is homotopic to f', then $f_* = f'_* : \pi_1^{ab}(X) \to \pi_1^{ab}(Y)$. Show that this map has the functorial properties:

(1) $(1_X)_* = 1_{\pi_1^{ab}(X)}$;

(2) $(gf)_* = g_* f_*$.

Use the functorial properties above to show that, if f is a homotopy equivalence, then $\pi_1^{ab}(X) \simeq \pi_1^{ab}(Y)$.

Returning to the regular fundamental group, use Proposition 3.10.1 to show the following result.

Proposition 3.10.2. *If X and Y are homotopy equivalent via $f : (X, x_0) \to (Y, y_0)$ and $g : (Y, y_0) \to (X, z_0)$, then $f_* : \pi_1(X, x_0) \to \pi_1(Y, y_0)$ is an isomorphism.*

In your proof it will be useful to note that $f_* : \pi_1(X, x_0) \to \pi(Y, y_0)$ and $g_* : \pi_1(Y, y_0) \to \pi_1(X, z_0)$. You can also use the induced homomorphism $f_* : \pi_1(X, z_0) \to \pi_1(Y, f(z_0))$. It is easiest to first show that g_* is an isomorphism and then use this to get the result.

3.11   Supplementary exercises

Exercise 3.11.1. Check whether the following sets together with their operations form groups:

(a) irrational numbers under addition;

(b) $S^3 = \{(z_1, z_2) : |z_1|^2 + |z_2|^2 = 1, z_1, z_2 \in \mathbb{C}\}$ with the operation

$$(z_1, z_2) \cdot (u_1, u_2) = (z_1 u_1 - z_2 \bar{u}_2, z_1 u_2 + z_2 \bar{u}_1),$$

where we are using complex multiplication and $\bar{u} = a - bi$ denotes the complex conjugate of $u = a + bi$. (Hint: $|z|^2 = z\bar{z}$.)

(c) $(\{1, 2, 3\}, \cdot)$, where $1 \cdot a = a \cdot 1 = a, 2 \cdot 2 = 3, 3 \cdot 2 = 2 \cdot 3 = 1$, and $3 \cdot 3 = 2$.

Exercise 3.11.2. Let $(\mathbb{R}/\sim, \cdot)$ have as its set the equivalence classes of real numbers, where $a \sim b$ if there is an integer n with $a = b + n$. The operation on equivalence classes is given by $[a] \cdot [b] = [a + b]$. Verify that this operation is well defined and that $(\mathbb{R}/\sim, \cdot)$ forms a group.

Exercise 3.11.3. Let $\bar{p} : \mathbb{R}/\sim \rightarrow S^1$ be induced from $p : \mathbb{R} \rightarrow S^1$, $p([a]) = (\cos 2\pi a, \sin 2\pi a)$. Show that $\bar{p}$ is well defined and is an isomorphism of groups.

Exercise 3.11.4. Suppose G is an abelian group and H is a nonabelian group (i.e. there exist $h_1, h_2 \in H$ with $h_1 \cdot h_2 \neq h_2 \cdot h_1$).

(a) Suppose $f : G \rightarrow H$ is a homomorphism. Show that f is not onto H. (Hint: Show that the image is an abelian group.)

(b) Suppose $g : H \rightarrow G$ is a homomorphism. Show that g is not 1–1. (Hint: Look at $h_1 \cdot h_2$ and $h_2 \cdot h_1$.)

Exercise 3.11.5. Show that if $H \subset G$ is closed under multiplication ($a, b \in H$ implies $a \cdot b \in H$) and taking inverses ($a \in H$ implies $a^{-1} \in H$), then H is a subgroup of G.

Exercise 3.11.6. Find all of the subgroups of $(\mathbb{Z}, +)$.

Exercise 3.11.7. Show that the relation f_0 homotopic to $f_1 \operatorname{rel} 0, 1$ is an equivalence relation on paths $\{f : (I, 0, 1) \rightarrow (X, x, y)\}$ connecting x to y.

Exercise 3.11.8. Denote the set of equivalence classes of paths $f : (I, 0, 1) \rightarrow (X, x, y)$, as above, by $\pi_1(X, x, y)$. Suppose X is path connected. Show that there is a bijection of sets $g : \pi_1(X, x, y) \rightarrow \pi_1(X, x)$. (Hint: Pick a fixed path α from x to y and use it to get from a path from x to y to a loop at x.)

Exercise 3.11.9.

(a) Compute $\pi_1(I, 0, 1)$ and give a representative for each equivalence class.

(b) Compute $\pi_1(S^1, 1, -1)$ and give a representative for each equivalence class.

Exercise 3.11.10.

(a) Show that addition of paths determines a well-defined map $A(x, y, z) :$ $\pi_1(X, x, y) \oplus \pi_1(X, y, z) \to \pi_1(X, x, z)$.

(b) Show that there is associativity; that is, there is a formula

$$A(x, z, w)(A(x, y, z)([\alpha], [\beta]), [\gamma]) = A(x, y, w)([\alpha], A(y, z, w)([\beta], [\gamma]))$$

and that this common value is represented by

$$h(s) = \begin{cases} \alpha(3s) & \text{if } 0 \le s \le \frac{1}{3}, \\ \beta(3s - 1) & \text{if } \frac{1}{3} \le s \le \frac{2}{3}, \\ \gamma(3s - 2) & \text{if } \frac{2}{3} \le s \le 1. \end{cases}$$

Definition 3.11.1. A *topological group* is a group $(G, \cdot)$ which is a topological space so that the multiplication $G \times G \to G, (g_1, g_2) \to g_1 \cdot g_2$, and the map taking inverses, $G \to G, g \to g^{-1}$, are continuous. Here $G \times G$ is given the product topology.

Exercise 3.11.11.

(a) Show that $\mathbb{R}$ with the usual addition forms a topological group.

(b) Show that S^1 with multiplication coming from complex multiplication is a topological group.

(c) Show that $p : \mathbb{R} \to S^1, p(t) = e^{2\pi i t}$, is a continuous group homomorphism.

Exercise 3.11.12. Consider the group of 2×2 real matrices with nonzero determinant

$$A = \begin{pmatrix} a_{11} & a_{12} \\ a_{21} & a_{22} \end{pmatrix}.$$

By identifying these matrices with a subset of $\mathbb{R}^4$ using the four coordinates, we can make this into a topological space. This space is denoted $GL(2, \mathbb{R})$ and is called the general linear group of 2×2 real matrices. Show that $GL(2, \mathbb{R})$ forms a topological group with the operation of matrix multiplication.

Exercise 3.11.13. (a) Continuing with the matrices in the last exercise, consider the matrices $O(2) \subset GL(2, \mathbb{R})$ which are the orthogonal matrices Q satisfying $QQ^t = Q^tQ = I$. Show that $O(2)$ is a subgroup of $GL(2, \mathbb{R})$ and is itself a topological group.

(b) Use the Gram–Schmidt orthogonalization process which writes a given element A in $GL(2, \mathbb{R})$ uniquely as a product $A = QR$, where $Q \in O(2)$ and R is an upper triangular matrix with positive diagonal entries to show that there is a deformation retraction of $GL(2, \mathbb{R})$ onto $O(2)$.

(c) Let $SO(2)$ denote the matrices in $O(2)$ with determinant 1. Show that $SO(2)$ is a topological group. Show that there is a homeomorphism between $O(2)$ and the product space $SO(2) \times \{\pm 1\}$.

(d) Show that $SO(2)$ is homeomorphic to S^1, and this homeomorphism is a group isomorphism.

(e) Compute the fundamental groups $\pi_1(SO(2), I)$, $\pi_1(O(2), I)$ and $\pi_1(GL(2, \mathbb{R}), I)$.

Exercise 3.11.14. Consider the set $GL(2, \mathbb{C})$ of 2×2 complex matrices with nonzero determinant. Give it a topology by identifying it with a subset of $\mathbb{C}^4$. Show that $GL(2, \mathbb{C})$ is a topological group.

Exercise 3.11.15. Consider the subset $U(2) \subset GL(2, \mathbb{C})$ of unitary matrices U satisfying $U^*U = UU^* = I$, where $U^* = \bar{U}^t$ is the adjoint of the matrix U. Show that $U(2)$ is a subgroup of $GL(2, \mathbb{C})$ and is a topological group. Use the complex Gram–Schmidt algorithm which decomposes a complex matrix $A = QR$ with $Q \in U(2)$ and R an upper triangular matrix with positive diagonal entries to show that $GL(2, \mathbb{C})$ deformation-retracts onto $U(2)$.

Exercise 3.11.16. Consider the subset $SU(2) \subset U(2)$ of unitary matrices which satisfy the additional condition $\det U = 1$. Show that $SU(2)$ is a topological group and that it is homeomorphic to S^3. (Hint: Identify S^3 as a subset of $\mathbb{C}^2$ of points (z_1, z_2) with $|z_1|^2 + |z_2|^2 = 1$. For such a point, consider the special unitary matrix

$$U(z_1, z_2) = \begin{pmatrix} z_1 & -\bar{z}_2 \\ z_2 & \bar{z}_1 \end{pmatrix}.$$

Show that this correspondence gives a homeomorphism.)

Exercise 3.11.17. Consider the unitary group $U(1)$, which is the set of 1×1 complex matrices which satisfy $U^*U = UU^* = I$. Show that $U(1)$ is a topological group which is homeomorphic to S^1 via a group isomorphism.

Exercise 3.11.18.

(a) Show that there is a continuous map $S^1 \times SU(2) \to U(2)$ which is also a group homomorphism given by $p(\zeta, A) = \zeta A$, where every element of the matrix A is multiplied by the unit complex number ζ.

(b) Show that the inverse image of I is the pair $\{(-1, -I), (1, I)\}$.

(c) Show that the map p is a covering map for the covering space $p : S^1 \times SU(2) \to U(2)$.

Exercise 3.11.19. Suppose $(G, \cdot)$ is a topological group with identity e. Define an operation on loops at e by $(f \circ g)(s) = f(s) \cdot g(s)$. Show that this respects homotopy classes in that if $f \sim f', g \sim g'$, then $f \circ g \sim f' \circ g'$. Hence we can define an operation on homotopy classes of loops by $[f]\bar{\circ}[g] = [f \circ g]$. Show that the homotopy classes of loops with this operation form a group, which we will denote by $\pi_1'(G, e)$. (Hint: Use the group properties of G to find identities and inverses, denoting by $E(s) = e$ the constant loop at e.)

Figure 3.42. Exercise 3.11.22(a).

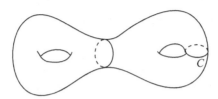

Figure 3.43. $T^{(2)}\backslash C$.

Exercise 3.11.20. Use the fact that $f \sim f * E \sim E * f, g \sim g * E \sim E * g$ to show that $\pi_1'(G, e)$ is abelian.

Exercise 3.11.21. Define a map from $\pi_1(G, e)$ to $\pi_1'(G, e)$ by the identity at the set level. This is necessarily a bijection since the underlying sets are the same and only the group operations are different. Prove that this is an isomorphism and thus the fundamental group of a topological group is abelian. (Hint: By Proposition 3.1, we need only show that it is a homomorphism, and for this we need to see why $f * g$ is homotopic to $f \circ g$. Consider the facts cited in the previous exercise for this.)

Exercise 3.11.22.

(a) Show that the region in Figure 3.42 is strongly contractible.

(b) Show that if X is contractible, then it is path connected.

(c) Show that if X is contractible, then $\pi_1(X, x) \simeq \{e\}$ for any $x \in X$.

Exercise 3.11.23.

(a) Let $T = S_a^1 \times S_b^1$. Show that $T\backslash(S_a^1 \times \{-1\})$ deformation-retracts to $S_a^1 \times \{1\}$.

(b) Show that $T\backslash(S_a^1 \times \{-1\})$ deformation retracts to a subset which is homeomorphic to $S_{(2)}$.

Exercise 3.11.24. Consider the curve C in Figure 3.43. Show that $T^{(2)}\backslash C$ deformation-retracts onto a subset which is homeomorphic to $T_{(2)}$.

Exercise 3.11.25. Find two disjoint circles C_1, C_2 in $T^{(2)}$ so that $T^{(2)} \backslash (C_1 \cup C_2)$ deformation retracts onto a subset homeomorphic to $S_{(4)}$. Identify the subset and give a description of the deformation retraction. (Hint: Use ideas from the previous two exercises.)

Exercise 3.11.26. Show that there are g disjoint circles $C_1, \ldots, C_g$ in $T^{(g)}$ so that $T^{(g)} \backslash (C_1 \cup \cdots \cup C_g)$ deformation-retracts onto a subset homeomorphic to $S_{(2g)}$. Identify the subset and give a description of the deformation retraction.

Exercise 3.11.27.

(a) If C denotes the center circle in the Möbius band M, then show that $M \backslash C$ deformation-retracts onto ∂M.

(b) Show that there is a circle C in the projective plane P so that $P \backslash C$ deformation-retracts onto a disk in P.

Exercise 3.11.28. Show that there is a circle C in the Klein bottle so that $K \backslash C$ deformation-retracts onto an annulus contained in the Klein bottle. Give a description of the subset and the deformation retraction.

Exercise 3.11.29. Show that there are h disjoint circles $C_1, \ldots, C_h$ in $P^{(h)}$ so that $P^{(h)} \backslash (C_1 \cup \cdots \cup C_h)$ deformation-retracts onto a subset which is homeomorphic to $S_{(h)}$. Give a description of the subset and the deformation retraction.

Exercise 3.11.30.

(a) By regarding $P^{(3)}$ as $P \# T$, show that there are two disjoint circles $C_1, C_2 \subset P^{(3)}$ so that $P^{(3)} \backslash (C_1 \cup C_2)$ deformation-retracts onto a subset homeomorphic to $S_{(3)}$. Give a description of the subset and the deformation retraction.

(b) Generalize the above argument to show that there are $k+1$ disjoint circles $C_1, \ldots, C_{(k+1)}$ so that $P^{(2k+1)} \backslash (C_1 \cup \cdots \cup C_{k+1})$ deformation-retracts onto a subset homeomorphic to $S_{(2k+1)}$.

(c) Show that there are k disjoint circles $C_1, \ldots, C_k$ so that $P^{(2k)} \backslash (C_1 \cup \cdots \cup C_k)$ deformation-retracts onto a subset homeomorphic to $S_{(2k)}$.

Exercise 3.11.31. Two paths $f_0, f_1 : I \to X$ are said to be *freely homotopic* if there is a continuous map $F : I \times I \to X$, $F(s, t) = F_t(s)$, with $F_0 = f_0$, $F_1 = f_1$. Show that any two paths in a path-connected space X are freely homotopic. (Hint: First show that $f : I \to X$ is freely homotopic to the map sending I to $f(0)$.)

Two loops $f, g : I \to X$ are called *freely homotopic loops* if there exists a continuous map $F : I \times I \to X$ with $G(s, 0) = f(s)$, $F(s, 1) = g(s)$, and $F(0, t) = F(1, t) = w(t)$ is a path between $f(0) = f(1) = a$ and $g(0) = g(1) = b$. Note that we are not requiring $a = b$ as in the definition of $\pi_1(X, a)$ and, even if $a = b$, we are not requiring the homotopy to keep the image of 0 and 1 fixed at a. We are requiring, however, that the image of 0 and 1 move in the same way during the homotopy—this distinguishes freely homotopic loops from freely homotopic paths as in Exercise 3.11.31. Denote by $\pi_1^f(X)$ the set of free homotopy classes of loops in X. We can no longer compose loops since they do not always begin and end at the same point—hence there is no group operation.

Exercise 3.11.32. Define a map $r : \pi_1(X, x_0) \to \pi_1^f(X)$ by ignoring the base point; that is, the equivalence class $[g]$ of $g : (I, \{0,1\}) \to (X, x_0)$ in $\pi_1(X, x_0)$ is sent to the equivalence class $[g]$ of $g : I \to X$ in $\pi_1^f(X)$. Show that any loop g is freely homotopic to a loop g' with $g'(0) = g'(1) = x_0$, and hence r is onto. (Hint: Suppose $g(0) = g(1) = a$. Let $\alpha : I \to X$ be a path joining x_0 and a, so $\alpha(0) = x_0, \alpha(1) = a$. Let $g' = \alpha * g * \bar{\alpha}$. The homotopy should gradually use more of α and $\bar{\alpha}$.)

Exercise 3.11.33. Suppose $f, g : (I, 0, 1) \to (X, x_0)$ are freely homotopic loops. Show that there exists a loop α at x_0 with $[f] = [\alpha][g][\alpha]^{-1} \in \pi_1(X, x_0)$. Show that the converse is also true; that is, $[f] = [\alpha][g][\alpha]^{-1}$ in $\pi_1(X, x_0)$ implies that f, g are freely homotopic. (Hint: Let $F : I \times I \to X$ be the free homotopy, with $F(s, 0) = f(s), F(s, 1) = g(s), F(0, t) = F(1, t) = \alpha(t)$. Now consider the proof of Proposition 3.10.1.)

Exercise 3.11.34. Show that $r : \pi_1(X, x_0) \to \pi_1^f(X)$ is a bijection iff $\pi_1(X, x_0)$ is abelian.

Suppose X, Y are path-connected spaces and they have base points $x_0 \in X$, $y_0 \in Y$. Consider the set of homotopy classes of continuous maps from (X, x_0) to (Y, y_0) which we denote as $[(X, x_0), (Y, y_0)]$. In the equivalence relation here, homotopies have to preserve the base point. If we ignore the base point, then there is a corresponding set $[X, Y]$.

Exercise 3.11.35. By using the identification $S^1 = I/0 \sim 1$, show that $\pi_1(Y, y_0)$ corresponds bijectively to $[(S^1, 1), (Y, y_0)]$.

Exercise 3.11.36. Show that $\pi_1^f(Y, y_0)$ corresponds bijectively to $[S^1, Y]$.

Exercise 3.11.37. Show that if X and Y are homotopy equivalent, then $\pi_1^f(X)$ and $\pi_1^f(Y)$ correspond bijectively.

Exercise 3.11.38. Show that the map $j : \pi_1(S^1, 1) \to \pi_1^f(\mathbb{R}^2 \backslash \{0\})$ is a bijection, where for $f : (I, \{0,1\}) \to (S^1, 1)$, the map j sends $[f]$ to the equivalence class of f, considered as a map $f : I \to S^1 \subset \mathbb{R}^2 \backslash \{0\}$.

Exercise 3.11.39. Show that a handlebody $H = h^0 \cup h^1$ is homotopy equivalent to a circle. (Hint: First collapse h^0 to a point, then push $h^1 = D^1 \times D^1$ to $D^1 \times \{0\}$.)

Exercise 3.11.40. Show that $H = h^0 \cup h_1^1 \cup \cdots \cup h_k^1$ (where we assume that the h_i^1 are all attached disjointly to ∂h^0) is homotopy equivalent to the one-point union of k circles (see Figure 3.44).

Exercise 3.11.41. Call a continuous map $f : S^1 \to S^1$ *regular* if

   (a) $f^{-1}\{1\} = \{a_1, \ldots, a_k\}$ consists of a finite number of points;

   (b) for each $a_i \in f^{-1}\{1\}$, there is a small arc A_i about a_i with $A_i \cap f^{-1}\{1\} = a_i$ and $A_i \backslash \{a_i\} = A_{i1} \bigsqcup A_{i2}$ so that if $f = (f_1, f_2)$, we have $f_2(A_{i1}) < 0$

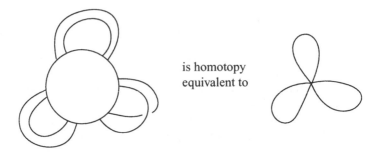

is homotopy
equivalent to

Figure 3.44. A homotopy equivalence.

and $f_2(A_{i2}) > 0$; that is, as we cross a_i the image crosses $\mathbf{1}$. If, when traversing S^1 in a counterclockwise direction, we encounter A_{i1} first and then A_{i2}, assign $e(a_i) = 1$. If we encounter A_{i2} first and then A_{i1}, assign $e(a_i) = -1$. In other words, if the image curve goes from the negative side of $\mathbf{1}$ to the positive side as we pass a_i, then $e(a_i) = 1$; if it goes from the positive side to the negative side, $e(a_i) = -1$.

Let $d(f) = \sum e(a_i)$. Show that $d(f) = \deg(f)$.

Exercise 3.11.42. Call a map $f : S^1 \to S^1$ *quasiregular* if (1) above is satisfied. Generalize the previous exercise to quasiregular maps.

For the next group of exercises we need to review some results from advanced calculus which we will be using. A differential 1-form $w = M(x,y)\, dx + N(x,y)\, dy$ defined in a region R of the plane is *closed* if $M_y = N_x$. It is *exact* if there is a function $F : R \to \mathbb{R}$ with $F_x = M, F_y = N$. Exact forms are closed, but closed forms need not be exact in general. The region R should be an open set, or at least the functions involved should extend to an open set containing R. The relationship between closed 1-forms and exact 1-forms measures something about R which is closely related to $\pi_1(R)$. If $\gamma = (\gamma_1, \gamma_2) : [a,b] \to R$ is a differentiable curve, then

$$\int_\gamma M\, dx + N\, dy = \int_a^b M(\gamma(t))\gamma_1'(t) + N(\gamma(t))\gamma_2'(t)\, dt.$$

Thus integrals over curves are evaluated by changing to one-variable integrals. The change of variables formula in one variable then implies that the integral does not depend upon the parametrization of the curve other than its orientation. If we reverse the direction in which we traverse the curve, then the integral is multiplied by -1. Thus the integral really only depends on the "oriented curve" γ. Another way to phrase this integral is that we are taking the integral over the defining interval of the dot product of the vector field $(M(\gamma(t)), N(\gamma(t)))$ with tangent vector $\gamma'(t)$.

This definition is extended to piecewise differentiable curves γ by defining

$$\int_\gamma w = \int_{\gamma_1} w + \cdots + \int_{\gamma_k} w,$$

where $[a, b]$ is partitioned into subintervals $[a_i, a_{i+1}]$ so that $\gamma|[a_i, a_{i+1}] = \gamma_i$. Green's theorem implies that if γ is a loop given by a union of paths which traverse the boundary of a compact R exactly once (oriented consistently on boundary circles so the exterior normal followed by γ' gives the orientation of the region or its negative), and w is closed on R, then $\int_\gamma w = 0$. For example, if R is a disk of radius r and $\gamma(t) = r(\cos t, \sin t)$, then $\int_\gamma w = 0$ if w is closed in R. This does *not* apply if R is the disk minus a point, however, as our exercises will show—R must be a compact region, which is a surface with boundary. However, when R is an annulus on which the closed form w is defined, it applies to say that the integral over the outer circle is the same as the integral over the inner circle, as long as each is oriented in the same way (clockwise or counterclockwise).

If w is exact in R and γ is any piecewise differentiable loop in R, then the fundamental theorem of integral calculus implies that $\int_\gamma w = 0$. Conversely, if $\int_\gamma w = 0$ for all differentiable loops γ in R, then w can be shown to be exact.

Suppose $f : S \to R$ is differentiable and γ is a differentiable curve in S. Suppose f is given by $x = f^1(u, v)$, $y = f^2(u, v)$. If $w = M\,dx + N\,dy$, is a 1-form in R, let f^*w be the 1-form in S defined by $f^*w = P\,du + Q\,dv$, with

$$P = (M \circ f)f_u^1 + (N \circ f)f_u^2, \quad Q = (M \circ f)f_v^1 + (N \circ f)f_v^2.$$

Then the definitions and the chain rule imply that $\int_{f \circ \gamma} w = \int_\gamma f^*w$ as each integral is being pulled back to the same integral over the parameter domain.

It is a fact that every continuous loop in R is freely homotopic to a differentiable one, and homotopic loops are differentiably homotopic. Henceforth, we will assume that all loops and paths are (piecewise) differentiable and our homotopies are differentiable as well.

Exercise 3.11.43. Show that if $w = M\,dx + N\,dy$ is closed in R, $f : S \to R$, then f^*w is closed in S.

Exercise 3.11.44. Suppose $\gamma : [a, b] \to R$ is a loop in R. Show that if $\alpha : [c, d] \to [a, b]$ is an affine linear homeomorphism, and $\beta = \gamma\alpha$, then $\int_\gamma w = \int_\beta w$ if α preserves order, and $\int_\gamma w = -\int_\beta w$ if α reverses order. Use this to show that $\int_{\gamma*\delta} = \int_\gamma w + \int_\delta w$, where $\gamma * \delta$ denotes the usual addition of loops. Also, show that $\int_{\bar\gamma} w = -\int_\gamma w$.

Exercise 3.11.45. Suppose that $f_0, f_1 : rS^1 \to \mathbb{R}^2$ are differentiable maps which are differentiably homotopic. Let $\alpha_i = f_i(rS^1)$ be the loops which come from composing f_i with the standard parametrization $\gamma_r(s) = re^{2\pi i s}$ of rS^1. Show that if w is a closed form which is defined on an open set containing the image of the homotopy, then $\int_{\alpha_0} w = \int_{\alpha_1} w$. (Hint: Use the homotopy to pull the problem back to integrals on the boundary circles of an annulus.)

Exercise 3.11.46. Suppose $\tau_1, \tau_2 : [0,1] \to R$ are differentiably homotopic loops in R and w is a closed 1-form in R. Show that $\int_{\tau_1} w = \int_{\tau_2} w$. (Hint: Use the homotopy to show that they can be computed as the integrals of a closed form F^*w over the boundary circles of an annulus.)

This last exercise says that the integral only depends on the differentiable homotopy class (hence on the free homotopy class) of the curve. The previous exercise allows us to define a function d from differentiable maps f from $rS^1 \to R \subset \mathbb{R}^2$ to $\mathbb{R}$ given by using a fixed closed 1-form w and defining $d(f) = \int_{\gamma_r} f^*(w) = \int_{f \circ \gamma_r} w$. Here R is some open set in $\mathbb{R}^2$. This has the property that if two maps f, g are differentiably homotopic, then $d(f) = d(g)$. Moreover, if the two maps are defined on the boundary circle of an annulus and they extend to a differentiable mapping of the annulus into R, then they have the same value. Also, Green's theorem also implies that if the map $f : rS^1 \to R$ extends to a differentiable map of rD^2 to R, then the integral of the closed form f^*w over rS^1 will be zero. These properties of d are similar to those where we defined the degree. We will see that if we choose $R = \mathbb{R}^2 \backslash \{\mathbf{0}\}$ and choose w appropriately, we will have $d(f) = \deg f$.

Exercise 3.11.47. Show that

$$w = \frac{-y}{x^2 + y^2}\,dx + \frac{x}{x^2 + y^2}\,dy$$

is a closed 1-form in $\mathbb{R}^2 \backslash \{\mathbf{0}\}$.

We will now restrict to $R = \mathbb{R}^2 \backslash \{\mathbf{0}\}$ and

$$w = \frac{-y}{x^2 + y^2}\,dx + \frac{x}{x^2 + y^2}\,dy.$$

Exercise 3.11.48. Evaluate $d(z^n)$ for $\gamma_r(t) = re^{2\pi i t} = (\cos 2\pi t, \sin 2\pi t)$.

Exercise 3.11.49. Suppose $\tilde{\alpha}(t)$ is a lift for $\alpha(t) = f(\gamma_r(t))$; that is,

$$\alpha(t) = (r \cos 2\pi \tilde{\alpha}(t), r \sin 2\pi \tilde{\alpha}(t)).$$

Show that $d(f) = \tilde{\alpha}(1) - \tilde{\alpha}(0)$. This integral thus gives the same definition of degree which we had earlier.

We now look at implications of these results about closed 1-forms in $\mathbb{R}^2 \backslash \{\mathbf{0}\}$. Note that all paths are freely homotopic to a multiple of γ_1. Thus a 1-form is exact precisely when it integrates to 0 on this loop since it would then integrate to 0 over any multiple of the loop and thus over any differentiable loop.

Exercise 3.11.50. Suppose $\eta = M\,dx + N\,dy$ is a closed differential 1-form in $\mathbb{R}^2 \backslash \{\mathbf{0}\}$. Let $\gamma(s) = (\cos 2\pi s, \sin 2\pi)$,

$$\omega = \frac{-y}{x^2 + y^2}\,dx + \frac{x}{x^2 + y^2}\,dy.$$

Show that if $(1/2\pi) \int_\gamma \eta = a$, then $\eta = a\omega + \nu$, where ν is an exact 1-form in $\mathbb{R}^2 \backslash \{\mathbf{0}\}$. Thus, modulo exact forms, all closed forms in $\mathbb{R}^2 \backslash \{\mathbf{0}\}$ are multiples of ω. (Hint: A 1-form ν in $\mathbb{R}^2 \backslash \{\mathbf{0}\}$ is exact iff $\int_\delta \nu = 0$ for all loops δ in $\mathbb{R}^2 \backslash \{\mathbf{0}\}$. Exercise 3.11.38 says that any loop in $\mathbb{R}^2 \backslash \{\mathbf{0}\}$ is freely homotopic to a multiple of γ.)

Exercise 3.11.51. Show that closed forms in $R \subset \mathbb{R}^2$ form a vector space $\mathcal{C}(S)$ under addition and scalar multiplication. Show that exact forms form a subspace $\mathcal{E}(S)$.

Exercise 3.11.52. Show that $I : \mathcal{C}(\mathbb{R}^2 \backslash \{\mathbf{0}\}) \to \mathbb{R}, I(\eta) = \int_\gamma \eta$, where $\gamma(t) = (\cos 2\pi t, \sin 2\pi t)$ is a surjective vector space homomorphism whose kernel is $\mathcal{E}(\mathbb{R}^2 \backslash \{\mathbf{0}\})$. Hence I induces a vector space isomorphism from the quotient vector space $\mathcal{C}(\mathbb{R}^2 \backslash \{\mathbf{0}\})/\mathcal{E}(\mathbb{R}^2 \backslash \{\mathbf{0}\})$ to $\mathbb{R}$.

Many of the applications of $\pi_1(S^1)$ rely only on a few basic functorial properties of the fundamental group. We illustrate this in the following exercises by asking you to prove analogous theorems in higher dimensions based on the existence of the appropriate functor. Such a functor is provided by singular homology theory, which we study in Chapter 6. Suppose h_n (hereafter just denoted h since we will assume n is fixed) is a functor from topological spaces and continuous maps to abelian groups and group homomorphisms; that is, for any topological space X, there is a corresponding abelian group $h(X)$, and for each continuous map $f : X \to Y$, there is a homomorphism $h(f) : h(X) \to h(Y)$ so that $h(1_X) = 1_{h(X)}$ and $h(gf) = h(g)h(f)$. Suppose that h also has the following properties.

(1) $h(S^n) \simeq \mathbb{Z}$ via an isomorphism k.

(2) $h(p) \simeq \{e\}$, where p is a point.

(3) If $f, g : X \to Y$ are homotopic maps, then $h(f) = h(g)$. Define the degree of $f : S^n \to S^n$ by using the composition

$$\mathbb{Z} \xrightarrow{k^{-1}} h(S^n) \xrightarrow{h(f)} h(S^n) \xrightarrow{k} \mathbb{Z}$$

by

$$\deg f = kh(f)k^{-1}(1).$$

Thus (3) says that homotopic maps have the same degree.

(4) If $f(-\boldsymbol{x}) = -f(\boldsymbol{x})$, then $\deg f$ is odd. Here $-\boldsymbol{x} = -(x_1, \dots, x_{n+1}) = (-x_1, \dots, -x_{n+1})$ is the antipodal point of $\boldsymbol{x}$.

Exercise 3.11.53. Show that if there are maps $f : X \to Y, g : Y \to X$ so that gf is homotopic to 1_X and fg is homotopic to 1_Y, then $h(f) : h(X) \to h(Y)$ and $h(g) : h(Y) \to h(X)$ are isomorphisms (and $h(f)$ and $h(g)$ are inverses to each other.) The hypotheses could have been stated in terms of X being homotopy equivalent to Y.

Exercise 3.11.54. Show that the inclusion $\mathbf{0} \to D^{n+1}$ and $g : D^{n+1} \to \mathbf{0}$ have $gi = 1_{\mathbf{0}}$ and ig homotopic to $1_{D^{n+1}}$. Conclude that $h(D^{n+1}) \simeq \{e\}$. (Hint: Contract D^{n+1} to $\mathbf{0}$ along radial lines.)

Exercise 3.11.55. Show that there does not exist a continuous map $g : D^{n+1} \rightarrow S^n$ with $gi = 1_{S^n}$.

Exercise 3.11.56. Show that a continuous map $f : D^{n+1} \rightarrow D^{n+1}$ must have a fixed point.

Exercise 3.11.57. Show that if $f : S^n \rightarrow S^n$ extends to a continuous map $F : D^{n+1} \rightarrow S^n$ (i.e. $f = Fi$), then $\deg(f) = 0$.

Exercise 3.11.58. Show that there does not exist a continuous map $f : S^{n+1} \rightarrow S^n$ satisfying $f(-x) = -f(x)$.

Exercise 3.11.59. Show that if $f : S^{n+1} \rightarrow R^{n+1}$, then there is a point $x \in S^{n+1}$ with $f(x) = f(-x)$.

Exercise 3.11.60. Assuming $f(z) = z^k : S^1 \rightarrow S^1$ has degree k, use h to prove the fundamental theorem of algebra.

Part II

Covering Spaces, CW Complexes and Homology

4

Covering spaces

4.1 Basic examples and properties

This chapter elaborates upon the ideas used in Chapter 3 to compute the fundamental group of the circle using the covering space $p : \mathbb{R} \to S^1$ to develop the theory of covering spaces. We will show that there is an intimate connection between the covering spaces of B and the subgroups of the fundamental group $\pi_1(B, b)$. As before, we are assuming that the spaces A, B in a covering space $p : A \to B$ are each path connected and locally path connected. We start with some basic examples and properties.

Our first examples and exercises involve making new covering spaces from old ones. For example, if we start with the covering $p : \mathbb{R} \to S^1, p(t) = e^{2\pi i t}$, we can get a covering space of the plane over the infinite cylinder by taking $P : \mathbb{R}^2 = \mathbb{R} \times \mathbb{R} \to S^1 \times \mathbb{R}$ with $P(t, s) = (p(t), s)$. To show that this is a covering space, we note that if $U \subset S^1$ is evenly covered so that $p^{-1}(U) = \bigsqcup_{j \in \mathbb{Z}} U_j$ with $p : U_j \to U$ a homeomorphism, then $U \times \mathbb{R} \subset S^1 \times \mathbb{R}$ is also evenly covered since $P^{-1}(U \times \mathbb{R}) = \bigsqcup_{j \in \mathbb{Z}} U_j \times \mathbb{R}$ and $P : U_j \times \mathbb{R} \to U \times \mathbb{R}$ is a homeomorphism. Here we can choose U to be any arc and get a covering of S^1 with two such arcs.

Exercise 4.1.1. Show that whenever $p : A \to B$ is a covering map and C is a path-connected, locally path-connected Hausdorff space, then $P : A \times C \to B \times C, P(a, c) = (p(a), c)$ is a covering map.

We can also take the product of a covering space with itself and get a new covering space. For example, we can take the product of the covering space of the reals over the circle with itself and get the covering space of the plane over the torus, $P : \mathbb{R}^2 \to S^1 \times S^1, P(s, t) = (p(s), p(t))$. We just use the fact that if $U, V \subset S^1$ are evenly covered open sets, then $U \times V \subset S^1 \times S^1$ is also evenly covered since $P^{-1}(U \times V) = \bigsqcup_{(j,k) \in \mathbb{Z} \times \mathbb{Z}} U_j \times U_k$ with $P : U_j \times U_k \to U \times V$ a homeomorphism. The next exercise generalizes this example.

Exercise 4.1.2. Show that if $p_1 : A_1 \to B_1, p_2 : A_2 \to B_2$ are covering maps, then so is $P : A_1 \times A_2 \to B_1 \times B_2, P(a_1, a_2) = (p_1(a_1), p_2(a_2))$.

The next exercise gives another important cover of the circle where the covering space is the circle itself.

Exercise 4.1.3. Show that p_m : S^1 → $S^1, p_m(z)$ = z^m (complex multiplication)—or, equivalently, $p(\cos x, \sin x)$ = $(\cos mx, \sin mx)$—is a covering map.

We now explore the properties of $p^{-1}\{x\}$ for different x.

Exercise 4.1.4. Show that if A is compact, and $p : A \to B$ is a covering map, then for any $b \in B, p^{-1}(b)$ consists of a finite number of points. (Hint: Use limit point compactness. See Exercise 1.9.19.)

Exercise 4.1.5. Using the framework of Exercise 4.1.4, show that since B is assumed path connected, then the number of points in $p^{-1}(b)$ does not depend on b. (Hint: Show that $\{y : |p^{-1}(b)| = n\}$ is both open and closed.)

Definition 4.1.1. We define the number of points in the inverse image $p^{-1}(x)$ to be the *order* of the cover. When the order is a finite number k, the covering is called a *k-fold* cover.

Exercise 4.1.3 can be crossed with another such map of the circle to get examples of the covering spaces of the torus by itself.

Exercise 4.1.6.

(a) Show that $P_{m,n} : S^1 \times S^1 \to S^1 \times S^1, P_{m,n}(z, w) = (z^m, w^n)$, is a covering space.

(b) Show that the order of this cover is mn.

Consider the cover $P_{2,1} : T \to T$ given by Exercise 4.1.6. We give a picture of this cover in Figure 4.1(a). Now we change the base space to $T^{(2)}$ by using the same construction we used to get T from S^2; that is, doing surgery by removing two disks and replacing them by a cylinder. In the top space T we now do this twice. This is pictured in Figure 4.1(b).

Exercise 4.1.7.

(a) Show that the construction described above leads to a double covering $P : T^{(3)} \to T^{(2)}$.

(b) By starting with $P_{m,1}$, construct an m-fold cover $p : T^{(m+1)} \to T^{(2)}$.

Exercise 4.1.8. Continuing with the ideas in the previous exercise and starting with $P_{m,1}$ but now performing n surgeries on the base space, construct an m-fold cover $p : T^{(mn+1)} \to T^{(n+1)}$.

We now consider some covering spaces of nonorientable surfaces. Recall that in Section 3.4 we discussed the examples of 2-fold covering spaces $p : S \to P$ and $p : T \to K$. We can get new covering spaces from these by doing surgery as in the preceding exercises.

Exercise 4.1.9. Show that there is a 2-fold covering space $p_n : T^{(2n)} \to P^{(2n+1)}$. (Hint: Do n surgeries in small evenly covered disks in P.)

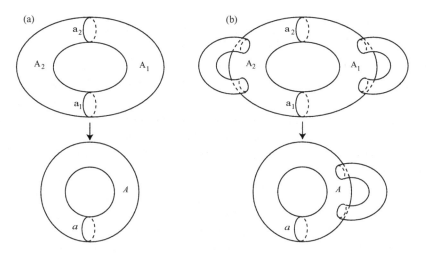

Figure 4.1. Constructing a cover $p : T^{(3)} \to T^{(2)}$.

Exercise 4.1.10. Show that there is a 2-fold covering space $p_n : T^{(2n+1)} \to P^{(2n+2)}$. (Hint: Start from the covering space $p : T \to K$ and do n surgeries to $K = P^{(2)}$ in small evenly covered disks in K.)

Here is another example dealing with surfaces which is motivated by a similar construction as the one forming $p : T \to K$. Start with $K = S^1 \times S^1/(z, w) \sim (-z, \bar{w})$. Form a new space K_n with a similar construction $K_n = S^1 \times S^1/(z, w) \sim (e^{\pi i/n}, \bar{w})$ when n is odd. Note that for $n = 1$, this is just the usual construction of K. There is a natural map from K to K_n which is induced from the identity map of T to itself. That this does induce a map uses the fact that n is odd so that $(z, w) \sim (-z, \bar{w})$ is part of the equivalence relation in the quotient construction of K_n. If we take a small disk in K_n, it will be evenly covered by n small disks in K where we get from one to the next by rotating by $e^{\pi i/n}$ in the first S^1 and then reflecting via $w \to \bar{w}$ in the second factor. That we get back to the first disk in n steps uses the equivalence relation in K. The next exercise asks you to show that K_n is homeomorphic to K, and so the construction gives an n-fold cover $p_n : K \to K$ for n odd.

Exercise 4.1.11. Show that K_n is homeomorphic to K. (Hint: Use the function $h : T \to T$ given by $h(z, w) = (z^n, w)$ and show that it induces a homeomorphism $\bar{h} : K_n \to K$.)

In each of the examples of finite coverings dealing with surfaces, we can compute for $p : A \to B$ how the Euler characteristics of A, B are related. In the next exercise you are asked to do this calculation to show that $\chi(A) = k\chi(B)$, where k is the order of the cover. This is a general result for finite covers of surfaces (actually much more generally).

Exercise 4.1.12. Verify in the last four exercises that there is a formula $\chi(A) = k\chi(B)$ where k is the order of the cover.

Exercise 4.1.13. Assuming the formula $\chi(A) = k\chi(B)$ for surfaces, show that the only surfaces without boundary that can cover themselves with order $k > 1$ are T and K.

One explanation for the Euler characteristic formula comes from equivariant handle decompositions. For example, consider the covering of the sphere over the projective plane. The projective plane is the union of a Möbius band and a disk. Lying above the Möbius band is an annulus about the equator of the sphere. Lying above the disk are two disks in the sphere, one in the upper hemisphere and one in the lower hemisphere. These will take the role of 2-handles in the equivariant handle decomposition of the sphere. For the Möbius band, it has a handle decomposition with a 0-handle and a 1-handle. Looking above this in the sphere, there will be two disks lying above the 0-handle, which can be considered 0-handles there that the antipodal map will interchange. Similarly, lying above the 1-handles are two disks, which take the role of 1-handles in the cover as well since they are each attached along a pair of intervals (up to homeomorphism). The next exercise generalizes this discussion.

Exercise 4.1.14. Suppose that $p : A \to B$ is a covering map of finite order k from one surface to another and we have a handle decomposition of B where each handle lies within an evenly covered neighborhood.

  (a) Then there is an equivariant handle decomposition of A with k i-handles for each i-handle of B.

  (b) Use (a) to show that $\chi(A) = k\chi(B)$.

We look next at some covering spaces of the wedge W of two circles. We determine how the covering spaces of order 2 should to look like (up to equivalence). We label the circles with a, b and label the cover with a_1, a_2, b_1, b_2. We picture one example of how to do this in Figure 4.2. In this example, each circle is double covered by a single circle.

Exercise 4.1.15. There are two other nonequivalent 2-fold covering spaces of W. Give pictures of them. (Hint: Consider cases where one of the circles is covered by two disjoint circles mapped homeomorphically.)

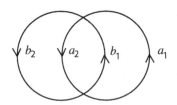

Figure 4.2. A double cover of $S^1 \vee S^1$.

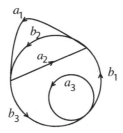

Figure 4.3. A three-fold cover of $S^1 \vee S^1$.

Figure 4.3 shows a one 3-fold cover of W. Here the edges labeled a are mapped to one circle and those labeled b are mapped to the other.

Exercise 4.1.16. Give pictures of two different nonequivalent 3-fold covering spaces of W from that given in Figure 4.3. (Note: There are more than two others, so there are many answers possible.)

The ability to lift paths in the base space to the covering space represents one of the most important properties. We have to first give a version of Theorem 3.3.1 to a general covering space.

Lemma 4.1.1. *Suppose $p : A \to B$ is a covering map and $U \subset B$ is a path-connected open set which is evenly covered with $p^{-1}(U) = \bigsqcup U_i, b \in U$. Let $a_i \in U_i$ satisfy $p(a_i) = b$. If X is connected and $f : X \to B$ is continuous with $f(X) \subset U, f(x) = b$, then there is a unique lift $\tilde{f}_i$ of f with $\tilde{f}_i(x) = a_i$. Moreover, $\tilde{f}_i(X) \subset U_i$.*

Exercise 4.1.17. Prove Lemma 4.1.1.

Theorem 4.1.2 (Unique path lifting theorem). *Suppose $p : A \to B$ is a covering map with $p(a) = b$. Given a path $f : I \to B$ with $f(0) = b$, there is a unique path $\tilde{f} : I \to A$ with $\tilde{f}(0) = a$ and $p\tilde{f} = f$.*

Exercise 4.1.18. Prove Theorem 4.1.2 by mimicking the proof of the unique path lifting property for the circle, Theorem 3.3.1.

Exercise 4.1.19. Suppose $r : I \to B$ is a path in B with $r(0) = b_1$ and $r(1) = b_2$. Define a map from $p^{-1}(b_1)$ to $p^{-1}(b_2)$ as follows: For each $a \in p^{-1}(b_1)$, let $\tilde{r}_a$ be the unique path lifting of r with $\tilde{r}_a(0) = a$. Then send a to $\tilde{r}_a(1)$. Show that this defines a bijection between $p^{-1}(b_1)$ and $p^{-1}(b_2)$. (Hint: Find an inverse for this map.)

We now want to extend the unique path lifting property to homotopies. The idea is to modify the argument given in Chapter 3 during the proof of the isomorphism $\pi_1(S^1, 1) \simeq \mathbb{Z}$ to get the following result.

Theorem 4.1.3. *Suppose that $F : (I \times I, \{0\} \times I, 1 \times I) \to (B, b_1, b_2)$ is a homotopy between f_0, f_1, and $p(a) = b_1$.*

(a) *There is a unique lifting $\widetilde{F}$ with $\widetilde{F}(0,0) = a$.*

(b) *If $\widetilde{f_0}, \widetilde{f_1}$ are liftings with $\widetilde{f_0}(0) = \widetilde{f_1}(0) = a$, then $\widetilde{f_0}(1) = \widetilde{f_1}(1)$.*

Exercise 4.1.20. Prove Theorem 4.1.3. (Hint: Lift the homotopy F connecting f_0, f_1 to $\widetilde{F}$ and utilize unique path lifting as in the proof that $\bar{h}$ is well defined in Theorem 3.3.3.)

Exercise 4.1.21. Utilize the previous exercise to define a map from $\pi_1(B, b)$ to $p^{-1}(b)$ as follows. Assign to $[f]$ the point $\widetilde{f}(1)$, where $\widetilde{f}$ is the unique path lifting of f with $\widetilde{f}(0) = a$.

   (a) Use the assumption that A is path connected to show that this map is onto.

   (b) Let $G \subset \pi_1(B, b)$ be the subset of $[f]$ with $\widetilde{f}(1) = a$. Show that G is a subgroup of $\pi_1(B, b)$.

4.2   Conjugate subgroups of π_1 and equivalent covering spaces

In this section we explore the relation between subgroups of the fundamental group of B and the possible covering spaces of B. We start by showing that whenever $p : (A, a) \to (B, b)$ is a covering map, then the fundamental group of A injects into the fundamental group of B.

Exercise 4.2.1. Show that $p_* : \pi_1(A, a) \to \pi_1(B, b)$ is 1–1. Thus $p_*(\pi_1(A, a))$ is a subgroup of $\pi_1(B, b)$ isomorphic to $\pi_1(A, a)$. (Hint: If $p\widetilde{f}$ is homotopic to a constant map at b, then lift the homotopy to a homotopy between $\widetilde{f}$ and the constant map at a.)

Exercise 4.2.2. Combine the ideas of Exercises 4.1.21 and 4.2.1 to show that $p_*(\pi_1(A, a)) \subset \pi_1(B, b)$ is the subgroup of classes of loops $[f]$ so that the lift of the loop to a path $\widetilde{f}$ with $\widetilde{f}(0) = a$ satisfies $\widetilde{f}(1) = a$. That is, it is the subgroup of loops at b which lift to loops at a.

   We now look at the influence of the base point a chosen in A on the subgroup $p_*(\pi_1(A, a))$ obtained. Ideas from Section 3.10 play a key role here. If a different base point is chosen, we do not necessarily get the same image subgroup, but we do get a subgroup which is conjugate to the original one.

Definition 4.2.1. Two subgroups $H_1, H_2 \subset G$ are called *conjugate* if there is an element $g \in G$ so that $H_1 = gH_2g^{-1}$; that is, each element of H_1 is of the form ghg^{-1}, where $h \in H_2$.

Exercise 4.2.3. Show that conjugacy is an equivalence relation on subgroups of G.

Theorem 4.2.1. *Suppose that a_0, a_1 satisfy $p(a_0) = p(a_1) = b$. Let $\widetilde{\alpha}$ be a path joining a_0 to a_1 with $\widetilde{\alpha}(0) = a_0, \widetilde{\alpha}(1) = a_1$. Let $\alpha = p\widetilde{\alpha}$ and $g = [\alpha] \in \pi_1(B, b)$. Let $G_0 = p_*(\pi_1(A, a_0)) \subset \pi_1(B, b)$ and $G_1 = p_*(\pi_1(A, a_1)) \subset \pi_1(B, b)$.*

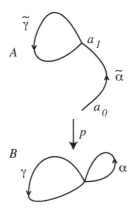

Figure 4.4. Conjugate loops.

(a) *Then the relation between the subgroups* G_0 *and* G_1 *is* $G_0 = gG_1g^{-1}$; *that is, the subgroups are conjugate and the element* g *that induces the conjugation is represented by the path* α *which lifts to a path connecting* a_0 *to* a_1 *in* A. *See Figure 4.4.*

(b) *If* $G = gHg^{-1} \subset \pi_1(B, b)$ *with* $g \in \pi_1(B, b), H = p_*(\pi_1(A, a_1))$, *then* $G = p_*(\pi_1(A, a_0))$ *for some* a_0 *with* $p(a_0) = p(a_1) = b$.

Exercise 4.2.4. Prove Theorem 4.2.1. (Hint: Recall from Section 3.10 that $\tilde{\alpha}_* : \pi_1(A, a_1) \to \pi_1(A, a_0)$ is an isomorphism.)

Exercise 4.2.5. For the covering map $p_m : S^1 \to S^1, p_m(z) = z^m$ of Exercise 4.1.3, compute $(p_m)_*(\pi_1(S^1, 1))$ and $(p_m)_*(\pi_1(S^1, e^{2\pi i/m}))$. Are they conjugate? Explain.

Exercise 4.2.6. Consider the covering space A of the wedge of two circles $B = S_x^1 \vee S_y^1$ which is formed from the union of the x-axis and the y-axis together with copies of the circle attached at each nonzero integer point $(n, 0)$ and $(0, m)$. Denote these circles as S_{xn}^1 and S_{ym}^1. The map $p : A \to B$ sends the x-axis to the circle S_x^1 by the usual covering map and similarly sends the y-axis to the circle S_y^1. The circle S_{xn}^1 is mapped via identification to the circle S_y^1 and the circle S_{ym}^1 is mapped via identification to the circle S_x^1. See Figure 4.5.

(a) Show that $p : A \to B$ is a covering map.

(b) Show that $p_*(\pi_1(A, (0, 0))) \neq p_*(\pi_1(A, (1, 0)))$ by finding an element of $p_*(\pi_1(A, (1, 0)))$ which does not lift to a loop starting at $(0, 0)$.

Exercise 4.2.7. Consider the 2-fold covering space of the three tangent circles A over the wedge of two circles B that is depicted in Figure 4.6. The arcs labeled a_i cover the first circle and the circles labeled b_i cover the second circle. Show that $G_1 = p_*(\pi_1(A, v_1)) = p_*(\pi_1(A, v_2)) = G_2$. (Hint: Consider the rotation about the center point of the middle circle in A.)

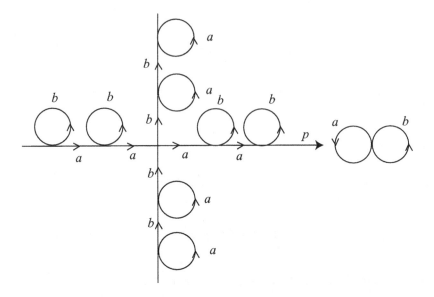

Figure 4.5. Covering space for Exercise 4.2.6.

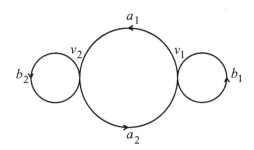

Figure 4.6. Covering space for Exercise 4.2.7.

Exercise 4.2.8. Consider the covering space A of the wedge of two circles $B = S_a^1 \vee S_b^1$ (with base point v where the two circles are joined) which is shown in Figure 4.7. In this covering space each arc a_i wraps around the circle S_a^1 once in the counterclockwise direction and each arc (or loop) b_i wraps around the circle S_b^1 once in the counterclockwise direction. We will denote the generators of $\pi_1(B, v)$ by a, b, which are equivalence classes of loops running around each of the circles once.

(a) Let $G_i = p_*(\pi_1(A, v_i))$. Show that $G_1 \neq G_2$ by considering how b lifts at different base points.

(b) Show that $G_2 \neq G_3$ but there is an isomorphism from G_2 to G_3 induced by the map sending $a \to a^{-1}, b \to b^{-1}$. (Hint: Consider the homeomorphism from A to itself which is induced by reflection through

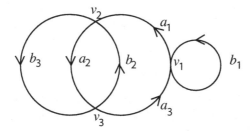

Figure 4.7. Covering space for Exercise 4.2.8.

a horizontal line of symmetry of A that interchanges v_2, v_3 and leaves v_1 fixed.)

Suppose that $p : A \to B$ is a covering map and $p(a) = b$. Let $G = p_*(\pi_1(A, a))$. Suppose X is a path-connected, locally path-connected space and $f : (X, x) \to (B, b)$ is continuous. We want to relate various covering spaces of B in terms of the subgroup G. To do this we first have to discuss lifting of maps from X into B to maps from X to the covering space A.

Definition 4.2.2. A continuous map $\tilde{f} : X \to A$ is called a *lifting* of f if $p\tilde{f} = f$.

$$\begin{array}{ccc} & & (A, a) \\ & \overset{\tilde{f}}{\nearrow} & \downarrow p \\ (X, x) & \overset{f}{\longrightarrow} & (B, b) \end{array}$$

Note that this is a generalization of a lifting of a path. The next theorem generalizes the path-lifting property to characterize when f has a lifting which sends x to a.

Theorem 4.2.2 (Lifting criterion). *Suppose that $p : A \to B$ is a covering map and $p(a) = b$. Let $G = p_*(\pi_1(A, a))$. Suppose X is a path-connected, locally path-connected space and $f : (X, x) \to (B, b)$ is continuous. There is a (unique) lifting of $f : X \to B$ to $\tilde{f} : X \to A$ with $\tilde{f}(x) = a$ iff*

$$f_*(\pi_1(X, x)) \subset p_*(\pi_1(A, a)).$$

Exercise 4.2.9. Follow the outline below to prove Theorem 4.2.2.

(a) Show that if there is a lifting $\tilde{f}$ with $\tilde{f}(x) = a$, then $f_*(\pi_1(X, x)) \subset p_*(\pi_1(A, a))$.

(b) Show that if there is a lifting $\tilde{f}$, then it is unique. (Hint: If $\tilde{f}_1, \tilde{f}_2$ are liftings with $\tilde{f}_1(x) = \tilde{f}_2(x) = a$, show that $S = \{y \in X : \tilde{f}_1(y) = \tilde{f}_2(y)\} = X$ by showing it is open and closed in X.)

Show that the converse of (a) is true by following the outline below. Try to define a lifting by defining $\tilde{f}(x) = a$ and then for each $y \in X$, choose a path α from x to y. Look at the image $\beta = f\alpha$ of this path, which is a path β in B starting at b. Then use unique path lifting to get a lifting $\tilde{\beta}$ of β with $\tilde{\beta}(0) = a$, and define $\tilde{f}(y) = \tilde{\beta}(1)$.

(c) Show that $f_*(\pi_1(X, x)) \subset p_*(\pi_1(A, a))$ implies that if α is a loop at x, then the lifting $\tilde{\beta}$ will be a loop at a.

(d) Suppose that α_1 and α_2 are two paths joining x to y which are homotopic relative to the end points. Then show that the liftings $\tilde{\beta}_1$ and $\tilde{\beta}_2$ are homotopic liftings and satisfy $\tilde{\beta}_1(1) = \tilde{\beta}_2(1)$.

(e) Use the fact that α_2 is homotopic relative to the end points to $\alpha_2 * \overline{\alpha_1} * \alpha_1$ and part (d) to show that the liftings $\tilde{\beta}_1, \tilde{\beta}_2$ of $\beta_i = f\alpha_i$ satisfy $\tilde{\beta}_1(1) = \tilde{\beta}_2(1)$. Deduce from this that the map $\tilde{f}$ as defined above is well defined, independent of the path used to join x to y.

(f) Let $\{U_i\}$ be a covering of B by path-connected open sets (using local path connectivity) so that $p^{-1}(U_i) = \bigsqcup_j \tilde{U}_{ij}$ and p maps each $\tilde{U}_{ij}$ homeomorphically to U_i. Use local path connectivity of X to show that the map $\tilde{f}$ is continuous at a given point $y \in X$. (Hint: Find a path-connected neighborhood V_y of y which is mapped into some U_i and show that the point $\tilde{f}(y)$ determines uniquely how that neighborhood V_y lifts and leads to a continuous function.)

We now apply these results to characterize equivalent covering spaces by replacing $f : X \to B$ by a covering map. We first relate two covering maps in terms of the image subgroups.

Exercise 4.2.10. Suppose $p_i : A_i \to B$ are covering maps, $p_i(a_i) = b$. Suppose

$$(p_1)_*(\pi_1(A_1, a_1)) \subset (p_2)_*(\pi_1(A_2, a_2)).$$

Then there is a continuous map $P : A_1 \to A_2$ with $P(a_1) = a_2, Pp_2 = p_1$, and P is a covering map. (Hint: Use Theorem 4.2.2 to get P. To show that it is a covering map, use the fact that we can select evenly covered sets that work for both p_1 and p_2.)

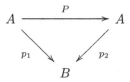

Exercise 4.2.11. Suppose $f : A \to A$ is a continuous map which is a lifting of $p : A \to B$ and $f(a) = a$. Then f is the identity.

Exercise 4.2.12. Suppose $p_1 : A_1 \to B, p_2 : A_2 \to B$ are equivalent covering spaces with equivalence $h : A_1 \to A_2$ and $h(a_1) = a_2$. Show that $(p_1)_*(\pi_1(A, a_1)) = (p_2)_*(\pi_1(A_2, a_2))$.

Exercise 4.2.13. Suppose $p_1 : A_1 \to B, p_2 : A_2 \to B$ are covering spaces with $p_i(a_i) = b$. Show that if $(p_1)_*(\pi_1(A_1, a_1)) = (p_2)_*(\pi_1(A_2, a_2))$, then the covering spaces are equivalent via an equivalence $F : A_1 \to A_2$ with $F(a_1) = a_2$. (Hint: Use Theorem 4.2.2.)

We now put these facts together to give the following characterization of when covering spaces are equivalent.

Theorem 4.2.3 (Characterization of equivalence of covering spaces). *Two covering spaces* $p_1 : A_1 \to B, p_2 : A_2 \to B$ *with* $p(a_i) = b$ *are equivalent iff* $(p_1)_*(\pi_1(A, a_1))$ *and* $(p_2)_*(\pi_1(A, a_2))$ *are conjugate subgroups of* $\pi_1(B, b)$.

Exercise 4.2.14. Combine the last exercises with Theorem 4.2.1 to prove Theorem 4.2.3.

Exercise 4.2.15.

(a) Find all of the subgroups of $\mathbb{Z}$.

(b) Show that any covering space of S^1 is equivalent to the covering space $p_m : S^1 \to S^1$, $p_m(z) = z^m$, $m \in \mathbb{N}$, or the covering space $p : \mathbb{R} \to S^1$.

Exercise 4.2.16. Show that there are only two covering spaces of $\mathbb{RP}^2$ up to equivalence, one of which is the identity covering space and the other $p : S^2 \to \mathbb{RP}^2$.

Exercise 4.2.17. Suppose $p : A \to B$ is a covering space, A is path connected, and $\pi_1(B, b) = \{e\}$. Show that p is a homeomorphism.

Exercise 4.2.18. Suppose $p_1 : A_1 \to B, p_2 : A_2 \to B$ are covering maps and the continuous map $h : A_1 \to A_2$ satisfies $p_1 = p_2 h, (p_1)_*(\pi_1(A, a)) = (p_2)_*(\pi_1(A_2, h(a))$. Then show that h is a homeomorphism and so is an equivalence between the covering spaces.

Exercise 4.2.19. A subgroup $G \subset F$ is said to be of *index* k if the set of right cosets Gf of G has k elements. Use Exercise 4.1.15 to find all of the conjugacy equivalence classes of index 2 subgroups G of the free group F_2.

Exercise 4.2.20.

(a) Suppose $p : A \to B$ is a covering map and the subgroup $G = p_*(\pi_1(A, a)) \subset \pi_1(B, b)$ is of index k, where $p(a) = b$. Show that $p^{-1}(b)$ has k points.

(b) Conversely, show that if $p^{-1}(b)$ has k points and $p(a) = b$, then there are k cosets Gf of $G = p_*(\pi_1(A, a)) \subset \pi_1(B, b)$; that is, the order $|\pi_1(B, b)/G| = k$.

4.3   Covering transformations

Definition 4.3.1. If $p : A \to B$ is a covering map, then a homeomorphism $T : A \to A$ with $pT = p$ is called a *covering transformation*.

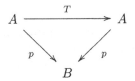

Exercise 4.3.1. Suppose $p : A \to B$ is a covering map, $p(a_1) = p(a_2) = b$ and $G_i = p_*(\pi_1(A, a_i)) \subset \pi_1(B, b)$. Suppose that $T : A \to A$ is a covering transformation with $T(a_1) = a_2$. Show that $G_1 = G_2$.

Exercise 4.3.2. Suppose $p : A \to B$ is a covering map with $p(a_1) = p(a_2) = b$. Show that there is a covering transformation $T : A \to A$ with $T(a_1) = a_2$ iff $p_*(\pi_1(A, a_1)) = p_*(\pi_1(A, a_2))$. (Hint: Use Theorem 4.2.2.)

Exercise 4.3.3. Show that the set of all covering transformations of $p : A \to B$ forms a group, which we denote G_p, under the operation of composition of homeomorphisms.

Exercise 4.3.4. Show that if T_1, T_2 are covering transformations with $T_1(a) = T_2(a)$ for some $a \in A$, then $T_1 = T_2$. (Hint: Use Theorem 4.2.2.)

Exercise 4.3.5. Consider the homeomorphism $T_n : \mathbb{R} \to \mathbb{R}$, $T_n(x) = x + n$, $n \in \mathbb{Z}$. Let $p : \mathbb{R} \to S^1$ be the standard covering map $p(t) = e^{2\pi i t}$.

   (a) Show that $pT_n = p$.
   (b) Show that if $\tilde{f_1}, \tilde{f_2}$ are two liftings of a loop $f : I \to S^1$, then for $n = \tilde{f_2}(0) - \tilde{f_1}(0)$ we have $T_n \tilde{f_1} = \tilde{f_2}$.

Exercise 4.3.6. Show that if $T : \mathbb{R} \to \mathbb{R}$ is a covering transformation, then $T = T_n$ for some n. Conclude that $G_p \simeq \mathbb{Z}$ for $(\mathbb{R}, p, S^1)$.

Exercise 4.3.7. Find the group of covering transformations of the covering $p : S^2 \to P$, $p(x) = [x]$ (where $P = S^2 / x \sim -x$ and $[x]$ denotes the equivalence class of x).

Exercise 4.3.8. Find the group of covering transformations of the covering $p_m : S^1 \to S^1$, $p_m(z) = z^m$ (see Exercise 4.1.3).

Exercise 4.3.9. Consider the covering space of $S^1 \vee S^1$ from Exercise 4.2.8. Show that there is no covering transformation sending v_1 to v_2. (Hint: Consider where the loop b_1 would have to be mapped.)

Theorem 4.3.1. *Suppose $p : A \to B$ is a covering space with $p(a) = b$ and $\pi_1(A, a) \simeq \{e\}$. Then the group of covering transformations G_p is isomorphic to the fundamental group $\pi_1(B, b)$.*

Exercise 4.3.10. Follow the outline below to prove Theorem 4.3.1. Define a map $r : G_p \to \pi_1(B, b)$ as follows. Given $T \in G_p$, let s be a path in A connecting a and $T(a)$, and let $r(T) = [ps]$. Prove that r is an isomorphism as follows.

(a) Show that if s is homotopic rel $0,1$ to s', then $ps \sim ps'$.

(b) Use $\pi_1(A, a) \simeq \{e\}$ to conclude that any two paths connecting a and $T(a)$ are homotopic rel $0,1$ and so r is well defined.

(c) Show that r is a homomorphism.

(d) Show that r is 1–1. (Hint: Show that if $r(T_1) = r(T_2)$, then $T_1(a) = T_2(a)$.)

(e) Show r is onto. (Hint: Let $[f] \in \pi_1(B, b)$. Lift f to $\widetilde{f}$ with $\widetilde{f}(0) = a$ and let $a_1 = \widetilde{f}(1)$. Then use Exercise 4.3.2.)

Exercise 4.3.11. (a) Suppose $p : A \to B$ is a covering space with $p(a_i) = b$, $i = 1, 2$. Let $G_i = p_*(\pi_1(A, a_i)) \subset \pi_1(B, b)$. Show that if there is a covering transformation $T : A \to A$ with $T(a_1) = a_2$, then $G_1 = G_2$ and if we call this common subgroup G, there is an element $g \in \pi_1(B, b)$ with $gGg^{-1} = G$, where g is represented by a loop which lifts to a path from a_1 to a_2.

(b) For the converse, suppose that $G = p_*(\pi_1(A, a_i))$, $i = 1, 2$ and $gGg^{-1} = G$, where $g \in \pi_1(B, b)$ is represented by a loop which lifts to a path from a_1 to a_2. Show that there is a covering transformation $T : A \to A$ which sends a_1 to a_2.

Exercise 4.3.12. Use the last exercise to give another argument for Exercise 4.3.9.

Definition 4.3.2. The *normalizer* $N(H)$ of a subgroup $H \subset G$ is the subgroup $\{g \in G : gHg^{-1} = H\}$.

The following theorem generalizes Theorem 4.3.1.

Theorem 4.3.2. *There is an isomorphism between the group of covering transformations G_p of $p : A \to B$ and the quotient $N(H)/H$ of $H = p_*(\pi_1(A, a)) \subset \pi_1(B, b)$.*

Exercise 4.3.13. Prove Theorem 4.3.2. (Hint: The argument should mimic the one given for the special case where $\pi_1(A, a) = \{e\}$ in Exercise 4.3.10.)

Exercise 4.3.14. Show that there is a bijection between

$$\{c \in p^{-1}(b): \text{there is a covering transformation } T \text{ with } T(a) = c\}$$

and $N(H)/H$. (Hint: Covering transformations are determined by their value at a by Exercise 4.3.4.)

Exercise 4.3.15. (a) Suppose that $H = p_*(\pi_1(A, a))$ is a normal subgroup of $\pi_1(B, b)$. Then show that for every pair of points a_1, a_2 in the pre-image $p^{-1}(b)$, there is a covering transformation T with $T(a_1) = a_2$.

(b) Show that the converse is true: whenever it is the case that for every pair of points a_1, a_2 in the pre-image $p^{-1}(b)$, there is a covering transformation T with $T(a_1) = a_2$, then $H = p_*(\pi_1(A, a))$ is a normal subgroup of $\pi_1(B, b)$.

Definition 4.3.3. A covering space is called *regular* if $p_*(\pi_1(A, a))$ is a normal subgroup of $\pi_1(B, b)$. By the previous exercise, this is equivalent to the group of covering transformations acting transitively on $p^{-1}(b)$.

Transitive actions of covering transformations on regular covering spaces have a special property. Suppose that U is a neighborhood of b which is evenly covered and $p^{-1}(U) = \bigsqcup_i \tilde{U}_i$, where $p : \tilde{U}_i \to U$ is a homeomorphism. There is one open set $\tilde{U}_i$ for each point a_i of $p^{-1}(b)$. If the covering space is regular, then there is a covering transformation T sending a_i to a_j for each pair i, j; T will then send $\tilde{U}_i$ homeomorphically to $\tilde{U}_j$.

Definition 4.3.4. A group G of homeomorphisms of a space A is called *properly discontinuous* if for every $a \in A$ there is an open set $\tilde{U}$ containing a such that $g(\tilde{U})$ is disjoint from $\tilde{U}$ whenever $g \neq e$.

Exercise 4.3.16. Show that if G is a group of homeomorphisms which is properly discontinuous, then for every $a \in A$ there is an open set $\tilde{U}$ containing a such that $g_0(\tilde{U}) \cap g_1(\tilde{U}) = \emptyset$ for all $g_0 \neq g_1 \in G$. (Hint: Look at $g_0^{-1}g_1$.)

Exercise 4.3.17. Suppose that A is path connected and locally path connected and G is a group of homeomorphisms of A. Let A/G denote the quotient space formed from A, where $a \sim g(a)$ for $g \in G$.

(a) Show that the quotient map $q : A \to A/G$ is a covering map iff the action of G is properly discontinuous.

(b) Show that when the conditions of (a) are satisfied, then the covering map $q : A \to A/G$ is regular and G is its group of covering transformations. (Hint: Use uniqueness of covering maps fixing sending one point to another and the characterization of regularity in terms of transitivity.)

Theorem 4.3.3 (Characterization of regular covering spaces). *Suppose $p : A \to B$ is a regular covering space and G is its group of covering transformations. Then $p = hq$, where $q : A \to A/G$ is the quotient covering space and h is a homeomorphism induced by p via the quotient construction.*

Exercise 4.3.18. Prove Theorem 4.3.3.

4.4   The universal covering space and quotient covering spaces

The previous exercises have shown how equivalence classes of covering spaces are related to conjugacy classes of subgroups of the fundamental group in terms of uniqueness. We next want to explore how existence of covering spaces is related to subgroups. The most basic question is whether there is a covering space $p : A \to B$ where $\pi_1(A, a)$ is the trivial group. Recall that a path-connected space A with $\pi_1(A, a)$ trivial is called simply connected.

Definition 4.4.1. A *universal covering space* of B is a simply connected covering space of B.

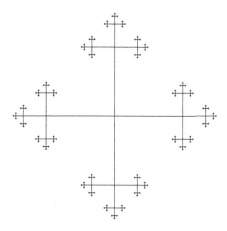

Figure 4.8. Start of universal cover of $S^1 \vee S^1$.

Exercise 4.4.1. Show that if a universal covering space exists, then it is unique up to equivalence. That is, if $p_i : A \to B$ are two simply connected covering spaces of A, then they are equivalent.

Universal covering spaces can be quite complicated, as different homotopy classes of loops have to lift to distinct end points in the universal cover. Figure 4.8 illustrates of a piece of the universal covering of $S^1 \vee S^1$. Note that the fundamental group is the free group on two generators. What the figure actually shows are paths which correspond to liftings of loops at the wedge point with word length ≤ 5. Each horizontal segment maps to the first S^1 in the counterclockwise direction as we move to the right, and each vertical segment maps to the second S^1 in the counterclockwise direction as we move upward. The vertices of the graph are all mapped to the wedge point. To get the next level in the universal cover, we would add a T at each end vertex point with the intersection of the T glued to the point. This pattern is repeated as we go to higher and higher levels. Note that there are 4×3^4 such T's that have to be added to get to the sixth level.

We next note the following property of a universal covering space $p : A \to B$. If $p(a) = b$ and U is a path-connected open set containing b which is evenly covered with $a \in \tilde{U}$ and $p : \tilde{U} \to U$ a homeomorphism, then the induced map $\pi_1(\tilde{U}, a) \to \pi_1(A, a) = \{e\}$ is necessarily trivial. But the diagram

$$
\begin{array}{ccc}
\pi_1(\tilde{U}, a) & \xrightarrow{\ i_* \ } & \pi_1(A, a) \\
{\scriptstyle p_*} \downarrow {\scriptstyle \simeq} & & \downarrow {\scriptstyle p_*} \\
\pi_1(U, b) & \xrightarrow{\ i_* \ } & \pi_1(B, b)
\end{array}
$$

shows that $i_* : \pi_1(U, b) \to \pi_1(B, b)$ is trivial.

Definition 4.4.2. We say that a space B is *semilocally simply connected* if, for each $b \in B$, there is a path-connected open set U containing b with the homomorphism $i_* : \pi_1(U, b) \to \pi_1(B, b)$ induced by inclusion being trivial.

The discussion above shows that a necessary condition for the existence of a universal covering space of B is that B is semilocally simply connected. The next set of exercises show that this condition is also sufficient by constructing a simply connected covering space for a semilocally simply connected space B.

Theorem 4.4.1 (Existence of universal covering space). *Suppose that B is a semilocally simply connected, locally path-connected, path-connected space. Then there is a universal covering space $p : A \to B$.*

To construct such a space, let $b \in B$ and form the space A of homotopy equivalence classes, relative to the end points, of paths $\alpha : I \to B$ with $\alpha(0) = b$. We need to put a topology on A, which we do by defining a basis. Let $[\alpha]$ denote an equivalence class and $\alpha(0) = b, \alpha(1) = b'$. Choose an open set U containing b' that is path connected so that $\pi_1(U, b') \to \pi_1(B, b')$ is trivial. Such a set will be shown to satisfy the evenly covered criterion for a covering space $p : A \to B$.

Exercise 4.4.2.

(a) Show that for any two paths $\beta, \gamma : I \to U$ with $\beta(0) = \gamma(0) = b', \beta(1) = \gamma(1) = b''$, β is homotopic to γ relative to the end points as maps to B.

(b) Show that $\alpha * \beta$ is homotopic to $\alpha * \gamma$ relative to the end points as maps to B.

(c) Conclude that if, for each point $u \in U$, we choose a path β in U from b' to u, then the equivalence class of $\alpha * \beta$ determines a unique point of A. Thus there is a 1–1 correspondence between points of U and points of the set $\widetilde{U}_{[\alpha]}$ of equivalence classes of paths from b which are represented by the juxtaposition of α with a path in U.

Exercise 4.4.3. Show that if $[\gamma] \in \widetilde{U}_{[\alpha]}$, then $\widetilde{U}_{[\gamma]} = \widetilde{U}_{[\alpha]}$.

Exercise 4.4.4. Show that if $\widetilde{U}_{[\alpha]} \cap \widetilde{U}_{[\beta]} \neq \emptyset$, then $\widetilde{U}_{[\alpha]} = \widetilde{U}_{[\beta]}$.

Exercise 4.4.5. Show that if $c \in U$, the distinct sets $\widetilde{U}_{[\alpha]}$ can be indexed by the distinct relative homotopy classes of paths from b to c, which is $\pi_1(B, b, c)$ and thus corresponds bijectively to $\pi_1(B, b)$, by Exercise 3.11.9.

Exercise 4.4.6. Show that if $x \in V \subset U$ and V is path connected and the inclusion $\pi_1(U, x) \to \pi_1(B, x)$ is trivial, then the inclusion $\pi_1(V, x) \to \pi_1(B, x)$ is trivial. Conclude that $\widetilde{V}_{[\alpha]} \subset \widetilde{U}_{[\alpha]}$.

Exercise 4.4.7. Show that the sets $\widetilde{U}_{[\alpha]}$ constructed above form the basis for a topology for A.

Exercise 4.4.8.

(a) Consider the function $p : A \to B$ given by $p([\alpha]) = \alpha(1)$. Show that p is well defined and continuous.

(b) Show that p is a covering map.

(c) Show that there is a bijection between $p^{-1}(c) \subset A$ and $\pi_1(B, c)$.

Exercise 4.4.9. Show that A is simply connected. (Hint: Since $p : A \to B$ is a covering space, it suffices to show that the image $p_*(\pi_1(A, a))$ is trivial, where a is the equivalence class of the constant path at b. Use the fact that p_* is injective, and interpret what it means to be a loop at a. Note that through unique path lifting we can describe the lifting of a loop s in B at b in terms of the initial segments $s|[0, t]$.)

Exercise 4.4.10. For the universal covering space A as constructed above, show how each element of $\pi_1(B, b)$ naturally leads to a covering transformation of the covering space. Show that this correspondence $\pi_1(B, b) \to G_p$ is an isomorphism.

Now suppose that H is a subgroup of the fundamental group $G = \pi_1(B, b)$. Using the previous exercise, we can identify H as a subgroup of covering transformations of A. Form the topological space A/H which is the quotient space of A by the equivalence relation that says $a \sim a'$ whenever there is a covering transformation $T \in H$ with $T(a) = a'$.

Exercise 4.4.11.

(a) Show that the map $p : A \to B$ factors as $p = p_2 p_1$, where $p_1 : A \to A/H$, $p_2 : A/H \to B$, and both p_1, p_2 are covering maps.

(b) Show that $\pi_1(A/H, [\alpha]) \simeq H$ and $(p_2)_*$ sends it to $H \subset G$.

Theorem 4.4.2 (Relation of subgroups of $\pi_1(B, b)$ and covering spaces of B). *Let B be a path-connected, locally path-connected, semilocally simply connected space. Then there is a bijection between the conjugacy classes of subgroups of $\pi_1(B, b)$ and the equivalence classes of covering spaces of B.*

Exercise 4.4.12. Put together the previous results to prove Theorem 4.4.2.

5

CW complexes

5.1 Examples of CW complexes

In Section 3.8, where we discussed homotopy equivalences, we showed that if we had a handle decomposition of a surface with one 0-handle and k 1-handles, then we could find a deformation retraction of this surface down to a wedge of k circles. This much of the space can be thought of as being built up from a central point corresponding to the 0-handle and then attaching intervals corresponding to the 1-handles where their end points are all identified to the central point. When the 2-handles are attached, we could compose their attaching maps with this homotopy equivalence and get a space which is built up from a point, some 1-disks attached, and then some 2-disks attached. It turns out that the original surface is homotopy equivalent to this space, which is an example of a two-dimensional CW complex. In this chapter we will develop the concepts of CW complexes and apply them to fundamental group calculations as well as discuss homotopy-theoretic ideas concerning them. Our discussion will include finite CW complexes of any dimension, but we will emphasize two-dimensional complexes where the geometry is easier to visualize. We will also discuss important special cases of simplicial complexes and Δ-complexes.

A finite two-dimensional *CW complex* structure for X expresses X as being built in stages as $X^0 \subset X^1 \subset X^2 = X$. Here X^0 is the disjoint union of a finite number of points, called 0-cells and denoted as $e_1^0, \ldots, e_{k_0}^0$, with the discrete topology. The space X^1 is built up from X^0 by taking a finite number of 1-disks, which are called 1-cells and denoted by $e_1^1, \ldots, e_{k_1}^1$, and then forming the quotient space X^1 from the disjoint union $X^0 \bigsqcup e_1^1 \bigsqcup \cdots \bigsqcup e_{k_1}^1$ by using continuous functions $f_i : \partial e_i^1 \to X^0$. These functions just identify the boundary points of a 1-cell to points in X^0—geometrically this is forming a one-dimensional graph. Finally, $X = X^2$ is built up from X^1 by taking a finite number of 2-disks, which are called the 2-cells and denoted by $e_1^2, \ldots, e_{k_2}^2$, and then forming the quotient space $X^2 = X$ from the disjoint union $X^1 \bigsqcup e_1^2 \bigsqcup \cdots \bigsqcup e_{k_2}^2$ by identifying the boundary points of a 2-cell to points in X^1 via continuous functions $g_i : \partial e_i^2 \to X^1$. The map g_i (resp., f_i) is called the *attaching map* of the cell e_i^2

(resp., e_i^1) The map $\phi_\alpha : e_\alpha \to X$ from a cell e_α to X is continuous, and the quotient topology has the property that a set A is closed iff the inverse image $\phi_\alpha^{-1}(A)$ of A is closed in the cell e_α for each α. The map ϕ_α is called the *characteristic map* of the cell e_α. Thus the space X can be regarded as a quotient space of the disjoint union of cells. The subsets X^0, X^1 are closed subsets of X, called the 0-skeleton and 1-skeleton of X, respectively. In a CW complex the cells are not embedded in general. However, the restriction of a characteristic map ϕ_α to the interior of the cell is embedded and the whole space can be expressed as the disjoint union of the images $\phi_\alpha(\text{int } e_\alpha)$ of all of the cells.

In our constructions of cells, we will use homeomorphs of D^i for the domains of our cells and not necessarily D^i itself. For example, when discussing simplicial complexes, our model cells will be simplices Δ^i. When we discuss surfaces, we will use polygons for the domains of our 2-cells. Although the topology is defined in terms of all of the cells, it is frequently the case that the whole space lies in the image of top-dimensional cells and the lower cells can be thought of as coming from first embedding a lower-dimensional cell into the top-dimensional one and then composing. In this case, the topology on the CW complex can be completely described in terms of the quotient topology from the map on the top-dimensional cells. This situation occurs in the CW decompositions we give for surfaces below.

The 2-sphere S^2 has the structure of a CW complex by taking X_0 as a single point e^0 and $X^1 = X^0$ (there are no 1-cells). Then $X^2 = X^0 \cup e^2$ is formed by using the constant map sending ∂e^2 to the point X^0. This CW complex with the quotient topology is just the same as taking a 2-disk D^2 (the 2-cell) and identifying all of its boundary points to a single point. This description can be used to show that the space is homeomorphic to S^2.

For the torus T, we can think of it as being formed from a square $D^1 \times D^1$ by identifying $(x, 1) \sim (x, 0), (1, y) \sim (0, y)$. Then the four corners are identified and determine a single 0-cell. The edge $D^1 \times \{0\}$ forms a 1-cell e_1^1—note that it is identified to the edge $D^1 \times \{1\}$ in the quotient space. Similarly, the edge $\{1\} \times D^1$ (which is identified to $\{0\} \times D^1$) forms a second 1-cell e_2^1. The space X^1 is homeomorphic to the one-point union of two copies of the circle. The point in common is e^0; the first circle comes from $e^0 \cup e_1^1$ and the second one from $e^0 \cup e_2^1$. Finally, $D^1 \times D^1$ forms a 2-cell e^2 whose boundary is identified to points in X^1. Basically, its boundary is divided into four parts which first run over the first circle, then run over the second circle, then run backwards over the first circle, and finally run backwards over the second circle. This can be expressed as the word $aba^{-1}b^{-1} \in \pi_1(S^1 \vee S^1, x)$.

The surface $T^{(g)}$ can be thought of as coming from a regular $4g$-gon whose boundary is identified via the pattern $a_1 b_1 a_1^{-1} b_1^{-1} \dots a_g b_g a_g^{-1} b_g^{-1}$. $T^{(g)}$ has the structure of a CW complex with one 0-cell (coming from all of the identified vertices), $2g$ 1-cells coming from the edges which are identified in pairs and whose boundaries are sent to the single 0-cell, and a single 2-cell coming from the $4g$-gon, whose boundary is mapped into X^1 according to the identification pattern. Here X^1 is homeomorphic to a wedge W_{2g} of $2g$ circles. Similarly, there is a decomposition of $P^{(h)}$ as a CW complex with a single 0-cell, h 1-cells, and a single 2-cell expressed as a regular $2h$-gon whose boundary is identified according to the pattern $a_1^2 \ldots a_h^2$. Note that X^1 in this case is homeomorphic to the wedge W_h of h circles.

Exercise 5.1.1. Verify that S^2 is homeomorphic to the CW complex $e^0 \cup e^2$.

Exercise 5.1.2.
  (a) Verify that T is homeomorphic to the CW complex $e^0 \cup e_1^1 \cup e_2^1 \cup e^2$ as described above.
  (b) Verify that $T^{(g)}$ is homeomorphic to the CW complex with a single 0-cell, $2g$ 1-cells, and a single 2-cell as described above.
  (c) Verify that $P^{(h)}$ is homeomorphic to the CW complex with a single 0-cell, h 1-cells, and a single 2-cell as described above.

Exercise 5.1.3. The projective plane is formed from a 2-disk by identifying x to $-x$ on the boundary circle. It has a CW decomposition $e^0 \cup e^1 \cup_{f_2} e^2$, where $f_2(z) = z^2$. The *pseudoprojective plane* P_p is formed from the disk by identifying x to $e^{2\pi i/p}x$ on the boundary circle. Give a similar CW decomposition for P_p with one cell in each dimension and identify the attaching map for the 2-cell.

Exercise 5.1.4. Consider the CW complex

$$X = e_1^0 \cup e_2^0 \cup e_1^1 \cup e_2^1 \cup_f e_1^2 \cup_g e_2^2.$$

Here each 1-cell e_i^1 is attached to X^0 by identifying its boundary to $e_1^0 \cup e_2^0$. We will orient each one cell and identify ∂e_1^1 by sending 1 to e_2^0 and -1 to e_1^0. Similarly, we identify ∂e_2^1 by sending 1 to e_1^0 and -1 to e_2^0. To attach the 2-cells, we identify the boundary with S^1 and divide it into the upper half arc S_+^1 and lower half arc S_-^1. We then identify the boundary of e_1^2 with X^1 by identifying the upper half arc with e_1^1 and identifying the lower half arc with e_2^1. Similarly, we identify the boundary of e_2^2 using the same map. We express this by saying that $\partial e_j^2 = e_1^1 + e_2^1$ (see Figure 5.1).

  (a) Show that X^1 is homeomorphic to S^1.
  (b) Show that $X^1 \cup_f e_1^2$ is homeomorphic to D^2.
  (c) Show that X is homeomorphic to S^2.

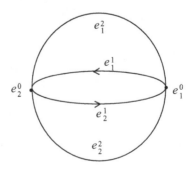

Figure 5.1. A CW decomposition of the sphere.

So far we have only considered two-dimensional CW complexes. More generally, a finite CW complex of dimension n is built up in stages $X^0 \subset X^1 \subset X^2 \subset X^3 \subset \cdots \subset X^n = X$, where X^i is obtained from X^{i-1} by attaching i-cells. Here an i-cell e^i is a homeomorph of the i-dimensional unit disk in $\mathbb{R}^i$ with boundary $\partial e^i \simeq S^{i-1}$ and there is a continuous map $f : \partial e^i \to X^{i-1}$. We then form X^i by taking the quotient space of the disjoint union of X^{i-1} and all of the i-cells and make identifications using the attaching maps of the i-cells. In the corresponding topology, a set is open (closed) iff its inverse image in each cell is an open (closed) set. The subset X^k is called the k-*skeleton* of X.

Exercise 5.1.5. Show that the k-skeleton X^k is a closed subset of X. (Hint: Use an inductive argument.)

Exercise 5.1.6. Show that a finite CW complex is a compact Hausdorff space. (Hint: Use an inductive argument to show that X is Hausdorff.)

In this book, we will restrict our attention to finite CW complexes. This will allow us to avoid discussing more subtle questions about point set topology which occur for a general CW complex. Thus we will always mean a finite CW complex when we use the term CW complex here. See [13] for a nice discussion of the topology of a CW complex with possibly infinitely many cells.

The k-skeleton $X^k \subset X$ is an example of a subcomplex of a CW complex. A *subcomplex* $Y \subset X$ is a collection of cells of X which is itself a CW complex. This requires that when we are attaching a cell of Y, then the image of the attaching map is within Y.

Exercise 5.1.7. Using the CW decomposition of S^2 from Exercise 5.1.4, show that the upper hemisphere is a subcomplex.

Exercise 5.1.8. Show that a subcomplex is a closed subset of X. (Hint: Show that it is compact and use the fact that X is Hausdorff.)

Here are some examples of higher-dimensional CW complexes.

The n-sphere has a CW complex description with one 0-cell and one n-cell, which is attached by the constant map from S^{n-1} to the 0-cell. Under the CW topology, the space just comes from D^n by identifying all of the boundary points to a single point.

Exercise 5.1.9. Verify that the n-sphere is homeomorphic to the space with the CW decomposition described above. (Hint: Construct the homeomorphism from $D^n/\sim$ to S^n analogously to what is done in the case $n = 2$.)

For a product space $X \times Y$, where X and Y each has a structure as a CW complex, $X \times Y$ inherits a structure as a CW complex since the product of an i-cell and a j-cell is homeomorphic to a $(i + j)$-cell. For example, think of $S^1 = e^0 \cup e^1$ and form the product

$$S_1^1 \times S_2^1 = (e_1^0 \cup e_1^1) \times (e_1^0 \cup e_2^1) = (e_1^0 \times e_2^0) \cup (e_1^1 \times e^0) \cup (e_1^0 \times e_2^1) \cup (e_1^1 \times e_2^1).$$

The only nontrivial attaching map to understand is for $e_1^1 \times e_2^1$. Here we use $\partial(e_1^1 \times e_2^1) = \partial e_1^1 \times e_2^1 \cup e_1^1 \times \partial e_2^1$. We map $\partial e_1^1 \times e_2^1$ to $e_1^0 \times e_2^1 \subset (X \times Y)^1 = X^1 \times Y^0 \cup X^0 \times Y^1$ by using the product of the attaching map $\partial e_1^1 \to X^0$ with the identity on the e_2^1 factor. Similarly, we map $e_1^1 \times \partial e_2^1$ to $e_1^1 \times e_2^0 \subset (X \times Y)^1$ by using the product of the attaching map $\partial e_2^1 \to Y^0$ with the identity on the e_1^1 factor. This leads to the same CW decomposition for the torus with a single 0-cell, two 1-cells, and a single 2-cell that we had before. Note that $X \times e^0 \subset X \times Y$ and $e^0 \times Y \subset X \times Y$ are each subcomplexes of $X \times Y$, as is their union.

Exercise 5.1.10. Give a CW decomposition for $S^1 \times S^2$ by using the product decomposition of our CW decompositions of S^1 and S^2. Your description should identify how the skeleta are built up in the product. (Hint: You should have a single cell in each dimension from 0 to 3.)

Exercise 5.1.11. Give a CW decomposition for $S^1 \times \mathbb{RP}^2$ by using the product decomposition of our CW decompositions of S^1 and $\mathbb{RP}^2$. Your description should identify how the skeleta are built up in the product. (Hint: You should have six cells.)

Exercise 5.1.12. Give a CW decomposition for $S^1 \times S^1 \times S^1$ by using the product decomposition of our CW decompositions of $S^1 \times S^1$ and S^1. Your description should identify how the skeleta are built up in the product. (Hint: You should have eight cells.)

Exercise 5.1.13. Give a CW decomposition for $S^2 \times S^2$ by using the product decomposition of our CW decomposition of S^2. Your description should identify how the skeleta are built up in the product. (Hint: You should have four cells.)

Exercise 5.1.14. Give a CW decomposition for $\mathbb{RP}^2 \times \mathbb{RP}^2$ by using the product decomposition of our CW decomposition of $\mathbb{RP}^2$. Your description should identify how the skeleta are built up in the product. (Hint: You should have nine cells.)

There is an n-dimensional generalization of the projective plane, which is n-dimensional projective space $\mathbb{RP}^n$. It is formed from the n-sphere by identifying antipodal points $\boldsymbol{x} = (x_0, \ldots, x_n) \sim (-x_0, \ldots, -x_n) = -\boldsymbol{x}$. Since S^n has trivial fundamental group for $n \geq 2$, it serves as the universal cover of $\mathbb{RP}^n$. From our work on covering spaces, this implies that $\mathbb{RP}^n$ has fundamental group $\mathbb{Z}_2$. We want to give a CW decomposition. The key idea is to see how we get from $\mathbb{RP}^{k-1}$ to $\mathbb{RP}^k$. In the covering space above, this will correspond to going from S^{k-1} to S^k. We do this by attaching the upper and lower hemispheres. But these hemispheres are each homeomorphic to D^k via vertical projection. Thus we could think of getting from S^{k-1} to S^k by attaching two k-cells. From the point of view of forming the quotient space, we want to attach these two cells in a consistent manner. The way to do this is to get the upper hemisphere by attaching an k-cell corresponding to vertical projection and then attach the lower hemisphere by composing this attaching map with the antipodal map which sends the upper hemisphere to the lower hemisphere. This leads to the description of $S^k = S^{k-1} \cup_{f_{k-1}} e^k \cup_{a_k f_{k-1}} e^k$. Here f_{k-1} is the identity map of S^{k-1} and a_k is the antipodal map of S^k. When we pass to the quotient, the two k-cells get identified to give $\mathbb{RP}^k = \mathbb{RP}^{k-1} \cup_{p_{k-1}} e^k$. Here $p_{k-1} : S^{k-1} \to \mathbb{RP}^{k-1}$ is the quotient map. By a simple inductive argument, this leads to a CW decomposition of S^n with $2(n+1)$-cells, two in each dimension between 0 and n, and a corresponding

CW decomposition of $\mathbb{RP}^n$ with $(n + 1)$-cells, one in each dimension between 0 and n. The k-skeleton for S^n is S^k and the k-skeleton for $\mathbb{RP}^n$ is $\mathbb{RP}^k$.

There is a complex analogue of $\mathbb{RP}^n$, which is called complex projective n-space and denoted $\mathbb{CP}^n$. To describe it, we start with the description of $\mathbb{RP}^n$ as the quotient space of S^n via $\{(x_0, x_1, \ldots, x_n) \in \mathbb{R}^{n+1} \colon \sum x_i^2 = 1\}/(x_0, \ldots, x_n) \sim \pm(x_0, \ldots, x_n)$. Note that the numbers ± 1 are the real numbers of length 1. For $\mathbb{CP}^n$, we form a similar quotient, but now we use the complex numbers instead of the real numbers. We start with $S^{2n+1} = \{(z_0, \ldots, z_n) \in \mathbb{C}^{n+1} \colon \sum |z_i|^2 = 1\}$. We then introduce the equivalence relation $(z_0, \ldots, z_n) \sim \zeta(z_0, \ldots, z_n) = (\zeta z_0, \ldots, \zeta z_n)$, where $\zeta \in \mathbb{C}$ is a unit complex number: $|\zeta| = 1$. Note that the set of ζ with $|\zeta| = 1$ is just the unit circle. We then form $\mathbb{CP}^n$ as the quotient space $S^{2n+1}/(z_0, \ldots, z_n) \sim \zeta(z_0, \ldots, z_n)$. When $n = 0$, the equivalence relation just identifies all points of the circle to each other, so $\mathbb{CP}^0$ is just a point. When $n = 1$, then each point $(z_0, z_1) \in S^3$ is equivalent to a point (w_0, r) with $r \geq 0$ and $|w_0|^2 + r^2 = 1$. These points form the upper hemisphere $S_+^2 \subset S^3$, which is homeomorphic to a disk. Moreover, the points of S_+^2 not in S^1 are mapped injectively to the quotient, and the boundary $z_1 = 0$ is sent to a single point of the quotient. But the points where $z_1 = 0$ just correspond to $\mathbb{CP}^0$. Thus we can express $\mathbb{CP}^1 = \mathbb{CP}^0 \cup_{p_0} S_+^2$, where $p_0 \colon S^1 = \partial S_+^2 \to \mathbb{CP}^0$ is the quotient map $p_0 \colon S^1 \to \mathbb{CP}^0$. When we identify $\mathbb{CP}^0$ to a point and S_+^2 to D^2, we then get the cell decomposition $\mathbb{CP}^1 = e^0 \cup e^2$, which is just the CW decomposition for the 2-sphere. Hence $\mathbb{CP}^1$ is homeomorphic to S^2. We can use the same idea to show that $\mathbb{CP}^k$ is built from $\mathbb{CP}^{k-1}$ by attaching a $2k$-cell, and hence inductively get a handle decomposition for $\mathbb{CP}^n$.

When $\mathbb{CP}^1$ is identified with S^2, the map $S^3 \to S^2 = \mathbb{CP}^1$ is called the Hopf map after Heinz Hopf. This is the lowest-dimensional example of a homotopically nontrivial (i.e. not homotopic to a constant) map of a sphere to a sphere of lower dimension, and played an important role in the early development of homotopy theory. The map $S^{2k+1} \to \mathbb{CP}^k$ is sometimes called a generalized Hopf map.

Exercise 5.1.15. Fill in the details of the above argument to show that $\mathbb{CP}^n$ has a CW decomposition with one cell in each even dimension between 0 and $2n$.

We close this section with another example, which is a different type of generalization of a projective space. We give a three-dimensional example, but there are examples in any odd dimension. We start with $S^3 = \{(z_0, z_1) \colon \sum |z_i|^2 = 1\}$. We could form $\mathbb{RP}^3$ from this by taking the quotient using $\{\pm 1\}$ or form $\mathbb{CP}^1 = S^2$ by taking the quotient using S^1. If we think of $\{\pm 1\} \subset S^1$, we could factor the quotient space $S^3 \to S^3/\{\pm 1\} = \mathbb{RP}^3 \to S^3/S^1 = \mathbb{CP}^1 \simeq S^2$. Instead of using $\{\pm 1\}$ we could use other finite groups of unit complex numbers. In particular, we could take a subgroup of S^1 consisting of pth roots of unity $G_p = \{1, \eta, \eta^2, \ldots, \eta^{p-1}\}$, where $\eta = e^{2\pi i/p}$. The case when $p = 2$ is just $\{\pm 1\}$. Now G_p acts on S^3 as a subset of the action of S^1. When we form the quotient space by this action, we are making $(z_0, z_1) \sim (\tau z_0, \tau z_1)$ for $\tau \in G_p$. Since each τ is just a power of η, the equivalence relation is generated from the case when $\tau = \eta$.

We first focus on a special case $p = 3$. Sitting inside S^3 is S^1 where the second coordinate is 0. When we take the quotient by G_3, we get S^1/G_3. This motivates us to give a CW decomposition of S^1 which is consistent with this action. For 0-cells, we use the points $1, \eta, \eta^2$. Then there are 1-cells e_0^1, e_1^1, e_2^1, where e_0^1 is mapped via the characteristic map ϕ_0^1 to the arc on the circle joining 1 and η. The characteristic maps for e_1^1, e_2^1 are the compositions of this with multiplication by η, η^2, respectively. Thus we get a CW decomposition for S^1 with three 0-cells and three 1-cells. When we take the quotient by the G_3 action, it identifies the 0-cells and identifies the 1-cells. Thus this quotient has a single 0-cell and a single 1-cell, and the quotient is homeomorphic to S^1. We next look at the rest of S^3. Consider the upper hemisphere $S_+^2 \subset S^3$. This can be thought of as a 2-cell e_0^2. It will be attached to S^1 so that its boundary is the sum of three 1-cells. When η acts on the 2-cell e_0^2, it sends its boundary to itself, but otherwise is disjoint from itself. Thus we can get three 2-cells by letting η, η^2 act on e_0^2. We call these new 2-cells e_1^2, e_2^2. When we take the quotient space by the G_3 action, these three 2-cells become equivalent and so the quotient space will have a single 0-cell, a single 1-cell, and a single 2-cell. The 2-cell will be attached so that it runs over the S^1 formed by the 0-cell and 1-cell three times. Thus this part of the quotient is just $S^1 \cup_{m_3} e^2$, where the map $m_3 : S^1 \to S^1$ can be taken as $m_3(z) = z^3$. This space is the pseudoprojective plane P_3 from Exercise 5.1.3. The 2-cells e_0^2 and e_1^2 intersect only along their boundary in a circle, and they are otherwise disjoint. Up to homeomorphism, they provide an S^2 in S^3. In fact, they enclose a region, which is characterized by the second coordinate being 0 (the S^1 already identified on the boundary) or lying between 1 and η when normalized by making it a unit vector. This region can be shown to be homeomorphic to a three-dimensional disk D^3, and so forms a 3-cell e_0^3. All points S^3 either lie in this 3-cell or in its image under the action of multiplying by η or η^2. Thus there are two more 3-cells e_1^3, e_2^3. When we take the quotient, then these 3-cells become identified. Thus the quotient space turns out to be described by a single cell in each dimension between 0 and 3. The standard notation for this quotient space is $L(3, 1)$ and it is called a *lens space*.

Exercise 5.1.16. By replacing 3 with p, give a description of the CW decomposition of the lens space $L(p, 1) = S^3/G_p$.

5.2 The fundamental group of a CW complex

In this section we consider a two-dimensional CW complex with a single 0-cell e^0. In computations of the fundamental group, the 0-cell will serve as the base point. The effect of adding the 1-cells will then be to form a wedge $X^1 = \bigvee_{j=1}^n S_j^1$ of circles. Each time we add a 2-cell, the homotopy type of the resulting space will be completely determined by the homotopy class of the attaching map $f_k : \partial e_k^2 \to X^1$. For simplicity of description, we will assume that 1 is mapped to the base point so that the attaching map can be described, up to homotopy, by a word in the free group $F_n = \pi_1(X^1, e^0)$.

Exercise 5.2.1. Suppose that there is a single 2-cell. Use the Seifert–van Kampen theorem to show that the fundamental group of X is isomorphic to the quotient of the free group F_n by the relation given by the word used in attaching the 2-cell which is given by $f_*(g)$, where g is the generator of $\pi_1(S^1, 1)$. (Hint: Use the decomposition of X into the two sets A, B, where A is the complement of the center point of the 2-cell and B is the interior of the 2-cell. Note that $A \cap B$ deformation retracts to the circle at radius $\frac{1}{2}$. Using that B is contractible and that there is a homotopy equivalence of A with X^1, show that the Seifert–van Kampen theorem leads to the above description of the fundamental group. First use the base point at $(\frac{1}{2}, 0)$ and then use the results of Section 3.10 to change the base point.)

Exercise 5.2.2. (a) Show that if Y is a two-dimensional CW complex and we attach a 2-cell to Y^1 by a map f sending $\mathbf{1}$ to e^0, then $\pi_1(Y \cup_f e^2, e^0) \simeq \pi_1(Y)/N$, where N is the normal subgroup generated by the word in $\pi_1(Y, e^0)$ determined by the image of the generator of $\pi_1(S^1, 1)$ under f_*.

(b) Show that the fundamental group of a two-dimensional CW complex with a single 0-cell is isomorphic to the group with one generator for each 1-cell and one relation for each 2-cell coming as in part (a) from the word in $F_n = \pi_1(X^1, e^0)$, which is the image of the generator under $(f_j)_*$.

The hypothesis that the attaching maps send $\mathbf{1}$ to e^0 is unnecessary but technically somewhat difficult to avoid. The easiest way to handle removing this assumption is to use a result in a later section that says that homotoping the attaching maps does not change the space up to homotopy type, and homotopy-equivalent spaces have isomorphic fundamental groups. We will always look at examples where the condition is satisfied. If there is more than one 0-cell and the space is connected, it is homotopy equivalent to a CW complex with a single 0-cell, so we can again reduce to this case in computing the fundamental group.

In the following exercises, we describe spaces as quotients of polygons by indicating identifications of edges. These spaces have natural decompositions as CW complexes. You are to use these decompositions to compute the fundamental groups. For those which are surfaces, identify the surface by abelianizing the fundamental group. Check your result using Euler characteristics. All should have a single 0-cell coming from the vertices.

Exercise 5.2.3. Find the fundamental groups of the two spaces in Figure 5.2. Find their abelianizations and from this determine which surface the space is homeomorphic to.

Exercise 5.2.4. Find the fundamental groups of the two spaces in Figure 5.3.

Exercise 5.2.5. Find the fundamental groups of the two spaces in Figure 5.4. Part (a) is a surface with boundary. Identify it.

Exercise 5.2.6. Find the fundamental groups of the two spaces in Figure 5.5.

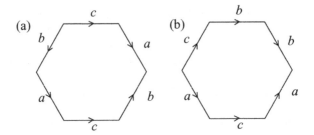

Figure 5.2. Figure for Exercise 5.2.3.

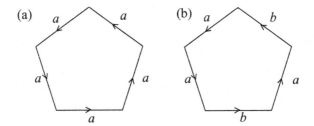

Figure 5.3. Figure for Exercise 5.2.4.

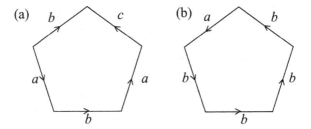

Figure 5.4. Figure for Exercise 5.2.5.

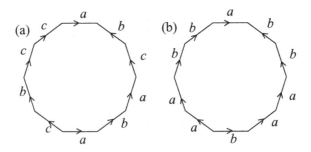

Figure 5.5. Figure for Exercise 5.2.6.

Exercise 5.2.7. Use the Seifert–van Kampen theorem to show that, for a connected CW complex X, $\pi_1(X, e^0) \simeq \pi_1(X^2, e^0)$. Thus the fundamental group only detects information about the 2-skeleton X^2.

Exercise 5.2.8. Use the last exercise to show that $\pi_1(S^k, x) \simeq \{e\}$ for $k \geq 3$ since it has a CW decomposition with one 0-cell and one k-cell.

Exercise 5.2.9. Use the CW decomposition for $S^1 \times S^2$ from Exercise 5.1.10 to compute the fundamental group. Compare your answer with the general result for the fundamental group of a product space.

Exercise 5.2.10. Use the CW decomposition for $S^1 \times \mathbb{RP}^2$ from Exercise 5.1.11 to compute the fundamental group. Compare your answer with the general result for the fundamental group of a product space.

Exercise 5.2.11. Use the CW decomposition for $S^1 \times S^1 \times S^1$ from Exercise 5.1.12 to compute the fundamental group. Compare your answer with the general result for the fundamental group of a product space.

Exercise 5.2.12. Use the CW decomposition for $S^2 \times S^2$ from Exercise 5.1.13 to compute the fundamental group. Compare your answer with the general result for the fundamental group of a product space.

Exercise 5.2.13. Use the CW decomposition for $\mathbb{RP}^2 \times \mathbb{RP}^2$ from Exercise 5.1.14 to compute the fundamental group. Compare your answer with the general result for the fundamental group of a product space.

Exercise 5.2.14. Give the fundamental group of $\mathbb{CP}^n$, $n \geq 1$, using its CW decomposition.

Exercise 5.2.15. Show that the fundamental group of the pseudoprojective plane of Exercise 5.1.3 is $\mathbb{Z}_p$.

Exercise 5.2.16. Show that S^3 is the universal covering space of $L(p, 1)$ and use this to give the fundamental group. Then use the CW decomposition from Exercise 5.1.16 to compute the fundamental group in another way.

5.3   Homotopy type and CW complexes

In this section we discuss some key homotopy-theoretic ideas about CW complexes. When we are attaching a cell, the only thing that matters up to homotopy type of the result is the homotopy class of the attaching map. Moreover, if we start off with homotopy equivalent spaces X, Y and form $X \cup_f e^n$ by attaching a cell to X, then composing the attaching map with the homotopy equivalence gives a homotopy-equivalent space $Y \cup_g e^n$. The next two exercises lead you through the proof of these two statements.

Exercise 5.3.1. Let X be a topological space and $f : S^{n-1} \to X$ a continuous function. Let $X_f = X \cup_f D^n$ be the quotient space formed from the disjoint union by identifying $x \in S^{n-1}$ with $f(x) \in X$.

(a) Show that if f is homotopic to f' via a homotopy $F : S^{n-1} \times I \to X$ with $F_0 = f', F_1 = f$, then there is a continuous function $\alpha : X_f \to X_{f'}$ which is the identity on X. (Hint: Regard D^n as being built from a smaller D^n and an annular set A which is homeomorphic to $S^{n-1} \times I$. Here $S^{n-1} \times 1$ is identified to the boundary ∂D^n and $S^{n-1} \times 0$ is identified to the boundary of the smaller disk. Use F on the annulus to map the annulus to X and then map the smaller disk to D^2, checking that these maps fit together to give a continuous map; see Figure 5.6.)

(b) Analogously, find a continuous function $\beta : X_{f'} \to X_f$.

(c) Show that α is a homotopy equivalence with homotopy inverse β. (Hint: Show that $\beta\alpha$ maps a small disk to D^n by radial expansion and an annular region to X by fitting together $\bar{F}$ and F. Show how to homotope this to the identity; see Figure 5.6.)

Exercise 5.3.2. Suppose that $h : X \to Y$ is a homotopy equivalence with homotopy inverse $g : Y \to X$. That is, gh is homotopic to 1_X and hg is homotopic to 1_Y. Name the homotopies $H : X \times I \to X$, $H_t(x) = H(x,t)$, $H_0(x) = gh(x)$, $H_1(x) = x$, and $G : Y \times I \to Y$, $G_t(y) = G(y,t)$, $G_0(y) = hg(y)$, $G_1(y) = y$.

(a) Consider the map $\alpha : X_f \to Y_{hf}$, which is given by h on X and the identity on D^n, and $\beta : Y_{hf} \to X_{ghf}$, which is defined by g on Y and the identity on D^n. Show that they are continuous (see Figure 5.7).

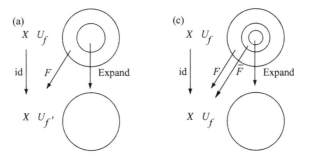

Figure 5.6. Figure for Exercise 5.3.1.

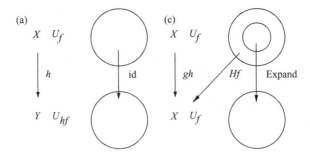

Figure 5.7. Figure for Exercise 5.3.2.

(b) Use the previous exercise to show that there is a homotopy equivalence $\epsilon : X_{ghf} \to X_f$.

(c) Show that $\gamma = \epsilon\beta$ is a left homotopy inverse to α; that is, $\epsilon\beta\alpha$ is homotopic to the identity. (Hint: $\epsilon\beta\alpha$ maps X to X via gh, maps an annular region via Hf, and expands a smaller disk in D^n; see Figure 5.7.)

(d) Show that $\beta : Y_{hf} \to Y_{ghf}$ has a left homotopy inverse. (Hint: Use $hg \sim 1_Y$ and the argument of (c).)

(e) In a group, show that whenever $ba = 1 = ac$, then $a = c$. That is, whenever there is a left inverse and a right inverse to a, then a is invertible and the inverse is $b = c$. Conclude that a left inverse for an invertible element is also a right inverse. (Hint: Start with $b(ac) = (ba)c$. In Section 3.1 there is a relevant discussion of this idea.)

(f) Starting with $\epsilon\beta\alpha \sim 1$, use the fact that ϵ is a homotopy equivalence to show that $\beta\alpha\epsilon \sim 1$ and conclude that β has a right homotopy inverse. From this and part (e), conclude that β is a homotopy equivalence, and then that α is a homotopy equivalence.

The outline in Exercise 5.3.2 is motivated by an argument in [19].

Exercise 5.3.3. (a) Show that the dunce hat (see Figure 3.34) is homeomorphic to a CW complex with one 0-cell, one 1-cell, and one 2-cell. Show that X^1 is homeomorphic to the circle, and the attaching map $f : S^1 \to S^1 = X^1$ of the 2-cell is homotopic to the reflection.

(b) Use Exercise 5.3.1 to show that the dunce hat is homotopy equivalent to the disk.

(c) Show that the dunce hat is not homeomorphic to the disk by examining neighborhoods of edge points and looking at the fundamental group when the point is deleted. (Hint: From the assumption that D is homeomorphic to a disk, find a nesting of neighborhoods $N_1 \subset N \subset N_2$, so that $N_1\backslash\{x\} \to N_2\backslash\{x\}$ is a homotopy equivalence with $\pi_1 \simeq F_2$ and $N\backslash\{x\}$ is homotopy equivalent to S^1.)

Exercise 5.3.4. (a) Suppose X is a two-dimensional CW complex with at least two 0-cells. Show that if X is connected, at least one of the 1-cells must have its boundary attached to two different 0-cells.

(b) Assume that e^1 is a 1-cell with boundary attached to two different 0-cells e_1^0 and e_2^0. Show that $e_1^0 \cup e_2^0 \cup e^1$ is homeomorphic to a 1-disk. Use Exercise 5.3.2 to show that X is homotopy equivalent to a CW complex where $e_1^0 \cup e_1^1 \cup e^1$ is replaced by a single 0-cell and there are corresponding 0-cells, 1-cells, and 2-cells to the other cells of X.

(c) Use induction to show that a connected two-dimensional CW complex is homotopy equivalent to a CW decomposition with a single 0-cell.

A somewhat more sophisticated way to approach the result of the last exercise is to find a subcomplex of a connected X^1 which contains all of the vertices and deformation-retracts to any one of its 0-cells through a process of collapsing 1-cells one edge at a time. Such a subcomplex is called a maximal tree in X^1. This can always be shown to exist by an inductive argument on the number of

0-cells of X. If there is only one 0-cell, then that 0-cell provides the maximal tree. Assuming that there exists a maximal tree with fewer than n 0-cells, suppose that there are n 0-cells in X. Choose one of the 0-cells e^0 of X and consider a maximal subcomplex K of X^1 which does not contain e^0. The subcomplex K does not have to be path connected; suppose it has p path components. By the inductive hypothesis, we can find a maximal tree in each path component that connects all of 0-cells in that path component. Then adding a 1-cell from e^0 to each path component of K gives a maximal tree in X^1. Let us call this maximal tree T.

As an example, Figure 5.8 shows two maximal trees (in bolder lines) in a 1-dimensional CW complex.

Exercise 5.3.5. Use Exercise 5.3.2 and a maximal tree T in X to show that there is a homotopy-equivalent CW complex Y with a single 0-cell.

Consider the CW complex $X = X^1$ given by the left-hand diagram in Figure 5.9. A maximal tree is given by the subcomplex including a, c and their endpoints. When we collapse it to a point, we get a homotopy-equivalent CW complex, which is shown is the figure at the right. We extend X by attaching two 2-cells, attached via maps of the circle which are described by words in terms of the 1-cells which are traversed. The two words used are abc and $c^{-1}d$. There is a homotopy-equivalent CW complex with 1-skeleton given by the right-hand diagram. For the 2-cells, each subinterval of the attaching circle which was mapped to a or c is now mapped to the 0-cell to which they were collapsed. Up to homotopy (which does not change the homotopy type of the CW complex), these can be described by words formed from the original words by deleting the symbols a, c where they occurred. Thus the attaching maps are given by b, d, which just means that the homotopy-equivalent complex is just two disks, joined at one point. This space is homotopy equivalent to a point.

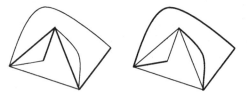

Figure 5.8. Examples of maximal trees.

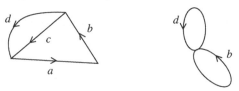

Figure 5.9. Collapsing a tree.

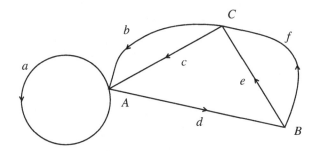

Figure 5.10. Figure for Exercise 5.3.6.

Exercise 5.3.6. Consider the CW complex whose 1-skeleton X^1 is pictured in Figure 5.10. Here A, B, C denote 0-cells and a, b, c, d, e, f are 1-cells. Suppose that there are two 2-cells attached to X^1, which can be described in terms of the 1-cells that are traversed by $adec$ and dfb.

(a) Find Y^1 homotopy equivalent to X^1 so Y^1 has a single 0-cell.

(b) Find Y homotopy equivalent to X so the 1-skeleton is Y^1 and use it to compute the fundamental group of X.

Exercise 5.3.7. Consider a handle decomposition of a surface. The 0-handles deformation-retract to a collection of points. When we attach a 1-handle, then the composition of the attaching maps with the deformation retraction will send these intervals to one or two points. Use this idea to show that the union of the 0-handles and 1-handles is homotopy equivalent to a CW complex with h_0 0-cells and h_1 1-cells, where h_i is the number of i-handles.

Exercise 5.3.8. Show that if a surface (possibly with boundary) has a handle decomposition with h_0 0-handles, h_1 1-handles, and h_2 2-handles, then there is a corresponding CW complex with h_0 0-cells, h_1 1-cells, and h_2 2-cells to which it is homotopy equivalent.

Exercise 5.3.9. Show that a connected surface (possibly with boundary) is homotopy equivalent to a CW complex with a single 0-cell.

Exercise 5.3.10. For Figure 5.11(a), give a CW decomposition and then get a homotopy-equivalent CW complex with a single 0-cell. Use this to compute the fundamental group.

Exercise 5.3.11. For Figure 5.11(b), give a CW decomposition and then get a homotopy-equivalent CW complex with a single 0-cell. Use this to compute the fundamental group.

Exercise 5.3.12. Using the notation established in Exercise 5.1.4, consider the CW complex Y where $Y^1 = X^1$ but $Y = X^1 \cup_f e^2$, and ∂e^2 is identified via sending S^1_+ to e^1_1 as before but sending S^1_- to e^1_1 by reversing the direction.

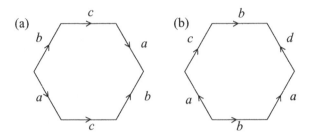

Figure 5.11. Figure for Exercises 5.3.10 and 5.3.11.

Geometrically, we can think of e_1^1 as being $[-1, 1]$ on the x-axis and the map is just a vertical projection. Identify geometrically what the quotient space will be. Show that Y is homotopy equivalent, but not homeomorphic, to $S^2 \vee S^1$, with the homotopy equivalence coming from collapsing e_1^1 to a point.

To close this section, we study one of the important properties that a CW complex has for homotopy theory, which is called the *homotopy extension property*. Its proof illustrates how the structure of a CW complex facilitates inductive arguments, reducing problems to problems on a disk. Suppose K is a CW complex and L is a subcomplex of K. Then K is built from L by the process of attaching k-cells for various k. This allows us to reduce arguments to the case of attaching a single cell.

Theorem 5.3.1 (Homotopy extension theorem). *Suppose $L \subset K$ is a subcomplex. Let $g : K \to Y$ and suppose there is a homotopy of the restriction $g|L : L \to Y$, that is, a map $H : L \times I \to Y$ with $H(x, 0) = g(x)$. Then there is a homotopy $H' : K \times I \to Y$ with $H'(x, t) = H(x, t)$ for $x \in L$ and $H'(x, 0) = g(x)$.*

To understand why this should be true, assume for the moment that $K = L \cup_f D^k$. Here $f : S^{k-1} \to K$ is a continuous map. The homotopy extension property asserts that if we have a map defined on $L \times I \cup K \times \{0\}$, we can extend it to a continuous map from $K \times I$. The key to the argument is to reduce to the case where $L = S^{k-1}$ and $K = D^k$.

Lemma 5.3.2. *Consider $S^{k-1} \times I \cup D^k \times \{0\} \subset D^k \times I$. There is a deformation retraction of $D^k \times I$ onto $S^{k-1} \times I \cup D^k \times \{0\}$.*

Here is the idea. We give an illustration for the case $k = 2$ in Figure 5.12. Let $p = (\mathbf{0}, 2) \in \mathbb{R}^k \times \mathbb{R} = \mathbb{R}^{k+1}$. Consider rays emanating from p. For each point $b \in B = S^{k-1} \times I \cup D^k \times \{0\}$, there is a unique ray from p to x. That ray will intersect $D^k \times I$ in an interval, with the first point of intersection on $D^k \times \{1\}$ and the last point of intersection being x. For points on $S^{k-1} \times \{1\}$, the intersection will consist of a single point. Now consider a point $x \in D^k \times I$. The ray from p through x will contain an interval which connects x to a point $b \in B$. We can define a deformation retraction H on $x \times I$ by just moving along this interval to contract it to the point b. If we take a point $x \in D^k \times I$, and look at $h(x) = H(x, 1)$, then $h : D^k \times I \to B$ gives a continuous map with $h|B = \mathrm{id}|B$.

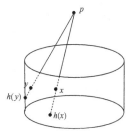

Figure 5.12. Deformation-retracting $D^2 \times I$ to $S^1 \times I \cup D^2 \times \{0\}$.

Suppose $g : K = L \cup_f D^k \to Y$ and there is a homotopy $H : L \times I \to Y$ with $H(x,0) = g(x)$ for $x \in L$. Consider the composition $J : S^{k-1} \times I \cup D^k \times \{0\} \to Y$ given by $J(u,t) = H(f(u),t)$, $u \in S^{k-1}$, $J(v,0) = g(v)$, $v \in D^k$. The lemma allows us to extend J to $D^k \times I$ by composing with the deformation retraction.

Exercise 5.3.13. By applying the lemma and using how $K \times I$ is formed as a quotient space, show that there is an extension $H' : K \times I \to Y$ of H.

Exercise 5.3.14. Use an inductive argument with induction over the number of cells of $K \backslash L$ to prove the homotopy extension theorem.

5.4 The Seifert–van Kampen theorem for CW complexes

In this section we explore applications of the Seifert–van Kampen theorem in terms of the hypothesis that the two sets A, B we use in $X = A \cup B$ have to be open. We show that this hypothesis can be replaced by one that is sometimes easier to apply. We say that a subset $A \subset X$ is a *neighborhood deformation retract* if there is an open set U containing A and a deformation retraction of U onto A. Recall that a deformation retraction is a map $h : U \to A$ with $h|A = \mathrm{id}|A$ so that h is homotopic to the identity with the homotopy restricting to the constant homotopy on A. If we have a pair of sets (A, B), then we say the pair is a neighborhood deformation retract if there are open sets U, V with $A \subset U$, $B \subset V$ so that U deformation-retracts to A, V deformation-retracts to B, and $U \cap V$ deformation-retracts to $A \cap B$.

Exercise 5.4.1. Show that if (A, B) is a neighborhood deformation retraction of (U, V), then $A, B, A \cap B$ are all path connected iff $U, V, U \cap V$ are all path connected.

Exercise 5.4.2. Show that if $A, B, A \cap B$ are path connected and (A, B) is a neighborhood deformation retraction of (U, V), then if $x_0 \in A \cap B$, there is an isomorphism between $\pi_1(A, x_0) *_{\pi_1(A \cap B, x_0)} \pi_1(B, x_0)$ and $\pi_1(U, x_0) *_{\pi_1(U \cap V, x_0)} \pi_1(V, x_0)$.

Exercise 5.4.3. Show that if $X = A \cup B$, where $A, B, A \cap B$ are path connected, $x_0 \in A \cap B$, and (A, B) is a neighborhood deformation retract, then there is an isomorphism $\pi_1(A, x_0) *_{\pi_1(A \cap B, x_0)} \pi_1(B, x_0) \simeq \pi_1(X, x_0)$.

Exercise 5.4.4. (a) Show that if $X = S_a^1 \vee S_b^1$ is the wedge of two copies of the circle and $A = S_a^1$, $B = S_b^1$, then (A, B) is a neighborhood deformation retract. Use Exercise 5.4.3 to show that $\pi_1(X, v) = F_2$. Here $\{v\}$ is the wedge point.

(b) Use induction to show that if W_k is the wedge of k copies of S^1, then $\pi_1(W_k, v) \simeq F_k$.

Exercise 5.4.5. Suppose M is a path-connected surface which is the union of two path-connected surfaces with boundary N_1, N_2 which intersect along a circle $C = \partial N_1 = \partial N_2$. Suppose there is a neighborhood H of C so that $(H, H \cap N_1, H \cap N_2, H \cap C)$ is homeomorphic to $(S^1 \times (-1, 1), S^1 \times (-1, 0], S^1 \times [0, 1), S^1 \times \{0\})$.

(a) Show that (N_1, N_2) is a neighborhood deformation retract.

(b) Show that $\pi_1(M, x) \simeq \pi_1(N_1, x) *_{\pi_1(C, x)} \pi_1(N_2, x)$.

Exercise 5.4.6. Apply Exercise 5.4.5 to compute

(a) $\pi_1(T \# T, x)$,

(b) $\pi_1(P \# P, x)$.

Exercise 5.4.7. Suppose K is a two-dimensional CW complex and L is a subcomplex K, so that we can form K from L by inductively attaching cells of dimensions 0,1,2. Using an inductive argument, show that L is a neighborhood deformation retract.

Exercise 5.4.8. Suppose that L, M are subcomplexes of the two-dimensional CW complex K. Show that (L, M) is a neighborhood deformation retract.

The above two results hold without the two-dimensional hypothesis, but that is all that we need for the next result.

Theorem 5.4.1 (Seifert–van Kampen theorem for CW complexes). *Suppose L, M are path-connected subcomplexes of the CW complex K so that $L \cap M$ is path connected with $x \in L \cap M$. Then $\pi_1(K, x) \simeq \pi_1(L, x) *_{\pi_1(L \cap M, x)} \pi_1(M, x)$.*

Exercise 5.4.9. Prove Theorem 5.4.1. (Hint: First reduce to the two-dimensional case using Exercise 5.2.7.)

5.5   Simplicial complexes and Δ-complexes

A special case of a CW complex is a *simplicial complex* where the i-cells are identified to an i-dimensional simplex, which is the affine span of the vertices $v_0, v_1, \ldots, v_i$ which are affinely independent in $\mathbb{R}^n$. From a CW point of view, we can identify these simplices to a standard i-simplex with vertices $e_0, e_1, \ldots,$

$e_i \in \mathbb{R}^i$ but we will frequently think of them as geometric simplices all living in a common $\mathbb{R}^n$. By a *face* of a simplex with vertices $v_0, \ldots, v_i$, we mean a simplex which has as its vertices a subset of the these vertices. For example, the 2-simplex $[v_0, v_1, v_2]$ with vertices v_0, v_1, v_2 has three one-dimensional faces, which are $[v_0, v_1], [v_0, v_2], [v_1, v_2]$. It has three zero-dimensional faces, which are the 0-simplices $[v_0], [v_1], [v_2]$. In a simplicial complex, each simplex is embedded in X and each $(i-1)$-dimensional face of an i-simplex obtained by taking the span of i of its vertices is identified with one of the $(i-1)$-simplices. The simplices can be thought of as independent entities that are glued together along faces or from embedding each of them in a common Euclidean space $\mathbb{R}^n$ so that any two simplices which intersect do so along a common face. Note that in a simplicial complex, the characteristic maps of the simplices are embeddings.

Historically, simplicial complexes came before CW complexes and dominated much of geometric topology, particularly in arguments in homotopy theory and homology. They still play an important role in topology, but CW complexes have become more important for many parts of homotopy theory and homology. In this section we introduce some of the basic concepts of simplicial complexes, and then discuss briefly the notion of a Δ-complex, which has many of the best features of both simplicial and CW complexes. We will use Δ-complexes in the next chapter on homology to give geometric examples where it is relatively simple to compute a form of homology.

In Figure 5.13(a) we show a collection of simplices which do not form a simplicial complex because they do not fit together along faces and in (b) an example of a simplicial complex. Note that (b) would not be a simplicial complex if the large right-hand blank simplex were filled in to be a 2-simplex, for now there would be two 2-simplices that did not intersect in a common face. As it is, this just represents a topological circle that is divided into four adjoining 1-simplices (and their faces), and one of these 1-simplices meets the other part of the complex. Frequently, it is easier to think of a simplicial complex in terms of the simplices and how they intersect rather than in terms of a specific embedding in some Euclidean space such as $\mathbb{R}^3$. In Figure 5.14 we give both views for the

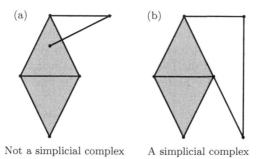

(a)   (b)

Not a simplicial complex   A simplicial complex

Figure 5.13. Simplices must intersect in a common face.

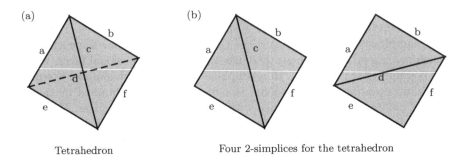

Tetrahedron   Four 2-simplices for the tetrahedron

Figure 5.14. Tetrahedron as a simplicial complex.

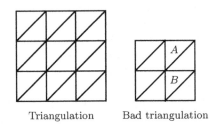

Triangulation   Bad triangulation

Figure 5.15. How to (and not to) triangulate the torus.

surface of a tetrahedron. When a space is given as a quotient space, such as our description of the torus as a quotient space of the square, we can get a simplicial decomposition by using a simplicial decomposition of the square that survives to give a simplicial decomposition of the torus after all of the identifications are made. We have to be careful to see that after the identifications that individual simplices are still embedded in the quotient space and they fit together along their faces. We illustrate this for the torus in Figure 5.15. The subdivision into eight triangles does not form a simplicial decomposition after the identifications are made since simplices A and B intersect in two vertices, not in a common face as they are supposed to. The second diagram using 18 triangles leads to a proper triangulation of the quotient torus after the identifications are made.

Exercise 5.5.1. Thinking of the projective plane as a quotient of a rectangle with identifications $abab$ as you go around the four edges in counterclockwise direction, subdivide this to make it a simplicial complex.

Exercise 5.5.2. For the surface $T\#T$ thought of as an octagon with identifications on the boundary $aba^{-1}b^{-1}cdc^{-1}d^{-1}$, give a decomposition as a simplicial

complex. (Hint: We need to separate identified edges. Construct a smaller octagon within the octagon and then subdivide so that it is the total space of a
subcomplex.)

Generally speaking, simplicial complexes have more structure than CW
complexes but also require many more cells. For many homotopy-theoretic computations such as the fundamental group and homology, CW complexes are
easier to work with than simplicial complexes. An intermediate structure that
has some of the advantages of both simplicial complexes and CW complexes is
a Δ-complex. Δ-complexes are developed more thoroughly in [13]. We introduce them here mainly for their use in homology computations in the next
chapter.

A Δ-complex is built up like a CW complex from cells, and the cells have
the structure of simplices as in a simplicial complex. Each simplex is determined by its vertices, and the vertices of the simplex are given an order. Each
i-simplex then can be thought of as $[v_0, v_1, \ldots, v_i]$, where the v_j are the vertices
in the chosen order $v_0 < v_1 < \cdots < v_i$. When the boundary of the i-simplex
is identified to points in K^{i-1}, the identification is required to use the unique
affine linear order-preserving maps from each face of the i-simplex to an $(i-1)$-
dimensional simplex. Any simplicial complex can be made into a special type
of Δ-complex by giving an ordering of all of its vertices. We give a couple of
examples.

The simplest nontrivial example is the circle with a Δ-complex structure with
one vertex and one 1-simplex, where both vertices are identified. This structure
just comes from the usual simplicial structure of an interval with two vertices
and one edge after we make the identification of the two vertices to get a circle
from the interval.

The torus has the structure of a Δ-complex with one vertex, three 1-simplices,
and two 2-simplices. We start with the usual picture of the torus as coming from
the square $D^1 \times D^1$ by identifying $(x, -1) \sim (x, 1)$, and $(-1, y) \sim (1, -y)$. The
four corner points then get identified to a single vertex. We first label the corner
points via $v_1 = (-1, -1)$, $v_2 = (1, -1)$, $v_3 = (-1, 1)$, $v_4 = (1, 1)$. The ordering
is given by the ordering of the vertices. In the quotient topology, these will give
a single vertex v. The 1-simplices are then $a = [v_1, v_2]$, which is also identified to
$[v_3, v_4]$, and $b = [v_1, v_3]$, which is identified to $[v_2, v_4]$, and $c = [v_1, v_4]$. There are
two 2-simplices, $A = [v_1, v_2, v_4]$, $B = [v_1, v_3, v_4]$. The labeling of the vertices then
determines completely how the attaching maps are given. If we want to give a
Δ-complex structure for the Klein bottle, we can again start with a rectangle, but
we now have the identifications $(x, -1) \sim (-x, 1)$ and $(-1, y) \sim (1, y)$. We can
give this a Δ-complex structure by ordering the vertices via $v_1 = (-1-1)$, $v_2 =
(-1, 1)$, $v_3 = (1, 1)$, $v_4 = (1, -1)$. These are identified to a single vertex v. Then
the 1-simplices are $a = [v_1, v_4] \sim [v_2, v_3]$, $b = [v_1, v_2] \sim [v_3, v_4]$, $c = [v_1, v_3]$. The
2-simplices are $A = [v_1, v_2, v_3]$, $B = [v_1, v_3, v_4]$. We illustrate these Δ-complex
decompositions in Figure 5.16.

Exercise 5.5.3. Give a Δ-complex structure to S using two triangles. (Hint:
Start with a rectangle and subdivide it into two triangles. Label the vertices in a

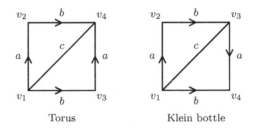

Figure 5.16. Δ-complex structures for T, K.

way to determine the Δ-complex structure and also consistent with the labeling of the edges as $abb^{-1}a^{-1}$ as we read around the edges in the counterclockwise direction. Note that identified edges must have the same ordering of vertices; that is, if we identify $[v_i, v_j]$ with $[v_k, v_l]$ and $i < j$, then $k < l$.)

Exercise 5.5.4. Give a Δ-complex structure to P using two triangles. (Hint: Label vertices to get the $ab^{-1}ab^{-1}$ pattern.)

6

Homology

6.1 Chain complexes and homology

In this chapter we study the concept of the homology of a topological space X as well as the homology of a pair (X, A) of spaces. We will also discuss related concepts of homology for Δ-complexes and for CW complexes. For each integer $k \geq 0$, there will be a group $H_k(X)$ for a topological space X. The group $H_0(X)$ will measure path connectivity of X, in the sense that $H_0(X)$ is a free abelian group with one generator for each path component of X. The group $H_1(X)$ of a path-connected space measures something like simple connectivity. It turns out to be isomorphic to the abelianization of the fundamental group of X. The first homology measures when certain one-dimensional objects such as loops which have no boundary are the boundaries of some two-dimensional objects. As an example, it measures when a loop in a surface is the boundary of a subsurface with boundary, such as how the equator in a sphere bounds the upper hemisphere or the circle where a connected sum in formed bounds the two summands. In general, n-dimensional homology measures when n-dimensional objects which have no boundary are boundaries of $(n + 1)$-dimensional objects. This is admittedly very vague, and making it precise requires introducing the notion of a chain complex and its homology as well as determining chain complexes which are associated to topological spaces, Δ-complexes, or CW complexes.

In this section we introduce the basic ideas of a chain complex. In succeeding sections, we will look at specific chain complexes related to topology and do some fundamental calculations. We then look at some properties that characterize homology for an important class of topological spaces and use these properties to do more calculations. We will give a number of important applications of homology. It turns out that although homology is somewhat complicated to define, it obeys some properties which make it relatively easy to compute. Some basic results which will be shown are:

$$H_k(D^p) = \begin{cases} \mathbb{Z} & k = 0, \\ 0 & \text{otherwise}, \end{cases} \qquad H_k(S^n) = \begin{cases} \mathbb{Z} \oplus \mathbb{Z} & n = k = 0, \\ \mathbb{Z} & n = k > 0 \text{ or } k = 0, n > 0, \\ 0 & \text{otherwise}, \end{cases}$$

$$H_k(D^{n+1}, S^n) = \begin{cases} \mathbb{Z} & k = n+1, \\ 0 & \text{otherwise}. \end{cases}$$

We will also show that $H_2(M)$ measures whether a compact connected surface is orientable or not, compute the homology for all compact connected surfaces, and discuss the relation between orientability and homology for manifolds of any dimension. We also apply homology to give a proof of the Jordan curve theorem as well as to develop its generalizations.

We now introduce the notion of a chain complex, on which homology will be based. We will always be dealing with a chain complex of abelian groups and so will embed this in our definition.

Definition 6.1.1. A *chain complex* $(C, \partial) = \{(C_n, \partial_n)\}$ consists of a sequence of abelian groups $C_n, n \geq 0$ together with homomorphisms $\partial_n : C_n \to C_{n-1}, n > 0$ and $\partial_0 = 0$ so that the composition $\partial_k \partial_{k+1} : C_{k+1} \to C_{k-1}$ is the zero homomorphism. Given a chain complex (C, ∂), the condition $\partial_k \partial_{k+1} = 0$ implies that $\text{im}(\partial_{k+1}) \subset \text{ker}(\partial_k)$. The elements of $\text{ker}(\partial_k)$ are called the *cycles* of dimension k and the elements of $\text{im}(\partial_{k+1})$ are called the *boundaries* of dimension k. The k-dimensional homology is the quotient $H_k(C) = \text{ker}(\partial_k)/\text{im}(\partial_{k+1})$.

We will look at specific chain complexes coming from topology in the next sections. Here we give a few calculations to familiarize ourselves with the basic concepts.

Example 6.1.1. Suppose $\partial_k = 0$ for all $k \geq 0$. Then $H_k(C) \simeq C_k$. For $\text{ker}(\partial_k) = C_k, \text{im}(\partial_{k+1}) = 0$, and hence

$$H_k(C) \simeq \text{ker}(\partial_k)/\text{im}(\partial_{k+1}) \simeq C_k/0 \simeq C_k.$$

This last example does occur when we compute the cellular homology of a sphere or a disk pair (D^{n+1}, S^n) as well as while computing the cellular homology of an orientable surface. Here is an example which occurs in computing the cellular homology of the projective plane.

Example 6.1.2. Suppose that

$$C_i = \begin{cases} \mathbb{Z} & 0 \leq i \leq 2, \\ 0 & \text{otherwise}, \end{cases}$$

and $\partial_0 = \partial_1 = 0, \partial_2(g_2) = 2g_1$, where g_i denotes the generators of the copies of $C_i \simeq \mathbb{Z}, i = 1, 2$. Then

$$H_0(C) = \text{ker}(\partial_0)/\text{im}(\partial_1) = \mathbb{Z}/0 \simeq \mathbb{Z}; H_1(C) = \text{ker}(\partial_1)/\text{im}(\partial_2) = \mathbb{Z}/2\mathbb{Z} \simeq \mathbb{Z}_2;$$
$$H_2(C) = \text{ker}(\partial_2)/0 = 0.$$

Since $C_i = 0$ for $i \geq 3$, we have $H_i(C) = 0, i \geq 3$.

We now give a few exercises to acquaint ourselves with basic computations using the definition.

Exercise 6.1.1. Suppose

$$C_i = \begin{cases} \mathbb{Z} & 0 \leq i \leq 2, \\ 0 & \text{otherwise,} \end{cases}$$

and $\partial_0 = \partial_1 = 0, \partial_2(1) = p > 0$. Show that

$$H_0(C) = \mathbb{Z}; \qquad H_1(C) = \mathbb{Z}_p; \qquad H_i(C) = 0, \quad i \geq 2.$$

Exercise 6.1.2. Suppose $C_i = 0, i > 2; C_2 = \mathbb{Z}; C_1 = \mathbb{Z} \oplus \mathbb{Z}; C_0 = \mathbb{Z}$; and $\partial_0 = \partial_1 = 0, \partial_2(1) = (2, 0)$. Show that

$$H_0(C) = \mathbb{Z}; \qquad H_1(C) = \mathbb{Z}_2 \oplus \mathbb{Z}; \qquad H_i(C) = 0, \quad i \geq 2.$$

Exercise 6.1.3. Suppose $C_i = 0, i > 2; C_i = \mathbb{Z} \oplus \mathbb{Z}, 0 \leq i \leq 2$, and $\partial_0 = 0, \partial_1(1,0) = (-1,1), \partial_1(0,1) = (1,-1), \partial_2(1,0) = (1,1) = \partial_2(0,1)$. Show that

$$H_i(C) = 0, \quad i > 2; \qquad H_0(C) = \mathbb{Z}; \qquad H_1(C) = 0; \qquad H_2(C) \simeq \mathbb{Z}.$$

Exercise 6.1.4. Suppose $C_i = 0, i > 3; C_i = \mathbb{Z}, 0 \leq i \leq 3$; and $\partial_i = 0, i = 0, 1, 3; \partial_2(1) = p$. Show that

$$H_0(C) \simeq H_3(C) \simeq \mathbb{Z}; \qquad H_1(C) \simeq \mathbb{Z}_p; \qquad H_2(C) = 0; \qquad H_i(C) = 0, \quad i \geq 4.$$

Exercise 6.1.5. Suppose $n \geq 1$ and $C_i = 0, i > n; C_i = \mathbb{Z}, 0 \leq i \leq n$; and $\partial_i = 0, 0 < i \leq n, i = 2k + 1$ or $i = 0$; and $\partial_i(1) = 2, 0 < i \leq n, i = 2k$.

(a) Show that

$$H_i(C) = 0, \quad i > n; \qquad H_0(C) \simeq \mathbb{Z}; \qquad H_i(C) = 0, \quad 0 < i < n, \ i = 2k;$$
$$H_i(C) \simeq \mathbb{Z}_2, \quad 0 < i < n, \ i = 2k + 1.$$

(b) Show that if n is odd, then $H_n(C) \simeq \mathbb{Z}$.

(c) Show that if n is even, then $H_n(C) = 0$.

6.2 Homology of a Δ-complex

For a Δ-complex K, we first define a chain complex $(C_i^\Delta(K), \partial_i)$ associated to K. This definition will also apply in the context of a simplicial complex when it is made into a Δ-complex by using an ordering of the simplices (e.g. one which comes from a total ordering of its vertices). Our viewpoint is to give a fairly concrete example of how to compute the homology of a chain complex which arises geometrically. The homology we compute agrees with the singular and cellular homologies we discuss later.

The chain complex consists of free abelian groups $C_i^\Delta(K)$ for each $i \geq 0$ and homomorphisms $\partial_i : C_i^\Delta(K) \to C_{i-1}^\Delta(K)$. The group $C_i^\Delta(K)$ is defined to be the 0-group if there are no simplices of dimension i. Otherwise, $C_i^\Delta(K)$ is the free abelian group with one generator for each i-simplex $[v_0, \ldots, v_i]$ in the Δ-complex structure. Note that these simplices automatically come with an order on the vertices. We define $\partial_0 = 0$. For $i > 0$, there is a homomorphism $\partial_i : C_i \to C_{i-1}$ which is defined on generators by

$$\partial[v_0, \ldots, v_i] = \sum_{k=0}^{i} (-1)^k [v_0, \ldots, \widehat{v_k}, \ldots, v_i].$$

The notation uses the identifications of a face of an i-simplex with an $(i-1)$-simplex, and $\widehat{v_k}$ indicates that the vertex v_k is omitted. For example, we have

$$\partial_1([v_0, v_1]) = [v_1] - [v_0], \qquad \partial_2([v_0, v_1, v_2]) = [v_1, v_2] - [v_0, v_2] + [v_0, v_1].$$

The definition of ∂_i is then extended from the generators to all of $C_i^\Delta(K)$ by extending it linearly, using the fact that the group is free abelian. For the case of a two-dimensional Δ-complex, the chain complex is depicted via the diagram

$$0 \longrightarrow C_2^\Delta(K) \overset{\partial_1}{\longrightarrow} C_1^\Delta(K) \overset{\partial_1}{\longrightarrow} C_0^\Delta(K) \overset{\partial_0}{\longrightarrow} 0.$$

Exercise 6.2.1.
   (a) Show that the composition $\partial_1 \partial_2([v_0, v_1, v_2]) = 0$.
   (b) Show that in general $\partial_i \partial_{i+1} = 0$. (Hint: It suffices to check this on a gener-
       ator $[v_0, \ldots, v_{i+1}]$. For any $(i-1)$-simplex formed from this by omitting
       two vertices, show that this term occurs twice with opposite signs in
       $\partial_i \partial_{i+1}([v_0, \ldots, v_{i+1}])$.)

We denote the image of ∂_{i+1} in $C_i^\Delta(K)$ by $\mathrm{im}(\partial_{i+1})$ and denote the kernel of ∂_i in $C_i^\Delta(K)$ by $\ker(\partial_i)$. The condition $\partial_i \partial_{i+1} = 0$ from the last exercise shows that $\mathrm{im}(\partial_{i+1}) \subset \ker(\partial_i)$.

Definition 6.2.1. We define the Δ-*homology groups*

$$H_i^\Delta(K) = \ker(\partial_i)/\mathrm{im}(\partial_{i+1}).$$

We carry out a couple of calculations as examples and then leave others as exercises.

   • The interval I has a Δ-complex structure with two 0-simplices $[0]$, $[1]$, and
     one 1-simplex $[0, 1]$, with $\partial_1[0, 1] = [1] - [0]$. Then $\ker(\partial_0) = C_0 = \mathbb{Z} \oplus \mathbb{Z}$.
     Since $\mathrm{im}(\partial_1)$ is generated by $[1] - [0]$, taking the quotient of $\ker(\partial_0)/\mathrm{im}(\partial_1)$
     identifies $[0]$ with $[1]$. The quotient is $\mathbb{Z}$, with generator the equivalence
     class of $[0] \sim [1]$. We then compute that $\ker(\partial_1) = 0$, and so we get
     $H_1^\Delta(I) = 0, H_0^\Delta(I) = \mathbb{Z}$, and all other H_i are 0 by definition since there
     are no simplices in dimensions besides 0, 1.

- For the circle with the Δ-complex structure arising as a quotient of I where we identify $0 \sim 1$, we have $C_1^\Delta(S^1) = \mathbb{Z}, C_0^\Delta(S^1) = \mathbb{Z}$, and $\partial_1 = 0$ since $\partial_1([0,1]) = [1] - [0] = 0$ (since $[0] = [1]$ here). Thus $H_0^\Delta(S^1) = \mathbb{Z}$, $H_1^\Delta(S^1) = \mathbb{Z}$.

- For a slightly more complicated example, we look at the torus with the Δ-complex structure of the torus in Figure 5.16. Here $C_2^\Delta(T) = \mathbb{Z} \oplus \mathbb{Z}$, with generators $[v_1, v_2, v_4], [v_1, v_3, v_4], C_1^\Delta(T) = \mathbb{Z} \oplus \mathbb{Z} \oplus \mathbb{Z}$, with generators a, b, c, and $C_0^\Delta(T) = \mathbb{Z}$, with generator $[v] = [v_i]$, noting that all of the vertices are identified. The boundary homomorphisms are

$$\partial_2([v_1, v_2, v_4]) = a - c + b, \qquad \partial_2([v_1, v_3, v_4]) = b - c + a,$$
$$\partial_1(a) = \partial_1(b) = \partial_1(c) = 0, \qquad \partial_0 = 0.$$

Thus $\ker(\partial_2)$ is the free abelian group generated by $[v_1, v_2, v_4] - [v_1, v_3, v_4]$ and so $H_2^\Delta(T) = \mathbb{Z}$. The kernel $\ker(\partial_1) = C_1^\Delta(T)$, and the image $\text{im}(\partial_2)$ is the free abelian group generated by $a + b - c$. Factoring out by it serves to identify c to $a + b$. Thus the quotient $H_1^\Delta(T) = \ker(\partial_1)/\text{im}(\partial_2)$ is the free abelian group $\mathbb{Z} \oplus \mathbb{Z}$ on two generators a, b. The group $H_0^\Delta(T) = \mathbb{Z}$ since $\text{im}(\partial_1) = 0$ and $\ker(\partial_0) = C_0^\Delta(T) = \mathbb{Z}$.

Note that in each of the examples we have found that $H_0^\Delta(K) \simeq \mathbb{Z}$. The reason for this is that each one is path connected and H_0 measures the number of path components. We outline a proof for this in the next exercise.

Exercise 6.2.2. Suppose K is a path connected Δ-complex.

(a) Show that K^1 is path connected, and in particular, given any two vertices v_1, v_2, there is a sequence of directed edges $e_1, e_2, \ldots, e_k$ (which give some of the 1-simplices, possibly with the opposite directions) which connect v_1 with v_2. (Hint: First show that adding higher-dimensional simplices does not change path connectivity, so that K^1 and K have the same path connectivity properties. Then show that a path connecting two vertices may be replaced by a simplicial path.)

(b) Show that factoring out by $\text{im}(\partial_1)$ serves to identify all vertices in $C_0^\Delta(K) = \ker(\partial_0)$ by showing that $[w] - [v] \in \text{im}(\partial_1)$ for any v_1, v_2. Conclude that $H_0^\Delta(K) = \mathbb{Z}$, with a generator coming from each vertex.

Exercise 6.2.3. Suppose that K is a Δ-complex with n path components $K_1, \ldots, K_n$. Show that each chain group $C_i^\Delta(K) = C_i^\Delta(K_1) \oplus \cdots \oplus C_i^\Delta(K_n)$ and the boundary homomorphisms respect this splitting since $\partial_i : C_i^\Delta(K_j) \to C_{i-1}^\Delta(K_j)$. Use this to show that $H_i^\Delta(K) = H_i^\Delta(K_1) \oplus \cdots \oplus H_i^\Delta(K_n)$. Thus the computation of homology just decomposes into the computation of the homology groups of the path components. Conclude that $H_0^\Delta(K)$ is the free abelian group with n generators.

Exercise 6.2.4. Use the Δ-complex structure of the Klein bottle K from Figure 5.16 to compute the Δ-homology groups of K.

Exercise 6.2.5. Use the Δ-complex structure of the sphere S from Exercise 5.5.3 to compute the Δ-homology groups of S.

Exercise 6.2.6. Use the Δ-complex structure of the projective plane P from Exercise 5.5.4 to compute the the Δ-homology groups of P.

In your computations above, you should have found that $H_1^\Delta(M)$ is the abelianization of $\pi_1(M, v)$. The second homology group $H_2^\Delta(M)$ of a surface M turns out to be $\mathbb{Z}$ whenever M is orientable and is 0 when M is nonorientable, as occurred in the examples and exercises above.

Orientability of a surface that is given as a simplicial complex can be phrased as follows. First, the condition that the space is a surface (possibly with boundary) is phrased by requiring that, for each 1-simplex, there are either exactly two 2-simplices with that simplex as a face (for an interior point) or there is exactly one 2-simplex with that 1-simplex as a face (for a boundary point). A simplicial complex is called orientable if we can choose an orientation for each 2-simplex so that, for each interior 1-simplex, the two 2-simplices with it as a face impose opposite orientations on that 1-simplex. In a Δ-complex where there are identifications on the boundary of a 2-simplex, it is also possible to have a single 2-simplex where two edges emanating from a vertex are identified. This occurred in the sphere example. In the Δ-complex case, each simplex has the orientation given by its ordered vertices, and this determines how boundaries are mapped and the corresponding boundary map for the Δ-chain complex. Then the condition required for the orientability of the surface is that orientations extend across identifications. This translates to the condition that we can select a sign ± 1 for each ordered 2-simplex so that, for any 1-simplex, it occurs with opposite signs in the boundary of two different signed 2-simplices *or* occurs with both signs in the boundary of a single 2-simplex. One can show that this condition is equivalent to our earlier notions of orientability. Later in the chapter, we will study orientability more thoroughly, so we do not pursue this point here. We use this definition of orientability in the following exercise.

Exercise 6.2.7. Show that a compact connected surface M without boundary with a Δ-complex structure satisfies $H_2^\Delta(M) \simeq \mathbb{Z}$ when it is orientable as given above, and $H_2^\Delta(M) \simeq 0$ when it is nonorientable. Here we are using 0 for the trivial abelian group with one element. (Hint: Show that the only way to get a 2-cycle is from a multiple of $\sum \epsilon(i)\sigma_i$, where $\epsilon(i) = \pm 1$ and we sum over all 2-simplices. You may use the fact that connectedness implies that, given any two 2-simplices, there is a chain of 2-simplices going from one to the other with a common face. Start with an arbitrary sum $\sum n_i \sigma_i$ of 2-simplices and show that the only way it could be a cycle is that all $|n_i|$ are equal.)

6.3 Singular homology $H_i(X)$ and the isomorphism $\pi_1^{ab}(X, x) \simeq H_1(X)$

We next define singular homology. One difficulty with the definition of homology of a Δ-complex is its dependence on the structure of the space as a Δ-complex.

It is nontrivial to show that if we express the space as a Δ-complex in two different ways, then the homology groups we get are the same. By their definition, the singular homology groups are invariant under homeomorphisms. The tradeoff is that they are more difficult to understand and compute directly from the definition. In singular homology theory, what is done is to prove general properties of the singular homology groups and use these to do the computations and not just rely on the definitions as we did above for Δ-homology. It is a fact that for a Δ-complex, the singular homology groups and the Δ-homology groups are isomorphic, leading to the invariance of the Δ-homology groups up to homeomorphism (or even homotopy equivalence) independent of the Δ-complex structure on a space.

As with Δ-homology, we start by defining a chain complex. Here the singular chain groups are defined by letting $S_i(X)$ be free abelian groups, with one generator for each singular i-simplex. By a *singular i-simplex*, we mean a continuous map $\sigma : \Delta_i \to X$, where Δ_i is the standard i-simplex with vertices $e_0, e_1, \ldots, e_i$. Note that these groups have an infinite number of generators in most cases. Thus the chain groups are more complicated than those that occur for Δ-complexes. The elements of $S_i(X)$ are finite linear combinations $\sum_{k=1}^{n} n_k \sigma_k$, where $n_k \in \mathbb{Z}$ and σ_k is a singular i-simplex. These linear combinations are called *singular i-chains*.

We define a boundary homomorphism $\partial_i : S_i(X) \to S_{i-1}(X), i > 0$ by defining it on generators by

$$\partial_i \sigma = \sum_{k=0}^{i} (-1)^k \sigma F_k,$$

where $F_k : \Delta_{i-1} \to \Delta_i$ is the affine linear map that is order preserving and whose image omits the kth vertex. As an example, $F_0 : \Delta_1 \to \Delta_2$ sends $[e_0, e_1]$ affine linearly to $[e_1, e_2]$, F_1 sends $[e_0, e_1]$ affine linearly to $[e_0, e_2]$, and F_2 sends $[e_0, e_1]$ affine linearly to $[e_0, e_1]$. As before, we define $\partial_0 = 0$. The argument in Exercise 6.2.1 extends to show that $\partial_i \partial_{i+1} = 0$.

Exercise 6.3.1. Verify that $\partial_i \partial_{i+1} = 0$.

Definition 6.3.1. The ith *singular homology group* $H_i(X) = \ker(\partial_i)/\operatorname{im}(\partial_{i+1})$.

Note that since Δ_0 is just a point, the chain group $S_0(X)$ can be identified with the free abelian group with one generator for each point of X. Since $\partial_0 = 0$, we have $\ker(\partial_0) = S_0(X)$. A singular 1-simplex is a continuous map $\sigma : \Delta_1 = I \to X$, and so it gives a path in X connecting the end points. When we form the quotient $H_1(X) = \ker(\partial_0)/\operatorname{im}(\partial_1)$, we will be identifying $[x] = [y] \in H_0(X)$ whenever there is a path joining x to y.

Exercise 6.3.2. Show that if X is path connected, then $H_0(X) \simeq \mathbb{Z}$.

Exercise 6.3.3. Suppose the path components of X are $X_j, j = 1, \ldots, n$.

(a) Show that $S_i(X) \simeq \bigoplus_{j=1}^{n} S_i(X_j)$ and $\partial_i : S_i(X) \to S_{i-1}(X)$ sends $S_i(X_j)$ to $S_{i-1}(X_j)$ in this direct sum decomposition.

(b) Show that $H_i(X) \simeq \bigoplus_{j=1}^{n} H_i(X_j)$.

(c) Show that $H_0(X)$ is the free abelian group with one generator for each path component.

It is somewhat inconsistent with later computations that path connectivity is measured by $H_0(X) \simeq \mathbb{Z}$, rather than being trivial. In Section 6.13 we will introduce the notion of reduced homology, which will adjust our chain complex so that path connectivity is measured by the reduced homology in dimension 0 being 0 in the path-connected case. Knowing that H_0 measures the number of path components, we now examine what H_1 measures. Because of the last result, we can restrict to the path-connected case.

Assume that X is path connected with base point x_0. We now concentrate on examining the relation of the fundamental group $\pi_1(X, x_0)$ and the first homology group $H_1(X)$. Note that the generators of $S_1(X)$ are similar to the maps used to form equivalence classes in π_1 in that they are maps from I to X; a major difference is that π_1 uses equivalence classes of loops and for S_1 we are just using paths. However, suppose that we have a singular 1-chain $C = \sum_{k=1}^m n_k \sigma_k$, where $\partial C = 0$. Such a 1-chain is called a *singular 1-cycle* and is similar to a loop in many ways. For each 1-simplex σ_k in C, there are two singular 0-simplices which map to the points $\sigma_k(1)$ and $\sigma_k(0)$, respectively. Let $x_i, i = 1, \ldots, p$ be the distinct points which occur as boundary points as we range over all singular 1-simplices in C. Since C is a 1-cycle, we will have $\sum n_k(\sigma_k(1) - \sigma_k(0)) = 0$, where we are abusing notation and identifying a singular 0-simplex with its image point. This means that the coefficient of each x_i in this sum is zero. For each $x \in X$, choose a singular 1-simplex $\tau_x : \Delta_1 \to X$ which is a path from the base point x_0 to x. Now suppose that $\sigma_k(0) = p, \sigma_k(1) = q$. Then define $c_k = \tau_p + \sigma_k - \tau_q$.

Exercise 6.3.4. Show that $\partial_1(C) = 0$ implies that $\sum_{k=1}^m n_k c_k = \sum_{k=1}^m n_k \sigma_k$.

Now look at the loop $\sigma'_k = \tau_p * \sigma_k * \bar{\tau}_q$ and consider it as a singular 1-simplex as well as a loop at x_0. There is a continuous map from the square to X which has σ'_k on $I \times \{0\}, \tau_p$ on $\{0\} \times I, \sigma_k$ on $I \times \{1\}$ and τ_q on $\{1\} \times I$. The square is divided into three pieces and the left piece is mapped using τ_p, the middle piece is mapped via σ_k, and the right piece is mapped via τ_q. We illustrate this with Figure 6.1.

Exercise 6.3.5. Use Figure 6.1 as a guide to construct the map of the square so that the boundary is mapped as described.

We now define a map h from continuous loops at x_0 to $S_1(X)$ by sending the loop γ to the singular 1-simplex $\gamma : \Delta_1 = I \to X$. Note that $\partial_1(h(\gamma)) = 0$. For the rest of this section, we will use $\sim$ to denote homotopy relative to the end points for paths (or loops) and $\sim_\partial$ for the equivalence relation on 1-chains (or 1-cycles) that $c_1 \sim_\partial c_2$ iff $c_1 = c_2 + \partial(D)$, where D is a 2-chain.

Exercise 6.3.6. Show that if $\gamma \sim \eta$ for cycles γ, η, then there is a singular 2-chain D so $h(\eta) - h(\gamma) = \partial_2(D)$; thus $h(\eta) \sim_\partial h(\gamma)$. (Hint: Take the homotopy as a map of the square and subdivide the square into two 2-simplices. Let D be the chain which is a difference of two singular 2-simplices $D = \alpha_1 - \alpha_2$ which first map into the two 2-simplices and then compose with the homotopy. See Figure 6.2 for an illustration.)

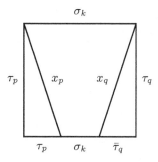

Figure 6.1. A homotopy.

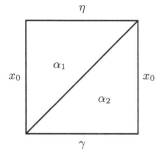

Figure 6.2. Constructing D.

Exercise 6.3.7. Using the previous exercise, show that h induces a map $\bar{h}$: $\pi_1(X, x_0) \to H_1(X)$.

Exercise 6.3.8. Show that $h(\alpha * \beta) \sim_\partial h(\alpha) + h(\beta)$ and so $\bar{h}$ is a homomorphism. (Hint: Consider Figure 6.3. First construct a map of the triangle with vertices $(0,0), (\frac{1}{2}, 1), (1,0)$ with $\alpha * \beta$ on the bottom, α on the left, and β (directed downward) on the right. Use it to show that $h(\alpha * \beta)$ differs from $h(\alpha) + h(\beta)$ by the boundary of a singular 2-simplex.)

Note that if e_{x_0} represents the trivial loop at x_0, the homomorphism property implies that $c_{x_0}^1 = h(e_{x_0})$ is a boundary.

Exercise 6.3.9. Show this directly by showing that if $c_{x_0}^2$ is the singular 2-simplex mapping to x_0, then $\partial_2(c_{x_0}^2) = c_{x_0}^1$.

The homomorphism property and the last exercise imply that $h(\bar{\alpha}) \sim_\partial -h(\alpha)$.

Exercise 6.3.10. Show this directly by showing that $h(\bar{\alpha}) + h(\alpha)$ is the boundary of a 2-chain.

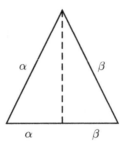

Figure 6.3. Diagram showing that $\bar{h}$ is a homomorphism.

Exercise 6.3.11. Use Exercises 6.3.4–6.3.8 to show that $\bar{h}$ is surjective. (Hint: Use the ideas in the previous exercises to show that if $C = \sum_{k=1}^{m} n_k \sigma_k$ is a 1-cycle in X, then, if $\sigma'_k = \tau_p * \sigma_k * \bar{\tau}_q$, we have $\bar{h}([\sigma'_k]) = [c_k]$. Then use the homomorphism property of $\bar{h}$ to show $[C]$ is in the image of $\bar{h}$.)

Since $H_1(X)$ is an abelian group, any element of the commutator subgroup of $\pi_1(X, x_0)$ must map to 0. Thus $\bar{h}$ factors through the abelianization $\pi_1^{ab}(X, x_0)$ and induces a map $\bar{h}' : \pi_1^{ab}(X, x_0) \to H_1(X)$. This induced map is surjective, so to show that it is an isomorphism it suffices to show that the kernel consists of the identity element.

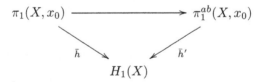

Theorem 6.3.1. *The map $\bar{h}' : \pi_1^{ab}(X, x_0) \to H_1(X)$ is an isomorphism.*

Theorem 6.3.1 states that the first homology group is the abelianization of the fundamental group. We give an argument that proves Theorem 6.3.1 through a series of exercises. Start with an element $[\alpha]$ in the kernel of $\bar{h}$. We have to show that $[\alpha]$ is in the commutator subgroup. Just as we replaced a singular 1-simplex σ with σ' which came from a loop, we want to replace a singular 2-simplex with one which maps the vertices to the base point. We can do this with the techniques developed in proving the Seifert–van Kampen theorem. We just take small triangular neighborhoods of the vertices that are not mapped to x_0 and first alter D so that the map is constant on these neighborhoods sending them to the same point that D sends the vertex. Then we take a map from the simplex to itself that sends the truncated neighborhood to the vertex and otherwise expands the rest of the simplex to fill up the simplex. This map can be constructed by defining it on the boundary of the truncated simplex and then

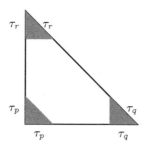

Figure 6.4. Constructing D' from D.

extending it to the whole truncated simplex by coning from the centroid. We define a new singular simplex D' by using the composition of our map from the truncated simplex to the simplex and composing with D. On the triangular neighborhoods of the vertices, we take the map that is the composition of τ_p with projection from the neighborhood to an interval (which we are identifying canonically with I). Basically, this map is defined so radial lines from the vertex formerly sent to p are mapped via a reparametrized τ_p. If we look at a term σ in $\partial_2 D$, then D' is constructed so that $\partial D'$ now contains the term σ' (at least up to homotopy relative to the end points).

We illustrate this in Figure 6.4.

Now suppose α is a loop so that $\bar{h}(\alpha) = 0$. Then $h(\alpha) = \sum_{i=1}^{k} n_i \partial_2 D_i$. But this means that the term $h(\alpha)$ occurs on the right-hand side and then all of the other terms which occur do so in pairs with canceling signs. This means that it will also be the case that $\partial D' = h(\alpha)$ since the term $h(\alpha)$ will be unchanged and any other term σ would be replaced by a corresponding term σ'. Hence we can assume from the outset that any term D_i is a singular 2-simplex whose vertices map to x_0. For any singular 2-simplex D whose vertices map to x_0, there is a loop at x_0 given by $D_2 * D_0 * \bar{D}_1$, where $D_i = DF_i$. The three components are just the face maps. This can also be written as $D(F_2 * F_0 * \bar{F}_1)$. The term $F_2 * F_0 * \bar{F}_1$ just represents a loop that runs counterclockwise around the standard 2-simplex.

Exercise 6.3.12. Show that $F_2 * F_0 * \bar{F}_1$ represents the trivial element of $\pi_1(\Delta_2, e_0)$ and that D represents the trivial element of $\pi_1(X, x_0)$.

For any singular 2-simplex D_i which occurs in the formula $h(\alpha) = \sum_{i=1}^{k} n_i \partial_2 D_i$ (with our assumption that D_i maps the vertices to x_0), look at the loop $D_{2i} * D_{0i} * \bar{D}_{1i} = \gamma_i$.

Exercise 6.3.13. Show that in the abelianization $\pi_1^{ab}(X, x_0)$ that $[\alpha]$ and $[\gamma_1]^{n_1} * \cdots * [\gamma_k]^{n_k}$ represent the same element. (Hint: Break the second term into the individual components D_{ji} so that each component occurs a net zero number of times except for α. Here $\bar{D}_{ji}$ counts as -1 when it occurs since it represents the inverse of D_{ji} in π_1.)

Exercise 6.3.14. Complete the proof of Theorem 6.3.1.

We can apply Theorem 6.3.1 to compute the first homology groups of surfaces.

Theorem 6.3.2.

(a) $H_1(T_{(p)}^{(g)})$ is the free abelian group on $2g + p - 1$ generators for $p \neq 0$ and the free abelian group on $2g$ generators when $p = 0$.

(b) $H_1 P_{(p)}^{(h)}$ is the free abelian group on $h + p - 1$ generators when $p \neq 0$ and is the direct sum of $\mathbb{Z}_2$ with the free abelian group with $h - 1$ generators when $p = 0$.

Exercise 6.3.15. Prove Theorem 6.3.2.

6.4 Cellular homology of a two-dimensional CW complex

We now consider a two-dimensional CW complex X and define a chain complex $(C_i^c(X), \partial_i)$, $i = 0, 1, 2$, of free abelian groups and boundary homomorphisms related to it. All higher-dimensional homology $H_i(C)$ is defined to be zero since there are no cells in these dimensions. In the two-dimensional case it is relatively simple to define this complex in an ad hoc manner not relying on the general theory of singular homology. Later on we will give the full definition which applies in all dimensions and serves to link cellular and singular homology. The justification for $\partial_1 \partial_2 = 0$ will be given then. Here we just focus on how computations are done in this low-dimensional situation.

We let $C_0^c(X)$ be the free abelian group with one generator g_i^0 for each 0-cell e_i^0. As before, we define $\partial_0 = 0$. For each 1-cell e_i^1, we choose the standard positive orientation of the interval and have a generator g_i^1 for this oriented 1-cell. If it is attached via f so that $f(1) = e_j^0, f(0) = e_k^0$, then we define $\partial_1(g_i^1) = g_j^0 - g_k^0$. For each 2-cell e_j^2, we choose the standard positive orientation of the cell and have a generator g_j^2. Corresponding to this orientation, there will be a positive orientation of the cell and a corresponding generator g of $H_1(S_j^1) \simeq \mathbb{Z}$. When the 2-cell is attached via $f_j : S_j^1 \to X^1$, there is an induced homomorphism $(f_j)_* : H_1(S_i^1) \to H_1(X^1)$. We can form the quotient space $q : X^1 \to X^1/X^0$ by identifying the points in X^0 to a single point. The quotient is homeomorphic to $\bigvee_{i=1}^p S_i^1$ with one circle for each 1-cell of X^1. By the results of the last section relating π_1 to H_1 as well as the Seifert–van Kampen theorem, $H_1(\bigvee_{i=1}^p S_i^1) \simeq \bigoplus_{i=1}^p \mathbb{Z} = F_p^{ab}$, the free abelian group on p generators, one for each 1-cell. We can thus identify the group to $C_1^c(X)$, and consider the image $(qf)_*(g)$ of the generator g of $H_1(S_j^1)$ in $H_1(\bigvee_{i=1}^p S_i^1)$ with an element of $C_1^c(X)$. This image is defined to be $\partial_2(g_i^2)$. Another way to think of this boundary map is to note that a map to a direct sum of copies of $\mathbb{Z}$ is determined by its image on each component. To determine the coefficient n_{ij} when $\partial_2(g_j^2) = \sum_{i=1}^p n_{ij} g_i^1$, we look at the composition $f_{ij} : S_j^1 \to S_i^1$ of the attaching map f_j with the quotient map $p_i : X^1 \to S_i^1$ coming from collapsing all of X^1 except the 1-cell e_i^1 to a point. The image of the generator will be n_{ij} times the generator, so we just need to compute the degree of this map between circles.

To help us understand this boundary homomorphism better, we look at some examples.

- For the torus T, the 2-cell is attached to X^1 via the commutator $aba^{-1}b^{-1}$, and so $\partial_2(g^2) = g_a^1 + g_b^1 - g_a^1 - g_b^1 = 0$. Since there is only one 0-cell, $\partial_1(g_a^1) = \partial_1(g_b^1) = 0$. We then form the cellular homology groups $H_i^c(X)$ via $\ker(\partial_i)/\mathrm{im}(\partial_{i+1})$. Thus all of the boundary maps are 0 and so $H_i^c = C_i^c$, giving that the homology of the torus is $H_2^c(X) \simeq \mathbb{Z}, H_1^c(X) \simeq \mathbb{Z} \oplus \mathbb{Z}, H_0^c(X) \simeq \mathbb{Z}$.

- The computation of the homology of the Klein bottle is similar, with the main difference being that the attaching map of the 2-cell uses the pattern $abab^{-1}$. This leads to $\partial_2(g^2) = g_a^1 + g_b^1 + g_a^1 - g_b^1 = 2g_a^1$. Thus $\ker(\partial_2) = 0$ and so $H_2^c(X) = 0$. Then

$$H_1^c(X) = \mathbb{Z}(g_a^1) \oplus \mathbb{Z}(g_b^1)/\{2g_a^1\} \simeq \mathbb{Z}_2 \oplus \mathbb{Z}.$$

As before, $H_0^c(X) \simeq \mathbb{Z}$.

As the two examples show, the way the 2-cells are attached determines the boundary map ∂_2, and it is largely a matter of reading off the boundary information and abelianizing it. We give an example which is not a surface; see Figure 6.5 for an illustration. Consider the space with three 0-cells, which we will call a,b,c. Then attach five 1-cells A,B,C,D,E so that their attaching maps determine $\partial_1(A) = a - b, \partial_1(B) = a - c, \partial_1(C) = c - c = 0, \partial_1(D) = b - c, \partial_1(E) = c - a$. To describe the attaching maps of the 2-cells, we think of dividing the boundary into subarcs and say where the image of consecutive subarcs is in terms of 1-cells with a superscript -1 to indicate running over the 1-cell in the opposite direction.

For our example, we will have two 2-cells which we label α, β, and are attached via the patterns $AECD$ and $BA^{-1}D^{-1}$. This leads to the chain groups $C_2^c = 2\mathbb{Z}, C_1^c = 5\mathbb{Z}, C_0^c = 3\mathbb{Z}$, with boundary map

$$\partial_2(\alpha) = A + E + C + D, \qquad \partial_2(\beta) = B - A - D.$$

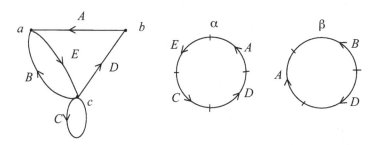

Figure 6.5. Computing the cellular homology.

The homology is then computed as follows. For H_0^c, factoring out by the image of ∂_1 identifies all of the generators of C_0^c, so $H_0^c(X) \simeq \mathbb{Z}$, which indicates that X is path connected. To compute H_1, we first find a basis for $\ker(\partial_1)$. To do this, we can write down a matrix which represents ∂_1 with respect to our basis of C_1^c. This matrix is

$$\begin{pmatrix} 1 & 1 & 0 & 0 & -1 \\ -1 & 0 & 0 & 1 & 0 \\ 0 & -1 & 0 & -1 & 1 \end{pmatrix}.$$

Using gaussian elimination, we can find a basis of $\ker(\partial_1)$ to be $C, A - B + D, B + E$. A basis of $\text{im}(\partial_2)$ is given by $A + E + C + D, -A + B - D$. These elements can be written in terms of the generators of $\ker(\partial_1)$ as $A + E + C + D = C - (A - B + D) + (B + E), -A + B - D = -(A - B + D)$. The quotient is $H_1^c(X) \simeq \mathbb{Z}$. One way of seeing this is to think of ∂_2 as identifying $(A - B + D)$ to 0 in the quotient and then identifying $(B + E)$ to C. Thus the quotient is generated by C with no further relations. Since $\partial_2(\alpha), \partial_2(\beta)$ are independent, $\ker(\partial_2) = 0$ and so $H_2^c(X) \simeq 0$.

Exercise 6.4.1. Compute the cellular homology groups for all surfaces without boundary, where we form the CW complex by thinking of the surface as a polygon with identifications.

Exercise 6.4.2. Compute the cellular homology groups for the dunce hat. (Hint: See Exercise 5.3.3.)

Exercise 6.4.3. Compute the cellular homology groups for the pseudoprojective plane P_q (Exercise 3.9.4) which has a CW complex decomposition as a 0-cell, one 1-cell, and one 2-cell, where the 1-cell is attached trivially to the 0-cell to form a circle for X^1, and the attaching map for the 2-cell is the map $z \to z^q$ which wraps around the circle q times.

Exercise 6.4.4. Compute the cellular homology groups for the CW complex with cellular decomposition given by a single 0-cell, two trivially attached 1-cells (labeled a, b), and three 2-cells attached via $a^2 b^{-1} a, ba^3, ababababab$.

Exercise 6.4.5. Find the cellular homology groups of the two spaces in Figure 5.2.

Exercise 6.4.6. Find the cellular homology groups of the two spaces in Figure 5.3.

Exercise 6.4.7. Find the cellular homology groups of the two spaces in Figure 5.4.

Exercise 6.4.8. Find the cellular groups of the two spaces in Figure 5.5.

6.5   Chain maps and homology

In this section we develop some basic results about chain complexes and homology which we will apply in studying singular homology more deeply.

We first look at the notion of a chain map between chain complexes.

Definition 6.5.1. Suppose $(C_k, \partial_k^C), (D_k, \partial_k^D)$ are chain complexes. A sequence of homomorphisms $h_k : C_k \to D_k$ is called a *chain map* if $\partial_k^D h_k = h_{k-1}\partial_k^C$.

This can be re-expressed in terms of the following commutative diagram:

$$
\begin{array}{ccc}
C_k & \xrightarrow{\partial_k^C} & C_{k-1} \\
\downarrow{\scriptstyle h_k} & & \downarrow{\scriptstyle h_{k-1}} \\
D_k & \xrightarrow{\partial_k^D} & D_{k-1}
\end{array}
$$

An important facet of chain maps is that they induce maps on homology by the formula $h_*([c]) = [h(c)]$. Here $[c]$ denotes the homology element which is represented by the chain c with $\partial(c) = 0$.

Exercise 6.5.1. Verify the claim above that chain maps induce homomorphisms on homology. You need to show that if $[c] = [c'] \in H_n(C)$, then $[h_n(c)] = [h_n(c')] \in H_n(D)$.

These induced maps behave well under composition.

Exercise 6.5.2. Show that if $A \xrightarrow{f} B \xrightarrow{g} C$ is a composition of chain maps, then $(gf)_* = g_* f_*$.

To simplify the notation from now on with chain maps and boundaries, we will delete the subscripts unless they are needed for clarity. Thus we express the chain map condition as $h\partial = \partial h$ and just use $h(c)$ for $h_n(c)$ when $c \in C_n$.

Definition 6.5.2. Suppose C, D, E are abelian groups and we have homomorphisms $f : C \to D$ and $g : D \to E$. The sequence of homomorphisms

$$ C \xrightarrow{f} D \xrightarrow{g} E $$

is said to be *exact* at D if $\ker(g) = \operatorname{im}(f)$. The sequence

$$ 0 \longrightarrow C \xrightarrow{f} D \xrightarrow{g} E \longrightarrow 0 $$

is called a *short exact sequence* if it is exact at C, D, E; this means that f is an injection and g is a surjection as well as $\ker(g) = \operatorname{im}(f)$.

When there is a short exact sequence as above, there is an isomorphism of abelian groups $E \simeq D/\operatorname{im}(f)$. Since f is an injection, $C \simeq \operatorname{im}(f)$ and so they are frequently identified. C is then thought of as a subgroup of D and the quotient $D/\operatorname{im}(f)$ is as written D/C. Two examples of short exact sequences are

- $0 \longrightarrow \mathbb{Z} \xrightarrow{f} \mathbb{Z} \oplus \mathbb{Z} \xrightarrow{g} \mathbb{Z} \longrightarrow 0$, where $f(n) = (n,0)$, $g(n,m) = m$.

- $0 \longrightarrow \mathbb{Z} \xrightarrow{f} \mathbb{Z} \xrightarrow{g} \mathbb{Z}_2 \longrightarrow 0$,   where $f(n) = 2n$ and $g(n) = n$ mod 2.

In the first example, the middle term is a direct sum of the first and third terms and the maps f, g have a standard form $f(n) = (n, 0), g(n, m) = m$. This is called a *split* short exact sequence. Any short exact sequence of abelian groups where the third term is a free abelian group can be shown to be equivalent to a split short exact sequence. By being equivalent to a split exact sequence, we mean that there is a commutative diagram where all the vertical maps are isomorphisms and the horizontal maps on the top row are $F(c) = (f(c), 0), G(c, e) = e$.

$$
\begin{array}{ccccccccc}
0 & \longrightarrow & C & \xrightarrow{\ F\ } & C \oplus E & \xrightarrow{\ G\ } & E & \longrightarrow & 0 \\
  &   & \Big\downarrow{=} &   & \Big\downarrow{\simeq} &   & \Big\downarrow{=} &   &   \\
0 & \longrightarrow & C & \xrightarrow{\ f\ } & D & \xrightarrow{\ g\ } & E & \longrightarrow & 0
\end{array}
$$

The second example shows that not all short exact sequences are split, and identifies the presence of torsion in the third term as a source of difficulty in splitting a short exact sequence.

Exercise 6.5.3. Show that if $E \simeq \mathbb{Z} \oplus \cdots \oplus \mathbb{Z}$ is a finitely generated free abelian group, then a short exact sequence $0 \longrightarrow C \xrightarrow{f} D \xrightarrow{g} E \longrightarrow 0$ is equivalent to a split short exact sequence. (Hint: Define a map $r : E \to D$ with $gr(e) = e$ by first defining it on generators and extending linearly. Then define a map $I : C \oplus E \to D$ by $I(c, e) = f(c) + r(e)$. Show that I gives an isomorphism and leads to the required commutative diagram. To show that I is onto, start with $d \in D$ and look at $d - r(g(d))$ and show that there is $c \in C$ with $d - r(g(d)) = f(c)$.)

More generally, two short exact sequences are equivalent if there are vertical isomorphisms so the following diagram is commutative:

$$
\begin{array}{ccccccccc}
0 & \longrightarrow & C & \xrightarrow{\ f\ } & D & \xrightarrow{\ g\ } & E & \longrightarrow & 0 \\
  &   & \Big\downarrow{\simeq} &   & \Big\downarrow{\simeq} &   & \Big\downarrow{\simeq} &   &   \\
0 & \longrightarrow & C' & \xrightarrow{\ f'\ } & D' & \xrightarrow{\ g'\ } & E' & \longrightarrow & 0
\end{array}
$$

We generalize the preceding exercise.

Exercise 6.5.4. Suppose that there is a commutative diagram with exact rows,

$$
\begin{array}{ccccccccc}
0 & \longrightarrow & C & \overset{f}{\longrightarrow} & D & \overset{g}{\longrightarrow} & E & \longrightarrow & 0 \\
& & \downarrow{\scriptstyle\gamma} & & \downarrow{\scriptstyle\delta} & & \downarrow{\scriptstyle\epsilon} & & \\
0 & \longrightarrow & C' & \overset{f'}{\longrightarrow} & D' & \overset{g'}{\longrightarrow} & E' & \longrightarrow & 0
\end{array}
$$

Show that if γ and ϵ are isomorphisms, then δ is an isomorphism and the two short exact sequences are equivalent. (Hint: The type of argument used is called diagram chasing. We have to show that δ is both surjective and injective. For surjectivity, start with $d' \in D'$. Take $g'(d') \in E'$ and use surjectivity of ϵ to find $e \in E$ with $\epsilon(e) = g'(d')$. Use surjectivity of g to find $d \in D$ with $g(d) = e$. Then look at $d' - \delta(d)$. If it were 0, then we would have shown that δ is surjective. It does not have to be 0, but you can use commutativity of the diagram to show that $g'(d' - \delta(d)) = 0$. Then use exactness at D' of the bottom row to show that there is a $c' \in C'$ with $f'(c') = d' - \delta(d)$. Use surjectivity of γ to find c with $\gamma(c) = c'$. Then use commutativity of the diagram to show that $d' - \delta(d) = \delta(f(c))$, from which you can conclude that $d' = \delta(d + f(c))$. Then argue that δ is injective by using a similar diagram chase.)

Now suppose C, D, E are chain groups of chain complexes. Let $f : C \to D, g : D \to E$ be chain maps.

Definition 6.5.3. A sequence of chain maps is a *short exact sequence* if, for each $i \geq 0$, the sequence $\quad 0 \longrightarrow C_i \overset{f}{\longrightarrow} D_i \overset{g}{\longrightarrow} E_i \longrightarrow 0 \quad$ is a short exact sequence of homomorphisms between abelian groups.

Theorem 6.5.1. *Suppose there is a short exact sequence*

$$
0 \longrightarrow C \overset{f}{\longrightarrow} D \overset{g}{\longrightarrow} E \longrightarrow 0
$$

of chain complexes. Then there is a corresponding long exact sequence in homology:

$$
\cdots \longrightarrow H_k(C) \overset{f_*}{\longrightarrow} H_k(D) \overset{g_*}{\longrightarrow} H_k(E) \overset{\partial}{\longrightarrow} H_{k-1}(C) \overset{f_*}{\longrightarrow} \cdots
$$

The sequence then continues to the right until we are in dimension 0 and ends with $H_0(E) \to 0$.

We break the proof up into a number of exercises. The induced maps f_*, g_* are all defined as in: $f_*([c]) = [f(c)]$, where $[c]$ is the homology class containing c. This is well defined by Exercise 6.5.1. We now indicate how the boundary map $H_k(E) \overset{\partial}{\longrightarrow} H_{k-1}(C)$ is defined. For an element $[e] \in H_k(E)$, we select an element $e \in E_k$ in the homology class. Note that $\partial e = 0$ since e represents a homology class. Since the map g is surjective, there is an element $d \in D_k$ with $g(d) = e$. We note $g(\partial(d)) = \partial(g(d)) = \partial(e) = 0$. Since the sequence is exact, there is an element $c \in C_{k-1}$ with $f(c) = \partial(d)$. Moreover, $f(\partial(c)) = \partial(f(c)) = \partial\partial(d) = 0$. Since f is an injection, this implies that $\partial(c) = 0$ and so represents a homology class. We define $\partial([e]) = [c]$. Note that we are using the same notation ∂ for the map on homology and the boundary maps in the chain complexes.

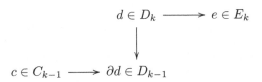

Exercise 6.5.5. Show that this definition is well defined. This means that if we choose another representative e' of $[e] \in H_k(E)$ and go through similar steps to define $[c'] \in H_{k-1}(C)$, then $[c'] = [c]$. (Hint: Start with $e - e' = \partial \epsilon$. Write $\epsilon = g(\delta)$. Then show $g(d - d' - \partial \delta) = 0$ and so $d - d' - \partial \delta = f(\gamma)$. Chase around the diagram to show that $c - c' = \partial \gamma$ and so $[c] = [c']$.)

Exercise 6.5.6. Show that the sequence is exact at $H_k(C)$. (Hint: This requires you to show that $\ker(f_*) = \mathrm{im}(\partial)$. Doing so requires showing inclusions in both directions. To show that $\mathrm{im}(\partial) \subset \ker(f_*)$, use the fact that, if $[c] = \partial([e])$, then we can assume $f(c) = \partial(d)$ with $g(d) = e$. For the other direction, if $f_*([c]) = 0$, then $f(c) = \partial d$. Use d to find e with $\partial(e) = 0$ and $\partial([e]) = [c]$.)

Exercise 6.5.7. Show that the sequence is exact at $H_k(D)$. (Hint: This requires you to show that $\ker(g_*) = \mathrm{im}(f_*)$. For one direction, you just need to use $gf = 0$. For the other direction, suppose that $g_*([d]) = 0$, which means that $g(d) = \partial(\epsilon)$. Choose δ so that $g(\delta) = \epsilon$. Then show that $g(d - \partial(\delta)) = 0$ and use this to show that there is a c with $\partial(c) = 0$ and $f_*([c]) = [d]$.)

Exercise 6.5.8. Show that the sequence is exact at $H_k(E)$. (Hint: This requires you to show that $\ker(\partial) = \mathrm{im}(g_*)$. To show that $\ker(\partial) \subset \mathrm{im}(g_*)$, suppose $[e] \in H_k(E)$ satisfies $\partial([e]) = 0$. Then there are elements d, c, γ with $g(d) = e$, $f(c) = \partial(d)$, $c = \partial(\gamma)$. Then show that $\partial(d - f(\gamma)) = 0$ and $g_*([d - f(\gamma)]) = [e]$.)

An important property of the long exact sequence is its "naturality".

Theorem 6.5.2 (Naturality of long exact sequence). *Suppose that there are two short exact sequences which are connected via chain maps as represented in the following commutative diagram:*

$$
\begin{array}{ccccccccc}
0 & \longrightarrow & C & \xrightarrow{f} & D & \xrightarrow{g} & E & \longrightarrow & 0 \\
  &   & \downarrow{\alpha} &   & \downarrow{\beta} &   & \downarrow{\gamma} &   &   \\
0 & \longrightarrow & C' & \xrightarrow{f'} & D' & \xrightarrow{g'} & E' & \longrightarrow & 0
\end{array}
$$

Then the corresponding long exact sequences are connected via the commutative diagram

$$
\begin{array}{ccccccccc}
\cdots \longrightarrow & H_k(C) & \xrightarrow{f_*} & H_k(D) & \xrightarrow{g_*} & H_k(E) & \xrightarrow{\partial} & H_{k-1}(C) & \xrightarrow{f_*} \cdots \\
& \downarrow{\scriptstyle\alpha_*} & & \downarrow{\scriptstyle\beta_*} & & \downarrow{\scriptstyle\gamma_*} & & \downarrow{\scriptstyle\alpha_*} & \\
\cdots \longrightarrow & H_k(C') & \xrightarrow{f'_*} & H_k(D') & \xrightarrow{g'_*} & H_k(E') & \xrightarrow{\partial} & H_{k-1}(C') & \xrightarrow{f'_*} \cdots
\end{array}
$$

Exercise 6.5.9. Prove Theorem 6.5.2.

We now give an algebraic result which we will apply to the long exact sequence in homology.

Exercise 6.5.10. Show that if there is a long exact sequence of abelian groups,

$$
A \xrightarrow{f} B \xrightarrow{g} C \xrightarrow{h} D \xrightarrow{k} E \ ,
$$

then there is a corresponding short exact sequence

$$
0 \longrightarrow \mathrm{coker}(f) \xrightarrow{\bar{g}} C \xrightarrow{\bar{h}} \ker(k) \longrightarrow 0
$$

Here $\mathrm{coker}(f) = B/\mathrm{im}(f)$.

This last exercise is used in computations as a means of determining C from $\mathrm{coker}(f)$ and $\ker(k)$. As important special cases, we note:

- if $B = D = 0$, then $C = 0$;
- if $B = E = 0$, then $h : C \to D$ is an isomorphism;
- if f is surjective, then $C \simeq \ker(k)$;
- if k is injective, then $C \simeq \mathrm{coker}(f)$;
- if D is a free abelian group, then $\ker(k)$ will also be free abelian and $C \simeq \mathrm{coker}(f) \oplus \ker(k)$.

The next result, the five lemma, allows us to compare two long exact sequences.

Exercise 6.5.11. (Five lemma) Suppose there is a commutative diagram of abelian groups,

$$
\begin{array}{ccccccccc}
A & \xrightarrow{f} & B & \xrightarrow{g} & C & \xrightarrow{h} & D & \xrightarrow{k} & E \\
{\scriptstyle\mathrm{surj}}\downarrow{\scriptstyle\alpha} & & {\scriptstyle\simeq}\downarrow{\scriptstyle\beta} & & \downarrow{\scriptstyle\gamma} & & {\scriptstyle\simeq}\downarrow{\scriptstyle\delta} & & {\scriptstyle\mathrm{inj}}\downarrow{\scriptstyle\epsilon} \\
A' & \xrightarrow{f'} & B' & \xrightarrow{g'} & C' & \xrightarrow{h'} & D' & \xrightarrow{k'} & E'
\end{array}
$$

where the horizontal rows are exact and the maps β, δ are isomorphisms, α is surjective, and ϵ is injective.

(a) Show that this diagram induces a commutative diagram

$$
\begin{array}{ccccccccc}
0 & \longrightarrow & \mathrm{coker}(f) & \xrightarrow{\bar{g}} & C & \xrightarrow{\bar{h}} & \ker(k) & \longrightarrow & 0 \\
& & \downarrow{\bar{\beta}} & & \downarrow{\gamma} & & \downarrow{\bar{\delta}} & & \\
0 & \longrightarrow & \mathrm{coker}(f') & \xrightarrow{\bar{g}'} & C' & \xrightarrow{\bar{h}'} & \ker(k') & \longrightarrow & 0
\end{array}
$$

(b) Show that $\bar{\beta}$ and $\bar{\delta}$ are isomorphisms.

(c) Use Exercise 6.5.4 to show that γ is an isomorphism.

We next define the notion of two chain maps as being chain homotopic. This definition is motivated by the notion of homotopic continuous maps. The goal is to show that if f, g are homotopic continuous maps, then they induce chain maps $f_\sharp, g_\sharp$ which are chain homotopic.

Definition 6.5.4. We say that chain maps $F, G : C \to D$ are *chain homotopic* if there is a map $H_k : C_k \to D_{k+1}$ for each $k \geq 0$ so that $G_k - F_k = \partial^D_{k+1} H_k + H_{k-1} \partial^C_k$.

$$
\begin{array}{ccccc}
C_{k+1} & \xrightarrow{\partial^C_{k+1}} & C_k & \xrightarrow{\partial^C_k} & C_{k-1} \\
\Big\downarrow{\scriptstyle F_{k+1}} \ \ \overset{H_k}{\underset{G_{k+1}}{\swarrow}} & & \Big\downarrow{\scriptstyle F_k} \ \ \overset{H_{k-1}}{\underset{G_k}{\swarrow}} F_{k-1} & & \Big\downarrow{\scriptstyle G_{k-1}} \\
D_{k+1} & \xrightarrow[\partial^D_{k+1}]{} & D_k & \xrightarrow[\partial^D_k]{} & D_{k-1}
\end{array}
$$

Exercise 6.5.12. Show that if the chain maps F, G are chain homotopic, then F and G induce the same map on homology. This involves showing that if c is a cycle in C_k, then $G_k(c) - F_k(c)$ is a boundary.

6.6 Axioms for singular homology

We have discussed a number of different homology theories. In this section, we discuss a set of properties which characterize singular homology theory for a CW complex. Because of these properties, it can be shown that cellular homology and singular homology coincide for a CW complex.

We start by listing these properties, which are called the Eilenberg–Steenrod axioms.

- (Functorial property) For each topological pair (X, A) and integer $k \geq 0$, there is an abelian group $H_k(X, A)$ so that if $f : (X, A) \to (Y, B)$ is continuous, then there is an induced homomorphism $f_* : H_k(X, A) \to H_k(Y, B)$ which is functorial. This means that the following two properties hold:
  — If we have a composition of continuous maps

  $$(X, A) \xrightarrow{f} (Y, B) \xrightarrow{g} (Z, C) \text{ , then } (gf)_* = g_* f_*.$$

— The identity induces the identity homomorphism: $(1_{(X,A)})_* = 1_{H_*(X,A)}$. When A is the empty set, we denote $H_k(X, \phi) = H_k(X)$.

Moreover, for each pair (X, A) and integer $k \geq 1$, there is a homomorphism $\partial : H_k(X, A) \to H_{k-1}(A)$ satisfying the property that, if $f : (X, A) \to (Y, B)$ is continuous, then there is a commutative diagram

$$
\begin{array}{ccc}
H_k(X, A) & \xrightarrow{\ \partial\ } & H_{k-1}(A) \\
\downarrow{\scriptstyle f_*} & & \downarrow{\scriptstyle f_*} \\
H_k(Y, B) & \xrightarrow{\ \partial\ } & H_{k-1}(B)
\end{array}
$$

- (Homotopy property) If $f, g : (X, A) \to (Y, B)$ are homotopic continuous maps, then $f_* = g_*$.

- (Exactness property) If $i_A : A \to X, j : X \to (X, A)$ are the inclusions, then there is a long exact sequence

$$
\cdots \longrightarrow H_k(A) \xrightarrow{\ i_*\ } H_k(X) \xrightarrow{\ j_*\ } H_k(X, A) \xrightarrow{\ \partial\ } H_{k-1}(A) \longrightarrow \cdots
$$

- (Excision property) If $\bar{U} \subset \mathrm{int}\, A$, then the inclusion $(X \backslash U, A \backslash U) \to (X, A)$ induces an isomorphism in homology.

- (Dimension property) If P is a one point space, then

$$
H_k(P) = \begin{cases} 0 & k \neq 0, \\ \mathbb{Z} & k = 0. \end{cases}
$$

Note that the invariance of homology under homeomorphisms follows from the functorial property since, if $f : (X, A) \to (Y, B)$ is a homeomorphism with inverse g, then $f_* : H_k(X, A) \to H_k(Y, B)$ is an isomorphism with inverse g_*. We note this result for future reference.

Theorem 6.6.1. *If $f : (X, A) \to (Y, B)$ is a homeomorphism of pairs, then $f_* : H_k(X, A) \to H_k(Y, B)$ is an isomorphism.*

Using the same argument, the homotopy property implies that a homotopy equivalence induces an isomorphism in homology.

Theorem 6.6.2. *Suppose $f : (X, A) \to (Y, B)$ is a homotopy equivalence of pairs. Then $f_* : H_k(X, A) \to H_k(Y, B)$ is an isomorphism.*

Exercise 6.6.1. Prove Theorem 6.6.2.

Definition 6.6.1. The induced chain map $f_\sharp : S(X) \to S(Y)$ coming from a continuous map $f : X \to Y$ is defined on a generating singular simplex $\sigma : \Delta_k \to X$ by $f_\sharp(\sigma) = f\sigma$. It is then extended to a general chain by linearity: if $c = \sum_i n_i \sigma_i$, then $f_\sharp(c) = \sum_i n_i f_\sharp(\sigma_i)$.

Exercise 6.6.2. Check that $f_\sharp$ is a chain map; that is, $f_\sharp \partial_k^X = \partial_k^Y f_\sharp$.

$$
\begin{array}{ccc}
S_k(X) & \xrightarrow{\ \partial_k^X\ } & S_{k-1}(X) \\
{\scriptstyle f_\sharp}\big\downarrow & & \big\downarrow{\scriptstyle f_\sharp} \\
S_k(Y) & \xrightarrow{\ \partial_k^Y\ } & S_{k-1}(Y)
\end{array}
$$

Exercise 6.6.3. Check that $f_\sharp$ satisfies the functorial property that, if we have

a composition $X \xrightarrow{\ f\ } Y \xrightarrow{\ g\ } Z$, then $(gf)_\sharp = g_\sharp f_\sharp$.

If $f : (X, A) \to (Y, B)$ is a continuous map, then this means that there is an equality $f i_A = i_B f$, where, on the left, $f : X \to Y$ and, on the right, we are regarding $f : A \to B$. Here i_A, i_B are the inclusion maps. Then the previous exercise implies that $f_\sharp (i_A)_\sharp = (i_B)_\sharp f_\sharp$.

For singular homology, we have defined chain groups and corresponding homology groups for a topological space X. We now extend this definition to a pair (X, A), where A is a subspace of X. We first define singular chain groups $S_k(X, A)$ as quotient groups $S_k(X, A) = S_k(X)/S_k(A)$. Here we are identifying $S_k(A)$ with its isomorphic image $(i_A)_\sharp(S_k(A))$. This quotient is isomorphic to the free abelian group with one generator for each singular simplex of X whose image does not lie in A. Each element of the quotient comes from a representative in $S_k(X)$. We denote by $\bar{c}$ the element in the quotient $S_k(X, A) = S_k(X)/S_k(A)$ with representative $c \in S_k(X)$. Then $f_\sharp (i_A)_\sharp = (i_B)_\sharp f_\sharp$ implies that, if $f : (X, A) \to (Y, B)$ is continuous, then $f_\sharp : S_k(X) \to S_k(Y)$ induces a homomorphism $f_\sharp : S_k(X, A) \to S_k(Y, B)$ via $f_\sharp(\bar{c}) = \overline{f_\sharp(c)}$.

Exercise 6.6.4. Check that this definition of $f_\sharp : S_k(X, A) \to S_k(Y, B)$ is well defined.

Exercise 6.6.5. Check that this induced map satisfies the functorial property

$(gf)_\sharp = g_\sharp f_\sharp$, where $(X, A) \xrightarrow{\ f\ } (Y, B) \xrightarrow{\ g\ } (Z, C)$ are continuous maps.

We then use the boundary map in $S(X)$ to induce a boundary map in $S(X, A)$. The construction used is a general construction for quotient groups. We define $\partial_k^{(X,A)}(\bar{c}) = \overline{\partial_k^X(c)}$. In order to see that this is well defined, we use the commutative diagram

$$
\begin{array}{ccc}
S_k(A) & \xrightarrow{\ \partial_k^A\ } & S_{k-1}(A) \\
{\scriptstyle i_\sharp}\big\downarrow & & \big\downarrow{\scriptstyle i_\sharp} \\
S_k(X) & \xrightarrow{\ \partial_k^X\ } & S_{k-1}(X)
\end{array}
$$

This commutes since $i_\sharp$ is a chain map.

Exercise 6.6.6. Show that the map $\partial_k^{(X,A)} : S_k(X, A) \to S_{k-1}(X, A)$ is well defined.

Exercise 6.6.7. Show that, if $f : (X, A) \to (Y, B)$, then $f_\sharp$ is a chain map; that is, $f_\sharp \partial_k^{(X,A)} = \partial_k^{(Y,B)} f_\sharp$.

Since $f_\sharp$ is a chain map, it induces a homomorphism in homology $f_* :$ $H_k(X, A) \to H_k(Y, B)$ via $f_*([c]) = [f_\sharp(c)]$. This map satisfies the functorial property since $f_\sharp$ does.

Exercise 6.6.8. Verify that $(gf)_* = g_* f_*$ and $(1_{(X,A)})_* = 1_{H(X,A)}$.

It is interesting to interpret what a cycle is in $S_k(X, A)$. If we take a representative to be a chain $c \in S_k(X)$, then $\partial_k^{(X,A)}(\bar{c}) = 0$ means that $\partial_k^X(c) \in S_{k-1}(A)$. This leads to a boundary homomorphism $\partial : H_k(X, A) \to H_{k-1}(A)$ being defined by $\partial[\bar{c}] = [\partial_k^X c]$. This is just the boundary homomorphism which is formed from having a short exact sequence $0 \to S(A) \to S(X) \to S(X, A) \to 0$ of chain complexes. For we are taking a cycle $\bar{c} \in S_k(X, A)$, using the fact that the map $S_k(X) \to S_k(X, A)$ is onto to pull the element $\bar{c}$ back to an element $c \in S_k(X)$, then take its boundary $\partial_k^X(c)$, and then pull it back to an element of $S_{k-1}(A)$. Here we are just identifying $S_{k-1}(A)$ with its image $(i_A)_\sharp(S_{k-1}(A))$. Then the work in the last section means that this is a well-defined homomorphism $\partial : H_k(X, A) \to H_{k-1}(A)$ given by this formula. Of course, we do need to establish that $0 \to S(A) \to S(X) \to S(X, A) \to 0$ is a short exact sequence of chain complexes.

Exercise 6.6.9. Show that $0 \to S(A) \to S(X) \to S(X, A) \to 0$ is a short exact sequence of chain complexes.

With this definition of $\partial : H_k(X, A) \to H_{k-1}(A)$, we then get the long exact sequence in singular homology as a consequence of Theorem 6.5.1. To establish the property that there is a commutative diagram

$$
\begin{array}{ccc}
H_k(X, A) & \xrightarrow{\partial} & H_{k-1}(A) \\
\downarrow{\scriptstyle f_*} & & \downarrow{\scriptstyle f_*} \\
H_k(Y, B) & \xrightarrow{\partial} & H_{k-1}(B)
\end{array}
$$

we can use Theorem 6.5.2 once we have established that, if $f : (X, A) \to (Y, B)$ is a continuous map, then there is a commutative diagram

$$
\begin{array}{ccccccccc}
0 & \longrightarrow & S(A) & \longrightarrow & S(X) & \longrightarrow & S(X, A) & \longrightarrow & 0 \\
& & \downarrow{\scriptstyle f_\sharp} & & \downarrow{\scriptstyle f_\sharp} & & \downarrow{\scriptstyle f_\sharp} & & \\
0 & \longrightarrow & S(B) & \longrightarrow & S(Y) & \longrightarrow & S(Y, B) & \longrightarrow & 0
\end{array}
$$

In fact, Theorem 6.5.2 will imply that there is a commutative diagram connecting the two long exact sequences.

Theorem 6.6.3. *Suppose $f : (X, A) \to (Y, B)$ is continuous. Then there is the following commutative diagram linking the long exact sequences:*

$$\cdots \longrightarrow H_k(A) \xrightarrow{i_*} H_k(X) \xrightarrow{j_*} H_k(X, A) \xrightarrow{\partial} H_{k-1}(A) \longrightarrow \cdots$$
$$\downarrow f_* \qquad\qquad \downarrow f_* \qquad\qquad \downarrow f_* \qquad\qquad \downarrow f_*$$
$$\cdots \longrightarrow H_k(B) \xrightarrow{i_*} H_k(Y) \xrightarrow{j_*} H_k(Y, B) \xrightarrow{\partial} H_{k-1}(A) \longrightarrow \cdots$$

Exercise 6.6.10. Prove Theorem 6.6.3 by showing that there is a commutative diagram connecting the two short exact sequences of chain complexes for $(X, A), (Y, B)$ as described above.

We next verify the dimension property.

Exercise 6.6.11.
 (a) Show that $S_k(P) \simeq \mathbb{Z}$ for $k \geq 0$ with generator the singular simplex σ_k which sends Δ_k to P.

 (b) Show that
$$\partial_k(\sigma_k) = \begin{cases} 0 & k \text{ odd or } k = 0, \\ \sigma_{k-1} & k \text{ even}, k \neq 0. \end{cases}$$

 (c) Prove the dimension property
$$H_k(P) = \begin{cases} 0 & k \neq 0, \\ \mathbb{Z} & k = 0. \end{cases}$$

We have shown the functorial, exactness, and dimension properties of singular homology. We postpone to later sections the homotopy and excision properties since they are substantially more difficult. However, we will be assuming them in the remainder of the chapter.

6.7   Reformulation of excision and the Mayer–Vietoris exact sequence

We now discuss a reformulation of the excision property of homology and the closely related Mayer–Vietoris exact sequence. As we stated it, excision concerns an isomorphism between the homology of (X, A) and $(X \backslash U, A \backslash U)$, where $\bar{U} \subset$ int A. We want to first rephrase the statement in terms of open covers and then show how the rephrased version leads to the excision theorem. For simplicity, we will use a cover with two sets A, B whose interiors cover X although the ideas work for a general cover by sets whose interiors cover X. We denote $S^{\{A,B\}}(X) = S(A) + S(B)$. By this notation we mean the sum as a subchain complex of $S(X)$, not the direct sum. There is a homomorphism $S^{\{A,B\}}(X) \to S(X)$ which is induced by the two inclusion maps $S(A), S(B) \to S(X)$. The key fact needed for excision is the following theorem.

Theorem 6.7.1. *If* $\{\text{int } A, \text{int } B\}$ *is an open cover of X, then the homomorphism* $S^{\{A,B\}}(X) \to S(X)$ *induces an isomorphism* $H_k^{\{A,B\}}(X) \to H_k(X)$ *in homology for all $k \geq 0$.*

To see how this relates to our statement of excision, we use the cover $\{A, X \setminus U\}$. The hypothesis in excision that $\bar{U} \subset \text{int } A$ means that $\{\text{int } A, \text{int}(X \setminus U)\}$ is an open cover of X. Then Theorem 6.7.1 says that the map $S(A) + S(X \setminus U) \to S(X)$ induces an isomorphism in homology. There is a similar statement for the quotient $(S(A) + S(X \setminus U))/S(A) \to S(X)/S(A)$, which can be proved via a relative version of the theorem or through some homological algebra. But the quotient $(S(A) + S(X \setminus U))/S(A)$ is naturally isomorphic to $S(X \setminus U)/S(A \setminus U)$ via a Noether isomorphism theorem from algebra. The combination of the isomorphism on homology induced by $S(X \setminus U)/S(A \setminus U) \to (S(A) + S(X \setminus U))/S(A)$ and the isomorphism on homology induced by $(S(A) + S(X \setminus U))/S(A) \to S(X)/S(A)$ gives the excision isomorphism $H_k(X \setminus U, A \setminus U) \to H_k(X, A)$. We will prove Theorem 6.7.1 in Section 6.17, and now pursue its implications and reformulations of excision.

Here is a useful restatement of excision.

Theorem 6.7.2. *Suppose $X = A \cup B$, where $\text{int } A$, $\text{int } B$ cover X. Then there is an isomorphism $H_k(B, A \cap B) \to H_k(X, A)$ induced by inclusion.*

Exercise 6.7.1. Show that Theorem 6.7.2 is equivalent to the original statement of excision. (Hint: Let $U = X \setminus B$. Be sure to check that the hypotheses of the excision property translate to the hypotheses of Theorem 6.7.2, and vice versa, with this substitution.)

We now look at the Mayer–Vietoris exact sequence, which follows from Theorem 6.7.1.

Theorem 6.7.3 (Mayer–Vietoris exact sequence). *Suppose $X = A \cup B$, where $\{\text{int } A, \text{int } B\}$ is an open cover of X. Then there is a long exact sequence*

$$\cdots \longrightarrow H_k(A \cap B) \overset{i_*}{\longrightarrow} H_k(A) \oplus H_k(B) \overset{j_*}{\longrightarrow} H_k(X) \overset{\delta}{\longrightarrow} H_{k-1}(A \cap B) \overset{i_*}{\longrightarrow} \cdots.$$

The map i_* is given by $i_*(x) = ((i_A)_*, -(i_B)_*)$ and the map $j_*(x, y) = (j_A)_*(x) + (j_B)_*(y)$. Here $i_A : A \cap B \to A, i_B : A \cap B \to B, j_A : A \to X, j_B : B \to X$ are the inclusions.

Exercise 6.7.2. Follow the outline to prove Theorem 6.7.3.

(a) First show that there is a short exact sequence

$$0 \longrightarrow S(A \cap B) \overset{i_\sharp}{\longrightarrow} S(A) \oplus S(B) \overset{j_\sharp}{\longrightarrow} S(A) + S(B) \longrightarrow 0.$$

(b) Deduce a long exact sequence in homology

$$\cdots \longrightarrow H_k(A \cap B) \overset{i_*}{\longrightarrow} H_k(A) \oplus H_k(B) \overset{j_*}{\longrightarrow} H_k^{\{A,B\}}(X) \overset{\partial}{\longrightarrow} H_{k-1}(A \cap B) \overset{i_*}{\longrightarrow}$$

from the above short exact sequence.

(c) Use Theorem 6.7.1 to transform the last exact sequence into the Mayer–Vietoris exact sequence.

Exercise 6.7.3. The map $H_k(X) \xrightarrow{\delta} H_{k-1}(A \cap B)$ in the Mayer–Vietoris exact sequence uses the composition $H_k(X) \to H_k^{\{A,B\}}(X) \to H_{k-1}(A \cap B)$. The first map is the isomorphism from Theorem 6.7.1 and the second is the boundary map ∂ from part (b) of the last exercise. Suppose that $x \in H_k(X)$ is represented by a cycle $\alpha + \beta$, where $\alpha \in S_k(A), \beta \in S_k(B)$. Show that

(a) $\partial(\alpha) = -\partial(\beta)$, and these give chains in $S_{k-1}(A \cap B) = S_{k-1}(A) \cap S_{k-1}(B)$;

(b) $\delta(x) = [\partial \alpha]$.

For many applications, it is useful to generalize slightly the hypotheses of the Mayer–Vietoris sequence to include situations where $X = A \cup B$, and int A, int B do not cover X but slight enlargements of them by homotopy equivalent sets do. For example, this will allow us to apply the Mayer–Vietoris sequence to decompositions of the sphere as the union of its hemispheres or to a surface which is a connected sum.

Exercise 6.7.4. Suppose $X = A \cup B = A' \cup B'$ where $A \subset A', B \subset B'$. Suppose the inclusions $A \cap B \to A' \cap B', A \to A', B \to B'$ induce isomorphisms in homology and $X = \text{int } A' \cup \text{int } B'$. By following the outline below, show that there is a Mayer–Vietoris exact sequence

$$\cdots \longrightarrow H_k(A \cap B) \xrightarrow{i_*} H_k(A) \oplus H_k(B) \xrightarrow{j_*} H_k(X) \xrightarrow{\delta} H_{k-1}(A \cap B) \xrightarrow{i_*} \cdots .$$

(a) Use the chain maps induced by inclusions between short exact sequences

$$0 \longrightarrow S(A \cap B) \longrightarrow S(A) \oplus S(B) \longrightarrow S(A) + S(B) \longrightarrow 0$$
$$\downarrow \qquad\qquad\qquad \downarrow \qquad\qquad\qquad \downarrow$$
$$0 \longrightarrow S(A' \cap B') \longrightarrow S(A') \oplus S(B') \longrightarrow S(A') + S(B') \longrightarrow 0$$

to get a diagram linking long exact sequences

$$\cdots \longrightarrow H_k(A \cap B) \xrightarrow{i_*} H_k(A) \oplus H_k(B) \xrightarrow{j_*} H_k^{\{A,B\}}(X) \xrightarrow{\delta} H_{k-1}(A \cap B) \xrightarrow{i_*} \cdots$$
$$\cdots \longrightarrow H_k(A' \cap B') \xrightarrow{i'_*,j'_*} H_k(A') \oplus H_k(B') \xrightarrow{j'_*} H_k^{\{A',B'\}}(X) \xrightarrow{\delta} H_{k-1}(A' \cap B') \xrightarrow{i'_*} \cdots$$

(b) Use the five lemma and the hypotheses to deduce that all vertical maps are isomorphisms, so in particular $H_k^{\{A,B\}}(X) \to H_k^{\{A',B'\}}(X)$ is an isomorphism.

(c) Combine (b) with the Mayer–Vietoris exact sequence for $\{A', B'\}$ to get the result.

Exercise 6.7.5. Suppose a surface $S = A \cup B$ is decomposed into two surfaces with boundary $(A, \partial A), (B, \partial B)$ with $A \cap B = \partial A = \partial B$. Suppose there is a closed neighborhood N of $A \cap B$ so that $(N, N \cap A, N \cap B, N \cap A \cap B)$ is homeomorphic to $(\partial A \times [-1, 1], \partial A \times [-1, 0], \partial A \times [0, 1], \partial A \times \{0\})$. The set N comes from collars in A, B and is called a bicollar neighborhood. Use the last exercise to show that there is a Mayer–Vietoris exact sequence based on $X = A \cup B$.

An n-manifold A with boundary ∂A possesses an interior collar $(C, \partial A) \simeq (\partial A \times [0, 1], \partial A \times \{0\})$. Frequently, A is a subset of a larger n-manifold M so that M splits as $M = A \cup B$, where A, B are n-manifolds with common boundary $\partial A = \partial B$ whose interior collars piece together to give a bicollar neighborhood N so that

$$(N, N \cap A, N \cap B, N \cap A \cap B) \simeq (\partial A \times [-1, 1], \partial A \times [-1, 0], \partial A \times [0, 1], \partial A \times \{0\}).$$

Exercise 6.7.6. Generalize the last exercise to decompositions of n-manifolds for arbitrary n.

There are also parallel results for both forms of excision. They are proved by similar techniques as the last exercises.

Exercise 6.7.7. Suppose $U' \subset U \subset A \subset X$

- $\bar{U}' \subset \operatorname{int} A$;
- the inclusion $(X \backslash U, A \backslash U) \to (X \backslash U', A \backslash U')$ induces an isomorphism in homology.

Then the inclusion $(X \backslash U, A \backslash U) \to (X, A)$ induces an isomorphism in homology.

As an example for the last exercise, consider $X = S^2, A$ equal to the lower hemisphere S_-^2, and $U = \operatorname{int} A$. We then choose for $U' = S_-^2 \backslash C$, where C is an interior collar within the lower hemisphere so that (C, S^1) is homeomorphic to $(S^1 \times [0, 1], S^1 \times 0)$, where $\partial S_-^2 = S^1$ is being sent to $S^1 \times 0$. Using spherical coordinates (θ, ϕ), we can specify U' by requiring ϕ to satisfy $\phi > 3\pi/4$, where the north pole corresponds to $\phi = 0$ and the south pole to $\phi = \pi$. We cannot excise U directly from S_+^2 since $\bar{U}$ is not contained in $\operatorname{int} S_+^2$. However, we can excise U'. Moreover, the inclusion $(S^2 \backslash U, S_-^2 \backslash U) \to (S^2 \backslash U', S_-^2 \backslash U')$ is a homotopy equivalence since $(S^2 \backslash U', S_-^2 \backslash U') = ((S^2 \backslash U) \cup C, (S_-^2 \backslash U) \cup C)$ and we can just deformation retract $C \simeq S^1 \times [0, 1]$ back to $S^1 \times \{0\}$, keeping the identity on $S^2 \backslash U = S_+^2$. Here the collar corresponds to $\phi \in [\pi/2, 3\pi/4]$ and we are using the linear homeomorphism between $[\pi/2, 3\pi/4]$ and $[0, 1]$ in expressing it as homeomorphic to $S^1 \times [0, 1]$.

Exercise 6.7.8. Suppose $M = A \cup B$ is an n-manifold and $A, B \subset M$ are embedded n-manifolds with common boundary. Assume that the interior collars of A, B piece together to give a bicollar neighborhood N so that $(N, N \cap A, N \cap B, N \cap A \cap B) \simeq (\partial A \times (-1, 1), \partial A \times (-1, 0], \partial A \times [0, 1), \partial A \times \{0\})$. Show that the inclusion maps induce an isomorphism $H_k(B, \partial B) \to H_k(M, A)$.

What is occurring is this last case is part of a more general result.

Exercise 6.7.9. Suppose $A \cup B = A' \cup B' = X$ and $A \subset A', B \subset B'$. Suppose int $A' \cup$ int $B' = X$ and the inclusion $(B, A \cap B) \to (B', A' \cap B')$ induces an isomorphism in homology. Then the inclusion $(B, A \cap B) \to (X, A)$ induces an isomorphism in homology.

These last exercises allow us to apply excision and the Mayer–Vietoris sequence more generally than its original statement. We will use these refined forms in our applications in succeeding sections.

6.8 Applications of singular homology

Most calculations using singular homology are a consequence of applying the basic properties rather than using the definition directly. We illustrate this in this section with a few basic applications.

We first use the basic properties to give the homology of a disk, a sphere, and a disk–sphere pair (D^{n+1}, S^n).

Exercise 6.8.1. Use Theorem 6.6.2 and the dimension property to show that

$$H_k(D^n) = \begin{cases} \mathbb{Z} & k = 0, \\ 0 & k > 0. \end{cases}$$

Exercise 6.8.2. Show that

$$H_k(S^0) = \begin{cases} \mathbb{Z} \oplus \mathbb{Z} & k = 0, \\ 0 & k > 0. \end{cases}$$

Exercise 6.8.3. Use the long exact sequence of the pair (D^1, S^0) to show that

$$H_k(D^1, S^0) = \begin{cases} \mathbb{Z} & k = 1, \\ 0 & k \neq 1. \end{cases}$$

Exercise 6.8.4. Give an explicit singular simplex $\sigma : \Delta_1 \to D^1$ which represents the generator of $H_1(D^1, S^0)$. (Hint: It should map via ∂ to a generator of the kernel of $i_*^0 : H_0(S^0) \to H_0(D^1)$.)

Exercise 6.8.5. Show that there is a homeomorphism $h : D^1 \to S_+^1 = \{x \in S^1 : x_2 \geq 0\}$ and use this to compute $H_k(S_+^1, S^0)$ for all $k \geq 0$.

Exercise 6.8.6. Suppose $B = S_+^1 \subset S^1$ is the upper semicircle $x_2 \geq 0$ and $A = S_-^1$ is the lower semicircle.

(a) Use Exercise 6.7.7 to show that there is an isomorphism $H_k(S^1_+, S^0) \to H_k(S^1, S^1_-)$. Indicate how to find the sets U, U' used in the exercise.

(b) Use Exercise 6.7.8 to show that there is an isomorphism $H_k(S^1_+, S^0) \to H_k(S^1, S^1_-)$. Indicate how to find the required bicollar neighborhood.

Exercise 6.8.7. Use the two previous exercises to compute $H_k(S^1, S^1_-)$.

Exercise 6.8.8. Give an explicit singular simplex $\sigma : \Delta_1 \to S^1$ which represents a generator of $H_1(S^1, S^1_-)$. (Hint: The essential ingredient of a generator is that it corresponds to a generator of $H_1(S^1_+, S^0)$, and this generator must map to a generator of $\ker(i^0_*) \subset H_0(S^0)$. From our earlier work, $\ker(i^0_*)$ is generated by $([1] - [-1])$. There is a simple trigonometric formula that works for σ.)

Exercise 6.8.9. Use the previous exercises to compute $H_k(S^1)$ by using the long exact sequence of the pair (S^1, S^1_-).

Exercise 6.8.10. Give an explicit singular chain which is a sum of two singular simplices, one of which is the answer to Exercise 6.8.8 and the other is a singular simplex in S^1_-, which represents a generator of $H_1(S^1)$. Explain how this relates to the isomorphism between $H_1(S^1)$ and $\pi_1(S^1, 1)$.

The argument so far has shown that there is a chain of isomorphisms

$$H_{k+1}(S^1) \simeq H_{k+1}(S^1, S^1_-) \simeq H_{k+1}(S^1_+, S^0) \simeq H_{k+1}(D^1, S^0).$$

When $k > 0$, there is also an isomorphism $H_{k+1}(D^1, S^0) \simeq H_k(S^0)$. When $k = 0$, the argument is slightly different since S^0 is not path connected, and we use the isomorphism $H_1(D^1, S^0) \simeq \ker(i^0_*)$ instead. These arguments all generalize to compute the homology of S^p and (D^{p+1}, S^p) for $p \geq 1$ via an inductive argument. Before starting with the general argument, note that S^p, D^{p+1} are path connected, so $H_0(S^p) \simeq H_0(D^{p+1}) \simeq \mathbb{Z}$. The portion of the exact sequence

$$H_0(S^p) \to H_0(D^{p+1}) \to H_0(D^{p+1}, S^p) \to 0$$

shows that $H_0(D^{p+1}, S^p) = 0$ since the first map is an isomorphism.

Exercise 6.8.11. Suppose $p \geq 1$. Show that $H_{k+1}(D^{p+1}, S^p) \simeq H_k(S^p), k > 0$. Show that $H_1(D^{p+1}, S^p) = 0$.

Exercise 6.8.12. Suppose $p \geq 1$. Show that $H_k(D^{p+1}, S^p) \simeq H_k(S^{p+1}_+, S^p), k \geq 0$. Here S^{p+1}_+ denotes the part of the sphere where the last coordinate is ≥ 0.

Exercise 6.8.13. Suppose $p \geq 1$. Show that $H_k(S^{p+1}_+, S^p) \simeq H_k(S^{p+1}, S^{p+1}_-), k \geq 0$.

Exercise 6.8.14. Suppose $p \geq 1$. Show that $H_k(S^{p+1}) \simeq H_k(S^{p+1}, S^{p+1}_-), k \geq 1$.

Exercise 6.8.15. Suppose $p \geq 1$. Combine the last exercises to show that $H_{k+1}(S^{p+1}) \simeq H_k(S^p)$ for $k \geq 1$, and $H_1(S^{p+1}) = 0$.

Exercise 6.8.16. Use an inductive argument to show that

(a)
$$H_k(S^n) = \begin{cases} \mathbb{Z} \oplus \mathbb{Z} & n = k = 0, \\ \mathbb{Z} & n = k > 0 \text{ or } n > 0, k = 0, \\ 0 & \text{otherwise;} \end{cases}$$

(b)
$$H_k(D^n, S^{n-1}) = \begin{cases} \mathbb{Z} & k = n \geq 0, \\ 0 & \text{otherwise.} \end{cases}$$

Later on, we will introduce reduced homology $\tilde{H}_k(X)$ and this will sim-plify the statements above for low-dimensional homology to give isomorphisms $H_{k+1}(D^{p+1}, S^p) \simeq \tilde{H}_k(S^p)$ and $\tilde{H}_{k+1}(S^{p+1}) \simeq \tilde{H}_k(S^p)$.

We now apply these calculations to generalize the result from Chapter 3 that a 2-disk cannot retract onto its boundary circle. We first recall the definition. If $A \subset X$ with inclusion map i, then a map $r : X \to A$ is called a retraction if $ri = 1_A$.

Exercise 6.8.17. Show that, if $r : X \to A$ is a retraction, then $r_* : H_k(X) \to H_k(A)$ is surjective for each k.

Exercise 6.8.18. Show that, if there is a k with $H_k(X) = 0$ and $H_k(A) \neq 0$, then there is no retraction of X onto A.

Exercise 6.8.19. Show that D^n does not retract onto S^{n-1}, $n \geq 1$.

Exercise 6.8.20. Show that if $f : D^n \to D^n$ is continuous, then there is a fixed point $x \in D^n$ with $f(x) = x$. (Hint: See the proof of Theorem 3.5.4.)

Exercise 6.8.21.
  (a) If M denotes the Möbius band with boundary ∂M, show that $H_1(\partial M)$ and $H_1(M)$ are each isomorphic to $\mathbb{Z}$, but the induced map i_* is given by multiplication by 2.
  (b) Use (a) to show that a Möbius band does not retract onto its boundary even though they have isomorphic homology.

6.9   The degree of a map $f : S^n \to S^n$

We discuss in this section the degree of a continuous map from S^n to S^n. Here we first take $n > 0$ and modify the definition for $n = 0$. We showed above that $H_n(S^n) = \mathbb{Z}$. Let g denote a generator. If $f : S^n \to S^n$ is continuous, then $f_*(g) = dg$ for some integer d. The *degree* of f is defined to be the integer $d = \deg(f)$ with $f_*(g) = \deg(f)g$. We note the following elementary facts about degree:

- $\deg(1) = 1$;
- $\deg(hf) = \deg(h)\deg(f)$;

- if f is homotopic to f', then $\deg(f) = \deg(f')$;
- if f is a homotopy equivalence, then $\deg(f) = \pm 1$.

Exercise 6.9.1. Prove the facts cited above about degree. Also, show that the degree does not depend on which generator of $H_n(S^n) = \mathbb{Z}$ you choose.

When $n = 0$, we have $H_0(S^0) = \mathbb{Z} \oplus \mathbb{Z}$, so the above definition of degree does not apply. However, there is a modification which can be made to define degree in this case. We look at the map induced by inclusion $i_*^0 : H_0(S^0) \to H_0(D^1)$. Then $\ker(i_*^0)$ is generated by $[1] - [-1]$, where $[x]$ denotes the homology class of the singular 0-simplex which maps to the point x. Moreover, any map $f : S^0 \to S^0$ must map this generator to a multiple of itself since if both points are sent to the same point, the induced map is 0, and if not, it is multiplication by ± 1. In defining degree in dimension 0, we replace $H_0(S^0)$ by $\ker(i_*^0) \simeq \mathbb{Z}$. Thus we can look at $g = [1] - [-1]$ and define $\deg(f)$ by the same formula, $f_*(g) = \deg(f)g$. The map $r(x) = -x$ is an important example of a map of degree -1.

When $f : (D^{n+1}, S^n) \to (D^{n+1}, S^n)$, there is a similar definition of degree which uses $H_{n+1}(D^{n+1}, S^n) \simeq \mathbb{Z}$. If g is a generator here, define degree by the same formula, $f_*(g) = \deg(f)g$.

Exercise 6.9.2. Show that $\deg(f)$ as defined above is the same as $\deg(f|S^n)$. (Hint: Use the exact sequence for the pair (D^{n+1}, S^n) to relate the two computations.)

Exercise 6.9.3. Suppose $f : (S^{n+1}, S_+^{n+1}, S_-^{n+1}, S^n) \to (S^{n+1}, S_+^{n+1}, S_-^{n+1}, S^n)$. We can define $\deg(f)$ and $\deg(f|S^n)$. Show that they are equal. (Hint: You need to relate the generators. Consider how we computed the homology of spheres inductively. Use isomorphisms

$$H_{n+1}(S^{n+1}) \simeq H_{n+1}(S^{n+1}, S_-^{n+1}) \simeq H_{n+1}(S_+^{n+1}, S^n) \simeq H_n(S^n),$$

when $n > 0$ and an appropriate modification when $n = 0$.)

Exercise 6.9.4. There is a reflection of any sphere of dimension greater than or equal to 0 given by $r_1 : S^n \to S^n$, where $r_1(x_1, x_2, \ldots, x_{n+1}) = (-x_1, x_2, \ldots, x_{n+1})$. By starting with the case of S^0 show that the degree of r_1 is -1 for all n.

Suppose $r : S^n \to S^n$ is a reflection. It is of the form $r(\boldsymbol{x}) = \boldsymbol{x} - 2(\boldsymbol{x} \cdot \boldsymbol{v})\boldsymbol{v}$, where $\boldsymbol{v} \in S^n$ is the normal vector of the plane through which you are reflecting. For example, for r_1 above, $\boldsymbol{v} = \boldsymbol{e}_1$. For $r_2(x_1, x_2, \ldots, x_{n+1}) = (x_1, -x_2, \ldots, x_{n+1})$, then $\boldsymbol{v} = \boldsymbol{e}_2$.

Exercise 6.9.5. Let $n > 0$.

(a) Use the fact that S^n is path connected to show that any reflection r is homotopic to r_1.

(b) Show that the degree of a reflection of S^n is -1.

The antipodal map $A : S^n \to S^n$ is $A(\boldsymbol{x}) = -\boldsymbol{x}$. It can be written as the composition of $n + 1$ reflections in the individual coordinates.

Exercise 6.9.6. Show that the degree of the antipodal map of S^n is $(-1)^{n+1}$.

Exercise 6.9.7. Show that the antipodal map of S^n is homotopic to the identity iff n is odd. (Hint: First consider the case of S^1 using a rotation to get a homotopy between the antipodal map and the identity. Then generalize this argument to take care of other odd-dimensional spheres. It is useful to write an element of an odd-dimensional sphere as $(z_1, \ldots, z_k)$, where $z_i = (x_{2i-1}, x_{2i})$ is identified with a complex number.)

We noted that we can also define degree in the context of maps $(D^n, S^{n-1}) \to (D^n, S^{n-1})$ since the nth homology is also $\mathbb{Z}$. This fact is useful in defining the notion of local degree of a map. Note that if D_x denotes a small disk about x, then $D'_x = S^n \backslash \text{int} \, D_x$ is also a disk that is contractible to a point. Then the long exact sequence of the pair says that $H_n(S^n) \simeq H_n(S^n, D'_x)$. Note that by excision $H_n(S^n, D'_x) \simeq H_n(D_x, \partial D_x)$. Alternatively, we can use the long exact sequence to show that $H_n(S^n) \simeq H_n(S^n, S^n \backslash \{x\})$, which is isomorphic to $H_n(D_x, \partial D_x)$ by excision. We will use the notation $S_x = S^n \backslash \{x\}$. We denote by g_x the image of g in $H_n(S^n, S_x)$.

Suppose that $f(x) = y$ and we have a small disk neighborhood D_x about x so that f restricts to give a map $f : (D_x, D_x \backslash \{x\}) \to (S^n, S_y)$. Then excision gives an isomorphism $H_n(D_x, D_x \backslash \{x\}) \simeq H_n(S^n, S_x)$. Then g_x determines via this isomorphism a generator $g_x^D \in H_n(D_x, D_x \backslash \{x\})$, and so we can form $f_*(g_x^D)$. It is a multiple of the generator g_y. We define this multiple as $\deg_x(f)$, and call it the *local degree* of f at x.

Exercise 6.9.8. Show that the local degree is well defined, independent of which disk D_x is chosen. (Hint: First show that if we choose a smaller disk, then there is a commutative diagram showing that the local degrees are the same. Then use the fact that any two disks contain a third disk in their intersection.)

Exercise 6.9.9. Show that the following alternative definition gives the same answer. Choose a small disk D_y about y so that $f(D_x) \subset D_y$ and $f : (D_x, D_x \backslash \{x\}) \to (D_y, D_y \backslash \{y\})$. Then define $\deg_x(f)$ by $f_*(g_x^D) = \deg_x(f) g_y^D$.

Exercise 6.9.10. Show that the local degree has the following properties analogous to the degree:

- $\deg_x(1) = 1$;
- if $f(x) = y, h(y) = z$, and there are disks D_x, D_y, D_z with $f : (D_x, D_x \backslash \{x\}) \to (D_y, D_y \backslash \{y\}), h : (D_y, D_y \backslash \{y\}) \to (D_z, D_z \backslash \{z\})$, then

$$\deg_x(hf) = \deg_y(h) \deg_x(f);$$

- if f is a local homeomorphism at x so that there is a disk D_x which is sent homeomorphically to a disk D_y, then $\deg_x(f) = \pm 1$.

In most applications of local degree, the original map will be modified up to homotopy so that it is a local homeomorphism when the local degree is being calculated. What the local degree is measuring there is whether the map is locally preserving or reversing the orientation. In a later section, we will discuss orientation in terms of homology.

We now begin to relate the definition of the local degree and the degree of the map. We first note a special case when $f^{-1}(\{y\}) = \{x\}$.

Exercise 6.9.11. Suppose $f : (S^n, S_x) \to (S^n, S_y)$. Then $f_*(g_x) = \deg(f)g_y$ and thus $\deg(f) = \deg_x(f)$.

It is usually not the case that $f^{-1}(\{y\}) = \{x\}$ but there are geometric techniques to homotope f so that $f^{-1}(\{y\})$ is a finite number of points $x_1, \ldots, x_k$. Then by choosing small disks about these points, we can define the local degrees $\deg_{x_i}(f)$.

Theorem 6.9.1. *When* $f^{-1}(\{y\}) = \{x_1, \ldots, x_k\}$, *then* $\deg(f) = \sum_{i=1}^{k} \deg_{x_i}(f)$.

Note that the hypothesis implies that $f : (S^n, S^n \backslash \{x_1, \ldots, x_k\}) \to (S^n, S_y)$. There is a commutative diagram

$$
\begin{array}{ccc}
H_n(S^n) & \xrightarrow{\ f_*\ } & H_n(S^n) \\
\downarrow{\scriptstyle e_x} & & \downarrow{\scriptstyle e_y} \\
H_n(S^n, S^n \backslash \{x_1, \ldots, x_k\}) & \xrightarrow{\ f_*\ } & H_n(S^n, S_y)
\end{array}
$$

Exercise 6.9.12. Show that $H_n(S^n, S^n \backslash \{x_1, \ldots, x_k\}) \simeq \bigoplus_{i=1}^{k} H_n(D_{x_i}, D_{x_i} \backslash \{x_i\})$ and, using the identification of these two, $e_x(g) = \sum_{i=1}^{k} g_{x_i}^D$.

Exercise 6.9.13. Use the last exercise and the commutative diagram preceding it to prove Theorem 6.9.1.

We now do a computation which we will use in the next section.

Exercise 6.9.14. Suppose that $T : S^n \to S^n$ is a homeomorphism and $f : S^n \to S^n$ satisfies $fT(x) = f(x)$. Suppose the local degree $\deg_x(f)$ of f is defined at x. Then show that $\deg_{T(x)}(f)$ is also defined and

$$
\deg_x(f) = \deg(T) \deg_{T(x)}(f).
$$

Exercise 6.9.15. Consider the quotient map $q : S^n \to S^n/(x \sim -x) = \mathbb{RP}^n$. We compose with the map $\mathbb{RP}^n \to \mathbb{RP}^n/\mathbb{RP}^{n-1}$ which identifies $\mathbb{RP}^{n-1}$ to a point, and identify this quotient as S^n since it is homeomorphic to a disk with its boundary sphere identified to a point. This composition gives a map $S^n \to S^n$. Use the preceding exercise and Theorem 6.9.1 to show that the degree of this map is $\pm(1 + (-1)^{n+1})$.

6.10 Cellular homology of a CW complex

In Section 6.4 we gave a description of the cellular homology of a two-dimensional CW complex. We now define the cellular homology more formally in terms of singular homology for any finite CW complex and will verify that our new definition agrees with the one given earlier in the two-dimensional case. In the next section we will show that cellular homology agrees with singular homology for a finite CW complex.

We denote by X^k the k-skeleton of X, which is the union of the cells of dimension less than or equal to k. Recall that the k-cell e_j^k is a unit disk (up to homeomorphism) which is attached to X^{k-1} via an attaching map $f_j : S^{k-1} \to X^{k-1}$. We can extend X^{k-1} to a larger set K^{k-1} in X^k by attaching $L^{k-1} = D^k \backslash (\text{int } \frac{1}{2} D^k)$ via f_i for each k-cell. We will use this set for excision purposes. There is a corresponding pair (D_j^k, L_j^{k-1}) for each k-cell and call the corresponding disjoint unions $(D_k, L_{k-1}) = (\bigsqcup_{j=1}^{c_k} D_j^k, \bigsqcup_{j=1}^{c_k} L_j^{k-1})$ as well as $(D_k, S_{k-1}) = (\bigsqcup_{j=1}^{c_k} D_j^k, \bigsqcup_{j=1}^{c_k} S_j^{k-1})$.

Exercise 6.10.1.
   (a) Show the inclusion $(D^k, S^{k-1}) \to (D^k, L^{k-1})$ induces an isomorphism in homology. (Hint: Show that there is a homotopy equivalence of pairs.)
   (b) Show that the inclusion $(X^k, X^{k-1}) \to (X^k, K^{k-1})$ induces an isomorphism in homology. (Hint: See the hint above.)

Exercise 6.10.2.
   (a) Show that if $(Y, B) = (\bigsqcup_{j=1}^N Y_j, \bigsqcup_{j=1}^N B_j)$ is a disjoint union of path components (Y_j, B_j), then $H_p(Y, B) = \oplus_{j=1}^N H_p(Y_j, B_j)$.
   (b) Show that $H_k(D_k, S_{k-1}) \simeq \oplus_{j=1}^{c_k} \mathbb{Z}$. Call the generator g_j^k that corresponds to a generator of $H_k(D_j^k, S_j^{k-1})$.
   (c) Show that the inclusion $H_p(D_k, S_{k-1}) \to H_p(D_k, L_{k-1})$ induces an isomorphism in homology. Conclude that
$$H_p(D_k, L_{k-1}) = \oplus_{j=1}^{c_k} H_p(D_j^k, L_j^{k-1}) \simeq \oplus_{j=1}^{c_k} \mathbb{Z}.$$

Exercise 6.10.3. Show that the restriction of the characteristic maps to the k-cells gives a homeomorphism of pairs $(D_k \backslash \text{int } L_{k-1}, L_{k-1} \backslash \text{int } L_{k-1}) \to (X^k \backslash K^{k-1}, K^{k-1} \backslash \text{int } K^{k-1})$. Conclude that there is an isomorphism in homology.

Exercise 6.10.4. Show that there is a commutative diagram where all maps are isomorphisms. The map $\phi = \bigsqcup \phi_j$ is the disjoint union of the characteristic maps of the k-cells which restricts to the attaching maps on the boundary spheres. The other horizontal maps are induced by restrictions of this map. The vertical maps are inclusions. Use the previous exercises to justify why each map is an isomorphism.

$$
\begin{array}{ccc}
H_p(D_k, S_{k-1}) & \xrightarrow{\ \phi_*\ } & H_p(X^k, X^{k-1}) \\
{\scriptstyle (i_L)_*}\downarrow & & \downarrow{\scriptstyle (i_K)_*} \\
H_p(D_k, L_{k-1}) & \xrightarrow{\ (\phi_L)_*\ } & H_p(X^k, K^{k-1}) \\
{\scriptstyle (j_L)_*}\uparrow & & \uparrow{\scriptstyle (j_K)_*} \\
H_p(D_k \backslash \text{int } L_{k-1}, L_{k-1} \backslash \text{int } L_{k-1}) & \xrightarrow{\ (\phi_L^-)_*\ } & H_p(X^k \backslash K^{k-1}, K^{k-1} \backslash \text{int } K^{k-1})
\end{array}
$$

When we combine the results of Exercises 6.10.3 and 6.10.4, they say that the characteristic maps induce an isomorphism $\oplus_{j=1}^{c_k} H_p(D_j^k, S_j^{k-1}) \to H_p(X^k, X^{k-1})$. In particular, all terms are 0 unless $p = k$. When $p = k$, the generator g_j^k of the term $H_p(D_j^k, S_j^{k-1})$ is mapped to a generator x_j^k of $H_k(X^k, X^{k-1})$. Thus $H_k(X^k, X^{k-1})$ is a free abelian group with generators $x_1^k, \ldots, x_{c_k}^k$.

For the next few exercises, we will be assuming that $k > 0$. Recall that the quotient space D^k/S^{k-1} is homeomorphic with S^k. When we start with the disjoint union (D_k, S_{k-1}) and form the quotient space D_k/S_{k-1}, we get the space formed from c_k copies of S^k where base points in each sphere (coming from the S_j^{k-1} terms which are collapsed to a point) are all identified to a single base point $*$. Thus the quotient space is $\vee_{j=1}^{c_k} S_j^k$. From a CW decomposition point of view, there is one 0-cell coming from the identified spheres and then c_k k-cells. When we form the quotient space X^k/X^{k-1}, this has the same CW decomposition and the map ϕ induces a homeomorphism $\bar{\phi}$. Thus there is an isomorphism $H_p(\vee_{j=1}^{c_k} S_j^k, *) \simeq H_p(X^k/X^{k-1}, *)$.

Exercise 6.10.5. Show that when $k > 0$, $H_k(\vee_{j=1}^{c_k} S_j^k, *) \simeq \oplus_{j=1}^{c_k} H_k(S_j^k, *)$. (Hint: Use an inductive argument with the Mayer–Vietoris sequence.)

Any element y of this direct sum is expressible as $\sum_{j=1}^{c_k} n_j y_j$, where y_j is a chosen generator of $H_k(S_j^k, *)$. Then the inclusion of the jth sphere into the wedge induces algebraically the inclusion of the term y_j in the direct sum. Let A_j represent the other spheres except for the jth one. When we form the quotient $\vee_{j=1}^{c_k} S_j^k/A_j$, we just get a homeomorphic copy of S_j^k. In this case the element y maps to $n_j y_j$. This uses the fact that the composition $S_j^k \to \vee_{j=1}^{c_k} S_j^k \to \vee_{j=1}^{c_k} S_j^k/A_j = S_j^k$ is the identity and the composition $S_p^k \to \vee_{j=1}^{c_k} S_j^k \to \vee_{j=1}^{c_k} S_j^k = S_j^k$ maps S_p^k to the base point $*$ when $p \neq j$. Since the identity induces the identity map and the constant map to the base point induces the zero map, this gives the algebraic result.

Exercise 6.10.6. Verify that the map $S_p^k \to S_j^k, p \neq j$ which sends the whole sphere to the base point induces the zero map in homology in dimension $k > 0$.

Exercise 6.10.7. Let $A_j \subset D_k, B_j \subset X^k$ denote the subcomplexes consisting of everything except the jth k-cell. We can form a quotient spaces $D_k/A_j, X^k/B_j$ by collapsing A_j and B_j to a point. Show that the characteristic map $D_j^k \to D_k \to X^k$ induces homeomorphisms between the quotient spaces $D_j^k/S_j^{k-1} \to D_k/A_j \to X^k/B_j$.

We next want to show that the maps $(D_k, S_{k-1}) \to (D_k/S_{k-1}, *)$ and $(X^k, X^{k-1}) \to (X^k/X^{k-1}, *)$ induce (consistent) isomorphisms in homology. This will use an excision argument. Each sphere S_j^k has a closed neighborhood $\bar{L}_j^{k-1}$ about the base point $*$ which comes from the image of L_j^{k-1} under the collapsing map. The union of all of these in $\vee_{j=1}^{c_k} S_j^k$ provides a closed neighborhood $\bar{L}^{k-1}$ of the base point $*$. Analogously, the set K^{k-1} provides a closed neighborhood $\bar{K}^{k-1}$ of the base point $* \in X^k/X^{k-1}$. The homeomorphism $\bar{\phi}$ sends $\bar{L}^{k-1}$ to $\bar{K}^{k-1}$.

Exercise 6.10.8. Show that $(D_k/S_{k-1}, \bar{L}^{k-1})$ is homotopy equivalent to $(D_k/S_{k-1}, *)$, and $(X^k/X^{k-1}, \bar{K}^{k-1})$ to $(X^k/X^{k-1}, *)$.

Exercise 6.10.9. Show that the restriction of the quotient map gives a homeomorphism $(X^k\backslash\text{int } K^{k-1}, K^{k-1}\backslash\text{int } K^{k-1}) \rightarrow (X^k/X^{k-1}\backslash\text{int } \bar{K}^{k-1}, \bar{K}^{k-1}\backslash\text{int } \bar{K}^{k-1})$.

Exercise 6.10.10.

(a) Suppose $k > 0$. Show that there is a commutative diagram where all maps are isomorphisms. Horizontal maps are induced by inclusions and vertical maps are induced by quotient maps.

$$\begin{array}{ccccc}
H_k(X^k, X^{k-1}) & \longrightarrow & H_k(X^k, K^{k-1}) & \longleftarrow & H_k(X^k\backslash\text{int } K^{k-1}, K^{k-1}\backslash\text{int } K^{k-1}) \\
\downarrow & & \downarrow & & \downarrow \\
H_k(X^k/X^{k-1}, *) & \longrightarrow & H_k(X^k/X^{k-1}, \bar{K}^{k-1}) & \longleftarrow & H_k(X^k/X^{k-1}\backslash\text{int } \bar{K}^{k-1}, \bar{K}^{k-1}\backslash\text{int } \bar{K}^{k-1})
\end{array}$$

(b) Deduce a similar result for (D_k, S_{k-1}) in place of (X^k, X^{k-1}).

(c) Show that there is a commutative diagram, where all maps are isomorphisms.

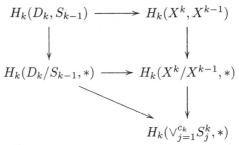

$$\begin{array}{ccc}
H_k(D_k, S_{k-1}) & \longrightarrow & H_k(X^k, X^{k-1}) \\
\downarrow & & \downarrow \\
H_k(D_k/S_{k-1}, *) & \longrightarrow & H_k(X^k/X^{k-1}, *) \\
& \searrow & \downarrow \\
& & H_k(\vee_{j=1}^{c_k} S_j^k, *)
\end{array}$$

We now define a generator y_j^k of $H_k(\vee_{j=1}^{c_k} S_j^k, *)$ coming from the image of the generator $g_j^k \in H_k(D_k, S_{k-1})$. Using the commutative diagram above, this is the image of x_j^k via the vertical maps on the right. If we name the map P_j that collapses all but the jth sphere of $\vee_{j=1}^{c_k} S_j^k$, then $(P_j)_*(y_j^k) = y_j^k$ and $(P_m)_*(y_j^k) = 0, m \neq j$ via the discussion before Exercise 6.10.6.

We define the cellular chain groups $C_k^c(X) = H_k(X^k, X^{k-1}), k \geq 0$, where $X^{-1} = \emptyset$ so $C_0^c(X) = H_0(X^0)$. This is a free abelian group with generators x_j^k, one for each k-cell. To define the boundary homomorphisms ∂_k^c we use the composition

$$H_k(X^k, X^{k-1}) \xrightarrow{\partial} H_{k-1}(X^{k-1}) \xrightarrow{j_*} H_{k-1}(X^{k-1}, X^{k-2}) \ .$$

Note that, when $k = 1$, we can just regard the definition as meaning $\partial_1^c = \partial : H_1(X^1, X^0) \to H_0(X^0)$. As usual, ∂_0^c is defined to be 0.

Exercise 6.10.11. Show that $\partial_k^c \partial_{k+1}^c = 0$. (Hint: Use the exactness of the long exact sequence of the pair (X^k, X^{k-1}).)

We now want to re-express this boundary map in a more computable form. Note that ∂_k^c is a mapping from the free abelian group with generators x_j^k to the free abelian group with generators x_i^{k-1}. Thus this homomorphism is determined by the formula $\partial_k^c(x_j^k) = \sum n_{ij} x_i^{k-1}$. To determine the coefficient n_{ij} of x_i^{k-1} we can collapse all cells of X^{k-1} except the ith one to get a sphere S_i^{k-1}. The generator x_i^{k-1} maps to y_i^{k-1} and the other generators map to 0. If we call this collapse p_i then we get $(p_i)_* \partial_k^c(x_j^k) = n_{ij} y_i^{k-1}$. We want to bring in the characteristic map of the k-cell that leads to the generator and so look at the following diagram:

$$
\begin{array}{ccc}
H_k(D_j^k, S^{k-1}) & \xrightarrow{(\phi_j)_*} & H_k(X^k, X^{k-1}) \\
\downarrow{\scriptstyle \partial} & & \downarrow{\scriptstyle \partial} \\
H_{k-1}(S_j^{k-1}) & \xrightarrow{(f_j)_*} & H_{k-1}(X^{k-1})
\end{array}
$$

Since $(\phi_j)_*(g_j^k) = x_j^k$, we can rewrite $\partial_k^c(x_j^k)$ as $i_*(f_j)_*(\partial g_j^k)$. Thus to find n_{ij} we just take $(p_i f_j)_*(\partial g_j^k)$, where $p_i : (X^{k-1}, X^{k-2}) \to (X^{k-1}/B_i, *)$. Another way to state this is to find n_{ij}, we take the induced map on the $(k-1)$st homology from

the composition $\quad S^{k-1} \xrightarrow{\ f_j\ } X^{k-1} \xrightarrow{\ p_i\ } X^{k-1}/B_i = S^{k-1}, \quad$ where the B_i

represents everything but the ith $(k-1)$-cell. Then we are computing the degree of the map $p_i f_j$ between these two spheres. In this case the generator for the first copy is coming from the chosen generator of $H_k(D^k, S^{k-1})$ coming from the jth k-cell after taking its boundary in $H_{k-1}(S^{k-1})$. The generator for the second copy is coming from the chosen generator of $H_{k-1}(D^{k-1}, S^{k-1})$ for the ith $(k-1)$-cell and then taking its image in $H_{k-1}(D^{k-1}, S^{k-1}) \to H_{k-1}(D^{k-1}/S^{k-1}, *) = H_{k-1}(S^{k-1}, *)$. The number n_{ij} then gives the degree of this map with respect to these generators. For the case when $k = 1$, this is simpler since we do not have to worry about collapsing a lower skeleton and so the boundary $\partial_1^c(x_j^1)$ is just computed by taking the composition $H_1(D_j^1, S_j^0) \to H_0(S_j^0) \to H_0(X^0)$. This just takes the difference $[f_j(1)] - [f_j(-1)]$. Note that our descriptions coincide with how we defined the cellular chain complex in the two-dimensional case.

We now summarize our results for future reference.

Theorem 6.10.1. *Let X be a finite CW complex, with X^k its k-skeleton. Then the map induced by the characteristic maps of the k-cells, $\phi_* : H_p(D_k, S_{k-1}) \to H_p(X^k, X^{k-1})$, induces an isomorphism. In particular, $H_p(X^k, X^{k-1}) = 0$ when $p \neq k$, and is the free abelian group on c_k generators $x_1^k, \ldots, x_{c_k}^k$ when $p = k$. The generator $x_j^k = (\phi_j^k)_*(g_j^k)$, where g_j^k generates $H_k(D_j^k, S_j^{k-1})$, and ϕ_j^k is the characteristic map of the jth k-cell. Moreover, for $k > 0$, there is an isomorphism $H_k(X^k, X^{k-1}) \to H_k(X^k/X^{k-1}, *) \simeq H_k(\vee_{j=1}^{c_k} S_j^k, *) \simeq \oplus_{j=1}^{c_k} H_k(S^k, *)$, and the*

*generator x_j^k maps to the generator y_j^k which corresponds to the image of g_j^k
under the map $(D_j^k, S_j^{k-1}) \to (D_j^k/S_j^{k-1}, *) \simeq (S_j^k, *)$.*

*In the cellular chain complex $C_k^c(X) = H_k(X^k, X^{k-1})$, the boundary map ∂_k^c
is characterized by $\partial_k^c(x_j^k) = \sum n_{ij} x_i^{k-1}$, where n_{ij} can be computed for $k > 1$
by taking the degree of the composition $p_i f_j$. Here $p_i : X^{k-1} \to S_i^{k-1}$ is the map
which collapses all but the ith $(k-1)$-cell to a point to form a sphere, and f_j is the
attaching map for the jth k-cell. The computation is done using the generator for
the first copy coming from the chosen generator of $H_k(D^k, S^{k-1})$ coming from the
jth k-cell after taking its boundary in $H_{k-1}(S^{k-1})$. The generator for the second
copy is coming from the chosen generator of $H_{k-1}(D^{k-1}, S^{k-1})$ for the ith $(k-1)$-
cell and then taking its image in $H_{k-1}(D^{k-1}, S^{k-1}) \to H_{k-1}(D^{k-1}/S^{k-1}, *) =
H_{k-1}(S^{k-1}, *)$. The map ∂_1^c is computed using the simpler formula $\partial_1^c(x_j^1) =
[f_j(1)] - [f_j(-1)] \in H_0(X^0)$.*

We have already given computations of the cellular homology in the two-
dimensional case. We look at some computations via the cellular chain complex
for higher-dimensional CW complexes. We first look at the case of $S^2 \times S^2$. Here
there are cells in dimensions $0, 2, 4$, and the cellular chain complex is $0 \to \mathbb{Z} \to
0 \to \mathbb{Z} \oplus \mathbb{Z} \to 0 \to \mathbb{Z}$ in low dimensions, 0–5. Since all boundary maps are 0, the
homology groups are the chain groups. Thus we get

$$H_k^c(S^2 \times S^2) \simeq \begin{cases} \mathbb{Z} & k = 0, 4, \\ \mathbb{Z} \oplus \mathbb{Z} & k = 2, \\ 0 & \text{otherwise.} \end{cases}$$

The computation for $\mathbb{CP}^n$ is similar and is left as an exercise.

Exercise 6.10.12. Show that

$$H_k^c(\mathbb{CP}^n) = \begin{cases} \mathbb{Z} & 0 \le k \le 2n, k \text{ even,} \\ 0 & \text{otherwise.} \end{cases}$$

We next look at the computation for $S^1 \times \mathbb{RP}^2$. The chain complex is

$$C_3^c = \mathbb{Z}(x^3) \to C_2^c = \mathbb{Z}(x_1^2) \oplus \mathbb{Z}(x_2^2) \to C_1^c = \mathbb{Z}(x_1^1) \oplus \mathbb{Z}(x_2^1) \to C_0^c = \mathbb{Z} \to 0.$$

The boundary map $\partial_1^c = 0$. We denote by x_1^1 the generator from the 1-cell
to form the circle $S^1 \times y$ and x_2^1 the generator from the 1-cell to form the circle
$x \times S^1 \subset \mathbb{RP}^2$. We denote by x_1^2 the generator from the 2-cell to form the
torus. We denote by x_2^2 the generator from the 2-cell used in forming $x \times \mathbb{RP}^2$.
Finally, x^3 corresponds to the generator for the 3-cell. From our methods for
two-dimensional complexes, we get that $\partial_2^c(x_1^2) = 0, \partial_2^c(x_2^2) = 2x_2^1$. To compute
the boundary map ∂_3^c, we look at the attaching map of the 3-cell, $f : S^2 \to X^2$.
We write $S^2 = \partial(D^1 \times D^2) = S^0 \times D^2 \cup D^1 \times S^1$. This map sends each copy
of $\{\pm1\} \times D^2$ via the characteristic map of the 2-cell in $\mathbb{RP}^2$. It sends $D^1 \times S^1$
via the characteristic map of the first circle crossed with the attaching map of
the 2-cell in $\mathbb{RP}^2$, which uses a map of degree 2. Since the two copies of D^2 in

the first part are mapped the same way, but inherit different orientations in S^2, they contribute 0 to $\partial_3(x^3)$. This can also be phrased in terms of local degree since they have opposite local degrees at two points in the inverse image of a point. For the second part, $D^1 \times S^1$ maps with a degree-2 map to the torus in X^2 which gives the x_1^2 generator, so $\partial_3^c(x_3) = 2x_1^2$. Thus we get $\ker(\partial_3) = 0$, so $H_3^c(S^1 \times \mathbb{RP}^2) = 0$. We also get $\ker(\partial_2^c)$ is generated by x_1^2, and $\mathrm{im}(\partial_3)$ is generated by $2x_1^2$. Hence $H_2(S^1 \times \mathbb{RP}^2) \simeq \mathbb{Z}_2$. This term is basically coming from the same source as the $\mathbb{Z}_2 = H_1(\mathbb{RP}^2)$. Finally, $\mathrm{im}(\partial_2^c)$ is generated by $2x_2^1$, so we get $H_1(S^1 \times \mathbb{RP}^2) = \mathbb{Z} \oplus \mathbb{Z}_2$. We have $H_0(S^1 \times \mathbb{RP}^2) = \mathbb{Z}$ and all other homology vanishes. The hardest part of the above argument to justify is the statement about orientation leading to 0 contribution from the $S^0 \times D^2$ part. What is geometrically clear is that each disk maps to a generator that comes from collapsing the lower skeleta to get $\mathbb{RP}^2/S^1 = S^2$. Thus the only other possibility would be that $\partial_3^c(x^3) = 2x_1^2 \pm 2x_2^2$ instead. But the fact that applying ∂_2^c to this element gives $\pm 4x_2^2 \neq 0$ and two successive applications of the boundary homomorphism have to give 0 verifies our earlier claim. This is a case where the algebra helps us out, so we do not have to figure out exactly what is happening geometrically. From the geometric point of view, we are computing the degree from the part of the map $S^0 \times D^2$ when we identify D^3 with $D^1 \times D^2$. Reflection in the first coordinate is a map of degree -1 and in S^2 it remains a map of degree -1 which interchanges the two copies of D^2 in $S^0 \times D^2$. Thus the local degree at a point in one copy of D^2 is the negative of the local degree at the other point. Thus these two contributions to the degree cancel one another.

Exercise 6.10.13. Compute the cellular homology of $S^1 \times S^2$.

Exercise 6.10.14. Compute the cellular homology of $\mathbb{RP}^2 \times \mathbb{RP}^2$.

We next look at $\mathbb{RP}^3$. The cellular chain complex is

$$0 \to \mathbb{Z} \to \mathbb{Z} \to \mathbb{Z} \to \mathbb{Z} \to 0.$$

The 2-skeleton is $\mathbb{RP}^2$ and we already computed its cellular homology, so we know that $\partial_2^c(x^2) = 2x^1, \partial_1^c(x^1) = 0$. To compute ∂_3^c, we can again rely on algebra as in the last example. For we must have $\mathrm{im}(\partial_3^c) \subset \ker(\partial_2^c) = 0$. Thus $\partial_3^c = 0$, and so we get

$$H_k^c(\mathbb{RP}^3) = \begin{cases} \mathbb{Z} & k = 0,3, \\ \mathbb{Z}_2 & k = 1, \\ 0 & \text{otherwise.} \end{cases}$$

This computation can also be done using our earlier result that the degree of the map $S^2 \to \mathbb{RP}^2 \to \mathbb{RP}^2/S^1 = S^2$ is $1 + (-1)^3 = 0$ from Exercise 6.9.15. If we try to continue to compute the cellular homology of higher-dimensional real projective spaces, we can no longer rely on algebra to help us, but can use Exercise 6.9.15. Since $\ker(\partial_3^c) = \mathbb{Z}$, it does not tell us anything about the map ∂_4^c. The cellular chain complex says to look at the attaching map of the 4-cell, which is just the standard double covering $S^3 \to \mathbb{RP}^3$. We then compose this with collapsing $\mathbb{RP}^2 \subset \mathbb{RP}^3$ to a point to get S^3. Each hemisphere covers the 3-sphere exactly once. The maps from the two hemispheres are related by $f(-x) = f(x)$. Thus $f = fA$, where $A(x) = -x$. $A : S^3 \to S^3$ has degree 1; that is, its induced map sends the generator of $H_3(S^3)$ to itself. We can express the degree of f as the sum of the local degrees. Since A has degree 1, the local degree is the same at two points in the inverse image of a point. The generator of S^3 is sent to twice the generator. This means that $\partial^c(x^4) = \pm 2x^3$. Hence $H_4(\mathbb{RP}^4) = 0, H_3(\mathbb{RP}^4) = \mathbb{Z}_2$. Since the 3-skeleton is the same for $\mathbb{RP}^3$ and $\mathbb{RP}^4$, they have the same cellular homology in dimensions 2 and lower. Thus

$$H_k(\mathbb{RP}^4) = \begin{cases} \mathbb{Z} & k = 0, \\ \mathbb{Z}_2 & k = 1, 3, \\ 0 & \text{otherwise.} \end{cases}$$

Exercise 6.10.15. Compute the cellular homology of $\mathbb{RP}^n$.

6.11   Cellular homology, singular homology, and Euler characteristic

We now want to show that the homology of the cellular chain complex gives the same homology as the singular homology of X. As a tool, we will use the long exact sequence of a triple (X, A, B), where $B \subset A \subset X$.

Exercise 6.11.1.
  (a) Show that there is a short exact sequence of chain complexes,

$$0 \to S(A)/S(B) \to S(X)/S(B) \to S(X)/S(A) \to 0.$$

  (b) Show that there is a long exact sequence of the triple (X, A, B) given by

$$\cdots \to H_{k+1}(X, A) \to H_k(A, B) \to H_k(X, B) \to H_k(X, A) \to \cdots .$$

Exercise 6.11.2.
  (a) Use the long exact sequence of the triple (X, X^p, X^{p-1}) to show that $H_k(X, X^p) \simeq H_k(X, X^{p-1})$ for $k > p + 1$.
  (b) Use (a) and induction to show that $H_k(X) \simeq H_k(X, X^{k-2})$.

Exercise 6.11.3.
  (a) Use the long exact sequence of the triple (X^{p+1}, X^p, A) to show that $H_k(X^p, A) \simeq H_k(X^{p+1}, A)$ for $k < p$.
  (b) Use (a) and induction to show that for a finite CW pair (X, A), we have $H_k(X, A) \simeq H_k(X^{k+1}, A)$.

Exercise 6.11.4. Combine the last two exercises to show that $H_k(X) \simeq H_k(X^{k+1}, X^{k-2})$.

By our definition of the chain complex (C^c, ∂^c) the kth homology is $\ker(\partial^c_k)/\mathrm{im}(\partial^c_{k+1})$. By the last exercise we need to identify this quotient with

$H_k(X^{k+1}, X^{k-2})$. First note that ∂_k^c occurs as a boundary map in the long exact sequence of the triple (X^k, X^{k-1}, X^{k-2}).

Exercise 6.11.5. Use the long exact sequence of the triple (X^k, X^{k-1}, X^{k-2}) to show that $H_k(X^k, X^{k-2})$ maps isomorphically onto $\ker(\partial_k^c)$ as a subgroup of $H_k(X^k, X^{k-1})$.

Exercise 6.11.6. Use the long exact sequence of the triple (X^{k+1}, X^k, X^{k-2}) to show that the quotient $H_k(X^k, X^{k-2})/\mathrm{im}(H_{k+1}(X^{k+1}, X^k) \to H_k(X^k, X^{k-2}))$ is isomorphic to $H_k(X^{k+1}, X^{k-2})$.

Exercise 6.11.7. Use the last two exercises and the commutative diagram

$$
\begin{array}{ccc}
H_{k+1}(X^{k+1}, X^k) & \xrightarrow{\quad\partial\quad} & H_k(X^k, X^{k-2}) \\
& \searrow{\scriptstyle\partial_{k+1}^c} & \downarrow{\scriptstyle(j_{12})_*} \\
& & H_k(X^k, X^{k-1})
\end{array}
$$

to complete the argument that the cellular homology $H_k^c(X)$ is isomorphic to the singular homology $H_k(X)$.

In Chapter 2 we used the Euler characteristic as defined in terms of a handle decomposition as a tool to distinguish surfaces up to homeomorphism. We stated there that the Euler characteristic was an invariant of the surface up to homeomorphism, independent of the chosen handle decomposition, but did not prove this claim. We will prove this by defining Euler characteristic in an invariant manner using singular homology and applying the result that the singular homology can be computed from the cellular chain complex. The connection to handle decompositions is that in Chapter 5 we showed that given a handle decomposition of a surface S (possibly with boundary), there is a corresponding CW complex K which is homotopy equivalent to the surface so that there is one k-cell of K for each k-handle of S.

Suppose that (C, ∂) is a chain complex of finitely generated free abelian groups so that there is a number n with $C_k = 0$ for $k > n$ (i.e. C is n-dimensional). The rank of C_k, which we denote by c_k, is the number of generators. In the case of the cellular chain complex, it is just the number of k-cells. The homology groups $H_i(C)$ will also vanish for $k > n$ and they will be finitely generated abelian groups of the form $H_i(C) \simeq h_i\mathbb{Z} \oplus T_i$. Here $h_i\mathbb{Z}$ denotes the free abelian group which is the direct sum of h_i copies of $\mathbb{Z}$, and T_i denotes a torsion group which is a direct sum of a finite number of copies of $\mathbb{Z}_{p_j}$. Here h_i is the rank of $H_i(C)$.

Definition 6.11.1. Define the Euler characteristic of C to be

$$
\chi(C) = \sum_{i=0}^{n}(-1)^i c_i
$$

and that of $H_*(C)$ to be

$$\chi(H_*(C)) = \sum_{i=0}^{n}(-1)^i h_i.$$

Our main result is

Theorem 6.11.1. $\chi(C) = \chi(H_*(C))$.

The starting point for our argument will be the result from algebra that whenever we have a short exact sequence of finitely generated abelian groups

$$0 \to A \to B \to C \to 0$$

then there is an equation

$$\mathrm{rk}(B) = \mathrm{rk}(A) + \mathrm{rk}(C).$$

Here rk denotes the rank of a finitely generated abelian group.

Now let $B_i \subset C_i$ denote the subgroup of boundaries and $Z_i \subset C_i$ be the subgroup of cycles.

Exercise 6.11.8.

(a) Show that there is a short exact sequence

$$0 \to Z_i \to C_i \to B_{i-1} \to 0.$$

(b) If $z_i = \mathrm{rk}(Z_i)$ and $b_i = \mathrm{rk}(B_i)$, show that $c_i = z_i + b_{i-1}$.

Exercise 6.11.9.

(a) Show that there is a short exact sequence

$$0 \to B_i \to Z_i \to H_i(C) \to 0.$$

(b) Show that $z_i = b_i + h_i$.

(c) Use the previous exercise to show that $c_i = h_i + b_i + b_{i-1}$.

Exercise 6.11.10.

(a) Show that $\sum_{i=0}^{n}(-1)^i(b_i + b_{i-1}) = 0$.

(b) Show that $\chi(C) = \chi(H_*(C))$.

Here is another algebraic result which is sometimes useful.

Exercise 6.11.11. Suppose there is a long exact sequence of finitely generated abelian groups:

$$0 \longrightarrow A_n \xrightarrow{a_n} A_{n-1} \xrightarrow{a_{n-1}} \cdots \xrightarrow{a_2} A_1 \xrightarrow{a_1} A_0 \longrightarrow 0$$

Show that $\sum_{i=0}^{n}(-1)^i \mathrm{rk}(A_i) = 0$. (Hint: Use induction on n starting with the case $n = 2$. For $n > 2$ rewrite the exact sequence in terms of two exact sequences

$$0 \longrightarrow A_n \xrightarrow{a_n} A_{n-1} \xrightarrow{a_{n-1}} \ker(a_{n-2}) \longrightarrow 0.$$

$$0 \longrightarrow \ker(a_{n-2}) \longrightarrow A_{n-2} \xrightarrow{a_{n-2}} \cdots \xrightarrow{a_2} A_1 \xrightarrow{a_1} A_0 \longrightarrow 0 \ .)$$

There is an alternative approach to equating the Euler characteristic given by a handle decomposition and the Euler characteristic coming from using the homology groups, that is, to use an inductive argument on the number of handles.

Exercise 6.11.12. Suppose that we have a handle decomposition of a surface S (possibly with boundary) with k_0 0-handles, k_1 1-handles, and k_2 2-handles. Let $h_i = \text{rk}(H_i(S))$.

(a) Show that if there is a single handle (i.e. $k_0 = 1, k_1 = k_2 = 0$), then $h_0 = 1, h_1 = h_2 = 0$.

(b) Show that if $S = H \cup h^i$ is formed from the handlebody H by attaching another handle h^i, then

$$H_p(S, H) = \begin{cases} \mathbb{Z} & p = i, \\ 0 & p \neq i. \end{cases}$$

(c) Use the exact sequence of the pair (S, H) and the Exercise 6.11.11 to show that

$$\chi(H_*(S)) = \chi(H_*(H)) + (-1)^i.$$

(d) Use induction to prove that $k_0 - k_1 + k_2 = h_0 - h_1 + h_2$.

Exercise 6.11.13. Use the technique of the last exercise to prove directly that if X is a finite CW complex of dimension n, and we define c_i to be the number of i-cells, then $\chi^c(X) = \sum_{i=0}^n (-1)^i c_i = \sum_{i=0}^n (-1)^i h_i = \chi(H(X))$.

6.12 Applications of the Mayer–Vietoris sequence

We next look at some applications of the Mayer–Vietoris exact sequence which use our refinements given earlier. We first use it to give a different derivation of the homology of spheres. We start with the basic computation

$$H_k(S^1) \simeq \begin{cases} \mathbb{Z} & k = 0, 1, \\ 0 & k > 1. \end{cases}$$

We know that $H_0(S^n) \simeq \mathbb{Z}$ for $n \geq 1$ by path connectivity, so we concentrate on higher-dimensional homology. What we want to show inductively is that $H_{k+1}(S^{n+1}) \simeq H_k(S^n), n \geq 1, k \geq 0$.

Exercise 6.12.1. (a) Use the decomposition $S^{n+1} = A \cup B$, where A is the upper hemisphere where $x_{n+2} \geq 0$, and B is the lower hemisphere where $x_{n+2} \leq 0$, and the refined Mayer–Vietoris sequence to show that $H_{k+1}(S^{n+1}) \simeq H_k(S^n), n \geq 0, k > 0$ and $H_0(S^n) \simeq H_1(S^{n+1}) \oplus \mathbb{Z}$.

(b) Show by induction that

$$
H_k(S^n) = \begin{cases} \mathbb{Z} \oplus \mathbb{Z} & n = k = 0, \\ \mathbb{Z} & n = k > 0 \text{ or } n > 0, k = 0, \\ 0 & \text{otherwise.} \end{cases}
$$

Note that the Mayer–Vietoris argument used for the sphere in Exercise 6.12.1 applies to suspensions of spaces. Suppose X is a path-connected topological space. Then the suspension $\sum X$ of X is the space formed from $X \times D^1$ by identifying $X \times \{1\}$ to a point and identifying $X \times \{-1\}$ to a point, and then using the quotient space topology.

Exercise 6.12.2. Show that ΣS^n is homeomorphic to S^{n+1}.

Exercise 6.12.3.
(a) Show that ΣX is path connected, and so $H_0(\Sigma X) \simeq \mathbb{Z}$.
(b) Use the Mayer–Vietoris exact sequence to show that $H_{k+1}(\Sigma X) \simeq H_k(X)$ for $k > 0$.
(c) Show that $H_1(\Sigma X) \oplus \mathbb{Z} \simeq H_0(X)$.

We now use the Mayer–Vietoris sequence to compute the homology of the torus. We divide the torus into two halves $T = T_+ \cup T_-$. We will think of T as arising as a quotient space of $D^1 \times D^1$ and let T_+ correspond to the quotient of points whose first coordinate is between $-\frac{1}{2}$ and $\frac{1}{2}$, with T_- corresponding to the quotient of those points whose first coordinate is either $\leq -\frac{1}{2}$ or $\geq \frac{1}{2}$. Note that each of T_+, T_- is homotopy equivalent to a circle as T_+ deformation retracts to the circle with first coordinate equal to 0 in the quotient space description, and T_- deformation retracts to the circle with first coordinate equal to 1 in the quotient space description. The intersection $T_- \cap T_+$ is the union of two circles $C_- \cup C_+$. Here C_- is chosen to be the circle with first coordinate equal to $-\frac{1}{2}$, and C_+ is the circle whose first coordinate is $\frac{1}{2}$.

Using standard identifications of the homology of T_+, T_-, C_+, C_- to the homology of the circles to which they each deformation-retract, we can look at the Mayer–Vietoris exact sequence.

Exercise 6.12.4. For $k \geq 2$, show that $H_k(T_+) = H_k(T_-) = H_k(T_+ \cap T_-) = 0$ and use this to show that $H_{k+1}(T) = 0$ for $k \geq 2$.

We thus concentrate on the lower part of the exact sequence.

Exercise 6.12.5. Using our earlier calculations, show that we can identify it to the following sequence:

$$
\begin{array}{ccccccc}
H_2(T_+) \oplus H_2(T_-) & \xrightarrow{j_2} & H_2(T) & \xrightarrow{\delta_2} & H_1(T_+ \cap T_-) & \xrightarrow{i_1} & H_1(T_+) \oplus H_1(T_-) \\
\Big\downarrow{\simeq} & & \Big\downarrow{=} & & \Big\downarrow{\simeq} & & \Big\downarrow{\simeq} \\
0 & \longrightarrow & H_2(T) & \xrightarrow{\delta_2} & \mathbb{Z} \oplus \mathbb{Z} & \xrightarrow{i_1} & \mathbb{Z} \oplus \mathbb{Z}
\end{array}
$$

$$\xrightarrow{\ j_1\ } H_1(T) \xrightarrow{\ \delta_1\ } H_0(T_+ \cap T_-) \xrightarrow{\ i_0\ } H_0(T_+) \oplus H_0(T_-) \xrightarrow{\ j_0\ } H_0(T) \longrightarrow 0$$

$$\downarrow{=} \qquad\qquad \downarrow{\simeq} \qquad\qquad \downarrow{\simeq} \qquad\qquad \downarrow{=}$$

$$\xrightarrow{\ j_1\ } \mathbb{Z} \oplus \mathbb{Z} \xrightarrow{\ \delta_1\ } \mathbb{Z} \oplus \mathbb{Z} \xrightarrow{\ i_0\ } \mathbb{Z} \oplus \mathbb{Z} \xrightarrow{\ j_0\ } \mathbb{Z} \longrightarrow 0$$

Exercise 6.12.6.

   (a) Show that $i_0(a,b) = (a+b, -a-b)$.

   (b) Show that $i_1(a,b) = (a+b, -a-b)$.

Exercise 6.12.7.

   (a) Show that $H_0(T) \simeq \mathbb{Z}$.

   (b) Use Exercise 6.5.11(a) to show that there are short exact sequences

$$0 \longrightarrow \operatorname{coker}(i_1) \longrightarrow H_1(T) \xrightarrow{\ \delta_1\ } ker(i_0) \longrightarrow 0$$

and

$$0 \longrightarrow H_2(T) \xrightarrow{\ \delta_2\ } ker(i_1) \longrightarrow 0.$$

Exercise 6.12.8.

   (a) Show that $ker(i_0) = \{(a, -a)\} \simeq \mathbb{Z}$.

   (b) Show that $\operatorname{coker}(i_1) \simeq \mathbb{Z}$.

   (c) Use (a),(b), and Exercise 6.5.3 to show that $H_1(T) \simeq \mathbb{Z} \oplus \mathbb{Z}$.

Exercise 6.12.9.

   (a) Show that $ker(i_1) = \{(a, -a)\} \simeq \mathbb{Z}$.

   (b) Use (a) to show that $H_2(T) \simeq \mathbb{Z}$.

Exercise 6.12.10. Give an explicit pair of singular 1-chains which generate $H_1(T) \simeq \mathbb{Z} \oplus \mathbb{Z}$.

Exercise 6.12.11. Give an explicit singular 2-chain which generates $H_2(T) \simeq \mathbb{Z}$. (Hint: First find chains in $S_2(T_+)$ and $S_2(T_-)$ whose boundaries give a representative of the generator of $ker(i_1) \subset H_1(T_+ \cap T_-)$.)

   We apply the same ideas used for the torus to compute the homology of the Klein bottle. As for our discussion of the torus, we regard K as a quotient space of $D^1 \times D^1$, where we make the identifications $(x, -1) \sim (x, 1), (-1, y) \sim (1, -y)$. We decompose K into $K_+ \cup K_-$, where K_+ corresponds to the first coordinate satisfying $-\frac{1}{2} \le x \le \frac{1}{2}$ and K_- corresponds to the first coordinate satisfying $-1 \le x \le -\frac{1}{2}$ or $\frac{1}{2} \le x \le 1$. Each of K_+, K_- is homeomorphic to an annulus and the intersection $K_+ \cap K_-$ is homeomorphic to the disjoint union of two circles.

Exercise 6.12.12. Use the Mayer–Vietoris sequence to show that $H_k(K) = 0$ for $k > 2$.

Exercise 6.12.13. Let $i_0 : H_0(K_+ \cap K_-) \simeq \mathbb{Z} \oplus \mathbb{Z} \to H_0(K_+) \oplus H_0(K_-) \simeq \mathbb{Z} \oplus \mathbb{Z}, i_1 : H_1(K_+ \cap K_-) \simeq \mathbb{Z} \oplus \mathbb{Z} \to H_0(K_+) \oplus H_0(K_-) \simeq \mathbb{Z} \oplus \mathbb{Z}$ be the maps in the Mayer–Vietoris sequence.

(a) Show that $i_0(a, b) = (a + b, -a - b)$.

(b) Show that $i_1(a, b) = (a + b, a - b)$. Here we are using the first coordinate of $\mathbb{Z} \oplus \mathbb{Z}$ to correspond to an upwardly oriented $\{-\frac{1}{2}\} \times D^1 / \sim$ and the second coordinate to correspond to an upwardly oriented $\{\frac{1}{2}\} \times D^1 / \sim$.

Exercise 6.12.14.
  (a) Show that $H_0(K) \simeq \mathbb{Z}$.

  (b) Use Exercise 6.5.11(a) to show that there are short exact sequences

$$0 \longrightarrow coker(i_1) \longrightarrow H_1(K) \xrightarrow{\delta_1} ker(i_0) \longrightarrow 0$$

and

$$0 \longrightarrow H_2(K) \xrightarrow{\delta_2} ker(i_1) \longrightarrow 0$$

Exercise 6.12.15.
  (a) Show that $ker(i_0) = \{(a, -a)\} \simeq \mathbb{Z}$.

  (b) Show that $coker(i_1) \simeq \mathbb{Z}_2$. (Hint: For $coker(i_1)$, show we have $[(1,0)] = [(0,1)] = [(-1,0)]$, which gives the relation $2[(1,0)] = 0$.)

  (c) Use (a),(b), and Exercise 6.5.3 to show that $H_1(K) \simeq \mathbb{Z}_2 \oplus \mathbb{Z}$.

Exercise 6.12.16.
(a) Show that $ker(i_1) = 0$.
(b) Use (a) to show that $H_2(K) \simeq 0$.

Exercise 6.12.17.
  (a) Show that reversing the direction of a singular 1-simplex σ to get $\bar{\sigma}(t) = \sigma(1 - t)$ satisfies $\sigma + \bar{\sigma} = \partial\tau$. Do this via a direct construction and by using $H_1(I) \simeq 0$.

  (b) Give an explicit pair of singular 1-chains which generate $H_1(K) \simeq \mathbb{Z}_2 \oplus \mathbb{Z}$. Explain how the 2-torsion arises for one of the generators by giving a 2-chain whose boundary represents twice the generator of $\mathbb{Z}_2$.

We now look at the general problem of computing the homology of a compact, connected surface. We first note how orientability is related to the inclusion map from the boundary circle to a surface with a disk removed. In Exercise 6.8.21(a), it is shown that the inclusion map from the boundary $\partial M \simeq S^1$ of a Möbius band to the Möbius band M induces multiplication by 2 on the first homology

$$H_1(\partial M) \simeq \mathbb{Z} \xrightarrow{2} H_1(M) \simeq \mathbb{Z}.$$

If P denotes the projective plane, then removing a disk from P to form $P_{(1)}$ gives the Möbius band. Thus the map $H_1(\partial P_{(1)}) \simeq \mathbb{Z} \to H_1(P_{(1)}) \simeq \mathbb{Z}$ is multiplication by 2. Now look at the Mayer–Vietoris sequence based on splitting P into $P_{(1)} \cup D^2$.

Exercise 6.12.18.
 (a) Show that $H_k(P) \simeq 0$ for $k \geq 3$.
 (b) Show that $H_2(P) = 0$ by using the fact that the map $i_1 : H_1(S^1) \to H_1(P_{(1)}) \oplus H_1(D^2)$ is injective.
 (c) Show that $H_1(P) \simeq \mathbb{Z}_2$ and $H_0(P) \simeq \mathbb{Z}$ by using the Mayer–Vietoris sequence.

We next consider a compact connected nonorientable surface N. Such a surface contains a Möbius band and so splits as a connected sum of P with another surface Q. Decompose $N = P_{(1)} \cup Q_{(1)}$. The intersection is the circle which is the common boundary of the two pieces.

Exercise 6.12.19.
 (a) Show that $H_k(N) = 0$ for $k \geq 3$.
 (b) Show that $H_2(N) = 0$ by using the fact that the map $i_1 : H_1(S^1) \to H_1(P_{(1)}) \oplus H_1(Q_{(1)})$ is injective.
 (c) Use the relation to the fundamental group to show that $H_1(P^{(k)}) \simeq \mathbb{Z}_2 \oplus (k-1)\mathbb{Z}$, where $(k-1)\mathbb{Z}$ denotes the connected sum of $k-1$ copies of $\mathbb{Z}$.
 (d) Use path connectivity to show that $H_0(P^{(k)}) \simeq \mathbb{Z}$.

We have seen that for the sphere and torus, the second homology is isomorphic to the integers. The other compact, connected and orientable surfaces are connected sums of copies of the torus. We want to show that their second homology is also the integers.

Exercise 6.12.20.
 (a) Show that the inclusion map $\partial T_{(1)} = S^1 \to T_{(1)}$ induces the zero map on H_1.
 (b) By using the decomposition $T = T_{(1)} \cup D^2$, use the Mayer–Vietoris sequence to show that $H_2(T) \simeq \mathbb{Z}$ and $H_k(T) \simeq 0$ for $k \geq 3$.
 (c) Use the relation to the fundamental group to show that $H_1(T) \simeq \mathbb{Z} \oplus \mathbb{Z}$. Use path connectivity to show that $H_0(T) \simeq \mathbb{Z}$.

We use Exercise 6.12.17(a) in the next exercise.

Exercise 6.12.21. (a) For the other oriented surfaces $T_{(1)}^{(k)}$ use the description of the $T^{(k)}$ as the quotient of a $4k$-gon with identifications to show that $H_1(\partial T_{(1)}^{(k)}) \to H_1(T_{(1)}^{(k)})$ is the zero map. (Hint: Think of the disk being removed as an interior $4k$-gon and its generator for H_1 as a sum of the edges and see where it maps via the standard deformation retraction to the boundary which is being identified.)
 (b) Use the Mayer–Vietoris sequence based on $T^{(k)} = T_{(1)}^{(k)} \cup D^2$ to show that $H_i(T^{(k)}) \simeq 0, i \geq 2$ and $H_2(T^{(k)}) \simeq \mathbb{Z}$.
 (c) Use the relation to the fundamental group to show that $H_1(T^{(k)}) \simeq 2k\mathbb{Z}$, where $2k\mathbb{Z}$ denotes the direct sum of $2k$ copies of $\mathbb{Z}$.
 (d) Use path connectivity to show that $H_0(T^{(k)}) \simeq \mathbb{Z}$.

Exercise 6.12.22. Consider $T^{(k)}$ as arising as a quotient of the $4k$-gon as in the previous exercise. Divide the $4k$-gon into $4k$ triangles with common point at the center of the $4k$-gon.

(a) Find a 2-cycle σ which is the signed sum of $4k$ singular simplices corresponding to these triangles.

(b) Let Δ represent one of the triangles. Show that the map which sends
$H_2(T^{(k)}) \to H_2(T^{(k)}, T^{(k)}\backslash \text{int}\, \Delta) \simeq H_2(\Delta, \partial\Delta) \simeq \mathbb{Z}$ sends $[\sigma]$ to a generator.

(c) Show that σ is a generator of $H_2(T^{(k)}) \simeq \mathbb{Z}$.

The calculations above show that H_2 detects orientability of compact connected surfaces, with oriented surfaces having integral second homology and nonorientable surfaces having zero second homology.

6.13   Reduced homology

In most of our calculations so far, the spaces have been path connected and so computation of H_0 has not been an issue. When it is a concern, it is useful to use reduced homology instead of homology. In the path-connected case, using reduced homology also makes the statements of results simpler. If we have a chain complex of free abelian groups $(C_i, \partial_i), i \geq 0$, we can extend it further by introducing $C_{-1} = \mathbb{Z}$. We define $\partial_0 = \epsilon$ by first selecting a basis of generators of C_0 and define $\epsilon(g) = 1$ for each generator g. This means that $\ker(\epsilon)$ is generated as a free group on a difference of generators. When ∂_1 has the property that, for a generator g of C_1, $\partial_1(g)$ is a difference of generators, we have $\text{im}(\partial_1) \subset \ker(\epsilon)$. Thus the extended chain complex is still a chain complex. The map $\partial_0 = \epsilon$ is called the *augmentation* of the original chain complex.

Definition 6.13.1. We define the reduced homology $\widetilde{H}_k(C)$ to be the homology of the augmented chain complex for $k \geq 0$. This is the same as $H_k(C)$ in dimensions greater than 0 but $\widetilde{H}_0(C) = \ker(\epsilon)/\text{im}(\partial_1)$, whereas $H_0(C) = C_0/\text{im}(\partial_1)$.

Exercise 6.13.1. Show that the short exact sequence

$$0 \longrightarrow \ker(\epsilon) \longrightarrow C_0 \overset{\epsilon}{\longrightarrow} \mathbb{Z} \longrightarrow 0,$$

leads to a short exact sequence

$$0 \longrightarrow \widetilde{H}_0(C) \longrightarrow H_0(C) \longrightarrow \mathbb{Z} \longrightarrow 0,$$

which splits to give

$$H_0(C) \simeq \widetilde{H}_0(C) \oplus \mathbb{Z}.$$

For the singular complex $S(X)$ of a topological space X we define the augmentation $\epsilon : S_0(X) \to \mathbb{Z}$ by $\epsilon(\sigma) = 1$ for a singular 0-simplex σ. Then the reduced homology $\widetilde{H}_k(X)$ is defined to be the homology of this augmented singular complex.

Exercise 6.13.2.

   (a) Verify that for $S(X)$ we have $\epsilon \partial_1 = 0$.

   (b) Show that $H_0(X) \simeq \widetilde{H}_0(X) \oplus \mathbb{Z}$.

   (c) Show that a topological space is path connected iff $\widetilde{H}_0(X) \simeq 0$.

Exercise 6.13.3. Show that the reduced homology of the sphere can be expressed as

$$\widetilde{H}_k(S^n) = \begin{cases} \mathbb{Z} & k = n, \\ 0 & k \neq n. \end{cases}$$

Exercise 6.13.4. Show that the homology of the suspension ΣX of a path-connected space can be expressed as $H_{k+1}(\Sigma X) \simeq \widetilde{H}_{k+1}(\Sigma X) = \widetilde{H}_k(X)$ for $k \geq 0$.

The long exact sequence of a pair (X, A) can be modified to replace homology by reduced homology, as can the Mayer–Vietoris exact sequence.

Exercise 6.13.5. By using the augmented complex in place of the original complex and using $S_{-1}(X, A) = S_{-1}(X)/S_{-1}(A) = \mathbb{Z}/\mathbb{Z} = 0$, show that there is a long exact sequence in reduced homology which ends in

$$H_1(X, A) \to \widetilde{H}_0(A) \to \widetilde{H}_0(X) \to H_0(X, A) \to 0.$$

The Mayer–Vietoris exact sequence can also be modified to use reduced homology as long as $A \cap B \neq \emptyset$. Here we need to augment the usual simplicial chains used by $S_{-1}(A \cap B) = \mathbb{Z}, S_{-1}(A) \oplus S_{-1}(B) = \mathbb{Z} \oplus \mathbb{Z}, S_{-1}(A) + S_{-1}(B) = \mathbb{Z}, S_{-1}(X) = \mathbb{Z}$.

Exercise 6.13.6. Assume that $X = A \cup B, A \cap B \neq \emptyset$ and the hypotheses giving the usual Mayer–Vietoris exact sequence hold. Show that there is a long exact sequence using reduced homology which ends in

$$H_1(X) \xrightarrow{\ \delta\ } \widetilde{H}_0(A \cap B) \xrightarrow{\ i_0\ } \widetilde{H}_0(A) \oplus \widetilde{H}_0(B) \xrightarrow{\ j_0\ } \widetilde{H}_0(X) \longrightarrow 0$$

Exercise 6.13.7. Use the above reduced sequence to show that, if A and B are path connected, $A \cap B \neq \emptyset, A \cup B = X$, and the hypotheses giving the Mayer–Vietoris exact sequence apply, then X is path connected.

6.14   The Jordan curve theorem and its generalizations

We now apply the reduced Mayer–Vietoris exact sequence to prove the Jordan curve theorem. Instead of $\mathbb{R}^2$, we consider S^2 and use the fact that there is a homeomorphism between $\mathbb{R}^2$ and $S^2 \backslash \{N\}$, where N is the north pole. Since there is a rotation sending any one point on the sphere to another (the sphere is homogeneous, as is any connected surface—see Exercise 2.9.34), $S^2 \backslash \{p\}$ is homeomorphic to $\mathbb{R}^2$.

Exercise 6.14.1. Suppose that C is a compact set in $\mathbb{R}^2$ and $h(C)$ its image under a homeomorphism between $\mathbb{R}^2$ and $S^2\backslash\{p\}$. Show that one of the path components of $\mathbb{R}^2\backslash C$ contains all points outside some large ball, and that there is a homeomorphism between this path component and $K\backslash\{p\}$, where K is the path component of $S^2\backslash h(C)$ which contains p. Show that for all other path components of $\mathbb{R}^2\backslash C$, there is a homeomorphism to a corresponding path component of $S^2\backslash h(C)$.

The previous exercise allows us to transfer the problem about a simple closed curve separating $\mathbb{R}^2$ into two path components to a problem in the sphere. What we first need to see is that an arc *does not* separate the sphere into multiple path components. By an arc we mean the homeomorphic image of an interval $[a, b]$.

Definition 6.14.1. We say a set C *disconnects* S^2 if $S^2\backslash C$ is not path connected, or, equivalently, $\widetilde{H}_0(S^2\backslash C) \not\simeq 0$.

Exercise 6.14.2. Let A be an arc $h([a, b])$ in S^2 and $c = \frac{1}{2}(a + b)$. Let $A_1 = h([a, c])$, $A_2 = h([c, b])$. Use the Mayer–Vietoris exact sequence with

$$S^2\backslash A = (S^2\backslash A_1) \cap (S^2\backslash A_2), \qquad S^2\backslash \{h(c)\} = (S^2\backslash A_1) \cup (S^2\backslash A_2)$$

to show that $\widetilde{H}_0(S^2\backslash A_1) \oplus \widetilde{H}_0(S^2\backslash A_2) \simeq \widetilde{H}_0(S^2\backslash A)$.

Exercise 6.14.3. Use the last exercise to show that if an arc A disconnects S^2, then there is a sequence of subarcs $\cdots \subset A_i \subset A_{i-1} \subset \cdots \subset A_1 \subset A$ so that the maps induced by inclusions $\widetilde{H}_0(S^2\backslash A) \to \widetilde{H}_0(S^2\backslash A_1) \to \cdots \to \widetilde{H}_0(S^2\backslash A_i)$ each map a given nontrivial element $[p] - [q]$ to a nontrivial element. Moreover, show that these subarcs intersect in a point.

Theorem 6.14.1. *If A is an arc in S^2, then $S^2\backslash A$ is path connected.*

Exercise 6.14.4. Use the previous exercise to prove Theorem 6.14.1 that an arc does not disconnect S^2. (Hint: Suppose it did. Choose p and q in different path components and use the previous exercise to show that there is a sequence of subarcs whose diameters tend to zero with limit a single point so that p and q remain in different path components in the complements of each of these subarcs. Show that this leads to a contradiction.)

Exercise 6.14.5. By using an analogous argument to the one used for $\widetilde{H}_0$, show that $H_1(S^2\backslash A) \simeq 0$, where A is an arc in S^2.

We now consider a simple closed curve $C \subset S^2$. Since $C = h(S^1)$ and S^1 is the union of two subarcs, we can write $C = A_1 \cup A_2$, where $A_1 \cap A_2 = \{p, q\}$ represent two points in the sphere.

Exercise 6.14.6.
(a) Show that $S^2\backslash\{p, q\}$ is homeomorphic to $\mathbb{R}^2\backslash\{\mathbf{0}\}$.
(b) Show that $\widetilde{H}_0(S^2\backslash\{p, q\}) \simeq 0$, $H_1(S^2\backslash\{p, q\}) \simeq \mathbb{Z}$.

Exercise 6.14.7.

(a) Let C be a simple closed curve in S^2. Use the above exercises and the reduced Mayer–Vietoris sequence to show that $\tilde{H}_0(S^2\backslash C) \simeq \mathbb{Z}$.

(b) Show that $S^2\backslash C$ has precisely two path components and that each path component is open.

Exercise 6.14.8. Suppose that p, q are points in different path components K_1, K_2 of $S^2\backslash C$, and $r \in C$. Let D be an open disk about r. Show that there is a point of each path component in D. (Hint: First find a subarc A_1 of C containing r that is in D. Then write $S^2\backslash C = (S^2\backslash A_1) \cap (S^2\backslash A_2)$, where $C = A_1 \cup A_2$ as before. Use the path connectivity of $S^2\backslash A_2$ to find a path connecting p to q which misses A_2. Show that this path must pass through D and use it to prove the result.)

Theorem 6.14.2 (Jordan curve theorem). *Suppose C is a simple closed curve in $\mathbb{R}^2$.*

(a) *$\mathbb{R}^2\backslash C$ has precisely two path components, which are also its two components and are open sets.*

(b) *One of these components is bounded and the other is not. Call the bounded component the interior I and the unbounded component the exterior E.*

(c) *$\bar{I} = I \cup C, \bar{E} = E \cup C$.*

Exercise 6.14.9. Prove Theorem 6.14.2.

The ideas that have gone into proving the Jordan curve theorem can be generalized to prove similar separation theorems. The natural context for proving these is for embedded subsets which are homeomorphic to a sphere of lower dimension.

We first use the Mayer–Vietoris sequence to show that the complement of an embedded disk of any dimension in a sphere has trivial reduced homology in all dimensions. The following lemma, which is implicit in our proof above, will be useful.

Lemma 6.14.3. *Suppose there is a closed set B which is expressed as the intersection of a nested family of closed sets $B_i, i = 0, \ldots, \infty$ with $B_{i+1} \subset B_i$ and a nonzero element $x \in \tilde{H}_p(S^n\backslash B_0)$ which maps injectively to $\tilde{H}_p(S^n\backslash B_i)$ for all i. Then x maps injectively to a nonzero element of $\tilde{H}_p(S^n\backslash B)$.*

Exercise 6.14.10. Prove Lemma 6.14.3. (Hint: If a chain $c \in S_p(S^n\backslash B_0)$ bounds a chain $d \in S_{p+1}(S^n\backslash B)$, show that d is actually a chain in $S_{p+1}(S^n\backslash B_i)$ for some i.)

Theorem 6.14.4. *Suppose $h : D^k \to S^n$ is an embedding. Then $\tilde{H}_i(S^n\backslash h(D^k)) \simeq 0, i \geq 0$.*

Exercise 6.14.11. Use induction and the Mayer–Vietoris sequence to prove Theorem 6.14.4.

We now apply the Mayer–Vietoris to relate the homology of the complement of an embedded S^k in S^n. Before doing this, we motivate the answer by looking at the model situation where S^k is embedded in a particularly nice fashion so that we can see directly what the complement is. This is basically the standard embedding, but we will look at it from a slightly different perspective. We start by writing $S^n = \partial D^{n+1}$. Then we note that there is a homeomorphism between D^{n+1} and $D^{k+1} \times D^{n-k}$. Using this homeomorphism, there is a corresponding homeomorphism between their boundaries given by

$$S^n \simeq \partial(D^{k+1} \times D^{n-k}) = S^k \times D^{n-k} \cup D^{k+1} \times S^{n-k-1}.$$

When $S^k = S^k \times \{\mathbf{0}\}$ is removed, $S^k \times (D^{n-k} \backslash \{\mathbf{0}\})$ first deformation-retracts to $S^k \times S^{n-k-1} = \partial(D^{k+1} \times S^{n-k-1})$ and then $D^{k+1} \times S^{n-k-1}$ deformation-retracts to S^{n-k-1}. Thus the complement $S^n \backslash S^k$ deformation-retracts to S^{n-k-1} and so

$$\tilde{H}_p(S^n \backslash S^k) = \begin{cases} \mathbb{Z} & p = n - k - 1, \\ 0 & \text{otherwise.} \end{cases}$$

The ideas used in the decomposition above are critical for understanding handle decompositions of higher-dimensional manifolds as well as generalizing the procedure of surgery for modifying manifolds of all dimensions. These ideas play a central role in the topology of manifolds.

We now show that the homology of the complement of an embedded sphere is independent of the embedding.

Theorem 6.14.5 (Alexander duality for spheres). *Suppose that* $h : S^k \to S^n, 0 \leq k < n$ *is an embedding. Then*

$$\tilde{H}_p(S^n \backslash h(S^k)) \simeq \begin{cases} \mathbb{Z} & p = n - k - 1, \\ 0 & \text{otherwise.} \end{cases}$$

The next three exercises lead to a proof of Theorem 6.14.5.

Exercise 6.14.12. Prove Theorem 6.14.5 when $k = 0$.

Exercise 6.14.13. Under the hypotheses of Theorem 6.14.5 with $k = m+1 \geq 1$, use the Mayer–Vietoris exact sequence based on the decomposition

$$S^n \backslash h(S^m) = (S^n \backslash h(S^k_+)) \cup (S^n \backslash h(S^k_-)), \qquad S^n \backslash S^k = (S^n \backslash h(S^k_+)) \cap (S^n \backslash h(S^k_-))$$

and the results above on the homology of complements of embedded disks to show that

$$\tilde{H}_{p+1}(S^n \backslash h(S^{m+1})) \simeq \tilde{H}_p(S^n \backslash h(S^m)).$$

Exercise 6.14.14. Use the previous two exercises and induction to prove Theorem 6.14.5.

Theorem 6.14.5 is part of a much more general phenomenon called Alexander duality. In the general case, the Alexander duality is concerned with the homology of a complement $M \backslash C$, where C is some subset a manifold M. The reader can find nice developments of this duality in [5, 13]. We now state a consequence of the special case $k = n - 1$ for further reference.

Theorem 6.14.6 (Jordan separation theorem). *Suppose that $h : S^{n-1} \to S^n$ is an embedding. Then $S^n \backslash h(S^{n-1})$ is the disjoint union of two open sets. The closure of each set is the union of the set and $h(S^{n-1})$.*

Exercise 6.14.15. Prove Theorem 6.14.6.

In Chapter 2 we used invariance of domain to prove a number of results. We now prove invariance of domain using the Jordan separation theorem.

Theorem 6.14.7 (Invariance of domain). *Suppose that U is an open subset of $\mathbb{R}^n$ and $f : U \to \mathbb{R}^n$ is 1–1 and continuous. Then f is an open map; that is, it maps open sets to open sets.*

By composing f with a homeomorphism between $\mathbb{R}^n$ and $S^n \backslash \{p\}$, we can assume that the image is contained in S^n and we just have to show that the image of an open set in U is an open set in S^n. The next exercise leads you through a proof of Theorem 6.14.7.

Exercise 6.14.16. Suppose $x \in U, y = f(x) \in S^n$ and let V be an open set in U containing x. Choose a small disk $D_x = D(x, \epsilon)$ about x which is contained in V with $S_x = \partial D_x$.

(a) Use the Jordan separation theorem to show that $S^n \backslash f(S_x)$ is the union of two disjoint connected open sets.

(b) Show that $f(\text{int } D_x)$ is one of these sets. (Hint: Show that $f(\text{int } D_x)$ and $S^n \backslash f(D_x)$ are each path components.)

(c) Conclude that $f(V)$ is open.

6.15 Orientation and homology

In this section we will discuss the concept of an orientation of an n-manifold in terms of homology. We will relate this to our earlier discussion of orientation of surfaces in Chapter 2 and isotopy classes of embedded disks. In particular, we will outline proofs of some statements made there. A key concept in this discussion is the relative homology group $H_n(M, M \backslash \{x\})$, so we will begin by discussing this group and then apply it to distinguish interior points and boundary points and to show that the dimension of an n-manifold is well defined.

Exercise 6.15.1. Suppose that M is an n-manifold, and $x \in \text{int } M$. Let $h : D^n \to M$ be the restriction of a homeomorphism h' from $\mathbb{R}^n$ onto an open set U' in int M so that if $U = h(\text{int } D^n)$, then $x = h(y) \in U$.

(a) Show that

$$H_k(M, M \backslash U) \simeq H_k(D^n, S^{n-1}) \simeq \begin{cases} \mathbb{Z} & k = n, \\ 0 & \text{otherwise.} \end{cases}$$

(b) Show that

$$H_k(M, M\backslash\{x\}) \simeq H_k(D^n, D^n\backslash\{y\}) \simeq \begin{cases} \mathbb{Z} & k = n, \\ 0 & \text{otherwise.} \end{cases}$$

(c) Show that the map $H_k(M, M\backslash U) \to H_k(M, M\backslash\{x\})$ is an isomorphism.

Exercise 6.15.2. Show that if D_+^n is the upper half disk and S_+^{n-1} is the upper half of the sphere, then $H_k(D_+^n, S_+^{n-1}) \simeq 0$.

Exercise 6.15.3. Show that a boundary point x and an interior point y of an n-manifold M with boundary are distinguished by their homology groups $H_n(M, M\backslash\{x\}) \simeq 0$ and $H_n(M, M\backslash\{y\}) \simeq \mathbb{Z}$.

Exercise 6.15.4. Show that if M is an n-manifold, then it is not an m-manifold for $m \neq n$.

Now suppose M is a connected n-manifold. We will assume that M has no boundary—if it does, then our first step is to replace M by $M\backslash\partial M = \text{int } M$ for this discussion. Note that we have shown that if $x \in M$ and U is an open set about x homeomorphic to the interior of an embedded n-disk which contains x, then if $r_{U,x} : H_n(M, M\backslash U) \to H_n(M, M\backslash\{x\})$ is the map induced by inclusion, then $r_{U,x}$ is an isomorphism of groups isomorphic to $\mathbb{Z}$. Moreover, $h : (D^n, S^{n-1}) \to (M, M\backslash U) \to (M, M\backslash\{x\})$ gives the corresponding generators $\mu_U \in H_n(M, M\backslash U), \mu_x \in H_n((M, M\backslash\{x\})$ with $r_{U,x}(\mu_U) = \mu_x$ as images of a generator of μ_D of $H_n(D^n, S^{n-1})$. This leads to the following definition.

Definition 6.15.1. An n-manifold M is *orientable* if there is a choice $\mu_x \in H_n(M, M\backslash\{x\})$ for each $x \in M$ so that

- μ_x is a generator of $H_n(M, M\backslash\{x\})$ for each $x \in M$;
- these choices are locally consistent in the sense that for each $x \in M$, there is an open set U containing x which is homeomorphic to the interior of a disk in an embedded $\mathbb{R}^n$ and a generator $\mu_U \in H_n(M, M\backslash U) \simeq \mathbb{Z}$ so that $r_{U,y}(\mu_U) = \mu_y$ for all $y \in U$. Here $r_{U,x}$ is the map induced by inclusion.

A choice of μ_x for all x as above is called a *homology orientation* or just an *orientation* for M.

Exercise 6.15.5. Show that if a connected n-manifold M is orientable, then the choice of homology orientation at a fixed x_0 determines the choice everywhere. (Hint: Use the fact that it determines it locally and path connectivity.)

Since we will be working extensively with the groups $H_i(M, M\backslash A)$, we introduce the shorthand notation $H_i^{M,A}$ for these groups. Whenever $A \subset B$, we have an inclusion $(M, M\backslash B) \subset (M, M\backslash A)$ and an induced homomorphism $r_{B,A} : H_i^{M,B} \to H_i^{M,A}$. Of particular importance to us are the homomorphisms $r_{A,x}$. Note that, if $x \in A \subset B$, then $r_{A,x} r_{B,A} = r_{B,x}$. We use the notation r_x for the $H_n(M) \to H_n^{M,x}$ induced by inclusion. For a manifold with boundary, there is also a map we denote by $r_x : H_n(M, \partial M) \to H_n^{M,x}$ when $x \in \text{int } M$.

We now show how to orient $\mathbb{R}^n$. Consider the disk of radius r about $\mathbf{0}$, which we denote D_r. When $r_1 < r_2$ there is an inclusion $D_{r_1} \subset D_{r_2}$. Let $C_r = \mathbb{R}^n \backslash D_r$. Now there is an inclusion $(\mathbb{R}^n, C_{r_2}) \subset (\mathbb{R}^n, C_{r_1})$. By excision, there is an isomorphism $H_n^{\mathbb{R}^n, D_r} \simeq H_n(\mathbb{R}^n, C_r) \simeq H_n(D^n, S^{n-1})$, where we are excising out the complement of a disk of larger radius and then deformation-retracting the annular region between the spheres to the outer sphere. Thus $H_n^{\mathbb{R}^n, D_r} \simeq \mathbb{Z}$ with a generator μ_r coming from a selected positive generator $\mu_D \in H_n(D^n, S^{n-1})$. Moreover, we can choose these generators consistently so that $r_{D_{r_2}, D_{r_1}}(\mu_{r_2}) = \mu_{r_1}$ by choosing μ_1 and getting the others using the isomorphisms induced by inclusion. Now let $x \in D_r$. Then the inclusion map $(\mathbb{R}^n, C_r) \to (\mathbb{R}^n, \mathbb{R}^n \backslash \{x\})$ induces an isomorphism in homology since $\mathbb{R}^n \backslash \{x\}$ deformation-retracts back to $\bar{C}_s$ for any $s > r$, as does C_r. Then $r_{D_r, x}(\mu_r)$ is a generator $\mu_x \in H_n^{\mathbb{R}^n, x}$ Moreover, if $r_1 < r_2$, and $x \in D_{r_1}$, then $r_{D_{r_2}, D_{r_1}}(\mu_{r_2}) = \mu_{r_1}$ and $r_{D_{r_2}, x} = r_{D_{r_1}, x} r_{D_{r_2}, D_{r_1}}$ imply that both μ_{r_2} and μ_{r_1} map to the same element μ_x. We now define an orientation μ_x by the following prescription. Choose a disk D_r with $x \in D_r$ and let $\mu_x = r_{D_r, x} \mu_r$. Note that, if U is an open disk about x with $U \subset D_r$, then $r_{U, x}$ and $r_{D_r, x}$ being isomorphisms imply that $r_{D_r, U}$ is as well. Thus these choices of μ_x give an orientation for $\mathbb{R}^n$ since they come from $r_{U, x}(\mu_U)$, where $\mu_U = r_{D_r, U}(\mu_r)$. Instead of the argument that $r_{D_r, U}$ is an isomorphism, we could use the convexity of U as follows. The radial deformation retraction of $\mathbb{R}^n \backslash \{x\}$ to $\bar{C}_s$ for $s > r$ will restrict to a deformation retraction of $\mathbb{R}^n \backslash U$ onto $\bar{C}_s$. This will work for any convex set K, giving consistent homotopy equivalences between $(\mathbb{R}^n, \bar{C}_s)$ and $(\mathbb{R}^n, \mathbb{R}^n \backslash D_r), (\mathbb{R}^n, \mathbb{R}^n \backslash K), (\mathbb{R}^n, \mathbb{R}^n \backslash \{x\})$ for $x \in K$. This implies that $H_i^{\mathbb{R}^n, K} = 0$ for $i \neq n$ and $H_n^{\mathbb{R}^n, K} \simeq \mathbb{Z}$ with a generator $r_{D_r, K}(\mu_r)$ so that $r_{K, x}(\mu_K) = \mu_x$ for each $x \in K$. If we start with a given orientation μ_x for each x, then this will determine consistent choices for μ_r for all r and a consistent μ_K. We will later use this for K an i-cube given by a product of closed intervals $I_1 \times \cdots \times I_n$ (some of which may be points). For future reference, we state the results of this discussion as a proposition.

Proposition 6.15.1. $\mathbb{R}^n$ *is oriented by choosing a consistent set of generators* $\mu_r \in H_n^{\mathbb{R}^n, D_r} \simeq \mathbb{Z}$ *so that, if* $r_1 < r_2$, *then* $r_{D_{r_2}, D_{r_1}}(\mu_{r_2}) = \mu_{r_1}$. *Then for any* $x \in \mathbb{R}^n$, *choosing* r *with* $x \in D_r$ *and defining* $\mu_x = r_{D_r, x}(\mu_r)$ *gives an orientation. Moreover, if* K *is a compact convex set contained in* D_r *and* $x \in K$, *then the inclusion maps* $\mathbb{R}^n \backslash D_r \to \mathbb{R}^n \backslash K \to \mathbb{R}^n \backslash \{x\}$ *are homotopy equivalences inducing isomorphisms for all* i. *The maps* $r_{D_r, K}, r_{K, x}$ *are each isomorphisms with composition the isomorphism* $r_{D_r, x}$. *Thus* $H_i^{\mathbb{R}^n, K} = 0$ *if* $i \neq n$ *and* $H_n^{\mathbb{R}^n, K} \simeq \mathbb{Z}$. *If we choose the generator* $\mu_K = r_{D_r, K}(\mu_r)$, *then it has the property that* $r_{K, x}(\mu_K) = \mu_x$ *for each* $x \in K$. *Given an orientation* $s(x) = \mu_x$, *the choice of* μ_x *for a single* x *will determine the choice of consistent* μ_r *for all* r *and hence a consistent* μ_K *whenever* $x \in K$.

Now suppose A is any nonempty compact set in $\mathbb{R}^n$. Choose r so that $A \subset D_r$. Then $r_{D_r, A}(\mu_r)$ is an element μ_A so that $r_{A, x}(\mu_A) = \mu_x$ for each $x \in A$. Thus the map $r_{A, x}$ is surjective for each $x \in A$, but it is not always an isomorphism. For example, if A consists of two points, then $H_n^{\mathbb{R}^n, A} \simeq \mathbb{Z} \oplus \mathbb{Z}$ by excision.

Here is another way to look at our orientation of $\mathbb{R}^n$. We can consider $\mathbb{R}^n$ as embedded in S^n as $S^n\backslash\{N\}$, where $N = (\mathbf{0}, 1)$, via stereographic projection. The pair $(\mathbb{R}^n, C_r)$ then corresponds to $(S^n\backslash\{N\}, B_s\backslash\{N\})$ where B_s is an open disk about N. Note that, as r increases, the diameter s decreases. By excision, $H_n(S^n\backslash\{N\}, B_s\backslash\{N\}) \simeq H_n(S^n, B_s)$. Thus our system of generators μ_r corresponds to consistent generators of $H_n(S^n, B_s)$. From the long exact sequence of the pair (S^n, B_s), the map $H_n(S^n) \to H_n(S^n, B_s)$ is an isomorphism. Thus our generators μ_s come from a single generator $\mu \in H_n(S^n)$. So an equivalent way to have found our orientation was to start with a generator $\mu \in H_n(S^n)$ and use it to induce generators of $\mu_x \in H_n(S^n, S^n\backslash\{x\})$ and then identify this group with $H_n(\mathbb{R}^n, \mathbb{R}^n\backslash\{x\})$ by excision whenever $x \neq N$.

Note that this discussion shows that S^n is oriented by choosing a generator $\mu \in H_n(S^n)$. Similarly, any compact connected n-manifold M without boundary with $H_n(M^n) \simeq \mathbb{Z}$ so that $r_x : H_n(M) \to H_n^{M,x}$ is an isomorphism for all $x \in M$ will be oriented in this fashion. For a manifold with boundary, we can get an orientation (which is a consistent choice of generators of $H_n^{M,x}$ for $x \in \operatorname{int} M$, or, equivalently, an orientation of $\operatorname{int} M$) by showing that $H_n(M, \partial M) \simeq \mathbb{Z}$ and $r_x : H_n(M, \partial M) \to H_n^{M,x}$ is an isomorphism for all $x \in \operatorname{int} M$.

Recall that for the case of compact connected surfaces without boundary, we found that the orientable surfaces were distinguished from the nonorientable ones in that $H_2(M) \simeq \mathbb{Z}$ when M is orientable and $H_2(M) \simeq 0$ when M is nonorientable. Since the surfaces with boundary come from ones without a boundary by removing disks, we could use long exact sequences of pairs and the fact that the disks are contractible to show that for an orientable surface with boundary $H_2(M, \partial M) \simeq \mathbb{Z}$. We could also have checked from our calculations via CW complexes that the map restricting to $(M, M\backslash\{x\})$ gives an isomorphism in homology.

Our main theorem is a parallel result for n-manifolds, where we are now using the definition of orientability in terms of homology.

Theorem 6.15.2. *Let M be a connected n-manifold with boundary ∂M, possibly empty.*

(1) $H_i(M) = 0, i > n$.

(2) If M is compact, nonorientable, or M is not compact, then $H_n(M) = 0$.

(3) If $\partial M = \emptyset$ and M is compact and orientable, then $r_x : H_n(M) \to H_n^{M,x}$ is an isomorphism for all $x \in M$.

(4) If M is compact with $\partial M \neq \emptyset$, then $H_n(M) = 0$.

(5) If M is compact with $\partial M \neq \emptyset$ and M is orientable, then $r_x : H_n(M, \partial M) \to H_n^{M,x}$ is an isomorphism for each $x \in \operatorname{int} M$.

In part (3), the homology orientation classes μ_x will all be in the image of a single generator $\mu \in H_n(M)$, which is called the orientation homology class of M corresponding to the orientation. Conversely, when there is a class $\mu \in H_n(M) \simeq \mathbb{Z}$ and the maps $r_x : H_n(M) \to H_n^{M,x}$ are isomorphisms for each x, then the classes $\mu_x = r_x(\mu)$ will determine a homology orientation. For if U is an open set homeomorphic to a disk about x and $y \in U$, then if we take the map $r_U : H_n(M) \to H_n^{M,U}$, then $r_U(\mu) = \mu_U$ gives a class so that $r_{U,y}(\mu_U) = r_{U,y}r_U(\mu) = r_y(\mu) = \mu_y$. Thus these classes satisfy the local consistency condition for a homology orientation. In part (5), the orientation class comes from the generator of $H_n(M, \partial M)$.

The proof of Theorem 6.15.2 will be quite involved. Before beginning our discussion of it, we want to look back at surfaces and see how to apply ideas of homology orientations to them. To distinguish different definitions of orientations, we will refer to the current form as homology-orientable. We will show that handle-orientable implies homology-orientable implies disk-orientable, providing another proof that handle-orientability implies disk-orientability. We also show that if r is a reflection of the disk, then ir and r are not ambient isotopic as maps of $\mathbb{R}^2$. We will work in dimension n and then apply this to surfaces.

Exercise 6.15.6. Suppose M is a compact and connected and homology-oriented n-manifold without boundary. Denote by $\mu \in H_n(M)$ a generator with $\mu \to \mu_z$ under $r_z : H_n(M) \to H_n^{M,z}$ and μ_z the class in the definition of homology orientation.

(a) Show that if $H_t : M \to M$ is an isotopy and $H_0 = \mathrm{id}$, then $(H_t)_*(\mu) = \mu$ for all $t \in I$.

(b) Suppose that $H_t(x) = y$. Show that $(H_t)_*(\mu_x) = \mu_y$. (Hint: Considering the commutative diagram

$$
\begin{array}{ccc}
H_n(M) & \xrightarrow{\;(H_t)_*\;} & H_n(M) \\
\downarrow{\scriptstyle r_x} & & \downarrow{\scriptstyle r_y} \\
H_n^{M,x} & \xrightarrow{\;(H_t)_*\;} & H_n^{M,y}
\end{array}
$$

is helpful.)

(c) Let $h_0, h_1 : D^n \to M$ be embedded disks which are ambient isotopic; that is, $h_1 = H_1 h_0$, where H_t is an isotopy of M. Let $\mu_D \in H_n(D^n, D^n \backslash \{0\})$ be the positive generator. If $(h_0)_* \mu_D = \epsilon \mu_{h_0(0)}$, show that $(h_1)_* \mu_D = \epsilon \mu_{h_1(0)}$.

(d) Show that h_0 is not ambient isotopic to $h_0 r$, where $r : D^n \to D^n$ is a reflection.

Note that if M is a compact, connected and homology-oriented n-manifold with boundary, then all of the previous exercise can be repeated virtually word for word where we now use the assumed class $\mu \in H_n(M, \partial M)$ instead.

We showed earlier that $H_n(S^n) \simeq \mathbb{Z}, n > 0$. Hence our theorem says that it is homology orientable, with homology orientation coming from $r_x : H_n(S^n) \to H_n^{S^n,x}$.

Exercise 6.15.7.

(a) Suppose that $n > 0$. Without using the theorem, show that $H_n(S^n) \simeq \mathbb{Z}$ with generator μ implies that $r_x : H_n(S^n) \to H_n^{S^n, x}$ is an isomorphism.

(b) If $\mu_x = r_x(\mu)$, show that μ_x satisfies the definition of a homology orientation.

Exercise 6.15.8. Suppose $H : \mathbb{R}^n \times I \to \mathbb{R}^n \times I$ is an isotopy. Extend H to a map $\widetilde{H} : S^n \times I \to S^n \times I$ by regarding $\mathbb{R}^n$ as embedded into S^n as $S^n \backslash \{N\}$, where $N = (\mathbf{0}, 1)$, and defining $\widetilde{H}(N, t) = (N, t)$. Show that $\widetilde{H}$ is an isotopy.

Exercise 6.15.9. Suppose that $h_0, h_1 : D^n \to \mathbb{R}^n$ are embedded disks which are ambient isotopic. By regarding $\mathbb{R}^n$ as embedded in S^n as $S^n \backslash \{N\}$ and taking the homology orientation of $\mathbb{R}^n$ as coming from the homology orientation of S^n, show that if $(h_0)_*(\mu_D) = \epsilon \mu_{h_0(0)}$ then $(h_1)_*(\mu_D) = \epsilon \mu_{h_1(0)}$. Conclude that h_0 is not ambient isotopic to $h_0 r$.

Here is a more direct approach to the last result, where we are now using ambient isotopies of $\mathbb{R}^n$ which are the identity outside a disk, as in Chapter 2.

Exercise 6.15.10. Let $G_t : \mathbb{R}^n \to \mathbb{R}^n$ be an isotopy which is the identity outside of a large disk B with $G_0 = \mathrm{id}$.

(a) Show that the map $(G_1)_* : H_n^{\mathbb{R}^n, \mathrm{int}\, B} \to H_n^{\mathbb{R}^n, \mathrm{int}\, B}$ is the identity.

(b) Use (a) to show that the map $(G_1)_* : H_n^{\mathbb{R}^n, x} \to H_n^{\mathbb{R}^n, G_1(x)}$ sends the local homology orientation at x determined via a choice of generator $g_B \in H_n^{\mathbb{R}^n, \mathrm{int}\, B}$ to the local homology orientation at $G_1(x)$ determined by g_B.

(c) Show that if $h_0, h_1 : D^n \to \mathbb{R}^n$ are ambient isotopic embedded disks with ambient isotopy G_t which is the identity outside of a large disk B (i.e. $G_0 = \mathrm{id}, G_1 h_0 = h_1$), and we choose the homology orientation consistent with h_0, then it is also consistent with h_1.

(d) Show that h and hr are not ambient isotopic.

Finally, we look at a direct approach without the assumption that the isotopy is the identity outside a disk.

Exercise 6.15.11. Let $G_t : \mathbb{R}^n \to \mathbb{R}^n$ be an isotopy with $G_0 = \mathrm{id}$.

(a) Show that if D_r is a disk about the origin in $\mathbb{R}^n$, then there is another disk D_s for $s \geq r$ about the origin so that $G_t(\mathbb{R}^n \backslash D_s) \subset \mathbb{R}^n \backslash D_r$ for all t. (Hint: Consider the image of $p_1 H^{-1}(D_r \times I)$, where $p_1 : \mathbb{R}^n \times I \to \mathbb{R}^n$ is projection onto the first coordinate. This is a compact set and so is contained in some D_s.)

(b) Show that the map $(G_1)_* : H_n^{\mathbb{R}^n, D_s} \to H_n^{\mathbb{R}^n, D_r}$ sends μ_s to μ_r.

(c) Use (b) to show that the map $(G_1)_* : H_n^{\mathbb{R}^n, x} \to H_n^{\mathbb{R}^n, G_1(x)}$ sends the local homology orientation at x determined via a choice of generator $g_s \in H_n^{\mathbb{R}^n, D_s}$ to the local homology orientation at $G_1(x)$ determined by g_r.

(d) Now suppose that $h_0, h_1 : D^n \to \mathbb{R}^2$ are ambient isotopic embedded disks (i.e. $G_0 = \text{id}, G_1 h_0 = h_1$). Show that if we choose the homology orientation consistent with h_0, then it is also consistent with h_1.

(e) Show that h and hr are not ambient isotopic.

We now apply this information to surfaces. First, any handle-orientable connected handlebody is homology-orientable. This follows from the classification theorem and homology calculations but can also be proved inductively. It suffices to assume that the handlebody is formed from a single 0-handle, disjointly attached 1-handles, and some 2-handles. The 0-handle is homology-oriented from its orientation as a handle. The assumption that the surface is oriented as a handlebody means that when we attach any 1-handle to the 0-handle, the result is identifiable to an annulus which has a homology orientation arising as a subset of the plane which is consistent with the homology orientations on the 0-handle and 1-handle. This then allows us to get a homology orientation of the sub-handlebody formed from the 0-handle and 1-handles. When a 2-handle is attached, then the collar where it is attached plus the 2-handle is identifiable with an open disk in the plane in a manner consistent with the homology orientation of the 2-handle and the one imposed on the collar by the rest of the surface.

For a compact connected surface without boundary, Exercise 6.15.6 implies that homology-orientability implies disk-orientability. If the surface has boundary, the bounded version of Exercise 6.15.6, which we discussed immediately following it, gives the result. Combining this with the last paragraph, this implies that handle-oriented surfaces are disk-oriented. This was used in the Chapter 2 supplementary exercises to show that all of the definitions of orientability are equivalent for handlebodies.

We now leave surfaces and return to begin the proof of Theorem 6.15.2. We first want to rephrase the choice of a consistent local orientation in terms of covering spaces. Suppose M is a connected n-manifold. We form two covering spaces of M, which we denote $\widetilde{M}_g, \widetilde{M}$. The points of $\widetilde{M}$ are elements of the group $H_n^{M,x}$, where x ranges over the points of M. To be a point of $\widetilde{M}_g$, the element must be a generator of the group. We are using the notion of covering space in a more general sense than we did earlier, in that we are not requiring $\widetilde{M}, \widetilde{M}_g$ to be path connected. In fact, the path connectivity of $\widetilde{M}_g$ will be shown to be equivalent to nonorientability. However, we are requiring the local triviality condition of a covering space. Thus we may think of these covering spaces as the disjoint union of covering spaces in the earlier sense. There is a natural inclusion $\widetilde{M}_g \subset \widetilde{M}$. We define a projection map $p : \widetilde{M} \to M$ by sending $k \in H_n^{M,x}$ to $x \in M$. By restriction we get $p_g : \widetilde{M}_g \to M$. We want to define a topology of $\widetilde{M}, \widetilde{M}_g$ so that these maps are continuous and are covering maps. To do this we use the idea of elements of $H_n^{M,x}$ being locally consistent. Thus we use as a basis of the topology the sets $\widetilde{U}_\alpha = \{r_{U,y}(\alpha)\}$, where U is the interior of an embedded disk in a Euclidean neighborhood and $\alpha \in H_n^{M,U}$ denotes a fixed element. Here the maps $r_{U,y}$ are induced by inclusion.

Exercise 6.15.12.

(a) Show that the elements of $\tilde{U}_\alpha$ are locally consistent in the sense that if we have $v_x \in \tilde{U}_\alpha$ with $v_x \in H_n^{M,x}$ and $W \subset U$ is a neighborhood of x homeomorphic to an open disk with v_W the element of $H_n^{M,W} \simeq H_n^{M,x}$ which maps to v_x under this isomorphism, then we have $r_{W,y}(v_W) \in \tilde{U}_\alpha$.

(b) Show that this does provide a basis for a topology.

(c) Show that $(\widetilde{M_g}, p_g, M)$ and $(\widetilde{M}, p, M)$ are covering spaces.

Exercise 6.15.13.

(a) Show that nearby points in $\widetilde{M}$ have the same divisibility as elements of groups isomorphic to $\mathbb{Z}$.

(b) Show that $\widetilde{M}$ consists of the disjoint union of one copy of the identity covering space (corresponding to the 0 element of $H_n^{M,x}$ for each x) and a natural number of copies of covering spaces equivalent to $(\widetilde{M_g}, p_g, M)$, one for each divisibility $d \in \mathbb{N}$.

Exercise 6.15.14.

(a) Show that the homology-orientability condition on M is equivalent to finding a continuous map $s : M \to \widetilde{M_g}$ so that $ps(x) = x$. Such a map s is called a *section* of the covering space.

(b) Show that if there is a nonzero section of $p : \widetilde{M} \to M$ (so that the image of x is *not* the zero element of $H_n^{M,x}$), then there is a section of $\widetilde{M_g}$.

(c) Show that M is nonorientable iff the only section of $p : \widetilde{M} \to M$ is the zero section.

We next look at properties of sections of $\widetilde{M}$, which we denote by Γ_M.

Exercise 6.15.15. Show that if s_1, s_2 are sections and $s_1(x) = s_2(x)$ for one x, then $s_1 = s_2$.

Exercise 6.15.16. Show that there exists a section $s : M \to \widetilde{M_g}$ iff the covering space $(\widetilde{M_g}, p_g, M)$ is equivalent to the two copies of the identity covering space, or, equivalently, $\widetilde{M_g}$ is the union of two components, each of which is sent homeomorphically to M via p_g.

Exercise 6.15.17. Show that M is orientable iff $\widetilde{M_g}$ is homeomorphic to the disjoint union of two copies of M.

Exercise 6.15.18. Use the theory of covering spaces to show that a simply connected manifold is orientable. (Hint: Nonorientability leads to a nontrivial connected double covering.)

Exercise 6.15.19.

(a) Show that if $\pi_1(M, x)$ does not have an index-2 subgroup, then M is orientable.

(b) Give an example of a surface which is orientable and whose fundamental group contains an index-2 subgroup.

Exercise 6.15.20. Suppose that M is connected. Show that Γ_M has the structure of a group, and the evaluation map $E : \Gamma_M \to H_n^{M,x} \simeq \mathbb{Z}$ given by $E(s) = s(x)$ is an injective homomorphism (monomorphism) for all $x \in M$.

Exercise 6.15.21.
    (a) Show that if M is connected and orientable, then $\Gamma_M \simeq \mathbb{Z}$ with isomorphism given by E.
    (b) Show that if M is connected and nonorientable, then $\Gamma_M = 0$.

For a manifold M with possibly empty boundary, there is a map $S_M : H_n(M, \partial M) \to \Gamma_{\text{int } M}$ so that $S_M(\alpha)(x) = r_x(\alpha)$.

Exercise 6.15.22.
    (a) Show that $S_M(\alpha)$ satisfies the local consistency condition required for a section.
    (b) Show that S_M is a homomorphism.

Our main tool in proving Theorem 6.15.2 is the following theorem.

Theorem 6.15.3. *Let M be a connected n-manifold, with $\partial M = \emptyset$, and let A be a compact subset of M.*

    (1) $H_i^{M,A} = 0, i > n$.
    (2) *If $\alpha \in H_n^{M,A}$ and $r_{A,x}(\alpha) = 0$ for all $x \in A$, then $\alpha = 0$.*
    (3) *If M is oriented via a section $s(x) = \mu_x$, then there exists $\mu_A \in H_n^{M,A}$ with $r_{A,x}(\mu_A) = \mu_x$.*

Proposition 6.15.1 asserts that Theorem 6.15.3 is true for $M = \mathbb{R}^n$ and A a compact convex set. The next set of exercises will deduce Theorem 6.15.2 from Theorem 6.15.3.

Exercise 6.15.23. Suppose that M is connected and compact, $\partial M = \emptyset$.

    (a) $H_i(M) = 0, i > n$.
    (b) If M is nonorientable, then $H_n(M) = 0$.
    (c) If M is orientable via a section $s(x) = \mu_x$, then there exists $\mu \in H_n(M)$ with $r_x(\mu) = \mu_x$. (Hint: Take $A = M$.)

Exercise 6.15.24. Suppose M is connected and is not compact and $\partial M = \emptyset$. Let $i \geq n$. Let $\alpha \in H_i(M)$ be represented by a chain c and suppose S is the support of c.

    (a) Show that there is an open set U with $S \subset U$ and $\bar{U}$ compact.
    (b) Let $V = M \setminus \bar{U}$. Show that $H_i(U) \simeq H_i(U \cup V, V)$.
    (c) Use the long exact sequence of the triple $(M, U \cup V, U)$ and Theorem 6.15.3(1) to show that $H_i(U \cup V, V) = 0$ for $i > n$.
    (d) Show that $\alpha = 0$ when $i > n$.
    For the remaining parts, let $i = n$.

(e) Show that $r_x(\alpha) = 0$ for all $x \in M$ by evaluating it at $x \in V$.

(f) By first regarding c as a chain in $(U \cup V, V)$ and using the long exact sequence of the triple, show that $\alpha = 0$.

The previous exercises have taken care of Theorem 6.15.2(1)–(3) when $\partial M = \emptyset$. The next exercises will assume $\partial M \neq \emptyset$. To deal with this case, we use the fact that there is a collar neighborhood C of ∂M which is homeomorphic to $\partial M \times [0, 1]$.

Exercise 6.15.25. Let $C \simeq_h \partial M \times [0, 1]$ be a collar. Let $C_t = h[0, t)$ and $A_t = M \backslash C_t, t \leq 1$.

(a) Show that $M, \mathrm{int}\, M$ and A_t deformation-retract to A_1.

(b) Show that (M, C_t) deformation-retracts to $(M, \partial M)$ for $t < 1$.

(c) Show that $H_i(M) \simeq H_i(\mathrm{int}\, M)$

(d) Show that the map induced by inclusion $H_i(M, \partial M) \to H_i^{M, A_t}$ is an isomorphism for all t.

We next take care of the remaining cases of Theorem 6.15.2(1)–(4) when $\partial M \neq \emptyset$.

Exercise 6.15.26. Show that $H_i(M) = 0, \ i \geq n$.

The next exercise takes care of Theorem 6.15.2(5).

Exercise 6.15.27. Suppose M is compact and orientable with $\partial M \neq \emptyset$. Let $s(x) = \mu_x \in \Gamma_{\mathrm{int}\, M}$ be a section for $x \in \mathrm{int}\, M$. Show that there exists a class $H_n(M, \partial M)$ which maps to μ_x under the map $r_x : H_n(M, \partial M) \to H_n^{M, x}$ for each $x \in \mathrm{int}\, M$.

We now return to prove Theorem 6.15.3. The key to the argument is to examine the relative Mayer–Vietoris sequence connecting the homology of $(M, M \backslash C)$ when C takes on the values $C = A, B, A \cap B, A \cup B$. Note that $(M \backslash A) \cup (M \backslash B) = M \backslash A \cap B$, $(M \backslash A) \cap (M \backslash B) = M \backslash A \cup B$. We leave it as an exercise to justify the necessary relative Mayer–Vietoris exact sequence since we have only discussed this in the absolute case.

Exercise 6.15.28 (Relative Mayer–Vietoris sequence). Show that if U, V are open sets in M, then there is a long exact sequence with segment

$$H_{k+1}(M, U \cup V) \xrightarrow{\delta} H_k(M, U \cap V) \xrightarrow{i_k} H_k(M, U) \oplus H_{k+1}(M, V) \xrightarrow{j_k} H_k(M, U \cup V)$$

(Hint: Start with the diagram

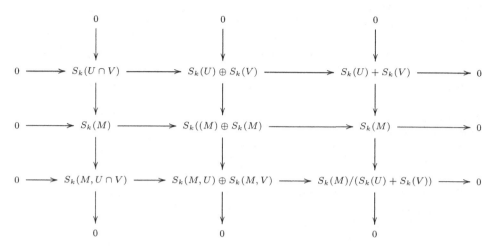

and consider the horizontal rows as chain complexes and the vertical columns as giving an exact sequence of chain complexes. Look at the long exact sequence in homology to see that it says that the bottom horizontal row is exact. Then take the long exact sequence arising from this exact row. Then show that the fact that there is an isomorphism induced by inclusion between $H_k^{\{U,V\}}(U \cup V)$ and $H_k(U \cup V)$ leads to an isomorphism between $H_k(M, U \cup V)$ and $H_k(S(M)/(S(U) + S(V)))$.)

Exercise 6.15.29. Use the previous exercise to show that if A, B are closed subsets of M, then there is long exact sequence

$$\cdots \longrightarrow H_{k+1}^{M,A\cap B} \overset{\delta}{\longrightarrow} H_k^{M,A\cup B} \overset{i_k}{\longrightarrow} H_k^{M,A} \oplus H_k^{M,B} \overset{j_k}{\longrightarrow} H_k^{M,A\cap B} \overset{\delta}{\longrightarrow} \cdots .$$

The exact sequence from Exercise 6.15.29 will be the basis for most of the remaining arguments. The next exercise shows that if Theorem 6.15.3 holds for $A, B, A \cap B$, then it holds for $A \cup B$ and is the basis of inductive proofs to show that it holds generally.

Exercise 6.15.30. Show that if $A, B, A \cap B$ satisfy Theorem 6.15.3, then so does $A \cup B$.

The point of Exercise 6.15.30 is that we can inductively build up sets C satisfying the conclusion of Theorem 6.15.3 by starting with pieces $A, B, A \cap B$ which have this property and using $C = A \cup B$. Note that if C is a compact subset of a coordinate neighborhood $U \subset M$ which is homeomorphic to $\mathbb{R}^n$, then $H_k^{M,C} \simeq H_k^{\mathbb{R}^n,C}$ by excision. Thus to prove our results for C it suffices to prove these statements when $M = \mathbb{R}^n$.

Our starting point for the argument will be when $M = \mathbb{R}^n$ and C is a cube of some dimension k with $0 \le k \le n$. By a cube we mean a product of intervals

$I_1 \times \cdots \times I_n$, and each interval is of the form $[a_i, b_i]$ with $a_i \leq b_i$. This case was done in Proposition 6.15.1.

Now consider the space C^s formed from an n-cube by subdividing it using subdivisions of each subinterval. This is built up from subcubes by adding them one step at a time, where the intersections are finite unions of cubes of lower dimension. We can think of this as a CW complex where all of the cells are embedded cubes of various dimensions. Let K be any subcomplex of C^s.

Exercise 6.15.31. Show that if K is a subcomplex of C^s, then show the conclusion of Theorem 6.15.3 holds for K. (Hint: Use induction on the number of cells in K and Exercises 6.15.30.)

Now suppose A is a compact subset of $\mathbb{R}^n$. It will be contained in a large cube C. By subdividing this cube into subcubes successively (say subdivide the edges in half, fourths, etc.) we can let K^i be the subcomplex of the ith subdivision consisting of all closed subcubes which intersect A somewhere in their interior. We will have $\cdots \subset K^2 \subset K^1 \subset K^0 = C$ and $A \subset K^i$ for all i. Moreover, $A = \cap_i K^i$.

Suppose $\alpha \in H_i^{\mathbb{R}^n, A}$ is represented by a chain c. Look at ∂c. It is a finite sum of singular simplices, each of which has image is $\mathbb{R}^n \backslash A$. Let $B \subset \mathbb{R}^n \backslash A$ be the compact set which consists of images of these simplices.

Exercise 6.15.32.
  (a) Show that there is a minimal distance between points of A and points of B. Use this to show that there is an integer p so that c also represents a cycle of $S_i(\mathbb{R}^n, \mathbb{R}^n \backslash K^p)$.

  (b) Show that (a) implies that $\alpha = 0$ when $i > n$ since it is in the image of $H_i(\mathbb{R}^n, \mathbb{R}^n \backslash K^p)$.

  (c) Suppose $i = n$ and $r_{A,x}(\alpha) = 0$ for $x \in A$. Show that this means that if α_K is the class represented by c, then $r_{K^p,x}(\alpha_K) = 0$ for $x \in K^p$. Then show that this implies that $\alpha_K = 0$ and hence $\alpha = 0$.

  (d) Show that the usual orientation of $\mathbb{R}^n$ provides the class μ_A that restricts to the orientation class μ_x for $x \in A$ in the statement of Theorem 6.15.3.

We now have proved Theorem 6.15.3 when $M = \mathbb{R}^n$ and A is any compact subset. We next consider a general connected n-manifold M.

Exercise 6.15.33. Show that Theorem 6.15.3 holds for M when A is a compact subset of a coordinate neighborhood U.

Exercise 6.15.34. Show that any compact set $A \subset M$ can be written as the union $A = A_1 \cup \cdots \cup A_m$ of a finite number of sets A_i which are compact sets in an Euclidean neighborhood $U_i \subset M$.

Exercise 6.15.35. By using induction on the number of sets, m, in the description above, complete the proof of Theorem 6.15.3.

6.16 Proof of homotopy invariance of homology

We next look at the homotopy invariance property. One of the important facets of singular homology is that it is a homotopy functor, which includes the fact that a homotopy equivalence of topological spaces $f : A \to X$ gives an isomorphism of homology groups $H_k(A) \to H_k(X)$ in each dimension. We first look at a proof of this when $k = 0, 1$, where the argument is simpler to understand. For H_0, we use the fact that H_0 just measures the path components, with one copy of $\mathbb{Z}$ for each path component. A representative cycle is a map from Δ_0 to a point in that path component.

Suppose that $f, g : X \to Y$ are homotopic maps. To show that they induce the same map in homology, we need to see that the chain maps $f_\sharp$ and $g_\sharp$ are chain homotopic.

We first look at dimension 0. Since all of the maps we are considering are linear, it suffices to look at a singular 0-simplex σ. Let F be the homotopy between f and g, so $F(x, 1) = g(x), F(x, 0) = f(x)$. The induced maps are $f_\sharp(\sigma) = f\sigma$ and $g_\sharp(\sigma) = g\sigma$. Suppose $\sigma(0) = p$. Let $\gamma : [0, 1] = \Delta_1 \to Y$ be the singular 1-simplex defined by the composition

$$[0, 1] \longrightarrow \{p\} \times [0, 1] \overset{F}{\longrightarrow} Y.$$

Exercise 6.16.1. Define $H_0(\sigma) = \gamma$ and show that H_0 provides a chain homotopy between $f\sigma$ and $g\sigma$.

We now look at dimension 1, noting that we have defined H_0 above. We need to define $H_1 : S_1(X) \to S_2(Y)$ so that, for a singular 1-simplex $\sigma : I \to X$, we have the formula

$$g_\sharp\sigma - f_\sharp\sigma = \partial_2 H_1(\sigma) + H_0\partial_1(\sigma).$$

We first use the homotopy F to define a map $K = F(\sigma \times \mathrm{id}) : I \times I \to Y$ where $\sigma \times \mathrm{id} : I \times I \to X \times I$. Note that $Ki_0 = f\sigma, Ki_1 = g\sigma$, where $i_0(s) = (s, 0), i_1(s) = (s, 1)$. Moreover, if $\tau_1(0) = \sigma(1), \tau_0(0) = \sigma(0)$ are the singular 0-simplices coming from restricting σ to its end points, then we have $\partial_1\sigma = \tau_1 - \tau_0$. Moreover, $H_0(\tau_1)(y) = K(1, y), H_0(\tau_0)(y) = K(0, y)$.

We now look at Figure 6.6.

Motivated by this figure, we can define two singular 2-simplices α, β in $I \times I = \Delta_1 \times I$ with positive orientation on α and negative orientation on β as depicted via the arrows on their boundaries, and then define $H_1(\sigma) = K\beta - K\alpha = K_\sharp(\beta - \alpha)$.

Exercise 6.16.2. Show that with this definition of H_1, we have the required formula

$$g_\sharp(\sigma) - f_\sharp(\sigma) = \partial_2 H_1(\sigma) + H_0\partial_1(\sigma).$$

Exercise 6.16.3. Use the previous exercises to show that if $f, g : X \to Y$ are homotopic maps then the induced maps $f_\sharp, g_\sharp$ are chain homotopic in dimensions 0 and 1. Deduce that $f_* = g_*$ in homology.

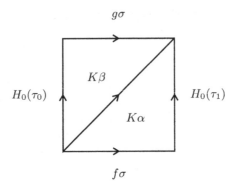

Figure 6.6. Constructing H_1.

In order to extend this to higher-dimensional chains, we need to analyze the process above. First, the homotopy F is used at the last step in each case and can be separated from the definition. The relation between f, g, F is $g = Fi_1^X$, $f = Fi_0^X$, where $i_j^X : X \to X \times I$ is $i_j^X(x) = (x, j)$. If we label the maps $H_k = H_k^{X,Y} : S_k(X) \to S_{k+1}(Y)$ that we have defined already, $k = 0, 1$, then there are corresponding maps related to the maps $i_j : X \to X \times I$, which we will label as H_k^X instead of the more cumbersome $H_k^{X, X \times I}$. Here we use the identity homotopy $I(x, t) = (x, t)$ as the map connecting i_0^X and i_1^X. The maps $H_k^{X,Y} : S_k(X) \to S_k(Y)$ then factor as $F_\sharp H_k^X$, where $H_k^X : S_k(X) \to S_{k+1}(X \times I)$. The term $(\partial_2 H_1^{X,Y} + H_0^{X,Y} \partial_1)(\sigma)$ just becomes $F_\sharp(\partial H_1^X + H_0^X \partial_1)(\sigma)$. On the right-hand side, we note that $g_\sharp = F_\sharp(i_1^X)_\sharp$, $f_\sharp = F_\sharp(i_0^X)_\sharp$, and so we can rewrite $g_\sharp(\sigma) - f_\sharp(\sigma) = F_\sharp((i_1^X)_\sharp(\sigma) - (i_0^X)_\sharp(\sigma))$. Thus our formula follows from the special case

$$(\partial_2 H_1^X + H_0^X \partial_1)(\sigma) = ((i_1^X)_\sharp - (i_0^X))(\sigma)$$

by taking composition with $F_\sharp$. Thus we now restrict to looking at what is happening in this case.

Let $i_j^{\Delta_1}$ be the corresponding inclusions for Δ_1. Then if we use the identity map $\iota_1 \in S_1(\Delta_1)$, we will have a formula

$$(\partial_2 H_1^{\Delta_1} + H_0^{\Delta_1} \partial_1)(\iota_1) = ((i_1^X)_\sharp - (i_0^X))(\iota_1).$$

Now apply $(\sigma \times \mathrm{id})_\sharp$ to both sides of this equation. On the left-hand side we note first that our definition of $H_1^X(\sigma)$ is just $(\sigma \times \mathrm{id})_\sharp(\beta - \alpha) = (\sigma \times \mathrm{id})_\sharp H_1^{\Delta_1}(\iota_1)$. When we take ∂_2 of this, we can use the fact that $(\sigma \times \mathrm{id})_\sharp$ is a chain map and commutes with ∂_2 to rewrite $\partial_2 H_1^X(\sigma) = (\sigma \times \mathrm{id})_\sharp \partial_2 H_1^{\Delta_1}(\iota_1)$. Moreover, $H_0^X \partial_1(\sigma) = (\sigma \times \mathrm{id})_\sharp \partial(\iota)$. by our construction. Thus the left-hand side of our fundamental equation becomes

$$(\sigma \times \mathrm{id})_\sharp((\partial_2 H_1^{\Delta_1} + H_0^{\Delta_1} \partial_1)(\iota)).$$

For the right-hand side, we note that $\sigma = \sigma_\sharp(\iota_1)$ and $i_j^X \sigma = (\sigma \times \text{id})i_j^{\Delta_1}$ implies $(i_j^X)_\sharp(\sigma) = (i_j^X)_\sharp \sigma_\sharp(\iota_1) = (\sigma \times \text{id})_\sharp(\iota_1)$. Thus the right-hand side can be rewritten as

$$(\sigma \times \text{id})_\sharp((i_1^{\Delta_1})_\sharp(\iota_1) - (i_0^{\Delta_1})_\sharp(\iota_1)).$$

Thus our formula for σ follows from the formula for ι by composing with $(\sigma \times \text{id})_\sharp$.

This discussion shows that generalizing this construction to higher dimensions can be done through first specializing to the case of Δ_k and the identity singular simplex ι_k and then defining our maps in general using this.

The key to extending the above argument to all dimensions is finding a way to systematically subdivide the product $\Delta_i \times I$ into $(i+1)$-simplices as we have done in the case $i = 1$ above. This can be done inductively on i. An important fact about $\Delta_i \times I$ is that it is a convex subset of $\mathbb{R}^i \times \mathbb{R} = \mathbb{R}^{i+1}$. We can redefine what we have done as follows. Whenever we have a convex space with points $u_0, \ldots, u_i$, we can define an affine linear singular i-simplex by using the affine linear map that sends e_j to u_j. We will denote this by $[u_0, u_1, \ldots, u_i]$. Now consider $\Delta_1 \times I$. Denote by v_j the point $(e_j, 0)$ and by w_j the point $(e_j, 1)$. Then $H_0([e_0]) = [v_0, w_0]$, and when we extend to Δ_1, we get $H_0([e_j]) = [v_j, w_j]$, $j = 0, 1$. Another way to state what we are doing for Δ_1 is that we are using the definition for Δ_0 to define H_0 on these particular singular simplices in Δ_1 by defining it on the face maps $F_j : \Delta_0 \to \Delta_1, j = 0, 1$, by $H_0(F_j) = (F_j \times \text{id})_\sharp H_0([e_0])$.

Now look at our definition of H_1. For simplicity, we will delete the superscript Δ_1. For the case when σ is just the identity map $\iota_1 = [e_0, e_1]$, then the simplex $\beta = [v_0, w_0, w_1]$ and $\alpha = [v_0, v_1, w_1]$. Thus our definition is $H_1([e_0, e_1]) = [v_0, w_0, w_1] - [v_0, v_1, w_1]$. Note that

$$H_0 \partial([e_0, e_1]) = H_0([e_1] - [e_0]) = [v_1, w_1] - [v_0, w_0].$$

When we take $\partial H_1([e_0, e_1])$, we get $[w_0, w_1] - [v_0, v_1] = (i_1)_\sharp([e_0, e_1]) - (i_0)_\sharp([e_0, e_1])$, as well as the terms coming from deleting a vertex besides v_0 in the first and w_1 in the last term. There are two types of terms here. The first type comes from deleting a nonrepeated index—here this gives $[v_0, w_0] - [v_1, w_1] = -H_0 \partial([e_0, e_1])$. The other type of terms come from deleting an interior repeated index. Here we get two terms such that $-[v_0, w_1] + [v_0, w_1] = 0$. Putting all of these together gives the formula

$$(\partial H_1 + H_0 \partial)([e_0, e_1]) = (i_1)_\sharp([e_0, e_1]) - (i_0)_\sharp([e_0, e_1]).$$

The formulas we gave above for H_0^X, H_1^X can now be rephrased in a more functorial way. We define them as maps from $S_i(X) \to S_{i+1}(X \times I)$ by using the formula $H_0^X(\sigma) = (\sigma \times \text{id})_\sharp H_0([e_0])$ and $H_1^X(\sigma) = (\sigma \times \text{id})_\sharp(H_1([e_0, e_1]))$. Then

$$\partial_1^X H_0(\sigma) = \partial_1^X(\sigma \times \text{id})_\sharp H_0([e_0]) = (\sigma \times \text{id})_\sharp \partial_1^{\Delta_0} H_0([e_0]))$$

$$= (\sigma \times \text{id})_\sharp((i_1^{\Delta_0})_\sharp([e_0]) - (i_0^{\Delta_0})_\sharp([e_0])) = (i_1^X)_\sharp \sigma_\sharp([e_0]) - (i_0^X)_\sharp \sigma_\sharp([e_0])$$

$$= (i_1^X)_\sharp(\sigma) - (i_0^X)_\sharp(\sigma).$$

Exercise 6.16.4. Verify by an argument analogous to the one given above that

$$(\partial_2^X H_1^X + H_0^X \partial_1^X)(\sigma) = (i_1^X)_\sharp(\sigma) - (i_0^X)_\sharp(\sigma)$$

follows from the similar formula

$$(\partial_2^{\Delta_1} H_1^{\Delta_1} + H_0^{\Delta_1} \partial_1^{\Delta_1})([e_0, e_1]) = (i_1^{\Delta_1})_\sharp([e_0, e_1]) - (i_0^{\Delta_1})_\sharp([e_0, e_1]).$$

We next note that our definition of H_0, H_1 satisfies the further property of naturality.

Exercise 6.16.5. Show that, if $f : X \rightarrow Y$ is a continuous map, then $H_i^Y(f_\sharp(\sigma)) = (f \times \mathrm{id})_\sharp(H_i^X(\sigma)), i = 0, 1$.

We now indicate what is involved in extending this to define $H_i^X(\sigma)$ for $i \geq 2$. We first need to define it in the case of $X = \Delta_i$ and $\sigma = [e_0, \ldots, e_i]$ and show that it satisfies the condition $\partial_{i+1} H_i([e_0, \ldots, e_i]) + H_{i-1}\partial([e_0, \ldots, e_i]) = (i_1)_\sharp([e_0, \ldots, e_i]) - (i_0)_\sharp([e_0, \ldots, e_i])$. We then extend it by using the formula $H_i^X(\sigma) = (\sigma \times \mathrm{id})_\sharp H_i^{\Delta_i}([e_0, \ldots, e_i])$ to show that it satisfies the required formula for X and that it has the same naturality property as in Exercise 6.16.5.

We extend the formula for H_i inductively where we have already defined H_j for $j < i$ on Δ_j and extended it over any X via

$$H_j^X(\sigma) = (\sigma \times \mathrm{id})_\sharp H_j^{\Delta_j}([e_0, \ldots, e_j]).$$

In particular, we will already have it defined for the face maps $F_k : \Delta_{i-1} \rightarrow \Delta_i$ and so we will have the definition of $H_{i-1}^{\Delta_i} \partial_i([e_0, \ldots, e_i])$ determined from this. The general definition of $H_i([e_0, \ldots, e_i])$ in any dimension is given by

$$H_i([e_0, \ldots, e_i]) = \sum_{j=0}^{i} (-1)^j [v_0, \ldots, v_j, w_j, \ldots, w_i].$$

Exercise 6.16.6.
(a) Check that this corresponds to the definition given for $i = 0, 1$ above.
(b) Check that our naturality property gives

$$H_{i-1}([e_0, \ldots, \widehat{e}_k, \ldots, e_i])$$

$$= \begin{cases} \sum_{j=0}^{i-1} (-1)^j [v_0, \ldots, v_j, w_j, \ldots, \widehat{w}_k, \ldots, w_i] & \text{if } k > j, \\ \sum_{j=0}^{i-1} (-1)^j [v_0, \ldots, \widehat{v}_k, \ldots, v_{j+1}, w_{j+1}, \ldots, w_i] & \text{if } k \leq j. \end{cases}$$

Using the last exercise, we can now compute

$$H_{i-1}\partial([e_0, \ldots, e_i])$$

$$= H_{i-1}\left(\sum_{k=0}^{i} (-1)^k [e_0, \ldots, \widehat{e}_k, \ldots, e_i]\right)$$

$$= \sum_{k=0}^{i} \left(\sum_{j=0, j<k}^{i-1} (-1)^{j+k} [v_0, \ldots, v_j, w_j, \ldots, \widehat{w}_k, \ldots, w_i]\right.$$

$$+ \sum_{j=0, j \geq k}^{i-1} (-1)^{j+k} [v_0, \ldots, \widehat{v}_k, \ldots, v_{j+1}, w_{j+1}, \ldots, w_i] \Bigg)$$

$$= \sum_{j,k=0, j<k}^{i} (-1)^{k+j} [v_0, \ldots, v_j, w_j, \ldots, \widehat{w}_k, \ldots, w_i]$$

$$+ \sum_{j,k=0, j>k}^{i} (-1)^{k+j+1} [v_0, \ldots, \widehat{v}_k, \ldots, v_j, w_j, \ldots, w_i].$$

We now look at $\partial_{i+1} H_i([e_0, \ldots, e_i])$.

Exercise 6.16.7. Show that

$$\partial_{i+1} H_i([e_0, \ldots, e_i]) = \sum_{j,k=0, j>k}^{i} (-1)^{j+k} [v_0, \ldots, \widehat{v}_k, \ldots, v_j, w_j, \ldots, w_i]$$

$$+ \sum_{j=0}^{i} [v_0, \ldots, v_{j-1}, w_j, \ldots, w_i] - \sum_{j=0}^{i} [v_0, \ldots, v_j, w_{j+1}, \ldots, w_i]$$

$$+ \sum_{j,k=0, j<k}^{i} (-1)^{j+k+1} [v_0, \ldots, v_j, w_j, \ldots, \widehat{w}_k, \ldots, w_i]$$

Exercise 6.16.8. Show that

$$\sum_{j=0}^{i} [v_0, \ldots, v_{j-1}, w_j, \ldots, w_i] - \sum_{j=0}^{i} [v_0, \ldots, v_j, w_{j+1}, \ldots, w_i]$$

$$= (i_1)_\sharp ([e_0, \ldots, e_i]) - (i_0)_\sharp ([e_0, \ldots, e_i]).$$

Exercise 6.16.9. Combine the calculations from the last exercises to show that

$$(\partial_{i+1} H_i + H_{i-1} \partial_i)([e_0, \ldots, e_i]) = ((i_1)_\sharp - (i_0)_\sharp)([e_0, \ldots, e_i]).$$

We now extend this result to a general space and singular simplex.

Exercise 6.16.10. Show that if we define $H_i^X(\sigma) = (\sigma \times \mathrm{id})_\sharp H_0([e_0, \ldots, e_i])$ and then extend linearly to chains, then this satisfies the condition

$$\partial_{i+1}^X H_i^X + H_{i-1}^X \partial_i^X = (i_1^X)_\sharp - (i_0^X)_\sharp.$$

Moreover, this satisfies the naturality property that when $f : X \to Y$, then $(f \times \mathrm{id})_\sharp H_i^X = H_i^Y f_\sharp$.

Note that the formula we have proved shows that $(i_1^X)_\sharp$ and $(i_0^X)_\sharp$ are chain-homotopic chain maps with the chain homotopy provided by H_i^X. Hence they induce the same map in homology. Thus $(i_0^X)_* = (i_1^X)_*$.

We now apply this to the situation of homotopic maps $f_0 \sim f_1$ via a homotopy $F : X \times I \to Y$. Then $(f_j)_* = F_*(i_j)_*$. Hence $(f_0)_* = (f_1)_*$ and we have proved the homotopy property in the absolute case.

Exercise 6.16.11. Prove that if $f : X \to Y$ is a homotopy equivalence, then $f_* : H_k(X) \to H_k(Y)$ is an isomorphism.

Exercise 6.16.12.
  (a) Adapt the previous exercises to prove the relative case: if $f, g : (X, A) \to (Y, B)$ are homotopic, then $f_* = g_*$.
  (b) Show that if $f : (X, A) \to (Y, B)$ is a homotopy equivalence, then f_* is an isomorphism.

6.17   Proof of the excision property

We now discuss the proof of the excision property of homology in terms of the reformulation Theorem 6.7.1, which we now restate for our convenience: If {int A, int B} is an open cover of X, then the homomorphism $S^{\{A,B\}}(X) \to S(X)$ induces an isomorphism in homology.

In discussing the proof of Theorem 6.7.1, we will mainly restrict our attention to the cases $k = 0, 1, 2$, where the geometric constructions are simpler, and then derive the general case based on these models. The fundamental idea is that since {int A, int B} is an open cover, then small enough images of singular simplices should lie in either int A or int B. Thus we want to replace a singular simplex up to homology by the sum of many small ones. To achieve this technically, we use the notion of barycentric subdivision of the domain. This is defined inductively, with no action on 0 simplices, subdividing a 1-simplex into two 1-simplices of $\frac{1}{2}$ the length, and subdividing a 2-simplex into six 2-simplices using the subdivision on the boundary. A basic estimate used is that, in subdividing an n-simplex, the maximal diameter of a subdivided simplex is reduced by a factor of $n/(n + 1)$. Another key idea, which is similar to that used in the last section, is that it suffices to deal with subdividing the domain Δ_k and then using naturality to define subdivision on a singular simplex which is defined on Δ_k.

We first show pictures of subdividing Δ_1 and Δ_2 (Figure 6.7). For a geometric 1-simplex e in the plane, the barycenter of e with vertices v, w is $\widetilde{e} = \frac{1}{2}(v + w)$. The new 1-simplices are then $[\widetilde{e}, v], [\widetilde{e}, w]$ and their length is half of the original length of $e = [v, w]$. For a 2-simplex $t = [u, v, w]$, we first subdivide the edges, introducing new vertices at the barycenters of $[v, w], [u, w], [u, v]$. Let us name these vertices $\widetilde{e}_0, \widetilde{e}_1, \widetilde{e}_2$, respectively. We introduce a new vertex $\widetilde{t}$ at the barycenter $\widetilde{t} = \frac{1}{3}(u + v + w)$ of t. We then have new 1-simplices

$$[\widetilde{e}_0, v], [\widetilde{e}_0, w], [\widetilde{e}_1, u], [\widetilde{e}_1, w], [\widetilde{e}_2, \widetilde{u}], [\widetilde{e}_2, \widetilde{v}]$$

coming from the subdivision of the boundary of t, new 1-simplices

$$[\widetilde{t}, \widetilde{e}_0], [\widetilde{t}, \widetilde{e}_1], [\widetilde{t}, \widetilde{e}_2], [\widetilde{t}, u], [\widetilde{t}, v], [\widetilde{t}, w]$$

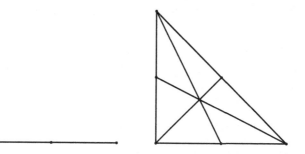

Figure 6.7. Barycentric subdivision of Δ_1 and Δ_2.

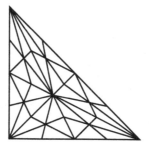

Figure 6.8. Second barycentric subdivision of Δ_2.

coming from joining $\tilde{t}$ to the 0-simplices of the subdivision of the boundary of t, and new 2-simplices

$$[\tilde{t}, \tilde{e}_0, v], [\tilde{t}, \tilde{e}_0, w], [\tilde{t}, \tilde{e}_1, u], [\tilde{t}, \tilde{e}_1, w], [\tilde{t}, \tilde{e}_2, \tilde{u}], [\tilde{t}, \tilde{e}_2, \tilde{v}]$$

coming from joining $\tilde{t}$ to the 1-simplices of the subdivision of the boundary of t. In this subdivision it can be shown that the diameter of any 2-simplex is less than two-thirds of the original diameter. Thus when the process is repeated again and again, the diameter of simplices goes to 0. Through repeated subdivision we can make the subdivided simplices lie within any open cover of the original simplex. We show in Figure 6.8 the result of subdividing Δ_2 twice.

In dimension 0, we know that $H_0(X)$ just measures the path components of X. We need to see that $H_0(S(A) + S(B))$ also measures the path components.

Exercise 6.17.1. Verify the conclusion of Theorem 6.7.1 in dimension 0 by showing the following.

(a) Each path component of X is represented by a point in A or a point in B.

(b) If two points x, y in X are in the same path component, then there is a 1-chain c in $S(A) + S(B)$ with $\partial(c) = x - y$. (Hint: Start with a path connecting x to y and subdivide it so that each subinterval is mapped to A or B. This should use the open cover $\{\text{int}\,A, \text{int}\,B\}$.)

(c) Use (a) and (b) to verify Theorem 6.7.1 in dimension 0.

We break the verification of Theorem 6.7.1 in low dimensions into a number of steps. We first introduce a subdivision operator $\text{Sd}: S_k(X) \to S_k(X)$ for $k = 0, 1, 2$. We start with the standard 2-simplex $\Delta_2 = [e_0, e_1, e_2]$. For a 0-simplex $[x]$, $\text{Sd}[x] = [x]$. For the identity 1-simplex $[e_0, e_1]$,

$$\text{Sd}[e_0, e_1] = [\tilde{e}, e_1] - [\tilde{e}, e_0] = \tilde{e}.\text{Sd}(\partial[e_0, e_1]),$$

where $\tilde{e}$ is the barycenter of $e = [e_0, e_1]$. Here $[a, b]$ denotes the affine linear map from Δ_1 to the segment joining a to b in that order. We are using the notation $x.w = [x, w]$ in forming $\tilde{e} \cdot \text{Sd}(\partial[e_0, e_1])$. More generally, when we start with an affine linear k-simplex $[v_0, \ldots, v_k]$, we form the affine linear $(k+1)$-simplex $v \cdot [v_0, \ldots, v_k] = [v, v_0, \ldots, v_k]$.

Exercise 6.17.2. Show that $\partial(v \cdot [v_0, \ldots, v_k]) = [v_0, \ldots, v_k] - v \cdot \partial[v_0, \ldots, v_k]$.

We then define $\text{Sd}(\sigma) = \sigma_\sharp(\text{Sd}[e_0, e_1])$. Geometrically, what this does is replace a singular simplex by sum of singular simplices (with signs ± 1) which represent the composition of the original singular simplex with affine linear maps into simplices that occur in the subdivision of the geometric simplex. By pulling back an open cover to the domain simplex, we can guarantee that after a finite number of subdivisions, the images of the new subdivisions are contained in elements of the open cover. We extend the definition of Sd linearly over 1-chains.

Exercise 6.17.3. Verify that

$$\partial(\text{Sd}[e_0, e_1]) = \partial(\tilde{e} \cdot \text{Sd}(\partial[e_0, e_1])) = \partial[e_0, e_1] = \text{Sd}(\partial[e_0, e_1]).$$

Extend this to show that for any 1-chain c, we have

$$\partial(\text{Sd}(c)) = \text{Sd}(\partial(c)) = \partial(c)$$

In particular, this says that Sd is a chain map in dimensions ≤ 1.

Exercise 6.17.4. Use the chain map Sd to give another proof that there is an isomorphism from $H_0(S(A) + S(B))$ to $H_0(X)$.

We now want to extend this construction to singular 2-simplices and 2-chains. We start with the identity map $[e_0, e_1, e_2]: \Delta_2 \to \Delta_2$. Using the notation established above for the barycentric subdivision, we define

$$\text{Sd}[e_0, e_1, e_2] = \tilde{t} \cdot \text{Sd}(\partial[e_0, e_1, e_2]).$$

We then define Sd on a singular 2-simplex σ by $\text{Sd}(\sigma) = \sigma_\sharp(\text{Sd}([e_0, e_1, e_2])$ and extend linearly to 2-chains.

Exercise 6.17.5. By starting with the identity singular 2-simplex and then extending to singular 2-simplices and 2-chains, show that there is the formula

$$\partial(\text{Sd}(c)) = \text{Sd}(\partial(c)).$$

Conclude that Sd is a chain map for dimensions ≤ 2.

Exercise 6.17.6. Extend the definition of Sd to a chain map in all dimensions.

We claim that subdividing a 1-chain should not change it up to homology. In order to see this, we need two more operators $H_0 : S_0(X) \to S_1(X)$, $H_1 : S_1(X) \to S_2(X)$, which essentially give a chain homotopy between Sd and the identity Id. The homomorphism $H_0(\sigma_0) = 0$. Note that we have the trivial formula $\partial H_0(c) = (\text{Id} - \text{Sd})(c) = 0$ for any 0-chain c. We want a map H_1 with a similar formula,

$$\partial H_1(c) + H_0 \partial(c) = \partial H_1(c) = (\text{Id} - \text{Sd})(c),$$

in dimension 1.

We start by defining H_1 on the identity singular 1-simplex by

$$H_1([e_0, e_1]) = \tilde{e} \cdot (\text{Id} - \text{Sd} - H_0 \partial)[e_0, e_1].$$

Then

$$\partial H_1([e_0, e_1]) = (\text{Id} - \text{Sd} - H_0 \partial)[e_0, e_1] - \tilde{e} \cdot \partial(\text{Id} - \text{Sd} - H_0 \partial)[e_0, e_1]$$
$$= (\text{Id} - \text{Sd} - H_0 \partial)[e_0, e_1].$$

Exercise 6.17.7. Extend the definition of H_1 to 1-chains so that the formula $\partial H_1 + H_0 \partial = \text{Id} - \text{Sd}$ holds.

Exercise 6.17.8. Suppose that $X = A \cup B$, where $\{\text{int } A, \text{int } B\}$ is an open cover of X. Show that for any singular 1-simplex σ, there is an integer k so that $\text{Sd}^k \sigma$ is a chain in $S(A) + S(B)$. Use this to show that the map $S(A) + S(B) \to S(X)$ induces a surjective map on H_1.

Exercise 6.17.9. Suppose that $X = A \cup B$, where $\{\text{int } A, \text{int } B\}$ is an open cover of X. Suppose $c \in S_1(A) + S_1(B)$ is a singular chain which is a cycle $(\partial c = 0)$, so that when it is considered as a chain in $S_1(X)$, it is a boundary: $c = \partial d$. Show that $\text{Sd}(c) = \partial \text{Sd}(d)$ is also a boundary in X and represents the same homology class as c in $H_1(S(A) + S(B))$. By taking enough subdivisions, show that the map $H_1(S(A) + S(B)) \to H_1(X)$ is injective.

Exercise 6.17.10. Prove Theorem 6.7.1 for dimension 1.

We want to define an analogous map $H_2 : S_2(X) \to S_3(X)$.

Exercise 6.17.11. Show that $\partial(\text{Id} - \text{Sd} - H_1 \partial)(c) = 0$. (Hint: Use the facts Id, Sd are chain maps and the formula proved earlier for $\partial H_1 + H_0 \partial$.)

Exercise 6.17.12. For the standard 2-simplex $[e_0, e_1, e_2]$, define

$$H_2[e_0, e_1, e_2] = \tilde{t}.((\text{Id} - \text{Sd} - H_1 \partial)[e_0, e_1, e_2]).$$

Show that

$$(\partial H_2 + H_1 \partial)[e_0, e_1, e_2] = (\text{Id} - \text{Sd})[e_0, e_1, e_2].$$

Exercise 6.17.13. By extending H_2 to a singular 2-simplex, and finally extending to 2-chains, define a map $H_2 : S_2(X) \to S_3(X)$ so that

$$\partial H_2 + H_1 \partial = \text{Id} - \text{Sd}.$$

Exercise 6.17.14. Prove Theorem 6.7.1 for dimension 2.

Exercise 6.17.15. Extend the definition of H_n to a chain map in all dimensions which satisfies $\partial H_n(c) + H_{n-1} \partial(c) = (\text{Id} - \text{Sd})(c)$.

Exercise 6.17.16. Extend the ideas of the previous exercises to prove Theorem 6.7.1.

Appendix

Selected solutions

1.1.2. (a) A is not open since any ball about $(0, y)$ will contain points with negative first coordinate.

(d) D is not open since $(0,0) \in D$, but every ball about $(0,0)$ contains points with negative first coordinate and so is not contained in D. It is open in A.

1.2.2. $d'((x,y),(u,v)) = |x - u| + |y - v| \geq 0$ and equals 0 iff $x = u$, $y = v$, so $(x,y) = (u,v)$. Since $|x - u| = |u - x|, |y - v| = |v - y|$, then $d'((x,y),(u,v)) = d'((u,v),(x,y))$. Finally,

$$d'((x,y),(w,z)) = |x - w| + |y - z| \leq (|x - u| + |u - w|) + (|y - v| + |v - z|)$$
$$= (|x - u| + |y - v|) + (|u - w| + |v - z|)$$
$$= d'((x,y),(u,v)) + d'((u,v),(w,z))$$

1.2.7. It is not closed since $b \in \mathbb{R} \backslash [a,b)$, but every interval about b contains a point of $[a,b)$, so $\mathbb{R} \backslash [a,b)$ is not open. The set $[a,b)$ is not open since every interval about a contains a point which is less than a and so is not in $[a,b)$.

1.2.11. Since $\bar{A}$ is defined as the intersection of closed sets, it is a closed set. Similarly, int A is open since it is a union of open sets.

1.2.12. (b) $\bar{A} = \{(x,y): y \geq 0\}$, int $A = \phi$, Bd $A = \bar{A}$.

1.3.3. (a) Suppose $\boldsymbol{a}_1 - \boldsymbol{a}_0, \boldsymbol{a}_2 - \boldsymbol{a}_0$ are linearly independent and

$$\lambda_0 \boldsymbol{a}_0 + \lambda_1 \boldsymbol{a}_1 + \lambda_2 \boldsymbol{a}_2 = \boldsymbol{0}, \quad \lambda_0 + \lambda_1 + \lambda_2 = 0.$$

Then $\lambda_0 \boldsymbol{a}_0 = (-\lambda_1 - \lambda_2) \boldsymbol{a}_0$ and we may rewrite $\lambda_0 \boldsymbol{a}_0 + \lambda_1 \boldsymbol{a}_1 + \lambda_2 \boldsymbol{a}_2 = \boldsymbol{0}$ as

$$\lambda_1 (\boldsymbol{a}_1 - \boldsymbol{a}_0) + \lambda_2 (\boldsymbol{a}_2 - \boldsymbol{a}_0) = \boldsymbol{0}.$$

Then linear independence of $\boldsymbol{a}_1 - \boldsymbol{a}_0$, $\boldsymbol{a}_2 - \boldsymbol{a}_0$ implies $\lambda_1 = \lambda_2 = 0$. Combining this with $\lambda_0 + \lambda_1 + \lambda_2 = 0$ implies $\lambda_0 = 0$ as well.

For the converse, suppose $\lambda_0 a_0 + \lambda_1 a_1 + \lambda_2 a_2 = 0$, $\lambda_0 + \lambda_1 + \lambda_2 = 0$ implies $\lambda_0 = \lambda_1 = \lambda_2 = 0$. Then suppose

$$c_1(a_1 - a_0) + c_2(a_2 - a_0) = 0.$$

Rewriting this as

$$(-c_1 - c_2)a_0 + c_1 a_1 + c_2 a_2 = 0,$$

note that the coefficients sum to 0. Thus our assumption on a_0, a_1, a_2 implies that the coefficients all vanish and hence $c_1 = c_2 = 0$.

1.3.5. We just need to see that for each map and its inverse, the inverse image of an open set is open. But each open set in the square is a union of the basic open sets described and the inverse images of these are the basic open sets we gave for the disk. Thus the inverse image of an arbitrary open set is the union of basic open sets and so is open. Continuity in the other direction is proved similarly.

1.4.2. Let $A = A_1 \cup \cdots \cup A_n$, where A_j is compact. Let $\mathcal{U} = \{U_i : i \in I\}$ be an open cover of A. In particular, it gives an open cover of each A_j. Since A_j is compact, there is a finite subcover for each j. The union of these subcovers give a finite subcover for A.

1.4.3. (c) This is not compact since the open cover $\{(1/n, 2)\}$ has no finite subcover.

1.5.2. To see that this subsequence is convergent to x, let U be an open set about x. Since U is open, we may find $\epsilon > 0$ so that the ball $B(x, \epsilon) \subset U$. Then choose N such that $1/N < \epsilon$. Then $n > N$ implies that $s_n \in B(x, 1/N) \subset B(x, \epsilon) \subset U$, so the sequence converges to x.

1.5.4. If f is uniformly continuous and $x \in X$, then given $\epsilon > 0$, uniform continuity implies there exists $\delta > 0$ so that $d(y, z) < \delta$ implies that $d(f(y), f(z)) < \epsilon$. Taking $z = x$ in this definition proves continuity at x.

A counterexample is $f(x) = 1/x$ on the interval $(0, 1]$. For given $\epsilon = 1$, to get $d(f(x), f(y)) < 1$ requires $(1/xy)d(x, y) < 1$ whenever $d(x, y) < \delta$. But the sequence $x_n = 1/n$, $y_n = 1/2n$ has distance $d(x_n, y_n) = 1/2n$, which tends toward zero yet $d(f(x_n), f(y_n)) = n \geq 1$ for all n.

1.6.2. X is connected iff it is not separated iff there do not exist open sets $U, V \subset X$ with $A \subset U \cup V$, $U \cap V \cap A = \emptyset$, $U \cap A \neq \emptyset$, $V \cap A \neq \emptyset$. But this means that whenever U, V are open sets in X with $U \cap V \cap A = \emptyset$, $A \subset U \cap V$, then we must have $A \cap U = \emptyset$ or $A \cap V = \emptyset$. The first case is equivalent to $A \subset V$ and the second case is equivalent to $A \subset U$.

1.6.6. (1) $x \sim x$: use the path $f(t) = x$;

(2) If $x \sim y$, then there is a path $f : I \to X$ with $f(0) = x, f(1) = y$. Then $g(t) = f(1-t)$ is a path connecting y to x since $g(0) = f(1) = y$, $g(1) = f(0) = x$;

(3) If $x \sim y$, $y \sim z$, then there exists paths f, g with $f(0) = x$, $f(1) = g(0) = y$, $g(1) = z$. Then

$$h(t) = \begin{cases} f(2t) & 0 \leq t \leq \frac{1}{2}, \\ g(2t - 1) & \frac{1}{2} \leq t \leq 1 \end{cases}$$

is a path connecting x to z, so $x \sim z$.

1.6.10. $S^1 \backslash \{x\}$ is path connected. If there were a homeomorphism with $h(x) = y$, then $\mathbb{R} \backslash \{y\}$ would have to be path connected, but it is not, since the intermediate value theorem says that there is no path in $\mathbb{R} \backslash \{y\}$ connecting $x < y$ and $z > y$.

1.7.4. The hint describes a continuous map from the square to the triangle which is a bijection except on the bottom edge of the square, which is sent to the bottom vertex of the triangle. This induces a continuous bijection from the quotient space of the square where the bottom edge is identified to a point to the triangle. This map is a homeomorphism, by Proposition 1.7.3.

1.7.9. Take the function which is defined by $f(z) = z^2$ on the upper half of the circle and sends the lower half to 1. This then induces a homeomorphism of the quotient space to S^1.

1.9.3. Following the hint, for each $x \in U$, choose a basis element $B_{i(x)}$ with $x \in B_{i(x)} \subset U$. Then $U = \bigcup_{x \in U} B_{i(x)}$.

1.9.8. Since A is open in B, then $A = B \cap V$, where V is open in X. Since B is open in X and the intersection of two open sets is open, then A is open in X.

1.9.12. (b) $\mathbb{R}$.

1.9.15. (a) Since $\bar{A}$ is the intersection of all closed sets containing A, we have $\bar{A} \subset C$.

1.9.21. Let U be an open set about $f(x)$. Then $f^{-1}(U)$ is an open set about x. Hence there exists N so that $n > N$ implies $x_n \in f^{-1}(U)$. This implies that $f(x_n) \in U$, so the sequence $\{f(x_n)\}$ converges to $f(x)$.

1.9.23. If $x \neq y$, then there are disjoint open sets U, V with $x \in U$, $y \in V$. If $x_n \to x$, then there exists N so that $n > N$ implies $x_n \in U$. This contradicts the sequence converging to y since the tail of the sequence is not in V.

1.9.27. Following the hint, let V, W be disjoint open sets about $x, X \backslash U$, respectively. Note that $X \backslash U \subset W$ implies $X \backslash W \subset U$. Then $D = X \backslash W$ is a closed set with $V \subset D$. Hence $\bar{V} \subset D \subset U$. Combining these statements, we have $x \in V \subset \bar{V} \subset U$.

1.9.32. Let $y \in C_\epsilon$. Then $d(y, C) < \epsilon$, so there exists $c \in C$ with $d(y, c) < \epsilon$. Let $r = \epsilon - d(y, c)$. Then, if $x \in B(y, r)$, the triangle inequality gives $d(x, c) \leq d(x, y) + d(y, c) < r + d(y, c) = \epsilon$. Hence $d(x, C) < \epsilon$ and so $x \in C_\epsilon$.

1.9.41. We use the fact that there is a countable neighborhood basis C_k at each point of X with $C_{k+1} \subset C_k$ as constructed in the last exercise. Then Exercise 1.9.19 says that X is limit point compact and Exercise 1.9.14 says that every neighborhood of a limit point of a set contains infinitely many points of the set. If we have a sequence with only a finite number of values, then we can find a constant, hence convergent, subsequence. Thus we may assume the sequence $\{s_n\}$ assumes an infinite number of values. Then the set of values will have a limit point x. Then if C_k is a neighborhood basis at x, each C_k will contain an infinite number of values. This allows us to select $n_1 < n_2 < \cdots$, so that $s_{n_k} \in C_k$. If U is an open set about x, there exists N so that $x \in C_N \subset U$. If $k > N$, then $s_{n_k} \in C_k \subset C_N \subset U$ so the subsequence converges to x.

1.9.45. Suppose Y is a connected subset of Z, and $\bar{Y} \subset U \cup V$, where $\bar{Y} \cap U \cap V = \emptyset$. Thus $Y \subset U \cup V$, $Y \cap U \cap V = \emptyset$. Since Y is connected, then $Y \subset U$ or $Y \subset V$. Suppose $Y \subset U$. Then let $x \in \bar{Y}$. If $x \in V$, then there is a point of Y in V, which contradicts $Y \subset U$ and $Y \cap U \cap V = \emptyset$. Then $x \in U$ and so $\bar{Y}$ is connected.

1.9.49. If U is the open set and $x \in U$, then there a ball $B(x, r) \subset U$. The ball is path connected since it is convex and straight line segments joining two points in the ball stay in the ball.

1.9.53. (b) A point y is in the component C containing x if there is a connected set C_y which contains both x, y. Then the set C_y also is in the component containing x for the same reason. Thus $C = \bigcup_{y \in C} C_y$. But then C is union of connected sets with the point x in common, so is connected. If two components have a point in common, then their union is connected and every point lies in each component. But this means the two components must be equal. Every point $x \in X$ is in the component of points equivalent to it.

1.9.59. From the definition, choose an open set U and a compact set C with $x \in U \subset C$. Since X is Hausdorff, then C compact implies C is closed. Since C is a closed set containing U, then the closure $\bar{U} \subset C$, since it is the intersection of all closed sets containing U. But $\bar{U}$ is then a closed subset of a compact set, so it is compact.

1.9.63. (a) Define a homeomorphism from $\mathbb{R}$ to $S^1 \backslash \{(0, -1)\}$ as a composition of arctan x with e^{2it}. Alternatively, stereographic projection from $(0, 1)$ gives a homeomorphism from $S^1 \backslash \{(0, 1)\}$ to $\mathbb{R} \subset \mathbb{R}^2$. Then apply the last exercise.

1.9.68. When we remove the crossing point of the X, the space separates into four components. When we remove any point in the Y, the space separates into at most three components. A homeomorphism from Y to T comes from sending the vertical parts to each other and then sending the upper prongs of the Y to the top of the T.

1.9.73. (a) Connected, path connected; (d) connected, path connected, open.

1.9.75. The subset A would have to be connected and compact and miss some point p. Since $S^1 \backslash \{p\}$ is homeomorphic to $\mathbb{R}$, the set would be homeomorphic to a compact connected subset of $\mathbb{R}$, which must be a closed interval. But the circle is not homeomorphic to a closed interval since removing a midpoint of an interval disconnects it but removing any point of the circle does not disconnect it.

1.9.81. Each homeomorphism can come from vertical projection $(x_1, x_2, x_3) \rightarrow (x_1, x_2)$. The map g is the identity since it is induced by taking the circle, sending it to the upper hemisphere and then projecting it back to the circle from the projection of the lower hemisphere.

1.9.87. The inside of the circle is path connected, so it has to be sent to a path-connected set and so the image must lie entirely in the outside component if there were such a homeomorphism. The homeomorphism would have to send the exterior component to the interior component then, or points in the interior component would not be in the image. But the image of the interior plus the circle would be a compact set in the exterior component and thus not all points in the exterior component would be in the image.

1.9.94. The two sets represent the inside of the triangle and the exterior. The first is convex, so path connected via straight line paths. It is bounded. The exterior is path connected as well. To connect two points, go radially on a ray until hitting a point on the circle of radius 2 and then connect the two points on the circle by an arc of the circle. If the two points already lie on the same radial arc, we can just use the segment of the arc connecting them. We still have to show that there is no path connecting the exterior to the interior missing the triangle. But the complement of the triangle is the union of two disjoint nonempty open sets in the plane and, for open sets, the components and the path components are the same. Thus these must be the path components as well.

2.1.1. (c) We use the restriction $g : [0, 1) \rightarrow [0, \infty)$ of part (a). Define $k : B(\mathbf{0}, 1) \rightarrow \mathbb{R}^n$ by

$$k(\mathbf{x}) = \begin{cases} g(|\mathbf{x}|)\mathbf{x}/|\mathbf{x}| & \mathbf{x} \neq \mathbf{0}, \\ \mathbf{0} & \mathbf{x} = \mathbf{0}, \end{cases}$$

with inverse

$$k^{-1}(\mathbf{y}) = \begin{cases} g^{-1}(|\mathbf{y}|)\mathbf{y}/|\mathbf{y}| & \mathbf{y} \neq \mathbf{0}, \\ \mathbf{0} & \mathbf{y} = \mathbf{0}. \end{cases}$$

2.1.6. The homeomorphism must send interior points to interior points and boundary points to boundary points. Thus it must restrict to a homeomorphism between ∂M and ∂N (as well as one between the interiors of the two manifolds).

2.1.11. Suppose that $(x, y) \in M \times N$ and there are homeomorphisms $h_M : U_x \rightarrow \mathbb{R}^m$, $h_N : V_y \rightarrow \mathbb{R}^n$, where U_x, V_y are open sets in M, N, respectively. Then $h : U_x \times V_y \rightarrow \mathbb{R}^m \times \mathbb{R}^n = \mathbb{R}^{m+n}$ with $h(u, v) = (h_M(u), h_N(v))$ is a homeomorphism. The torus $T = S^1 \times S^1$ is then a 2-manifold since it is the product of the 1-manifold S^1 with itself. It is compact and connected since S^1 is and products of compact, connected sets are compact and connected.

2.2.2. (a) Since f is continuous and S^2 is compact and connected, so is P. For any point $x \in S^2$, the map f is a local homeomorphism near x. This means that there is a small open set U about x so that f sends U homeomorphically onto an open set $\bar{U}$ about $f(x)$. We just have to choose U small enough so that it does not contain any pair of antipodal points.

2.2.3. We start with $\mathbb{R}^2$ and identify (x, y) with $(x, y + 2)$. After this identification, the quotient space is the infinite cylinder. To form the Klein bottle, we identify $(x, y) \sim (x + 2, -y)$. After making these identifications, each point will be identified with a point in the rectangle $D^1 \times D^1$, the upper and lower edges will be identified via $(x, -1) \sim (x, 1)$, and the right and left edges will be identified via $(-1, y) \sim (1, -y)$. The map from the plane to the quotient space is a local homeomorphism, so the quotient K is a surface.

2.3.1. Let $x = t + n$, $t \in [a, a+1)$. Then $x + 1 = t + (n+1)$ and $f(x+1) = f'(t) + (n+1) = f(x) + 1$, so f is periodic. The map $f|[a+n, a+n+1]$ is just $T_n f' T_{-n}$ and so is a homeomorphism since it is a composition of homeomorphisms. The piecing lemma then shows that f is a homeomorphism.

2.3.5. By (b) fr is isotopic to the identity via an isotopy F_t, so $f = frr$ is isotopic to r via the isotopy $G_t = F_t r$.

2.4.1. For the Möbius band, think of it as $D^1 \times D^1/(-1, y) \sim (1, -y)$. When we remove the center circle $D^1 \times \{0\}/(-1, 0) \sim (1, 0)$, the space can be divided into the equivalence classes of points M_1 with second coordinate > 0 or M_2 with second coordinate < 0, each of which is path connected from $D^1 \times D^1 \backslash D^1 \times \{0\}$. But the equivalence relation says that some points in M_1 are in M_2, so the whole complement is path connected. In fact, the complement can be shown to be homeomorphic to $S^1 \times (0, 1]$. For the annulus, the complement is just the disjoint union $S^1 \times [-1, 0) \cup S^1 \times (0, 1]$, which is separated.

2.4.4. (a) and (b) are orientable, but (c) is nonorientable.

2.5.1. The first connected sum is formed from $M \backslash i_M(\mathbf{0}) \bigsqcup N \backslash i_N(\mathbf{0})$ using the map $i_N R i_M^{-1}$, and the second connected sum is formed similarly using $i_N' R(i_M')^{-1}$. What is required for consistency is that $h_N i_N R i_M^{-1} = i_N' R(i_M')^{-1} h_M$ on $i_M(\text{int } D^2 \backslash \{0\})$. Composing with i_M on both sides and using $i_M' = k_M i_M$, both sides simplify to give the equivalent equation $h_N i_N R = i_N' R$, which holds since $h_N i_N = i_N'$.

2.5.2. This can be constructed by a coning construction. We divide each region into six triangles and map the triangles to each other via affine linear homeomorphisms determined by the maps on vertices. For each region, we choose the interior point $v_0 = (0, \frac{1}{2})$. For the larger rectangle we use the vertices $v_1 = (1, 0), v_2 = (1, -1), v_3 = (-1, -1), v_4 = (-1, 0), v_5 = (-1, 1), v_6 = (1, 1)$. For the smaller rectangle, we use $v_i' = v_i$, $i = 0, \dots, 4$, but have $v_5' = (-\frac{1}{3}, 0), v_6' = (\frac{1}{3}, 0)$. The map then is determined from sending v_i to v_i' and extending affine linearly on the six triangles. See Figure A.1.

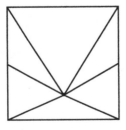 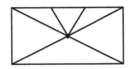

Figure A.1. Sending a big rectangle to a small one.

2.5.7. When we form $M \amalg D^2$, we are starting with M, adding a disjoint disk, and then adding a 1-handle to join the disk to M via the boundary sum. However, the disk is homeomorphic to a rectangle and the union of the disk and the rectangle used in forming the boundary sum is homeomorphic to the union of two rectangles along a common edge, which is a rectangle. Thus our operation of boundary sum is homeomorphic to adding a rectangle to M along an interval in the boundary. Then using an internal rectangle from either a collar on the boundary or just the local structure of a region near the boundary homeomorphic to $\mathbb{H}^2_+$ allows us to use the argument of Exercise 2.5.2 to absorb the external rectangle into M.

2.6.1. First construct a rectangular strip (longer than wide) and cylinder out of fairly stiff cloth. Cut two holes in the strip and attach the cylinder on one side about one-fourth and three-fourths of the way across lengthwise. Now form the Möbius band connected sum the torus by making a half twist and gluing the ends. Then note that when we look at the disk that includes the two handles and the glued edge, the two ends of the handle are glued to opposite sides of this disk.

2.6.5. We look at the 0-handle and the torus pair. This forms $T_{(1)}$, which is a torus with a disk removed. As boundary this has an oriented circle to which the other handles are attached. We replace $T_{(1)}$ with a disk as 0-handle with the other handles attached to it in the same way they were attached to $\partial T_{(1)}$. This is an oriented surface which has two fewer 1-handles and so by induction is of the form $T^{(g)}_{(p)}$. We get our surface from it by removing the disk and gluing to the boundary of $T_{(1)} = T \backslash D^2$. But this is the operation of connected sum, so we get $T^{(g+1)}_{(p)}$.

2.7.2. (a) We prove this by induction using the basic formula $\chi(A \amalg B) = \chi(A) + \chi(B) - 1$, which is the case $n = 2$. We suppose it is true for $n - 1 \geq 2$ and prove it for n:

$$\begin{aligned}
\chi(A_1 \amalg \cdots \amalg A_n) &= \chi((A_1 \amalg \cdots \amalg A_{n-1}) \amalg A_n) = \chi(A_1 \amalg \cdots \amalg A_{n-1}) + \chi(A_n) - 1 \\
&= (\chi(A_1) + \cdots + \chi(A_{n-1}) - (n-2)) + \chi(A_n) - 1 \\
&= \chi(A_1) + \cdots + \chi(A_n) - (n-1).
\end{aligned}$$

2.7.7. (a) The surface is nonorientable with one boundary circle and $\chi = -1$, so it is $P^{(2)}_{(1)}$.

2.9.3. Given $x \in U$, let $b = \sup\{y \in U : y > x, [x, y] \subset U\}$. Since U is open, this set is nonempty. If it is not bounded, then $[x, \infty) \subset U$. If it is bounded, then b will be its supremum or least upper bound. Given any y with $x < y < b$, the definition of b shows that there exists c with $y < c \leq b$ with $[x, c] \in U$. This implies that $[x, b) \subset U$. We claim that $b \notin U$. For if it were, then we would have some interval about b in U, and this would contradict the definition of b as the least upper bound. Thus we can find a maximal interval $[x, b) \subset U$. Analogously, we can find a maximal interval $(a, x] \subset U$ and so, given $x \in U$, there is a largest interval $I_x \subset U$. Any two such intervals are either equal or disjoint. These sets can be labeled by any point in them, and so we can choose a rational point in each set to label it. Since the rationals are countable, the labels are countable, so there will be a countable (possibly finite) number of these sets I_x.

2.9.9. An n-manifold, being locally homeomorphic to $\mathbb{R}^n$, is locally path connected. The result follows from the fact that a locally path connected and connected space is path connected. See Exercise 1.9.51.

2.9.15. Following the hint, let a denote the point on the right side, where three segments join. Take a connected neighborhood of it so that removing a disconnects the neighborhood into three components. If this were 1-manifold, then this neighborhood would be a connected open set in $\mathbb{R}$ and so it is homeomorphic to an open interval, open ray or $\mathbb{R}$. But this is a contradiction since removing a point from one of these only gives two components.

2.9.19. We think of the torus as a quotient of a square $D^1 \times D^1$, where opposite edges are identified via translation and the sphere as the quotient of the the square where all boundary points are identified to the same point. This latter description uses a similar description using the standard disk together with the homeomorphism of the disk to a square. Now the identity map on $D^1 \times D^1$ induces the desired map $T \to S$.

2.9.20. We think of a Möbius band as a quotient of $D^1 \times D^1$ where we identify $(-1, y)$ with $(1, -y)$. We remove the image of $D^1 \times (-\epsilon, \epsilon)$ in the quotient. What remains is the union $D^1 \times [\epsilon, 1] \bigsqcup [-1, -\epsilon]$. When the second piece is turned over vertically, translated by 2, and glued to the first according to identification of its left edge with the right edge of the first, then we can reexpress the complement as a strip $[-1, 3] \times [\epsilon, 1]$, where the vertical edges are identified via translation by 4. This gives a description of the annulus.

2.9.25. We apply the strong form of the Schönflies theorem to the embedded circle $C = f(\partial D^2)$. The ambient isotopy there will send C to S^1 and $f(D^2)$ to D^2 since $f(D^2)$ must map to the closure of the component which is compact. Then $g = G_1 f$ is ambient isotopic to f with $g(D^2) = D^2$ and $g(S^1) = S^1$.

2.9.29. The formula is

$$c_\epsilon(s) = \begin{cases} \epsilon s & s \in [0, 1], \\ \epsilon + (2 - \epsilon)(s - 1) & s \in [0, 2], \\ s & s \in [2, \infty). \end{cases}$$

The isotopy is given by

$$k_t(s) = \begin{cases} [(1 - t) + t\epsilon]s & s \in [0, 1], \\ [(1 - t) + t\epsilon] + (2 - [(1 - t) + t\epsilon])(s - 1) & s \in [1, 2], \\ s & s \in [0, \infty). \end{cases}$$

2.9.37. Note first that if C is coming from the outer circle of our model, we are using the discussion preceding Theorem 2.6.5 and Figure 2.44 to identify with the model above. We first identify the upper annulus to a rectangle with a hole and then use the argument from Exercise 2.5.2 to pull the exterior rectangle with a hole in it into the surface. We then apply our construction of an orientation reversing isotopy of Möbius band to get a self-homeomorphism which is the reverses the orientation of the circle bounding the hole and is the identity on the boundary of this Möbius band. Going back to the original set W, this gives a homeomorphism of W which reverses the orientation on C and is the identity on the rest of ∂W. We then can extend this by the identity outside of W to get a homeomorphism of the nonorientable handlebody that reverses orientation on C and preserves the other boundary circles.

2.9.39. The proof of the classification theorem implies that if a handlebody is nonorientable as a handlebody, then it possesses an embedded Möbius band B. For the standard form with a single 0-handle and the twisted 1-handle attached to it will have a Möbius band inside the 1-handle and a collar about the 0-handle. There is an isotopy of B which connects f and fr. If the Möbius band is embedded in a surface M, then we can extend this isotopy by the identity outside the Möbius band to get that M is disk-nonorientable. From Chapter 6, any handle-oriented surface is disk-oriented. The contrapositive says that a disk-nonorientable handlebody is handle-nonorientable.

2.9.44. We form a disk (the new 0-handle) with the three 0-handles and h_2^1, h_3^1. The 1-handles h_1^1, h_4^1 are attached to the boundary of this 0-handle.

2.9.50. We use the hint. This map is continuous and surjective. The only points which are sent to the same points are the points on the circle which are identified as indicated by the identifications of a. Since the domain is compact and the range is Hausdorff, this gives a homeomorphism of the quotient space with S.

2.9.56. These are the standard ways to write $P^{(3)}$, $P\#K$, and $P\#T$. The fundamental lemma of surface theory says that the last two are homeomorphic and the fact $K = P\#P$ says that the first two are homeomorphic.

2.9.61. The torus. Think of the handle as being the cylinder used to form T from S when two disks are removed.

2.9.66. (c) $S_{(3)}$.

2.9.70. This is nonorientable, has three boundary circles, and has $\chi = -3$. Thus $h = 2 + 3 - 3 = 2$. The surface is $P_{(3)}^{(2)}$.

2.9.75. Think of T as coming from revolving the circle in the yz-plane given by $(y - 2)^2 + z^2 = 1$ about the z-axis. Then remove a small disk near the point $(0, 3, 0)$ by slicing by a plane parallel to the xz-plane. Then reflection through the yz-plane will reverse orientation on the boundary circle of $T_{(1)}$.

2.9.80. The homotopy is given by $F_t(x) = (1 - t)x - tx = (1 - 2t)x$. An isotopy has to be a homeomorphism at each stage and so must either always preserve order or reverse it. In particular, if $F_0(1) = 1$, then $F_t(1)$ is either 1 or -1 and the only way the map can be continuous is for the $F_t(1) = 1$ for all t.

2.9.84. If there is a single 1-handle and M is nonorientable, then it must be attached to a single 0-handle. Otherwise, the union of two 0-handles and this 1-handle would be a disk, and then M would be orientable since there are no more 1-handles which would have to be oriented consistently. Note that the union of the 0-handle and 1-handle to which it is attached must give a Möbius band. Since there are no more 1-handles, the 2-handles do not change connectivity, and M is connected, there can be no more 0-handles. Since the boundary of the Möbius band has only one boundary circle, then there must be exactly one 2-handle. Thus M has three handles and is homeomorphic to a Möbius band with a disk attached to its boundary, which then can be used to give a homeomorphism of M to P by extending the attaching map to a homeomorphism the 2-handle, as in the previous exercise.

2.9.88. If the surface N is orientable and connected then the orientations of the disks that get embedded coming from $\partial(D^2 \times S^1)$ must not be consistent with the orientation of the rest of the surface would be orientable. Hence the result of removing those disks and replacing them with a cylinder will be forming a connected sum $M \simeq N \# K$. Now N is orientable with fewer 1-handles, so it is homeomorphic to $T^{(k)}$ with $2 - 2k = \chi(N)$. Again, we have the relation $\chi(M) = \chi(N) - 2$. But $M \simeq T^{(k)} \# K \simeq P^{(h)}$, where $h = 2k + 2$. Thus $\chi(M) = \chi(N) - 2 = 2 - 2k - 2 = 2 - h$.

2.9.92. For the case of a surgery of index 1, we define the map by using $H_1 : M \backslash f(\{-1, 1\} \times \text{int } D^2) \to M \backslash f'(\{-1, 1\} \times \text{int } D^2)$ and using the identity on $D^1 \times S^1$. A similar homeomorphism is used for the surgery of index 2.

2.9.96. (a) For (a) use the map $f(x, y) = (y, x)$.

2.9.100. That the genus of S is 0 is just the Jordan curve theorem, which implies that any embedded simple closed curve separates S into two disjoint open sets. When we have an embedded circle in T that does not separate T, we can do surgery on T using the extended embedding and write $T = T \# M$. From the classification theorem, we must have $M \simeq S$. An embedded circle in the complement of the first embedded circle in T would give an embedded circle in $S^2 \backslash D^2 \subset M$ and so must separate it.

2.9.104. Since M is nonorientable, there is an embedded Möbius band whose central curve C does not separate M or even a neighborhood of C. Thus there is no extension of C to a neighborhood $D^1 \times S^1$ since C would then separate this neighborhood.

3.1.3. Suppose that b, c are inverses of the group element a. Then

$$b = b(ac) = (ba)c = c.$$

3.1.6. (a) An isomorphism $f : S_2 \to \mathbb{Z}_2$ has $f([12]) = 0$, $f[21] = 1$, with inverse $f^{-1}(0) = [12]$, $f^{-1}(1) = [21]$.

(b) They are not isomorphic. For if $f : (\mathbb{Z}, +) \to (\mathbb{Q}, +)$ is a homomorphism and $f(1) = q$, then $f(n) = nq$. Thus the map is not surjective, as $q/2$ is not in the image, for example, unless $q = 0$ and then the whole image is just $\{0\}$.

3.1.9. If $h = [213]$ and $g = g^{-1} = [321]$, then $ghg^{-1} = [132]$. If $H = \{[123], [213]\}$, then $ghg^{-1} \notin H$.

3.2.3. We use the same basic formula as before, except reverse the roles of $\bar{f}, f$:

$$G(s,t) = \begin{cases} \bar{f}(2st) & \text{if } 0 \le s \le \frac{1}{2}, \\ f(2t(s-1)+1) & \text{if } \frac{1}{2} \le s \le 1. \end{cases}$$

3.3.4. We first check that this is well defined at $\frac{1}{2}$, that $\tilde{f}(1) = \tilde{g}_m(0)$. But this follows since $m = \tilde{f}(1)$. Then continuity follows from the piecing lemma. We compute

$$\widetilde{pf * g}(t) = \begin{cases} p\tilde{f}(2t) & \text{if } 0 \le t \le \frac{1}{2}, \\ p\tilde{g}_m(2t-1) & \text{if } \frac{1}{2} \le t \le 1 \end{cases}$$

$$= \begin{cases} f(2t) & \text{if } 0 \le t \le \frac{1}{2}, \\ g(2t-1) & \text{if } \frac{1}{2} \le t \le 1 \end{cases}$$

$$= f * g.$$

3.3.8. As sets they are the same, so it suffices to show that for representative classes $f_k(s) = e^{2\pi k s}$ used in the isomorphism $\pi_1(S^1, 1) \simeq \mathbb{Z}$, we have $f_m \circ f_n \sim f_{m+n}$. But $f_m \circ f_n(s) = p(ms + ns) = p((m+n)s) = f_{m+n}(s)$.

3.4.2. We use the straight line homotopy $G_t(s) = (1-t)g_0(s) + tg_1(s)$.

3.4.4. We just use the given homotopy on $[a, b]$ and the constant homotopy on its complement in $[0, 1]$.

3.4.8. First note that if we include $D^1 \times S^1$ into $T = S^1 \times S^1$ by sending $D^1 \to S^1_+$ by vertical projection, then this induces a map $D^1 \times S^1/(-1, w) \sim (1, \bar{w}) \to T/(z, w) \sim (-z, \bar{w})$ which is a homeomorphism. Then map $D^1 \times D^1 \to D^1 \times S^1$ by mapping the second D^1 factor via $e^{\pi i t}$ to wrap it once. Then this map induces a map of the quotient space $D^1 \times D^1/(-1, y) \sim (1, -y), (x, -1) \sim (x, 1) = K$ to $D^1 \times S^1/(-1, w) \sim (1, \bar{w})$, which is a homeomorphism. The composition of these two homeomorphisms gives our homeomorphism between K and $T/(z, w) \sim (-z, \bar{w})$. Alternatively, we could get the homeomorphism more directly by using the map $D^1 \times D^1 \to S^1 \times S^1$ sending $(x, y) \to (e^{\pi i x}, e^{\pi i y})$ and check that it induces an isomorphism of quotient spaces $D^1 \times D^1/(-1, y) \sim (1, -y), (x, -1) \sim (x, 1) = K \to S^1 \times S^1/(z, w) \sim (-z, \bar{w})$.

3.5.2. We solve $\langle x + tv, x + tv \rangle = 1$, where $v = x - f(x) \neq 0$, by the quadratic formula to get

$$t = \frac{-\langle x, v \rangle + \sqrt{\langle x, v \rangle^2 + |v|^2(1 - |x|^2)}}{|v|^2}.$$

Since v depends continuously on x, so does t.

3.5.5. (1) f homotopic to g implies $\bar{f}$ is homotopic to $\bar{g}$ and so $\deg f = \deg \bar{f} = \deg \bar{g} = \deg g$.

(2) The map m_r extends with the same definition to a homeomorphism from D^2 to rD^2, so the extension F of f determines an extension $\bar{F}$ of $\bar{f}$. Hence $\deg f = \deg \bar{f} = 0$, by Lemma 3.5.7.

(3) We let $S^1 \times [0,1] \to A(r_1, r_2)$ be defined by $M(z,t) = ((1-t)r_1 + tr_2)z$. Then $M_0 = m_{r_1}, M_1 = m_{r_2}$. The map uFM is a homotopy between $u(F|r_1 S^1)m_{r_1}$ and $u(F|r_2 S^1)m_{r_2}$, so $\deg(F|r_1 S^1) = \deg(F|r_2 S^1)$.

(4) The composition $u f m_r = f$, so this follows by Lemma 3.5.8.

3.6.1. When we take two different radii, then the annular region between the two circles allows us to find a homotopy between the two maps $\underline{v}m_{x,r_1}, \underline{v}m_{x,r_2}$. This is part (3) of Proposition 3.5.10, together with the translation from a neighborhood of $\mathbf{0}$ to a neighborhood of x. The fact that v only vanishes at x means that v defines a map into $\mathbb{R}^2\backslash\{\mathbf{0}\}$, which is required in defining the homotopy.

3.6.7. Following the hint, the composition FG gives a homotopy between f and h where $h(z) = F(\mathbf{0})$. By path connectivity of $\mathbb{C}\backslash\{\mathbf{0}\}$, there is a path $p : I \to \mathbb{C}\backslash\{\mathbf{0}\}$ connecting $F(\mathbf{0})$ and 1. Then $H(z,t) = p(t)$ gives a homotopy between $h(z)$ and the constant map $g(z) = 1$. Combining these two homotopies gives a homotopy between f and g.

3.6.12. The point is that since there are no singularities in the annular type region $A = B(x,r)\backslash\mathrm{int}\, B(z_1, r_1)$, then v extends to a map from A to $\mathbb{R}^2\backslash\{\mathbf{0}\}$. Moreover, there is a parametrization $P : S^1 \times I \to A$ which agrees with m_{z_1,r_1} and $m_{x,r}$ as pictured in Figure 3.19. The composition of P with the extension gives the homotopy, which when further composed with u, shows the two degrees are the same.

3.7.4. We can first homotope f by composing it with a rotation (which is homotopic to the identity) so that $f(1) = 1$. Then we can compose f with $p : I \to S^1$, $p(t) = e^{2\pi i t}$, to get a representative of $\pi_1(S^1, 1)$. This map is homotopic to the map $z^n p$ by our computation of $\pi_1(S^1, 1)$. Moreover, the homotopy preserves the base point 1 and thus induces a homotopy as maps of $(S^1, 1)$ to itself between our rotated f and z^n.

3.7.10. Let d_1, d_2 be the indices on the outer circles and $d_{11}, \ldots, d_{1k}, d_{21}, \ldots, d_{2k}$ be the indices of the vector fields on the inner circles. Then the way the corresponding circles are identified is the same locally as in the analysis of T, so there will again be a relationship $d_1 + d_2 = 2$, $d_{1j} + d_{2j} = 2$. But the sum of the indices of the singularities for each the two pieces is $I_i = d_i - (d_{i1} + \cdots + d_{ik})$. Thus the total index $I = I_1 + I_2 = d_1 + d_2 - \sum_{j=1}^{k}(d_{1j} + d_{2j}) = 2 - 2k$, which is the Euler characteristic of $T^{(k)}$.

3.8.3. (a), (b) The proof of (b) with the subspace the empty space gives (a). The identity map gives a homotopy equivalence of (X, A) to itself. If $f : (X, A) \to (Y, B)$ with homotopy inverse $g : (Y, B) \to (X, A)$, there are homotopies $F : (X, A) \times I \to (X, A)$, $G : (Y, B) \times I \to (Y, B)$ with $F_0 = 1_{(X,A)}, F_1 = gf, G_0 = 1_{(Y,B)}, G_1 = fg$. These same maps then show that there is a homotopy equivalence from (Y, B) to (X, A). Thus the relation is symmetric. Supposing that there is also a homotopy equivalence of pairs $h : (Y, B) \to (Z, C)$ with homotopy inverse k and homotopies H, K, then the map $hf : (X, A) \to (Z, C)$ will have homotopy inverse $gk : (Z, C) \to (X, A)$. The homotopy from $(gk)(hf) = g(kh)f$ will be given by first using H_t to homotope kh to the identity and thus homotoping $g(kh)f$ to gf and then using F_t to homotope gf to the identity. We similarly use G_t and K_t to homotope $(hf)(gk) = h(fg)k$ to the identity.

3.8.7. This just uses the same maps on $(I \cup S^1, 0)$ to $(S^1, (1, 0))$ and homotopy inverse on each factor of S^1. The deformation retractions used there fit together to give the result.

3.9.3. (a) This space deformation-retracts onto the boundary circle union an arc from y to the circle, and its fundamental group is just the fundamental group of the circle, which is $\mathbb{Z}$.

(d) $N \backslash \{x\}$ deformation-retracts onto two circles with a segment joining them, so the fundamental group is again F_2.

3.9.6. All are nonorientable. Part (a) has two boundary circles, and part (b) has one boundary circle. In part (a) the abelianized fundamental group is $\mathbb{Z} \oplus \mathbb{Z}$, while in part (b) it is $\mathbb{Z}$. The effect of adding the 2-handle in part (b) to add the relation $a = b^2$. The space is $P_{(2)}$ in part (a) and is $P_{(1)}$ in part (b).

3.11.1. (a) No, it does not contain an additive identity.

(b) Yes. The identity is $(1, 0)$. The inverse of (z_1, z_2) is $(\bar{z}_1, -z_2)$. We can check associativity. This group represents the unit quaternions.

(c) This is a group. The identity is 1. It is isomorphic to $(\mathbb{Z}_3, +)$ where a in this group corresponds to $a - 1 \in \mathbb{Z}_3$.

3.11.6. The subgroups are multiples $n\mathbb{Z} = \{nk : k \in \mathbb{Z}\}$.

3.11.11. (a) This follows since addition and taking the additive inverse are both continuous maps.

(b) Note first that complex multiplication is a continuous map. This can be shown by the same formula that shows that real multiplication is continuous:

$$|z_1 z_2 - w_1 w_2| \leq |z_1||z_2 - w_2| + |w_2||z_1 - w_1|.$$

To see that inverses are continuous, we note that the inverse is just given by the conjugate $z^{-1} = \bar{z}$ and $|\bar{z} - \bar{w}| = |z - w|$.

(c) Continuity comes from continuity of cosine and sine. That it is a group homomorphism just is the formula $e^{2\pi i(t+u)} = e^{2\pi it}e^{2\pi iu}$ for the exponential (or equivalently, trigonometric formulas for the addition of angles).

3.11.13. (b) This deformation retraction comes from deformation-retracting the matrices R to the identity. Write R as $\left(\begin{smallmatrix} a & b \\ 0 & c \end{smallmatrix}\right)$. Then

$$R_t = \begin{pmatrix} (1-t)a + t & (1-t)b \\ 0 & (1-t)c + t \end{pmatrix}$$

satisfies $R_0 = R$ and $R_1 = I$. The map $F(R,t) = R_t$ gives a deformation retraction of the upper triangular matrices with positive diagonal entries onto the identity matrix. Then $F(QR,t) = QR_t$ gives a deformation retraction of $GL(2,\mathbb{R})$ onto $O(2)$.

3.11.21. $f \circ g \sim (f * E) \circ (E * g) = f * g$.

3.11.23. (a) This just uses the deformation retraction of $S_b^1 \backslash \{-1\}$ to the point 1 which can be written as $F(e^{is},t) = e^{ist}$ for $-\pi < s < \pi$, $0 \leq t \leq 1$. On T we use $G(x,y,t) = (x, F(y,t))$.

3.11.27. (a) If we think of $M = D^1 \times D^1/(-1, y) \sim (1, -y)$, then we can take $C = D^1 \times \{0\}/\sim$. This deformation-retracts onto the boundary circle by deformation-retracting $D^1 \times (D^1 \backslash \{0\})$ onto $D^1 \times \{\pm 1\}$ by using the straight line homotopy in the second coordinate and noting that this is consistent with the identifications.

3.11.31. Following the hint, we define $F(s,t) = f(ts)$. Then if f, g are two paths, let F be the free homotopy to $f(0)$ and G the free homotopy to $g(0)$. Let $h(t)$ be a path connecting $f(0)$ to $g(0)$. Then the free homotopy H between f and g is given by

$$H(s,t) = \begin{cases} F(s, 3t) & 0 \leq t \leq \frac{1}{3}, \\ h(3t - 1) & 1/3 \leq t \leq \frac{2}{3}, \\ G(s, 3 - 3t) & 2/3 \leq t \leq 1. \end{cases}$$

3.11.38. Since $\pi_1(\mathbb{R}^2 \backslash \{0\}, 1)$ is abelian, the map $\pi_1(\mathbb{R}^2 \backslash \{0\}, 1) \to \pi_1^f(\mathbb{R}^2 \backslash \{0\})$ is bijective. Since the inclusion induces an isomorphism between $\pi_1(S^1, 1)$ and $\pi_1(\mathbb{R}^2 \backslash \{0\}, 1)$, the map j is bijective.

3.11.43. We compute

$$P_v = (M_x \circ f) f_u^1 f_v^1 + (M_y \circ f) f_u^1 f_v^2 + (N_x \circ f) f_u^2 f_v^1 + (N_y \circ f) f_u^2 f_v^2$$
$$\quad + (M \circ f) f_{uv}^1 + (N \circ f) f_{uv}^2,$$
$$Q_u = (M_x \circ f) f_u^1 f_v^1 + (M_y \circ f) f_u^2 f_v^1 + (N_x \circ f) f_u^1 f_v^2 + (N_y \circ f) f_u^2 f_v^2$$
$$\quad + (M \circ f) f_{uv}^1 + (N \circ f) f_{uv}^1.$$

Comparing these using $M_y = N_x$ gives $P_v = Q_u$.

3.11.48. This calculation is equivalent to substituting $x = r^n \cos 2\pi nt$, $y = r^n \sin 2\pi nt$ and integrating from 0 to 1. We get

$$d(z^n) = \frac{1}{2\pi} \int_0^1 2\pi n \, dt = n.$$

3.11.55. If there were such a map, then $h(g)h(i) = 1_{h(S^n)}$. But this is impossible since this map factors through a map of the trivial group.

4.1.1. Let $\{U_i\}$ be a covering of B by path-connected open sets that are evenly covered. That is, $p^{-1}(U_i) = \bigsqcup_{j \in \mathcal{J}_i} U_{ij}$ and $p : U_{ij} \to U_i$ is a homeomorphism. Then $B \times C$ has a covering by the path-connected open sets $\{U_i \times C\}$ since $P^{-1}(U_i \times C) = \bigsqcup_{j \in \mathcal{J}_i} U_{ij} \times C$ and $P|U_{ij} \times C : U_{ij} \times Y \to U_i \times C$ is a homeomorphism, being a product of homeomorphisms.

4.1.7. (a) The inverse image of the added cylinder is the disjoint union of two cylinders, so we can find evenly covered open sets for points in its interior. For points outside the added cylinder, we find the evenly covered open sets from the original cover. For points on the boundary of the added cylinder, we get our evenly covered open set by taking a disk (up to homeomorphism) which comes from a half disk in the cylinder and a half disk in the complement of the interior of the cylinder.

4.1.13. If a surface Σ covers itself, we get $\chi(\Sigma) = k\chi(\Sigma)$, $k > 1$ and so $\chi(\Sigma) = 0$. Hence Σ must be either T or K.

4.1.18. Let $f : I \to B$ with $f(0) = b$ be a given path. Cover B by path-connected open sets $\{U_i\}$ which are evenly covered. Then $\{f^{-1}(U_i)\}$ is an open cover of I. Choose a Lebesgue number for this cover and subdivide I into subintervals of equal length less than this Lebesgue number. Note that for a map from a path-connected set into some U_i, there is a unique lift to the cover once we specify where one point lifts. We then write $I = I_1 \cup I_2 \cup \cdots \cup I_n$ as the union of adjacent subintervals in the subdivision. Starting with $f|I_1$, we lift it to the unique map to the cover g_1 with $g_1(0) = a$. Letting $g(1/n) = a_1$, we next lift $f|I_2$ to the unique map to the cover g_2 with $g_2(1/n) = a_1$. Inductively, after we have defined lifts $g_1, \ldots, g_k$, we then lift $f|I_{k+1}$ to $g_{k+1} : I_{k+1} \to A$ with $g_{k+1}(k/n) = g_k(k/n)$. We define $\tilde{f}$ by $\tilde{f}|I_j = g_j$. Continuity follows from the piecing lemma and uniqueness follows from uniqueness over each subinterval.

4.2.1. Let e_a, e_b denote the constant maps at a, b, respectively. Following the hint, we start with a class $[\tilde{f}] \in \pi_1(A, a)$ which maps to $[e_b]$, and so there is a homotopy F between $p\tilde{f}$ and e_b. Use Theorem 4.1.3 to lift this to a homotopy $\tilde{F} : I \times I \to A$ with $\tilde{F}(0,0) = a$. By unique path lifting, this must be a homotopy relative to the end points between $\tilde{f}$ and e_a. The map p_* is a homomorphism, so it is 1–1 since its kernel consists of the identity element. Thus the image $p_*(\pi_1(A, a))$ is a subgroup of $\pi_1(B, b)$ which is isomorphic to $\pi_1(A, a)$.

4.2.6. (a) To show that this is a covering space, we must show that it is evenly covered. Near the wedge point, we see an X-like neighborhood which is covered by homeomorphic open sets at each point $(n, 0), (0, m)$ in the cover. For each point on the circle labeled a away from this wedge point, take an arc that does not include the wedge point. Above it in the cover are homeomorphic intervals along the x-axis and homeomorphic arcs in the a circles which are attached to the y-axis. The argument is similar for the b circle with the roles of the axes reversed.

(b) If we start with the loop in $p_*(\pi_1(A,(1,0))$ represented by the projection of the loop that runs around S^1_y in the counterclockwise direction (denoted b in the figure), then when it is lifted starting at $(0,0)$, it lifts to the line segment from $(0,0)$ to $(0,1)$, which is not a loop.

4.2.9. (a) We use $p\widetilde{f} = f$ to get $f_*(\pi_1(X,x)) = p_*(\widetilde{f}_*(\pi_1(X,x))) \subset p_*(\pi_1(A,a))$ since $\widetilde{f}_*(\pi_1(X,x)) \subset \pi_1(A,a)$.
   (d) This follows from Theorem 4.1.3.

4.2.13. By Theorem 4.2.2, there is a continuous lifting $g : A_1 \to A_2$ of p_1 so that $g(a_1) = a_2$. Similarly, there is a lifting $h : A_2 \to A_1$ so that $h(a_2) = a_1$. Then the composition hg is a lift of p_1 with $hg(a_1) = a_1$, and so, by Exercise 4.2.11, $hg = 1_{A_1}$. Similarly, we get $gh = 1_{A_2}$ and so g gives the equivalence we seek.

4.2.18. h is a lifting of p_2 and the lifting is unique sending a to $h(a)$. This lifting is an equivalence, by the proof of Theorem 4.2.2.

4.3.1. Since T is a homeomorphism, it induces an isomorphism between $\pi_1(A,a_1)$ and $\pi_1(A,a_2)$ and

$$G_1 = (p_1)_*(\pi_1(A,a_1)) = (p_2T)_*(\pi_1(A,a_1)) = (p_2)_*(T_*(\pi_1(A,a_1)))$$
$$= (p_2)_*(\pi_1(A,a_2)) = G_2.$$

4.3.6. Suppose $T(0) = n$. Then T and T_n are covering transformations which agree at a point, and so they are equal, by Exercise 4.3.4.

4.3.10. (a) If S is the homotopy between s and s', then pS gives the homotopy between ps and ps'.
   (d) Suppose $r(T_1) = r(T_2)$. This means that the loops representing $r(T_1)$ and $r(T_2)$ lift to paths joining a and the same point $T_1(a) = T_2(a)$. But two covering transformations which agree a point are equal, by Exercise 4.3.4.

4.3.14. From the hint, there is a 1–1 correspondence between

$$\{c \in p^{-1}(b)\colon \text{there is a covering transformation } T \text{ with } T(a) = c\}$$

and G_p. Combining this with Theorem 4.3.2 gives the result.

4.4.1. This is just a special case of Theorem 4.2.3.

4.4.5. With this restriction on α, if $[\alpha_1] \neq [\alpha_2]$ then $\widetilde{U}_{[\alpha_1]} \cap \widetilde{U}_{[\alpha_2]} = \emptyset$. For if they had a point in common, they would be equal and then $\alpha_2 = \alpha_1 * \beta$, where β is a loop at c in U. But the construction of U implies that β is homotopic to a constant in B. This implies that $\alpha_1 \sim \alpha_2$.

4.4.11. (a) The map $p : A \to B$ factors as the composition of $q : A \to A/G$, and the homeomorphism $\bar{p} : A/G \to B$, by Theorem 4.3.3. Now the factorization being alluded to just uses the factorization $A \to A/H \to A/G$. That $A \to A/H$ is a covering map just uses that G acts properly discontinuously on A and so does any subgroup H. Thus $p_1 : A \to A/H$ is also a regular covering space. That

$p_2 : A/H \to B$ is a covering space uses the same evenly covered open sets $U \subset B$ as for p, but now we identify in A/H all of the homeomomorphic copies of U in the inverse image in A which differ by the action of covering transformations in H.

5.1.4 (a) As described above, X^1 can be identified with the circle by identifying e_1^0 to 1 and e_2^0 to -1. Then the attaching maps of the 1-cells are consistent with sending e_1^1 to the upper half of the circle and e_2^1 to the lower half as depicted.

5.1.8. The subcomplex is the image of a finite number of cells, so it is compact. It is closed since a compact set is closed in a Hausdorff space.

5.1.13. We start with the CW decompositions $S_i^2 = e_i^0 \cup e_i^2$, $i = 1, 2$. Then there are four cells

$$E^0 = e_1^0 \times e_2^0, \qquad E_1^2 = e_1^2 \times e_2^0, \qquad E_2^2 = e_1^0 \times e_2^2, \qquad E^4 = e_1^2 \times e_2^2.$$

The 2-cells are attached trivially and $X^2 = S_1^2 \vee S_2^2$. The characteristic map for the 4-cell is the product of the characteristic maps for the two 2-cells. Its attaching map uses $\partial E^4 = \partial e_1^2 \times e_2^2 \cup e_1^2 \times \partial e_2^2$. The first term is mapped first to $e_1^0 \times e_2^2$ and then mapped via the characteristic map to $e_1^0 \times S_2^2$. The second term is mapped similarly onto $S_1^2 \times e_2^0$.

5.1.15. We prove the result by induction on n. The case $n = 1$ is established above. For the inductive step, we have to see that $\mathbb{CP}^k$ is built from $\mathbb{CP}^{k-1}$ by attaching a $2k$-cell as claimed. Any point in S^{2k+1} is equivalent to a point $(w_0, \ldots, w_{k-1}, r)$ with $r \geq 0$. Write the last coordinate as $z_k = r\zeta$, where $r = |z_k| \geq 0$. When $r > 0$, we can take $\zeta = z_k/|z_k|$. Then multiplying each coordinate by ζ^{-1} gets an equivalent representative of the required form. When $r > 0$, no two elements of this special form are equivalent. However, when $r = 0$, we could choose any unit complex number ζ. In this case, the point in the quotient space lies in $\mathbb{CP}^{k-1}$. Thus $\mathbb{CP}^k = \mathbb{CP}^{k-1} \cup_{p_{k-1}} S_+^{2k} \simeq \mathbb{CP}^{k-1} \cup_{p_{k-1}} e^{2k}$. Here we are identifying the upper hemisphere to a $2k$-cell via vertical projection.

5.2.1. Let $x = (1, 0) \sim e^0$ be our base point and $y = (1/2, 0)$. The Seifert–van Kampen theorem says that $\pi_1(X, y) \simeq \pi_1(A, y) *_{\pi_1(A \cap B, y)} \pi_1(B, y)$. Here $\pi_1(B, y)$ is trivial since the open disk is contractible, so we get $\pi_1(X, y) \simeq \pi_1(A, y)/N(\text{im}(\pi_1(A \cap B, y)))$. Here $N(\text{im}(\pi_1(A \cap B, y))$ denotes the normal subgroup generated by the image. But $A \cap B$ is just $D^2 \backslash \{0\}$ and its fundamental group has a generator running around a circle at radius $\frac{1}{2}$. When we take its image in $\pi_1(A, y)$, it is homotopic to the path $\gamma * \alpha * \bar{\gamma}$, where α runs once around the unit circle and γ is the linear path running from $(\frac{1}{2}, 0)$ to $(1, 0)$. Its image in $\pi_1(A, y)$ is represented by $\gamma * f\alpha * \bar{\gamma}$. Using the isomorphism $\gamma_* : \pi_1(X, x) \to \pi_1(X, y)$, this quotient is isomorphic to $\pi_1(X, x)/N(f_*[\alpha])$. The term $N(f_*[\alpha])$ is the term $f_*(g)$ referred to in the statement of the exercise.

5.2.7. We show that when we attach a cell of dimension > 2, then the fundamental group does not change up to isomorphism. Let $X = Y \cup e^{n+1}$, $n \geq 2$. Let $A = X \backslash \{0\}$, $B = \text{int } e^{n+1}$. Then $A \cap B$ is homeomorphic to $D^{n+1} \backslash \{0\}$, which deformation retracts to S^n. The Seifert–van Kampen theorem implies $\pi_1(S^n)$ is trivial for $n \geq 2$ since it can be written as the union of two sets homeomorphic to $\mathbb{R}^n$ with path-connected intersection. Returning to our calculation, then $\pi_1(B) \simeq \pi_1(A \cap B)$ being trivial and the Seifert–van Kampen theorem imply $\pi_1(X) \simeq \pi_1(Y)$. Inductively, this implies that attaching cells of dimension greater than 2 does not change the fundamental group and so $\pi_1(X, e^0) \simeq \pi_1(X^2, e^0)$.

5.2.10. The 2-skeleton is $S^1 \times S^1 \cup e^2$, where the 2-cell is attached via the map $f(z) = z^2$ running around the second S^1 (from $\mathbb{RP}^2$). Thus its fundamental group is $\langle a, b | ab = ba, b^2 = 1 \rangle$, which is $\mathbb{Z} \oplus \mathbb{Z}_2$. This agrees with the calculation as a product space since $\pi_1(S^1) \simeq \mathbb{Z}$ and $\pi_1(\mathbb{RP}^2) \simeq \mathbb{Z}_2$.

5.2.16. S^3 is simply connected. It is a covering space of $L(p, 1)$ since the homeomorphisms which we are using to form the quotient act in a properly discontinuous manner. Thus the fundamental group is the cyclic group $\mathbb{Z}_p$. From the CW decomposition, the 2-skeleton is the pseudoprojective plane, whose fundamental group is $\mathbb{Z}_p$.

5.3.1. (a) We follow the hint to define the map $\alpha : X_f \to X_{f'}$. It will be induced from $G : X \bigsqcup D^n \to X \cup_{f'} D^n$. The map $G|X$ includes $X \subset X \cup_{f'} D^n$. To define $G|D^n$, we first write $D^n = 0.5D^n \cup (D^n \backslash \text{int } 0.5D^n)$. Here $0.5D^n$ denotes the disk of radius 0.5. We can identify $D^n \backslash \text{int } 0.5D^n$ with $S^{n-1} \times [0, 1]$ by using the inverse of the homeomorphism $H(\boldsymbol{x}, t) = 0.5(1 + t)\boldsymbol{x}$. We then define $G|D^n \backslash \text{int } 0.5D^n :$ $D^n \backslash \text{int } 0.5D^n \to X \subset X \cup_{f'} D^n$ by using the composition FH^{-1}. Note that this sends S^{n-1} to X via f and $0.5S^{n-1}$ to X via f'. Then map $0.5D^n$ to D^n via $\boldsymbol{x} \to 2\boldsymbol{x}$ and compose with the quotient map $D^n \to X \cup_{f'} D^n$. Note that on $0.5S^{n-1}$ this agrees with f'. We then apply Proposition 1.7.5 to get the map α induced by G to be continuous. We have the following diagram of continuous maps:

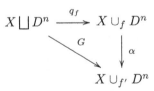

5.3.2. (a) X_f is a quotient space of $X \bigsqcup D^n$ and Y_{hf} is a quotient space of $Y \bigsqcup D^n$. The map $K : X \bigsqcup D^n \to Y \bigsqcup D^n$ given by $K|X = h, K|D^n = \text{id}$ induces a map $\alpha : X \cup_f D^n \to Y \cup_{hf} D^n$ which is continuous by Proposition 1.7.5. We have the following diagram of continuous maps:

$$
\begin{array}{ccc}
X \bigsqcup D^n & \xrightarrow{\ K\ } & Y \bigsqcup D^n \\
\downarrow{\scriptstyle q_f} & & \downarrow{\scriptstyle q_{hf}} \\
X \cup_f D^n & \xrightarrow{\ \alpha\ } & Y \cup_{hf} D^n
\end{array}
$$

(f) To use (e) here, we consider maps up to homotopy and note that it says that if a map has a left homotopy inverse and a right homotopy inverse, then it is invertible and the two partial inverses are equal up to homotopy and are each homotopy inverses of the map. Note that $\gamma = \epsilon\beta$ and so $\epsilon\beta\alpha \sim$ id. Since ϵ is a homotopy equivalence and $\beta\alpha$ is a right homotopy inverse, it is also a left homotopy inverse. Hence $\beta\alpha\epsilon \sim 1$. Hence β has a right homotopy inverse. However, the same argument which shows $\alpha : X_f \to X_{hf}$ has a left homotopy inverse shows that $\beta : Y_{hf} \to Y_{ghf}$ has a left homotopy inverse. Thus β is a homotopy equivalence, and so is γ. Since α is a right homotopy inverse to γ, it must also be a left homotopy inverse and so α is also a homotopy equivalence.

5.3.4. (a) Since X is connected, the 1-skeleton X^1 must also be connected since 2-cells are attached along connected sets and cannot change a separated set to a connected set. If each 1-cell was attached to a single 0-cell, then the number of components would not change in going from X^0 to X^1, which contradicts X^0 being separated and X^1 connected.

5.3.8. By the previous exercise, the 0- and 1-handles are homotopy equivalent to a CW complex K with h_0 0-cells and h_0 1-cells. Exercise 5.3.2 implies that the surface is homotopy equivalent to K with h_2 2-cells attached, which is the required CW complex.

5.3.13. If $J : S^{k-1} \times I \cup D^k \times \{0\} \to Y$ is given by $J(u,t) = H(f(u),t)$, $u \in S^{k-1}$, $J(v,0) = g(v)$, $v \in D^k$, then the lemma says that there is an extension $J' : D^k \times I \to Y$. Then the map $H' : K \times I \to Y$ defined by using H on $L \times I$ and J' on $D^k \times I$ gives a continuous extension of H.

5.4.2. The inclusions induce a commutative diagram with vertical isomorphisms:

$$
\begin{array}{ccccc}
\pi_1(A, x_0) & \longleftarrow & \pi_1(A \cap B, x_0) & \longrightarrow & \pi_1(B, x_0) \\
\downarrow & & \downarrow & & \downarrow \\
\pi_1(U, x_0) & \longleftarrow & \pi_1(U \cap V, x_0) & \longrightarrow & \pi_1(V, x_0)
\end{array}
$$

This implies that the inclusions fit together to induce an isomorphism between $\pi_1(A, x_0) *_{\pi_1(A \cap B, x_0)} \pi_1(B, x_0)$ and $\pi_1(U, x_0) *_{\pi_1(U \cap V, x_0)} \pi_1(V, x_0)$.

5.4.6. (a) Here we use that $T_{(1)} = T \backslash D^2$ is homotopy equivalent to a wedge of two circles and the $\pi_1(\partial T_{(1)})$ is mapped homotopically to the class $aba^{-1}b^{-1} \in F_2 = \pi_1(T_{(1)}, x)$. Using this on both factors, we get

$$
\pi_1(T \# T, x) \simeq \langle a, b, c, d | aba^{-1}b^{-1}cdc^{-1}d^{-1} \rangle.
$$

5.5.1. The subdivision of the rectangle on the right-hand side of Figure 5.15 leads to the structure of a simplicial complex for the quotient space giving the projective plane.

6.1.1. We have $H_0(C) = \ker(\partial_0)/\operatorname{im}(\partial_1) = \mathbb{Z}/0 \simeq \mathbb{Z}$; $H_1(C) = \ker(\partial_1)/\operatorname{im}(\partial_2) = \mathbb{Z}/p\mathbb{Z} \simeq \mathbb{Z}_p$; and $H_2(C) = \ker(\partial_2)/0 = 0$. Since $C_i = 0$, we have $H_i(C) = 0$, $i \geq 3$.

6.1.5. (a) $H_0(C) = \ker(\partial_0)/\mathrm{im}(\partial_1) = \mathbb{Z}$. For $0 < i = 2k+1 < n$, then $H_i(C) = \ker(\partial_i)/\mathrm{im}(\partial_{i+1}) = \mathbb{Z}/2\mathbb{Z} \simeq \mathbb{Z}_2$. For $0 < i = 2k < n$, then $H_i(C) = \ker(\partial_i)/\mathrm{im}(\partial_{i+1}) = 0$. Since $C_i = 0$ for $i > n$, we have $H_i(C) = 0$, $i > n$.

6.2.1. (a) We have $\partial_1\partial_2([v_0, v_1, v_2]) = \partial_1([v_1, v_2] - [v_0, v_2] + [v_0, v_1]) = [v_2] - [v_1] - [v_2] + [v_0] + [v_1] - [v_0] = 0$.

6.2.4. We have $\partial_1(a) = \partial_1(b) = \partial_1(c) = 0, \partial_2([v_1, v_2, v_3]) = a + b - c, \partial_2([v_1, v_3, v_4]) = a - b + c$. Then

$$\partial_2(n_1[v_1, v_2, v_3] + n_2[v_1, v_3, v_4]) = (n_1 + n_2)a + (n_1 - n_2)(b - c).$$

Then $n_1[v_1, v_2, v_3] + n_2[v_1, v_3, v_4] \in \ker(\partial_2)$ implies $n_1 = n_2 = 0$. Thus $H_2^\Delta(K) = 0$. Since $\mathrm{im}(\partial_2)$ is generated by $a + (b - c), a - (b - c)$, this introduces relations in the quotient $\ker(\partial_1)/\mathrm{im}(\partial_2)$, so $[a] = [b - c] = -[b - c]$. Then the quotient is generated by $[a], [b]$ with $2[a] = 0$. Hence $H_1^\Delta(K) \simeq \mathbb{Z}_2 \oplus \mathbb{Z}$. Since $\partial_1 = 0, H_0^\Delta(K) = \mathbb{Z}$. All $H_i^\Delta(K) = 0$ for $i > 2$ since there are no i-simplices.

6.3.2. If X is path connected, then $\mathrm{im}(\partial_1)$ is the subgroup generated by the differences $[x] - [y]$, where $x, y \in X$. Thus the quotient $H_0(X)$ is the free abelian group generated by $[x]$, where x is any chosen point in X.

6.3.7. The map h is defined on loops at x_0. The previous exercise says that homotopic loops are mapped to cycles which differ by a boundary, hence $[\eta] = [\gamma] \in \pi_1(X, x_0)$ implies $[h(\eta)] = [h(\gamma)] \in H_1(X)$. Hence h induces a well defined map $\bar{h}([\gamma]) = [h(\gamma)]$.

6.3.12. For $F_2 * F_0 * \bar{F}_1$ this holds, since Δ_2 contractible implies $\pi_1(\Delta_2, e_0)$ is trivial. The statement for $D_2 * D_0 * \bar{D}_1$ follows since this represents $D_*([F_2 * F_0 * \bar{F}_1])$.

6.4.2. There is one cell in dimensions 0,1,2. We have $\partial_0 = \partial_1 = 0$ and ∂_2 sends the generator to the generator. Thus $H_0^c(D) = \mathbb{Z}; H_1^c(D) = 0 = H_2^c(D)$.

6.4.7. (a) Here there is one 0-cell, three 1-cells with corresponding generators a, b, c, and one 2-cell, with ∂_2 sending the generator to $2a + c$. Then $H_0^c(X) = \mathbb{Z}$; $H_1^c(X) = 2\mathbb{Z}; H_2^c(X) = 0$.

6.5.3. Let $e_1, \ldots, e_n$ be generators of E. Then choose $d_1, \ldots, d_n$ so that $g(d_i) = e_i$. We define $r(e_i) = d_i$. Since E is free abelian, we can extend r uniquely to a homomorphism $r : E \to D$. Since any element e of E is expressible uniquely as $e = \sum n_i e_i$, then $r(e) = \sum n_i d_i$ and $g(r(e)) = e$. Thus gr is the identity map. To show that I is an isomorphism, we show that it is 1–1 and onto. Let $d \in D$. Look at $d - r(g(d))$. Then $g(d - rg(d)) = g(d) - gr(g(d)) = g(d) - g(d) = 0$. Hence there is an element $c \in C$ with $f(c) = d - r(g(d))$, and so $d = f(c) + r(g(d)) = I(c, g(d))$. Thus I is onto. Next suppose that $I(c, e) = f(c) + r(e) = 0$. Then $0 = g(f(c) + r(e)) = gr(e) = e$ and so $e = 0$. Then $f(c) = 0$. Since f is 1–1, this means $c = 0$ as well, so $(c, e) = 0$. Hence I is 1–1. To see that the diagram commutes, note that $I(c, 0) = f(c)$, and $g(I(c, e)) = g(f(c) + r(e)) = gr(e) = e$.

6.5.8. First suppose $[e] \in \operatorname{im}(g_*)$. Then $[e] = g_*([d]) = [g(d)]$ with $\partial(d) = 0$. Since the definition of ∂ does not depend on the representative of $[e]$ chosen, we start with $g(d)$ and apply the definition to get $\partial([e]) = [c]$, where $f(c) = \partial(d)$. But then $f_*([c]) = [f(c)] = [\partial(d)] = 0$. Thus $[e] \in \ker(\partial)$. Next suppose that $[e] \in \ker(\partial)$. Then there are elements d, c, γ with $g(d) = e, f(c) = \partial(d), c = \partial(\gamma)$. We compute $\partial(d - f(\gamma)) = \partial(d) - \partial f(\gamma) = \partial(d) - f(\partial(\gamma)) = \partial(d) - f(c) = 0$. Thus $d - f(\gamma)$ represents a homology class and $g_*([d - f(\gamma)]) = [g(d) - gf(\gamma)] = [g(d)] = [e]$. Thus $[e] \in \operatorname{im}(g_*)$.

6.5.12. Suppose $c \in C_k$ and $\partial_k^C(c) = 0$. Then $G_k(c) - F_k(c) = \partial_{k+1}^D H_k(c) + H_{k-1}\partial_k^C(c) = \partial_{k+1}^D(H_k(c))$, so $G_*([c]) = [G_k(c)] = [F_k(c)] = F_*([c])$.

6.6.2. It suffices to check this on a generator. Then $f_\sharp \partial_k^X(\sigma) = f_\sharp(\sum_{i=0}^k (-1)^i \sigma F_i) = \sum_{i=0}^k (-1)^i f(\sigma F_i) = \sum_{i=0}^k (-1)^i (f\sigma) F_i = \partial_k^Y(f\sigma) = \partial_k^Y(f_\sharp(\sigma)) = \partial_k^Y f_\sharp(\sigma)$.

6.6.8.

$$(gf)_*([c]) = [(gf)_\sharp(c)] = [g_\sharp(f_\sharp(c))] = g_*([f_\sharp(c)]) = g_* f_*([c]),$$
$$(1_{(X,A)})_*([c]) = [(1_{(X,A)})_\sharp(c)] = [c].$$

6.7.2. (a) Define $i_\sharp : S(A \cap B) \to S(A) \oplus S(B)$ by $i_\sharp(c) = ((i_A)_\sharp(c), -(i_B)_\sharp(c))$, where $i_A : A \cap B \to A$, $i_B : A \cap B \to B$ are the inclusions. Make $S(A) \oplus S(B)$ into a chain complex via the direct sum construction from the chain complexes $S(A), S(B)$. Then $i_\sharp$ becomes a chain homomorphism since $(i_A)_\sharp, (i_B)_\sharp$ are chain homomorphisms. $S(A) + S(B)$ is the subchain complex of $S(X)$ consisting of chains of the form $c_A + c_B$, where c_A is a chain in $S(A)$ and c_B is a chain in $S(B)$. It is a subchain complex since $S(A), S(B)$ are. The map $j_\sharp : S(A) \oplus S(B) \to S(A) + S(B)$ is $j_\sharp(c_A, c_B) = (j_A)_\sharp(c_A) + (j_B)_\sharp(c_B)$. Here $j_A : A \to X$, $j_B : B \to X$ are the inclusions. By its definition, $j_\sharp$ is surjective. $i_\sharp$ is injective since $(i_A)_\sharp, (i_B)_\sharp$ are. Note that $j_\sharp i_\sharp(c) = j_\sharp((i_A)_\sharp(c), -(i_B)_\sharp(c)) = (j_A)_\sharp(i_A)_\sharp(c) - (j_B)_\sharp(i_B)_\sharp(c) = (j_{A \cap B})_\sharp(c) - (j_{A \cap B})_\sharp(c) = 0$. Here $j_{A \cap B} : A \cap B \to X$ is the inclusion. We are using $j_A i_A = j_{A \cap B} = j_B i_B$. Thus $\operatorname{im}(i_\sharp) \subset \ker(j_\sharp)$. Suppose $(c_A, c_B) \in \ker j_\sharp$. Then $(j_A)_\sharp(c_A) = -(j_B)_\sharp(c_B)$. But $S(A) \cap S(B) = S(A \cap B)$ so $c_A = (i_A)_\sharp(c), c_B = -(i_B)_\sharp(c)$, where $c \in S(A \cap B)$. Hence $(c_A, c_B) \in \operatorname{im}(i_\sharp)$ and so $\ker(j_\sharp) \subset \operatorname{im}(i_\sharp)$. Thus the sequence is exact.

6.7.4. (b) Take a string of five terms with $H_k^{\{A,B\}}(X), H_k^{\{A',B'\}}(X)$ in the middle in the top and bottom rows. The outer four vertical maps are isomorphisms by our hypotheses. The five lemma then implies that the middle map $H_k^{\{A,B\}}(X) \to H_k^{\{A',B'\}}(X)$ is also an isomorphism.

6.7.9. This follows from the hypothesis and Theorem 6.7.2, since A', B' now satisfy the hypotheses of that theorem.

6.8.1. The disk deformation-retracts to a point, so Theorem 6.6.2 implies the homology of the disk is isomorphic to the homology of a point. The result follows from the dimension property.

6.8.6. (a) For the set U we take the points on the circle in the interior of the lower semicircle. For U' we take the smaller set where the angle $\theta \in (5\pi/4, 7\pi/4)$. Then $(S^1 \backslash U', A \backslash U')$ deformation-retracts to $(S^2 \backslash U, A \backslash U)$ since $S^1 \backslash U = S^1 \backslash U' \cup C, A \backslash U = A \backslash U' \cup C$, where C is the collar specified by letting the angle θ lie in the range $[0, -\pi/4]$ or $[\pi, 5\pi/4]$. Identifying each to the intervals to $[0, 1]$ linearly leads to the homeomorphism between C and $S^0 \times [0, 1]$ needed for the deformation retraction.

6.8.11. We use the long exact sequence of the pair (D^{p+1}, S^p):

$$H_{k+1}(D^{p+1}) \to H_{k+1}(D^{p+1}, S^p) \to H_k(S^p) \to H_k(D^{p+1}).$$

For $k > 0$, the first and last terms are 0, so the middle map is an isomorphism. For $k = 0$, the last map is an isomorphism so its kernel is 0. The first term is 0, so the long exact sequence implies that $H_1(D^{p+1}, S^p) = 0$.

6.8.17. $ri = 1_A$ implies $r_*(i_*(x)) = x$ for any $x \in H_k(A)$.

6.9.2. Following the hint, we use the diagram

$$
\begin{array}{ccc}
H_{n+1}(D^{n+1}, S^n) & \xrightarrow{\partial} & H_n(S^n) \\
\downarrow{\scriptstyle f_*} & & \downarrow{\scriptstyle (f|S^n)_*} \\
H_{n+1}(D^{n+1}, S^n) & \xrightarrow{\partial} & H_n(S^n)
\end{array}
$$

For $n > 0$, we start with a generator G of $H_{n+1}(D^{n+1}, S^n)$ and choose the generator $g = \partial G$ of $H_n(S^n)$. Then the commutative diagram and $\deg(f)G = f_*(G)$ imply that

$$\deg(f)\partial G = \partial(\deg(f)G) = \partial f_*(G) = (f|S^n)_*(\partial G) = \deg(f|S^n)(\partial G),$$

so $\deg(f) = \deg(f|S^n)$. For $n = 0$, the argument still works with our modified definition of degree and the isomorphism $H_1(D^1, S^0) \simeq \ker(i_*^0)$, replacing $H_0(S^0)$ by $\ker(i_*^0)$.

6.9.5. (a) Any reflection r is of the form $r_v(x) = x - 2(x \cdot v)v$ for some unit vector v. Choose a path $p(t)$ connecting $p(0) = v$ and e_1. Then let $R_t(x) = x - 2(x \cdot p(t))p(t)$. This gives a homotopy between r_v and $r_{e_1} = r_1$.

6.9.11. This follows from the commutative diagram

$$
\begin{array}{ccc}
H_n(S^n) & \xrightarrow{f_*} & H_n(S^n) \\
\downarrow{\scriptstyle j_*^x} & & \downarrow{\scriptstyle j_*^y} \\
H_n(S^n, S_x) & \xrightarrow{f_*} & H_n(S^n, S_y)
\end{array}
$$

Note that $j_*^y f_*(g) = f_* j_*^x(g)$, $j_*^y f_*(g) = j_*^y(\deg(f)g) = \deg(f)g_y$, and $f_* j_*^x(g) = f_*(g_x) = \deg_x(f)(g_y)$.

6.9.15. Choose a point $y \in \mathbb{RP}^n/\mathbb{RP}^{n-1} \simeq S^n$ away from the point coming from collapsing $\mathbb{RP}^{n-1}$. Then there are two points in the inverse image of the form $x, T(x)$, where T is the antipodal map. In identifying the quotient to S^n, we use the generator that corresponds under the local homeomorphism at $T(x)$. Then the local degree at $T(x)$ is, by definition, 1. The local degree at x is given by the last exercise as $\deg_x(f) = \deg(T)\deg_{T(x)}(f) = (-1)^{n+1}$. Then Theorem 6.9.1 gives the result.

6.10.1. (a) We use a map $r : (D^k, L^{k-1}) \to (D^k, S^{k-1})$ which maps L^{k-1} radially to S^{k-1} and maps $D^k\backslash\text{int } L^{k-1}$ onto D^k along radial lines using $[0.5, 1] \to [0, 1]$, $r \to 2(r - 0.5)$ on the radial parameter. The map $ir : (D^k, L^{k-1}) \to (D^k, L^{k-1})$ is homotopic to the identity via maps that use $[0.5t, 1] \to [0, 1]$. Thus r provides a deformation retraction of (D^k, L^{k-1}) to (D^k, S^{k-1}).

6.10.7. The maps induce maps of the quotient spaces which are continuous and bijective. Since the domain space of the quotient is compact and the range is Hausdorff, the maps are homeomorphisms.

6.10.11. $\partial_k^c \partial_{k+1}^c$ includes the composition $H_k(X^k) \to H_k(X^k, X^{k-1}) \to H_{k-1}(X^{k-1})$, which is two steps in the long exact sequence of the pair (X^k, X^{k-1}) and so is 0.

6.10.13. There is one cell in each dimension from 0 to 3. The 1-cell and 2-cell are attached trivially, so the boundary maps in dimensions 1 and 2 are both 0. To determine ∂_3, we look at the attaching map of the 3-cell, again regarding the boundary 2-sphere as $S^0 \times D^2 \cup D^1 \times S^1$. Then the argument we gave for $S^1 \times \mathbb{RP}^2$ again shows that the contributions from the two copies of D^2 cancel one another and so we look at the contribution from $D^1 \times S^1$. But this uses the trivial attaching map for the 2-sphere and so the image lies in the 1-skeleton which is being collapsed. Thus the degree is 0, and $\partial_3 = 0$ as well. Hence

$$H_k^c(S^1 \times S^2) = \begin{cases} \mathbb{Z} & 0 \leq k \leq 3, \\ 0 & \text{otherwise.} \end{cases}$$

6.11.1. (a) This follows from $S_k(X)/S_k(A)$ being the quotient of $(S_k(X)/S_k(B)/(S_k(A)/S_k(B))$ and all of the boundary maps being induced via quotient constructions from the consistent boundary maps in $S(B), S(A), S(X)$.

6.11.4. The last exercise says that $H_k(X^{k+1}, X^{k-2}) \simeq H_k(X, X^{k-2})$ and the previous exercise says that $H_k(X, X^{k-2}) \simeq H_k(X)$.

6.11.10. (a) The sum is $b_0 - (b_0 + b_1) + (b_1 + b_2) + \cdots + (-1)^n(b_{n-1} + b_n) = (-1)^n b_n$. Because of the signs, all of the middle terms cancel and we are left with $(-1)^n b_n = 0$ since $c_{n+1} = 0$.

6.11.12. (c) When we look at the long exact sequence of the pair and take the alternating sum of the ranks, we get $\chi(H_*(S)) = \chi(H_*(H)) + \chi(H_*(S, H))$. This uses the fact each term for $S, H, (S, H)$, respectively, occurs every third term in the long exact sequence and so the signs alternate as in forming the individual

Euler characteristics. However, $\chi(H_*(S,H)) = (-1)^i$ by part (b) and the result follows.

6.12.4. We can use a bicollar neighborhood of $T_+ \cap T_-$ coming from $([0.4, 0.6] \cup [0, 0.1] \cup [0.9, 1]) \times D^1 / \sim$ to justify the application of the Mayer–Vietoris sequence. Since each of T_+, T_- deformation retract to circles and $T_+ \cap T_-$ is a pair of circles, their kth homology is zero for $k \geq 2$. Then the Mayer–Vietoris sequence gives

$$0 = H_{k+1}(T_+) \oplus H_{k+1}(T_-) \to H_{k+1}(T) \to H_k(T_+ \cap T_-) = 0$$

for $k \geq 2$. Hence $H_{k+1}(T) = 0$ for $k \geq 2$.

6.12.10. In terms of the quotient description above, we can use $\sigma_1(t) = [(0, 2t - 1)], \sigma_2(t) = [(2t - 1, 0)]$. In terms of thinking of T as $S^1 \times S^1$, we can use $\sigma_1(t) = (1, e^{2\pi i t}), \sigma_2(t) = (e^{2\pi i t}, 1)$.

6.12.15. (b) We showed $i_1(a, b) = (a + b, a - b)$. In particular, $(1, 1)$ and $(1, -1)$ generate the image, so in the cokernel, we have $[(1, 0)] = [(0, 1)] = [(-1, 0)]$. But this means the cokernel is generated by $[(1, 0)]$ and $2[1, 0] = 0$. Thus $\operatorname{coker}(i_1) \simeq \mathbb{Z}_2$.

6.12.19. (b) The Mayer–Vietoris sequence gives

$$0 \to H_2(N) \to H_1(S^1) \to H_1(P_{(1)}) \oplus H_1(Q_{(1)}).$$

The map $H_1(S^1) \to H_1(P_{(1)})$ is the map shown above to be multiplication by 2, so is injective. Hence the map $i_1 : H_1(S^1) \to H_1(P_{(1)}) \oplus H_1(Q_{(1)})$ is injective as well. Thus $H_2(N) \simeq \ker(i_1) \simeq 0$.

6.13.1. The boundary map ∂_1 has $\operatorname{im}(\partial_1) \subset \ker(\epsilon)$. Thus the short exact sequence induces map of the quotient spaces

$$\widetilde{H}_0(C) = \ker(\epsilon)/\operatorname{im}(\partial_1) \xrightarrow{\ \bar{i}\ } H_0(C) = C_0/\operatorname{im}(\partial_1) \xrightarrow{\ \bar{\epsilon}\ } \mathbb{Z}. \quad \text{The first map is}$$

still injective since it was originally and we factored out by the same subgroup. The last map is still surjective since we factored C_0 by its kernel. Thus we get the desired short exact sequence, which splits since $\mathbb{Z}$ is free.

6.13.5. When we take the augmented complex, which we will call S^ϵ in each case, we again have a short exact sequence

$$0 \to S^\epsilon(A) \to S^\epsilon(X) \to S^\epsilon(X, A) \to 0,$$

which, in dimension -1, becomes $0 \to \mathbb{Z} \to \mathbb{Z} \to 0 \to 0$. This leads to a long exact sequence. Further, $H_{-1}(S^\epsilon(A)) = H_{-1}(S^\epsilon(X)) = 0$ since ϵ is surjective, and $H_{-1}(S^\epsilon(X, A)) = 0$ since the chain group is 0. Hence the sequence ends as claimed.

6.14.2. We note that $S^2 \backslash \{h(c)\}$ is homeomorphic to $\mathbb{R}^2$ and $\widetilde{H}_0(S^2 \backslash \{h(c)\}) = 0 = H_1(S^2 \backslash \{h(c)\})$. Then the end of the Mayer–Vietoris sequence is

$$0 \to \widetilde{H}_0(S^2 \backslash A) \to \widetilde{H}_0(S^2 \backslash A_1) \oplus \widetilde{H}_0(S^2 \backslash A_2) \to 0,$$

which gives the isomorphism.

6.14.7. (b) Using (a) $H_0(S^2\backslash C) \simeq \mathbb{Z}\oplus\mathbb{Z}$, and so there are two path components. Since C is compact, hence closed, $S^2\backslash C$ is an open subset of S^2, and so is itself are surface. Since it is locally path connected, the path components are open.

6.14.13. We subdivide the embedded sphere into the image of the upper and lower hemispheres. Applying Theorem 6.14.4, the exact sequence gives the isomorphism

$$\widetilde{H}_{p+1}(S^n\backslash h(S^{m+1})) \simeq \widetilde{H}_p(S^n\backslash h(S^m)),$$

which gives the result.

6.15.1. (c) This follows from the commutative diagram

$$
\begin{array}{ccc}
H_n(D^n, S^{n-1}) & \longrightarrow & H_n(M, M\backslash U) \\
\downarrow & & \downarrow \\
H_n(D^n, D^n\backslash\{y\}) & \longrightarrow & H_n(M, M\backslash\{x\})
\end{array}
$$

The horizontal maps are isomorphisms from above, and the left vertical map is an isomorphism, so the right vertical map is also one. This could also be shown from using the deformation retraction of $(D^n, D^n\backslash\{y\})$ onto (D^n, S^{n-1}) and the embedding h to get a deformation retraction of $(M, M\backslash\{x\})$ onto $(M, M\backslash U)$.

6.15.5. Since the manifold is assumed orientable, the homology orientation at a point determines the choice in an open set about the point. Suppose $y \in M$ is another point. Choose a path $f : I \to M$ connecting x_0 to y, using the fact that a connected n-manifold is path connected. Then the image of the path is compact and so is covered by a finite number of interiors of disks in Euclidean neighborhoods. By taking the inverse image of embedded disks about points of M, we get a covering of I. We can then take a finite subcovering and thus find a sequence $D_1, \ldots, D_k$ of embedded disks so that $D_{i+1} \cap D_i \neq \emptyset$ and $x_0 \in D_1$, $y \in D_k$. Since the homology orientation at one point in a disk determines the homology orientation at all other points, we inductively see that the homology orientation at x_0 determines it at y.

6.15.10. (a) Since G_1 and $G_0 = $ id are homotopic maps from $(\mathbb{R}^n, \mathbb{R}^n\backslash\text{int } B)$ to itself, they induce induce the same map on homology.
(b) Here we use the diagram

$$
\begin{array}{ccc}
H_n^{\mathbb{R}^n,\text{int } B} & \xrightarrow{(G_1)_*} & H_n^{\mathbb{R}^n,\text{int } B} \\
\downarrow & & \downarrow \\
H_n^{\mathbb{R}^n, x} & \xrightarrow{(G_1)_*} & H_n^{\mathbb{R}^n, G_1(x)}
\end{array}
$$

Here the vertical maps are induced by inclusion and are isomorphisms. The result follows from (a) and this commutative diagram.

6.15.14. (a) If μ_x is a homology orientation, then the map $s : M \to \widetilde{M_g}$ given by $s(x) = \mu_x$ is a nonzero section. Conversely, a nonzero section s gives the homology orientation by $\mu_x = s(x)$. We have constructed the basis for the topology on $\widetilde{M_g}$ so that this section is continuous—it is just a local homeomorphism using the disk neighborhood of the point.

6.15.18. A simply connected covering space has no nontrivial connected double coverings since they correspond to equivalence classes of index-2 subgroups.

6.15.23. We apply Theorem 6.15.3 with $A = M$. Note that $H_n^{M,A} = H_n(M)$.
  (a) This follows from Theorem 6.15.3(1).
  (b),(c) Here we look at the homomorphism $S_M : H_n(M) \to \Gamma_M$ which sends $\alpha \in H_n(M)$ to $s(x) = r_x(\alpha)$. Theorem 6.15.3(2) says this is injective. Since $\Gamma_M = 0$ when M is nonorientable, this implies $H_n(M) = 0$ then. Theorem 6.15.3(3) implies that S_M is an isomorphism onto $\Gamma_M \simeq \mathbb{Z}$ when M is orientable.

6.15.25. (a) These deformation retractions just use the deformation retractions of $[0, 1], (0, 1], [t, 1]$ to $\{1\}$, which leads to deformation retractions of the product of ∂M times each of these intervals to $\partial M \times \{1\}$. These extend to the claimed deformation retractions by using the identity map on A_1.

6.15.30. (1) We use the exact sequence from the previous exercise. Suppose $i > n$. Then $H_i^{M,A \cup B} = 0$ since it is squeezed between two terms which are 0 in the exact sequence.

6.15.33. This follows from the excision isomorphism $H_i^{M,A} \simeq H_i^{U,A} = H_i^{\mathbb{R}^n,A}$ when U is the coordinate neighborhood.

6.16.2. We compute

$$\partial_2 H_1(\sigma) = \partial(\beta - \alpha) = H_0(\tau_0) - H_0(\tau_1) + g\sigma - f\sigma = -H_0\partial_1(\sigma) + g_\sharp(\sigma) - f_\sharp(\sigma).$$

6.16.7. We first compute

$$\partial_{i+1} H_i([e_0, \ldots, e_i]) = \partial_{i+1} \sum_{j=0}^{i} (-1)^j [v_0, \ldots, v_j, w_j, \ldots, w_i].$$

When taking the boundary, there are three types of terms: (1) the ones where v_k is deleted, $k < j$; (2) the two terms where v_j, w_j are deleted; (3) the ones where w_k is deleted, where $k > j$. These three types of terms correspond to the three terms above.

6.17.1 (b) Following the hint, we start with a path $\sigma : I \to X$, which connects y to x. Pulling the open cover $\{\text{int } A, \text{int } B\}$ back to I via σ gives an open cover of I. By using enough subdivisions of I we can find a subdivision of I so that each subinterval is sent to int A or int B by σ. Then replace σ by the chain which is the sum of the reparametrizations of restrictions of σ to these subintervals. This gives a chain in $S(A) + S(B)$, which has boundary $x - y$.

6.17.6. The extension is inductive, assuming the map is defined and shown to be a chain map in dimensions less than n. We first define it on the identity n-simplex $[e_0, \ldots, e_n]$ by letting $v = (1/(n+1)) \sum e_i$ denote the barycenter and then defining $Sd([e_0, \ldots, e_n]) = v.Sd\partial[e_0, \ldots, e_n]$. Then check

$$\partial Sd([e_0, \ldots, e_n]) = Sd\partial([e_0, \ldots, e_n]) - v.\partial Sd\partial([e_0, \ldots, e_n])$$
$$= Sd\partial([e_0, \ldots, e_n]) - v.Sd\partial^2([e_0, \ldots, e_n]) = Sd\partial([e_0, \ldots, e_n]).$$

Now define $Sd(\sigma) = \sigma_\sharp(Sd([e_1, \ldots, e_n])$. We then compute

$$\partial Sd(\sigma) = \partial \sigma_\sharp Sd([e_1, \ldots, e_n]) = \sigma_\sharp \partial Sd([e_1, \ldots, e_n]) = \sigma_\sharp Sd\partial([e_1, \ldots, e_n])$$
$$= Sd\sigma_\sharp \partial([e_1, \ldots, e_n]) = Sd\, \partial\sigma.$$

Finally, extend Sd to a chain map in dimension n by linearity.

6.17.10. The previous two exercises have shown that the map $H_1^{\{A,B\}}(X) \to H_1(X)$ is both surjective and injective.

References

[1] M. Armstrong, *Basic Topology*, Springer, New York, 1983.

[2] S. Barr, *Experiments in Topology*, Dover, New York, 1989.

[3] R.H. Bing, *The Geometric Topology of 3-Manifolds*, AMS Colloquium Publications 40, American Mathematical Society, Providence, RI, 1983.

[4] R.H. Bing, An alternative proof that 3-manifolds can be triangulated, *Ann. Math.* 69 (1959) 37–65.

[5] G. Bredon, *Geometry and Topology*, Springer, New York, 1993.

[6] M. Brown, A proof of the generalized Schoenflies theorem, *Bull. Amer. Math. Soc.* 66 (1960) 74–6.

[7] S. Cairns, *Introductory Topology*, Ronald Press, New York, 1968.

[8] S. Cairns, Homeomorphisms between topological manifolds and analytic manifolds, *Ann. Math.* 41 (1940) 796–808.

[9] W. Chinn and N. Steenrod, *First Concepts of Topology*, The New Mathematical Library, Mathematical Association of America, Random House, 1966.

[10] P. Doyle and D. Moran, A short proof that compact 2-manifolds can be triangulated, *Invent. Math.* 5 (1968) 180–2.

[11] A. Gramain, *Topology of Surfaces*, BCS Associates, Moscow, ID, 1984.

[12] H.B. Griffiths, *Surfaces*, Cambridge University Press, Cambridge, 1976.

[13] A. Hatcher, *Algebraic Topology*, Cambridge University Press, Cambridge, 2001.

[14] M. Henle, *A Combinatorial Introduction to Topology*, Dover, New York, 1979.

[15] M. Hirsch, *Differential Topology*, Graduate Texts in Mathematics 33, Springer, New York, 1976.

[16] R. Kirby, Stable homeomorphisms and the annulus conjecture, *Ann. Math.* 89 (1969) 575–82.

[17] T. Lawson, *Linear Algebra*, Wiley, New York, 1996.

[18] W. Massey, *Algebraic Topology: An Introduction*, Harcourt, Brace and World, New York, 1967.

[19] J. Milnor, *Morse Theory*, Annals of Mathematics Studies, No. 51, Princeton University Press, Princeton, NJ, 1963.

[20] J. Milnor, *Lectures on the h-cobordism Theorem*, Princeton Mathematical Notes, Princeton University Press, Princeton, NJ, 1965.

[21] J. Milnor, *Topology from the Differentiable Viewpoint*, U. Virginia Press, Charlottesville, VA, 1965.

[22] E. Moise, *Geometric Topology in Dimensions 2 and 3*, Springer, New York, 1977.

[23] E. Moise, Affine structures in 3-manifolds. V. The triangulation theorem and the Hauptvermutung, *Ann. Math.* 55 (1952) 96–114.

[24] J. Munkres, *Topology*, Prentice-Hall, Englewood Cliffs, NJ, 2000.

[25] J. Munkres, *Elementary Differential Topology*, Princeton University Press, Princeton, NJ, 1961.

[26] J. Munkres, Obstructions to smoothing piecewise-differentiable homeomorphisms, *Ann. Math.* 72 (1960) 521–54.

[27] C. Papakkyriakopolous, A new proof of the invariance of homology groups of a complex, *Bull. Soc. Math. Greece* 22 (1943) 1–154.

[28] F. Quinn, Ends of maps. III. Dimensions 4 and 5, *J. Diff. Geom.* 17 (1982) 503–21.

[29] T. Radó, Über den Begriff die Riemann Fläche, *Acta. Univ. Szeged* 2 (1924–26) 101–21.

[30] D. Rolfsen, *Knots and Links*, Publish or Perish, Boston, MA, 1976.

[31] C. Rourke and J. Sanderson, *Introduction to Piecewise-Linear Topology*, Springer, New York, 1972.

[32] J. Vick, *Homology Theory. An Introduction to Algebraic Topology*, Pure and Applied Mathematics, Vol. 53, Academic Press, New York, 1973.

[33] J.H.C. Whitehead, Manifolds with transverse fields in Euclidean space, *Ann. Math.* 73 (1961) 154–212.

Index

Printed in the United States
By Bookmasters